Applications of Mathematical Models in Engineering

Applications of Mathematical Models in Engineering

Editors

Eva H. Dulf
Cristina I. Muresan

MDPI • Basel • Beijing • Wuhan • Barcelona • Belgrade • Manchester • Tokyo • Cluj • Tianjin

Editors
Eva H. Dulf
Technical University of
Cluj-Napoca
Romania

Cristina I. Muresan
Technical University of
Cluj-Napoca
Romania

Editorial Office
MDPI
St. Alban-Anlage 66
4052 Basel, Switzerland

This is a reprint of articles from the Special Issue published online in the open access journal *Mathematics* (ISSN 2227-7390) (available at: https://www.mdpi.com/journal/mathematics/special_issues/math_model_eng).

For citation purposes, cite each article independently as indicated on the article page online and as indicated below:

LastName, A.A.; LastName, B.B.; LastName, C.C. Article Title. *Journal Name* **Year**, *Volume Number*, Page Range.

ISBN 978-3-0365-3847-1 (Hbk)
ISBN 978-3-0365-3848-8 (PDF)

© 2022 by the authors. Articles in this book are Open Access and distributed under the Creative Commons Attribution (CC BY) license, which allows users to download, copy and build upon published articles, as long as the author and publisher are properly credited, which ensures maximum dissemination and a wider impact of our publications.

The book as a whole is distributed by MDPI under the terms and conditions of the Creative Commons license CC BY-NC-ND.

Contents

About the Editors ix

Preface to "Applications of Mathematical Models in Engineering" xi

Vlad Mihaly, Mircea Şuşcă and Eva H. Dulf
μ-Synthesis FO-PID for Twin Rotor Aerodynamic System
Reprinted from: *Mathematics* **2021**, *9*, 2504, doi:10.3390/math9192504 1

Jožef Ritonja
Adaptive Control of CO_2 Production during Milk Fermentation in a Batch Bioreactor
Reprinted from: *Mathematics* **2021**, *9*, 1712, doi:10.3390/math9151712 19

Vlad Mihaly, Mircea Şuşcă, Dora Morar, Mihai Stănese and Petru Dobra
μ-Synthesis for Fractional-Order Robust Controllers
Reprinted from: *Mathematics* **2021**, *9*, 911, doi:10.3390/math9080911 45

Frank A Plua, Francisco-Javier Sánchez-Romero, Victor Hidalgo, P. Amparo López-Jiménez and Modesto Pérez-Sánchez
New Expressions to Apply the Variation Operation Strategy in Engineering Tools Using Pumps Working as Turbines
Reprinted from: *Mathematics* **2021**, *9*, 860, doi:10.3390/math9080860 67

Lloyd Ling, Zulkifli Yusop and Joan Lucille Ling
Statistical and Type II Error Assessment of a Runoff Predictive Model in Peninsula Malaysia
Reprinted from: *Mathematics* **2021**, *9*, 812, doi:10.3390/math9080812 85

Mircea Şuşcă, Vlad Mihaly, Mihai Stănese, Dora Morar and Petru Dobra
Unified CACSD Toolbox for Hybrid Simulation and Robust Controller Synthesis with Applications in DC-to-DC Power Converter Control
Reprinted from: *Mathematics* **2021**, *9*, 731, doi:10.3390/math9070731 109

Jorge Muñoz, Francesco Piqué, Concepción A. Monje and Egidio Falotico
Robust Fractional-Order Control Using a Decoupled Pitch and Roll Actuation Strategy for the I-Support Soft Robot
Reprinted from: *Mathematics* **2021**, *9*, 702, doi:10.3390/math9070702 141

Blanca Viviana Martínez, Javier Sanchis, Sergio García-Nieto and Miguel Martínez
Tuning Rules for Active Disturbance Rejection Controllers via Multiobjective Optimization: A Guide for Parameters Computation Based on Robustness
Reprinted from: *Mathematics* **2021**, *9*, 517, doi:10.3390/math9050517 157

Mingming Zhang, Shurong Hao and Anping Hou
Study on the Intelligent Modeling of the Blade Aerodynamic Force in Compressors Based on Machine Learning
Reprinted from: *Mathematics* **2021**, *9*, 476, doi:10.3390/math9050476 191

Iskandar Waini, Anuar Ishak and Ioan Pop
Flow towards a Stagnation Region of a Vertical Plate in a Hybrid Nanofluid: Assisting and Opposing Flows
Reprinted from: *Mathematics* **2021**, *9*, 448, doi:10.3390/math9040448 205

Marjan Goodarzi, Ali Mohades and Majid Forghani-elahabad
Improving the Gridshells' Regularity by Using Evolutionary Techniques
Reprinted from: *Mathematics* **2021**, *9*, 440, doi:10.3390/math9040440 221

Serguei Maximov, Manuel A. Corona-Sánchez, Juan C. Olivares-Galvan, Enrique Melgoza-Vazquez, Rafael Escarela-Perez and Victor M. Jimenez-Mondragon
Mathematical Calculation of Stray Losses in Transformer Tanks with a Stainless Steel Insert
Reprinted from: *Mathematics* **2021**, *9*, 184, doi:10.3390/math9020184 239

Taehak Kang and Jaiyoung Ryu
Determination of Aircraft Cruise Altitude with Minimum Fuel Consumption and Time-to-Climb: An Approach with Terminal Residual Analysis
Reprinted from: *Mathematics* **2021**, *9*, 147, doi:10.3390/math9020147 253

Rasikh Tariq, Jacinto Torres Jimenez, Nadeem Ahmed Sheikh and Sohail Khan
Mathematical Approach to Improve the Thermoeconomics of a Humidification Dehumidification Solar Desalination System
Reprinted from: *Mathematics* **2021**, *9*, 33, doi:10.3390/math9010033 275

Seyed Nasrollah Mousavi and Daniele Bocchiola
A Novel Comparative Statistical and Experimental Modeling of Pressure Field in Free Jumps along the Apron of USBR Type I and II Dissipation Basins
Reprinted from: *Mathematics* **2020**, *8*, 2155, doi:10.3390/math8122155 307

Javier Velasco, Isidro Calvo, Oscar Barambones, Pablo Venegas and Cristian Napole
Experimental Validation of a Sliding Mode Control for a Stewart Platform Used in Aerospace Inspection Applications
Reprinted from: *Mathematics* **2020**, *8*, 2051, doi:10.3390/math8112051 325

Eva H. Dulf, Mihnea Saila, Cristina I. Muresan and Liviu C. Miclea
An Efficient Design and Implementation of a Quadrotor Unmanned Aerial Vehicle Using Quaternion-Based Estimator
Reprinted from: *Mathematics* **2020**, *8*, 1829, doi:10.3390/math8101829 341

Zakieh Avazzadeh, Omid Nikan and José A. Tenreiro Machado
Solitary Wave Solutions of the Generalized Rosenau-KdV-RLW Equation
Reprinted from: *Mathematics* **2020**, *8*, 1601, doi:10.3390/math8091601 365

Cristina I. Muresan, Isabela R. Birs and Eva H. Dulf
Event-Based Implementation of Fractional Order IMC Controllers for Simple FOPDT Processes
Reprinted from: *Mathematics* **2020**, *8*, 1378, doi:10.3390/math8081378 385

About the Editors

Eva H. Dulf, Professor, received her Ph.D. degree from the Technical University of Cluj-Napoca, Cluj-Napoca, Romania, in 2006, where she is currently Professor in the Automation Department. She has published more than 150 papers and received 39 awards at prestigious International Exhibitions of Inventions. She has been and currently is involved in more than 30 research grants, all dealing with the modeling and control of complex processes. Her research interests include modern control strategies and the modeling of biochemical and medical processes.

Cristina I. Muresan, PhD, received a degree in control systems and her Ph.D. degree from the Technical University of Cluj-Napoca, Cluj-Napoca, Romania, in 2007 and 2011, respectively, where she is currently an Associate Professor in the Automation Department. She has published more than 150 papers. She has been and currently is involved in more than ten research grants, all dealing with multivariable and fractional order control. Her research interests include modern control strategies, such as predictive algorithms, robust nonlinear control, fractional-order control, time-delay compensation methods and multivariable systems.

Preface to "Applications of Mathematical Models in Engineering"

The role of applied mathematics has continued to become increasingly important with the advancement of science and technology, ranging from the modeling and analysis of natural phenomena to the simulation, design, control and optimization of systems. With the advancements in computing technology, larger and more complex problems can now be tackled and analyzed in a very timely fashion. As a sample for this, the present book comprises 19 chapters that present a series of contributions in the field.

The manuscripts cover a wide spectrum in terms of the type of problems, methodologies and applications discussed. Different mathematical models are discussed: models for pumps working as turbines considering the modified affinity laws; a reassessment of the rainfall-runoff model and model calibration with inferential statistics; a three-dimensional unsteady aerodynamic force reduction model based on the eXterme Gradient Boosting algorithm in machine learning; a hybrid nanofluid flow model towards a stagnation region of a vertical plate with radiation effects; analytical formulations to calculate magnetic field distribution and stray losses in the transformer region where bushings are mounted, considering a stainless steel insert in the transformer tank; a mathematical model to enhance the freshwater productivity rate of a solar-assisted humidification–dehumidification type of desalination system; statistical modeling of the pressure field at the centerline of the apron along the USBR Type I and II basins; solitary wave solutions of the generalized Rosenau-Korteweg-de Vries-regularized-long wave equation. In the control engineering field, different innovative strategies are presented: an event-based algorithm for fractional-order IMCs for first-order plus dead-time processes, including delay- and lag-dominant ones; the integration of a fixed-structure, multiple-input-multiple-output, fractional-order, proportional-integral-derivative controller in the μ-synthesis optimization problem for different engineering applications; the design, implementation, validation and use of a Computer-Aided Control System Design (CACSD) toolbox for nonlinear and hybrid systems; uncertainty modeling, simulation, and control using μ-synthesis; robust fractional-order control using a decoupled pitch and roll actuation strategy for the I-Support soft robot; a set of tuning rules for Linear Active Disturbance Rejection Controller (LADRC) with three different levels of compromise between disturbance rejection and robustness; a sliding mode control for the precise positioning of a Stewart platform used as a mobile platform in non-destructive inspection applications. The discussed applications range from the development of a control system for fermentation production in batch bioreactors to the design and optimization of gridshell structures or the determination of aircraft cruise altitudes with minimum fuel consumption and time-to-climb, or even the design of a low-cost, performing quadrotor unmanned aerial vehicle using a quaternion-based estimator.

DWe would like to thank the MDPI publishing editorial team, the scientific peer reviewers and all of the authors who have contributed to this volume. We hope that the manuscripts are of value to researchers, academics and professionals involved in the resolution and optimization of real-world engineering problems.

Eva H. Dulf and Cristina I. Muresan
Editors

Article

µ-Synthesis FO-PID for Twin Rotor Aerodynamic System

Vlad Mihaly, Mircea Şuşcă and Eva H. Dulf *

Department of Automation, Technical University of Cluj-Napoca, Str. G. Barițiu nr. 26-28, 400027 Cluj-Napoca, Romania; vlad.mihaly@aut.utcluj.ro (V.M.); mircea.susca@aut.utcluj.ro (M.Ș.)
* Correspondence: eva.dulf@aut.utcluj.ro

Abstract: µ-synthesis is a NP-hard optimization problem based on the generalized Robust Control framework which manages to find a controller which fulfills both robust stability and robust performance. In order to solve such problems, nonsmooth optimization techniques are employed to find nearly-optimal parameters values. However, the free parameters available for tuning must be involved only in classical arithmetic operations, which leads to a problem for the fractional-order operator or for its integer-order approximation, exponential operations being involved. The main goal of the current article consists of presenting a possibility to integrate a fixed-structure multiple-input-multiple-output (MIMO) fractional-order proportional-integral-derivative (FO-PID) controller in the µ-synthesis optimization problem. The solution consists in a possibility to find a set of tunable parameters isomorphic with the fractional-order such that the coefficients involved in the approximation of the fractional element, along with the formulation of a fixed-structure mixed-sensitivity loop shaping µ-synthesis control problem. The proposed design procedure is applied to a twin rotor aerodynamic system (TRAS) using both MATLAB numerical simulation and practical experiments on laboratory scale equipment. Moreover, a comparison with the unstructured µ-synthesis is performed, highlighting the advantages of the proposed solution: simpler form and guaranteed robust stability and performance.

Keywords: robust control; mixed-sensitivity; µ-synthesis; fractional-order control; FO-PID; twin rotor aerodynamic system

Citation: Mihaly, V.; Şuşcă, M.; Dulf, E.H. µ-Synthesis FO-PID for Twin Rotor Aerodynamic System. *Mathematics* **2021**, *9*, 2504. https://doi.org/10.3390/math9192504

Academic Editor: Ali Farajpour

Received: 8 September 2021
Accepted: 2 October 2021
Published: 6 October 2021

Publisher's Note: MDPI stays neutral with regard to jurisdictional claims in published maps and institutional affiliations.

Copyright: © 2021 by the authors. Licensee MDPI, Basel, Switzerland. This article is an open access article distributed under the terms and conditions of the Creative Commons Attribution (CC BY) license (https://creativecommons.org/licenses/by/4.0/).

1. Introduction

One of the fundamental problems studied in Control Engineering concerns robustness, which characterizes the sensitivity of the closed loop system to the variation of plant parameters. One of the most used performance measures is the $\mathcal{H}_\infty$ norm. Starting from the approach of synthesizing a $\mathcal{H}_\infty$ controller by solving two Algebraic Riccati Equations (AREs) as in [1], a more numerically stable solution can be obtained using Popov triplets [2]. Alternatively, due to the limitations of this approach represented by the impossibility of solving singular problems, the AREs were replaced with Algebraic Riccati Inequalities (ARIs) and were solved using Linear Matrix Inequalities (LMIs) [3]. The last two approaches have been recently implemented in open-source manners in [4,5]. However, the classical $\mathcal{H}_\infty$ control problem manages to ensure nominal stability and nominal performance only. In order to consider dynamic and parametric uncertainties, the plant is formulated as an upper linear fractional transform with such an uncertainty block and the µ-synthesis can be used for computing a robust controller based on the classical D–K iterations [6]. The major concern about these methods consists of the fact that the controller is usually of high order. However, imposing a fixed structure leads to a non-convex problem which cannot be approximated as in the case of µ-synthesis. The solutions, initially proposed for $\mathcal{H}_\infty$ problem [7], and then for µ-synthesis [8] as well, are based on nonsmooth optimization techniques. A CACSD toolbox that manages to offer an end-to-end solution for designing a robust controller starting from a given plant is presented in [9].

The most well-known controller structure, which is highly used in industry, is the proportional-integral-derivative (PID) regulator. Its form is generally given as an example for fixed-structure controllers and nonsmooth optimization methods were designed around it [10]. An extension with two extra degrees of freedom is represented by fractional-order PID (FO-PID), which improves the robustness of the closed loop system. As tuning methods, the well-known methods used for designing integer-order controllers were extended for fractional-order controllers as well. As such, two generalized versions of Kessler's magnitude methods were presented in [11,12], while a fractional-order internal model controller with event-based implementation was developed in [13]. A fractional-order integrator was used as a model for the servo problem in [14], while the same structure was used as a speed controller for a DC motor in [15]. In [16] crone control methodologies were presented, along with LMI formulation for the $\mathcal{H}_\infty$ fractional-order control problem. An artificial bee colony optimization for a MIMO FO-PID controller design by solving the mixed-sensitivity μ-synthesis control problem is presented in [17].

The twin rotor aerodynamic system (TRAS) is a well-known benchmark system used to illustrate the control methods designed in literature. A two degrees of freedom (2-DOF) discrete-time μ-synthesis controller of order 24 was presented in [18]. A decentralized fixed-structure PID controller designed using $\mathcal{H}_\infty$ is presented in [19], along with a comparison between the full-order $\mathcal{H}_\infty$ controller. After the linearization and decoupling steps, 2-DOF continuous and discrete-time controllers were designed using $\mathcal{H}_\infty$ in [20]. A hybrid architecture using both $\mathcal{H}_\infty$ and Iterative Learning Control is described in [21]. A linear quadratic regulator (LQR) for MIMO TRAS problem was designed using particle swarm optimization in [22], while a frequency-based PID controller was combined with a lead compensator designed using root locus in [23]. An approach that further details the controller implementation with quantization aspects taken into consideration for the same family of processes is presented in [24].

In this paper, we present a design procedure that manages to optimize the controller parameters instead of tuning them. As such, we present a method for finding the parameters of a MIMO fractional-order PID (FO-PID) robust controller by solving a fixed-structure mixed-sensitivity loop shaping μ-synthesis control problem. Although the resulting control problem is nonconvex in terms of the controller's free parameters, the nonsmooth optimization techniques implemented in MATLAB's Robust Control Toolbox can be used. However, the realp object used for these free parameters does not support exponential operations necessary in the approximation of a fractional-order element. Therefore, we present in this paper an algorithm to construct the approximation function of a fractional-order element using integer-order elements and supported arithmetic operations applied on a free parameter isomorphic to the desired fractional order. As such, we successfully manage to formulate the problem of optimizing the parameters of a MIMO FO-PID such that the available techniques can be used. Moreover, we illustrate our design method on the twin rotor aerodynamic system stand, having both MATLAB simulations and physical experiments.

The remainder of this paper is organized as follows: Section 2 summarizes the main mathematical background in terms of available results in both Robust and Fractional-Order Control, along with the description of the proposed method in terms of the algorithm for approximation of the fractional order element and of the optimization problem; Section 3 starts with the presentation of the simplified nonlinear mathematical model of the TRAS system, the linearized mathematical model around an equilibrium point, and a list of parameters with their numerical values and tolerances which manages to encompass the nonlinearities; in Section 4, the numerical results are presented, starting from the augmentation step, followed by the proposed structure of the controller and the obtain results in MATLAB and on the experimental stand; Section 5 presents the discussions of the obtained results and a comparison with another method for solving the optimization problem, while in Section 6 there are some conclusions and possible research directions.

2. Proposed Method

In this section, the mathematical background for the proposed controller design method in terms of Robust Control Framework in Section 2.1 and Fractional-Order Control Framework in Section 2.2 is firstly presented, while in Section 2.3 the method for optimizing the controller parameters using a different approach as against the procedure presented in [17] is described.

2.1. Robust Control

The generalized Robust Control Framework [25] has, besides the control input vector $\mathbf{u} \in \mathbb{R}^{n_u}$, two extra inputs: the exogenous input vector $\mathbf{w} \in \mathbb{R}^{n_w}$ and disturbance input vector $\mathbf{d} \in \mathbb{R}^{n_d}$. Additionally, besides the output vector $\mathbf{y} \in \mathbb{R}^{n_y}$, the generalized plant contains two extra outputs: the performance vector $\mathbf{z} \in \mathbb{R}^{n_z}$ and the disturbance output $\mathbf{v} \in \mathbb{R}^{n_v}$. The input and output disturbance vectors encompass both parametric and unstructured uncertainties, which are generally modeled by the following set:

$$\Delta = \left\{ \mathrm{diag}\left(\delta_1 I_{n_1}, \ldots, \delta_s I_{n_s}, \Delta_1, \ldots, \Delta_f\right) \mid \delta_k \in \mathbb{R},\ \Delta_j \in \mathbb{R}^{m_j \times m_j},\ k = \overline{1,s},\ j = \overline{1,f} \right\}, \quad (1)$$

where I_n denotes the identity matrix of order n.

The uncertainty block Δ is interconnected with the generalized plant P_Δ via an upper linear fractional transformation (ULFT), while the controller K is interconnected via a lower linear fractional transformation (LLFT) with P_Δ, as noticed in Figure 1.

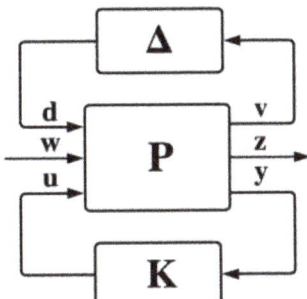

Figure 1. Generalized plant interconnection with the controller and uncertainty blocks [17].

The state-space representation of the generalized plant P_Δ is:

$$P_\Delta : \begin{pmatrix} \dot{\mathbf{x}}(t) \\ \mathbf{v}(t) \\ \mathbf{z}(t) \\ \mathbf{y}(t) \end{pmatrix} = \left(\begin{array}{c|ccc} A & B_d & B_w & B_u \\ \hline C_v & D_{vd} & D_{vw} & D_{vu} \\ C_z & D_{zd} & D_{zw} & D_{zu} \\ C_y & D_{yd} & D_{yw} & D_{yu} \end{array} \right) \begin{pmatrix} \mathbf{x}(t) \\ \mathbf{d}(t) \\ \mathbf{w}(t) \\ \mathbf{u}(t) \end{pmatrix}. \quad (2)$$

For robustness analysis, the singular value notion used for $\mathcal{H}_\infty$ synthesis was extended to the *structural singular value*, defined for the LLFT interconnection between the plant P_Δ and the controller K according to the uncertainty block Δ as:

$$\mu_\Delta(\mathrm{LLFT}(P_\Delta, K)) = \sup_{\omega \in \mathbb{R}_+} \frac{1}{\min_{\Delta \in \Delta}\{\overline{\sigma}(\Delta) \mid \det(I - \mathrm{LLFT}(P_\Delta, K)(j\omega)\Delta) = 0\}}. \quad (3)$$

Given that the problem of explicitly computing such structural singular values is NP-hard, an approximation must be used. The classical $\mathcal{H}_\infty$ control problem can be extended to the following optimization problem:

$$\inf_{K\ \mathrm{stab.}} \sup_{\omega \in \mathbb{R}_+} \mu_\Delta(\mathrm{LLFT}(P_\Delta, K)(j\omega)), \quad (4)$$

which can be considered solved if there is a controller K such that $\mu_\Delta(\mathrm{LLFT}(P_\Delta, K)) < 1$, according to main loop theorem. As such, an upper bound is necessary for $\mu_\Delta(\cdot)$ [6]:

$$\mu_\Delta(\mathrm{LLFT}(P_\Delta, K)(j\omega)) \leq \inf_{D \in \mathcal{D}} \overline{\sigma}\Big(D \cdot \mathrm{LLFT}(P_\Delta, K)(j\omega) \cdot D^{-1}\Big), \tag{5}$$

where the set $\mathcal{D}$ is defined according to the uncertainty block Δ as follows [6]:

$$\mathcal{D} = \Big\{ \mathrm{diag}\Big(D_1, \ldots, D_s, d_1 I_{m_1}, \ldots, d_f I_{m_f}\Big) | D_k = D_k^\top \in \mathbb{R}^{n_k \times n_k}, \ d_j > 0, \ k = \overline{1,s}, \ j = \overline{1,f} \Big\}. \tag{6}$$

Summarizing, robust stability and robust performance are achieved through a controller K obtained as a solution of the optimization problem (4) which manages to obtain an objective value lower than 1. But this NP-hard problem can be approximated by the following quasi-convex problem:

$$\inf_{K \text{ stab.}} \sup_{\omega \in \mathbb{R}_+} \inf_{D \in \mathcal{D}} \overline{\sigma}\Big(D(j\omega) \cdot \mathrm{LLFT}(P_\Delta, K)(j\omega) \cdot (D(j\omega))^{-1}\Big). \tag{7}$$

As already known, the last optimization problem can be solved using the so-called D–K iteration [9,17]. This iterative procedure starts with a fixed D (usually considered the unitary system) and alternatively computes the controller K, by solving the $\mathcal{H}_\infty$ control problem with fixed D, and the D-scale factor, by solving the Parrot problem, as defined in [6], for each point from a frequency set $\Omega = \{\omega_l = \omega_1 < \cdots < \omega_N = \omega_u\}$ followed an approximation of the obtained solutions with a minimum phase system. Therefore, after setting the initial D-scale step as $D = I$, the following steps are successively applied:

1: The D-scale step is fixed and the controller can be computed as:

$$K = \arg \inf_{K \text{ stab.}} \|\mathrm{LLFT}(P_\Delta, K)\|_\infty. \tag{8}$$

2: The controller K is fixed and the following set of convex problems must be solved:

$$D(j\omega) = \arg \inf_{D \in \mathcal{D}} \overline{\sigma}\Big(D \cdot \mathrm{LLFT}(P_\Delta, K)(j\omega) \cdot D^{-1}\Big), \tag{9}$$

for a given frequency range Ω and, then, a stable minimum phase transfer matrix $D(s)$ is fitted.

Steps **1** and **2** are executed in a loop sequence until the difference between two consecutive $\mathcal{H}_\infty$ norms is less than a prescribed tolerance, the maximum number of iterations is reached, or the improvement after a prescribed number of steps is under an imposed tolerance.

2.2. Fractional-Order Control

The domain of Fractional-Order Control has recently gained more attention due to their robustness. The fractional integral operator used in Control Engineering is [26]:

$$\mathcal{I}_\alpha\{f(t)\} = \frac{1}{\Gamma(\alpha)} \int_0^t (t-\tau)^{\alpha-1} f(\tau) d\tau, \ t > 0, \ \alpha \in \mathbb{R}_+, \tag{10}$$

where $\Gamma(\cdot) : \mathbb{C}_+ \to \mathbb{C}$ is the Euler Gamma function. In a similar manner with the integer order integral operator, the fractional order integral operator $\mathcal{I}_\alpha$ has the following Laplace transform:

$$\mathcal{L}\{\mathcal{I}_\alpha\{f(t)\}\}(s) = s^{-\alpha} \mathcal{L}\{f(t)\}(s). \tag{11}$$

As previously stated, the fractional-order calculus can be used to extend the classical 3-DOF proportional-integral-derivative (PID) controller to a fractional-order PID (FO-PID) having two extra DOF $\lambda, \mu \in \mathbb{R}_+$ – the order of the integral operator and the order of the derivative operator, respectively. As such, based on the error signal $\varepsilon(t)$, the command signal $c(t)$ has the following expression:

$$c(t) = K_P \cdot \varepsilon(t) + K_I \cdot \mathcal{I}_\lambda\{\varepsilon(t)\} + K_D \cdot \mathcal{I}_{-\mu}\{\varepsilon(t)\}, \quad (12)$$

where $c(t)$ would be $u(t)$ and $\varepsilon(t)$ would be $r(t) - y(t)$ according to the generalized framework from Figure 1, while the differences between the transfer functions of these two controllers are:

$$H_{PID}(s) = K_P + \frac{K_I}{s} + K_D s \;\Rightarrow\; H_{FO-PID}(s) = K_P + \frac{K_I}{s^\lambda} + K_D s^\mu. \quad (13)$$

The main drawback of the FO-PID revolves around the implementation of the fractional order elements. One possible solution is the Oustaloup recursive approximation (ORA) introduced in crone toolbox [16]. The approximation of a fractional-order element s^λ with an integer-order one is detailed for $\lambda \in (0,1)$, but it can be easily extended for $\lambda \in \mathbb{R}$. The ORA representation receives as inputs three parameters: the order N of the LTI system which approximates the fractional-order element, along with the lower bound ω_l and the upper bound ω_u of the frequency range where the approximation is valid. The LTI approximation is:

$$s^\lambda \approx \prod_{k=1}^{N} \frac{1 + s/\mathring{\omega}_k}{1 + s/\hat{\omega}_k}, \quad (14)$$

where the poles and zeros frequencies can be computed using two coefficients:

$$\varepsilon = \left(\frac{\omega_u}{\omega_l}\right)^{\frac{\lambda}{N}} \text{ and } \eta = \left(\frac{\omega_u}{\omega_l}\right)^{\frac{1-\lambda}{N}}, \quad (15)$$

followed by the recursive relations:

$$\mathring{\omega}_1 = \omega_l \sqrt{\eta}, \quad (16a)$$
$$\hat{\omega}_k = \mathring{\omega}_k \cdot \varepsilon, \; k = \overline{1, N}, \quad (16b)$$
$$\mathring{\omega}_{k+1} = \hat{\omega}_k \cdot \eta, \; k = \overline{1, N-1}. \quad (16c)$$

The MATLAB object `realp` used for fixed-structure robust synthesis does not allow the use of operations other than classical arithmetic operations. Therefore, the recursive fractional-order approximation (14) cannot be used as is in order to compute the fractional-order of the integrative and derivative effects. In Section 2.3 we will give a possible implementation in order to use the `realp` object for optimizing the controller parameters.

2.3. Controller Design Procedure

Although the controller which results by solving the quasi-convex problem (7) manages to fulfill the robust stability and robust performance, the major drawback consists in the fact that the controller is of high-order and cannot be easily implemented. As such, the problem should be constrained to use a specific controller structure. After imposing a fixed-structure family $\mathcal{K}$, the problem (7) can be written as:

$$\inf_{\substack{K \in \mathcal{K} \\ K \text{ stab}}} \sup_{\omega \in \mathbb{R}_+} \inf_{D \in \mathcal{D}} \overline{\sigma}\Big(D(j\omega) \cdot \text{LLFT}(P_\Delta, K)(j\omega) \cdot (D(j\omega))^{-1}\Big). \quad (17)$$

The above problem is non-convex in terms of the free tuning parameters of the controller $K \in \mathcal{K}$. However, the problem (17) can also be solved using the D–K iteration approach, where the K step from (8) is replaced with the following $K_\mathcal{K}$ step:

$$K = \arg \inf_{\substack{K \in \mathcal{K} \\ K \text{ stab}}} \|\text{LLFT}(P_\Delta, K)\|_\infty. \quad (18)$$

In the MATLAB environment there exists the `realp` object which can be used to construct a desired family of controllers $\mathcal{K}$ and then the closed loop system contains both uncertainties and free tunable parameters alike. Using nonsmooth optimization techniques presented in [8] and implemented in [25], the fixed-structure μ-synthesis control problem can be solved. For the purpose of this paper, we consider the fixed structure controller family:

$$\mathcal{K} = \left\{ K_\theta(s) = \begin{pmatrix} K_{1,1}(s) & K_{1,2}(s) & \cdots & K_{1,n_y}(s) \\ K_{2,1}(s) & K_{2,2}(s) & \cdots & K_{2,n_y}(s) \\ \vdots & \vdots & \ddots & \vdots \\ K_{n_u,1}(s) & K_{n_u,2}(s) & \cdots & K_{n_u,n_y}(s) \end{pmatrix} \mid \theta \in D \right\}, \quad (19)$$

where each controller $K_{i,j}$ has the form:

$$K_{i,j}(s) = K_P^{(i,j)} + \frac{K_I^{(i,j)}}{s^{\lambda^{(i,j)}}} + K_D^{(i,j)} s^{\mu^{(i,j)}}, \quad (20)$$

having the free parameters:

$$\theta_{i,j} = \begin{pmatrix} K_P^{(i,j)} & K_I^{(i,j)} & K_D^{(i,j)} & \lambda^{(i,j)} & \mu^{(i,j)} \end{pmatrix} \in \mathbb{R}^5. \quad (21)$$

However, the tunable parameters $\lambda^{(i,j)}$ and $\mu^{(i,j)}$ cannot be used as `realp` objects, due to exponential operations not supported. As a solution, ORA is used with the tunable parameter being $\theta_\lambda \equiv \sqrt{\eta}$ from (15). The transfer function (14) can be implemented using θ_λ as in Algorithm 1.

Algorithm 1: Construct Fractional-Order Element

Input: $\theta_\lambda, N, \omega_u, \omega_l$
Output: $H_{s^\lambda}(s)$

1 $\varepsilon = \left(\frac{\omega_u}{\omega_l} \right)^{\frac{1}{N}} \cdot \frac{1}{\theta_\lambda^2}$
2 $\mathring{\omega}_1 = \omega_l \cdot \theta_\lambda$
3 $H_{s^\lambda}(s) = 1$
4 **for** $k = \overline{1, N-1}$ **do**
5 $\quad \mathring{\omega}_k = \mathring{\omega}_k \cdot \varepsilon$
6 $\quad \mathring{\omega}_{k+1} = \mathring{\omega}_k \cdot \theta_\lambda^2$
7 **end**
8 $\mathring{\omega}_N = \mathring{\omega}_N \cdot \varepsilon$
9 $H_{s^\lambda}(s) = \prod_{k=1}^{N} \frac{s/\mathring{\omega}_k + 1}{s/\mathring{\omega}_k + 1}$

Therefore, the tunable parameters for each controller $K_{i,j}(s)$ are:

$$\hat{\theta}_{i,j} = \begin{pmatrix} K_P^{(i,j)} & K_I^{(i,j)} & K_D^{(i,j)} & \theta_{\lambda^{(i,j)}} & \theta_{\mu^{(i,j)}} \end{pmatrix} \in \mathbb{R}^5, \quad (22)$$

with the special mention that the parameters $\theta_{\lambda^{(i,j)}}$ and $\theta_{\mu^{(i,j)}}$ must be in the domain $\left[1, \left(\frac{\omega_u}{\omega_l} \right)^{\frac{1}{N}} \right]$. If a desired fractional order λ is out of the admissible domain, extra integrator/derivative terms can be added. Therefore, the fixed-structure μ-synthesis control problem can be solved in MATLAB from the desired family $\mathcal{K}$ from (19).

Additionally, the control problem will be posed in a mixed-sensitivity loop shaping μ-synthesis formulation. The main reason for this choice consists in the fact that the mixed-sensitivity loop shaping allows an adequate trade-off between robustness and performance. In the optimization process, the following functions will be used for the loop shaping procedure: the sensitivity function S, the complementary sensitivity function T, and the control effort KS. For each performance function, a set of performance outputs are considered, while the performance inputs are considered as the references.

On one hand, large magnitude in the open loop system implies good reference tracking, disturbance rejection, and unstable plant stabilization. On the other hand, small magnitude of the open loop system ensures robust stability and mitigation of measurement noise. Moreover, a small magnitude of the control effort is necessary to relieve actuator stress. Although all these magnitude requirements seem to lead to an impossible combination, the target frequency ranges for each component are disjunctive. Through the loop shaping mechanism, the engineer is supposed to find three weighting functions, one for each of the previously-mentioned closed loop performances and the frequency performance imposed by the weighting functions is strongly correlated to the corresponding time performance.

For the sensitivity function, the frequency performance indicators of the weighting function are the minimum bandwidth frequency ω_B, which is inversely proportional with the rise time, the maximum magnitude A_S at low frequencies, which imposes the maximum steady-state error, the peak magnitude M_S, which limits the overshoot of the system, along with the imposed slope n_S of the sensitivity function at low and medium frequencies [9]:

$$W_S(s) = \left(\frac{\frac{1}{M_S^{1/n_S}} s + \omega_B}{s + \omega_B A_S^{1/n_S}} \right)^{n_S}. \tag{23}$$

Similarly, the complementary sensitivity's weighting function can be constructed using the peak amplitude M_T, the maximum magnitude at high frequencies A_T, the minimum bandwidth ω_{BT} and the roll-off n_T:

$$W_T(s) = \left(\frac{s + \omega_{BT}}{A_T^{1/n_T} s + \omega_{BT} M_T^{1/n_T}} \right)^{n_T}. \tag{24}$$

The control effort is generally weighted by imposing the magnitude at low and high frequencies, along with an intermediate point of interest. However, the main goal is to maintain the control effort in the range given by the saturation of the physical actuator. For MIMO systems, the weighting matrices are diagonal concatenations of the weighting functions described above. Now the optimization problem that needs to be solved for the proposed method is the mixed-sensitivity fixed-structure loop shaping μ-synthesis:

$$\min_{\substack{K \in \mathcal{K} \\ K \text{ stab}}} \sup_{\omega \in \mathbb{R}_+} \inf_{D \in \mathcal{D}} \overline{\sigma} \Big(D(j\omega) \cdot \text{LLFT}(P,K)(j\omega) \cdot (D(j\omega))^{-1} \Big) \tag{25}$$

$$\text{s.t. } \| (W_S S \quad W_T T \quad W_{KS} KS) \|_\infty < 1$$

3. Mathematical Model of a TRAS

The TRAS model is of sixth order with four inputs and two outputs. The state variables considered are the rotational speed of the tail rotor (ω_h), the rotational speed of the main rotor (ω_v), the azimuth velocity of TRAS beam (Ω_h), the pitch velocity of TRAS beam (Ω_v), the azimuth position (α_h), and the pitch position (α_v), the state vector being:

$$\mathbf{x} = \begin{pmatrix} \omega_h & \omega_v & \Omega_h & \Omega_v & \alpha_h & \alpha_v \end{pmatrix}^\top \in \mathbb{R}^6. \tag{26}$$

There are two control inputs, u_h and u_v, representing the normalized horizontal and vertical DC-motor PWM duty cycles, while the considered outputs will be the azimuth and pitch positions of the TRAS beam:

$$\mathbf{u} = \begin{pmatrix} u_h & u_v \end{pmatrix} \in \mathbb{R}^2, \quad \mathbf{y} = \begin{pmatrix} \alpha_h & \alpha_v \end{pmatrix} \in \mathbb{R}^2. \tag{27}$$

The TRAS model is strongly nonlinear even under some simplifying assumptions, as stated in [27]. One simplification is regarding to the characteristics of the two rotors: their models are supposed to be of first order containing the moment of inertia and the velocity gain for each rotor. Moreover, the angular velocities of the TRAS beam is influenced by the aerodynamic force of each rotor, which is nonlinear in terms of its rotational speed, by the aerodynamic damping torque and by the cross momentum. Moreover, the azimuth velocity is strongly influenced by the pitch angle position, while the pitch velocity is influenced by the pitch angle as well by the return torque. The nonlinear model after some simplifying assumptions can be written as:

$$\dot{\omega}_h = -\frac{1}{I_h}f_1(\omega_h) + \frac{1}{I_h}u_h \tag{28}$$

$$\dot{\Omega}_h = \frac{l_t}{k_1 \cdot \cos^2(\alpha_v) + k_2}f_2(\omega_h) \cdot \cos(\alpha_v) - \frac{k_{fh}}{k_1 \cdot \cos^2(\alpha_v) + k_2}\Omega_h - \frac{k_{vh}}{k_1 \cdot \cos^2(\alpha_v) + k_2}\cos(\alpha_v) \cdot u_v \tag{29}$$

$$\dot{\alpha}_h = \Omega_h \tag{30}$$

$$\dot{\omega}_v = -\frac{1}{I_v}f_3(\omega_v) + \frac{1}{I_v}u_v \tag{31}$$

$$\dot{\Omega}_v = \frac{l_m}{J_v}f_4(\omega_v) - \frac{k_{fv}}{J_v}\Omega_v - \frac{k_3\cos(\alpha_v) + k_4\sin(\alpha_v) + k_5\sin(\alpha_v)\cos(\alpha_v)}{J_v} + \frac{k_{hv}}{J_v}u_h \tag{32}$$

$$\dot{\alpha}_v = \Omega_v \tag{33}$$

All parameters of both linearized an nonlinear systems are described in Table 1. The first step of the linearization process is to find approximations for the functions f_1 and f_3 such that the two systems from inputs to rotational speeds of the rotors are of first order. In order to obtain this scenario, these functions are estimated as $f_1(\omega_h) = k_{Hh} \cdot \omega_h$ and $f_3(\omega_v) = k_{Hv} \cdot \omega_v$, while the nonlinerity is treated using the sector bound technique, being included in the tolerance of each velocity gain. Moreover, the forces developed by each axis are also nonlinear in terms of rotational speeds of the rotors and can be approximated $f_2(\omega_h) = k_{Fh} \cdot \omega_h$ and $f_4(\omega_v) = k_{Fv} \cdot \omega_v$, where the trust coefficients encompass the nonlinearities in their tolerances. All sector bound nonlinearites described above are depicted in Figure 2.

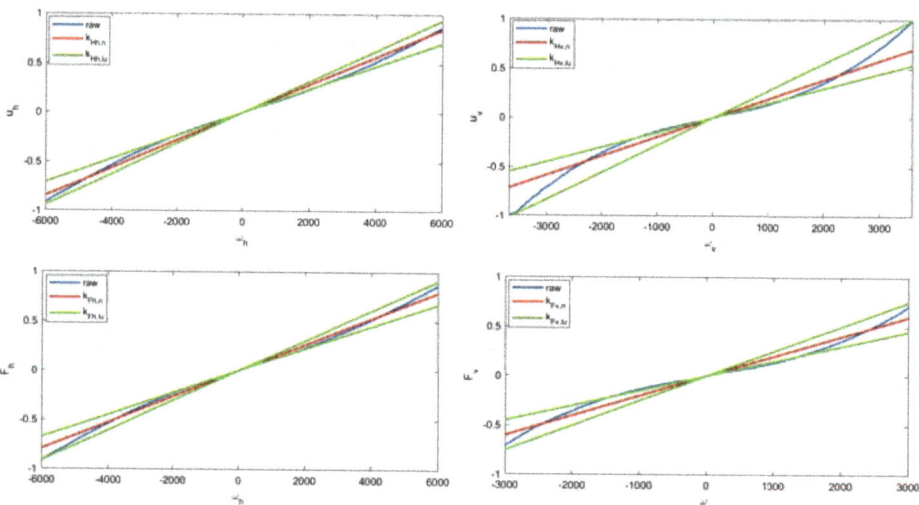

Figure 2. The tolerances of the parameters $k_{Hh}, k_{Hv}, k_{Fh}, k_{Fv}$ which encompass the behaviour of the nonlinear functions f_1, f_2, f_3 and f_4 into sector bound nonlinearities.

Table 1. Twin rotor aerodynamic system physical parameters, values and tolerances.

Symbols	Description	Nominal Value	Tolerance
I_h	moment of inertia of the tail rotor	1/37,000 (kg·m^2)	-
I_v	moment of inertia of the main rotor	1/6100 (kg·m^2)	-
J_h	moment of inertia with respect to the vertical axis	0.0268 (kg·m^2)	±10[%]
J_v	moment of inertia with respect to the horizontal axis	0.0268 (kg·m^2)	-
k_{H_h}	velocity gain of the tail rotor	7.0742 × 10^3 (rad/s)	±10[%]
k_{H_v}	velocity gain of the main rotor	5.1574 × 10^3 (rad/s)	±10[%]
k_{F_h}	thrust coefficient of the tail rotor	1.3218 × 10^{-4} (Ns/rad)	±10[%]
k_{F_v}	thrust coefficient of the main rotor	2.0124 × 10^{-4} (Ns/rad)	±10[%]
k_{f_h}	friction coefficient in the vertical axis	5.889 × 10^{-3} (Nms/rad)	±5[%]
k_{f_v}	friction coefficient in the horizontal axis	1.271 × 10^{-2} (Nms/rad)	±5[%]
k_{hv}	coefficient of the cross moment from tail rotor to pitch angle	4.175 × 10^{-3} (Nm)	±5[%]
k_{vh}	coefficient of the cross moment from main rotor to azimuth angle	−1.782 × 10^{-2}	±5[%]
R_v	coefficient of the return torque	9.360078 × 10^{-2} (Nm)	±10[%]
l_t	length of the tail part of the beam	0.2165 (m)	-
l_m	length of the main part of the beam	0.202 (m)	-
k_1	coefficient of J_h	2.379 × 10^{-2} (kg·m^2)	-
k_2	coefficient of J_h	3.009 × 10^{-3} (kg·m^2)	-
k_3	coefficient of R_v	5.006 × 10^{-2} (Nm)	-
k_4	coefficient of R_v	9.361 × 10^{-2} (Nm)	-
k_5	auxiliary coefficient	0.010624 (Nm)	-

After this first step, the nonlinear model can be now linearized around an equilibrium point. The forced equilibrium point has been chosen such that the outputs are $\bar{\alpha}_h = \bar{\alpha}_v = 0$ [rad], i.e., plant stabilization problem. In order to obtain this point, the state vector has the rest of the components $\bar{\omega}_h = -1336$ [rad/s], $\bar{\omega}_v = 1803.45$ [rad/s], $\bar{\Omega}_h = 0$ [rad/s], $\bar{\Omega}_v = 0$ [rad/s], while the input vector has the components $\bar{u}_h = -0.1492$ and $\bar{u}_v = 0.30559$. According to [27], the moment of inertia with respect to the horizontal axis is constant, while around the vertical axis the moment of inertia is nonlinear, having the expression $J_h = k_1 \cdot \cos^2(\alpha_v) + k_2$. In practice, we will consider this parameter uncertain, having the nominal value $\bar{J}_h = k_1 \cdot \cos(\bar{\alpha}_v) + k_2$, along with a tolerance of ±10[%]. The uncertainties from the thrust coefficients of the tail and the main rotors are necessary in order to compensate the nonlinearity of the aerodynamic forces from these rotors. The friction coefficients in the axes and the cross moments coefficients also present uncertainties in order to compensate the nonlinearities presented in the angular velocity parts and the interconnections between the two rotations. The return torque coefficient is a nonlinear function in terms of pitch position and velocity, which can be approximated by an uncertain parameter having the nominal value $\bar{R}_v = k_3 \sin(\bar{\alpha}_v) - k_4 \cos(\bar{\alpha}_v)$, and a tolerance of ±10%. As such, the linearized state-space model can now be written as:

$$\dot{x}(t) = \begin{pmatrix} -\frac{1}{I_h \cdot k_{H_h}} & 0 & 0 & 0 & 0 & 0 \\ \frac{l_t \cdot k_{F_h} \cdot \cos(\bar{\alpha}_v)}{\bar{J}_h} & -\frac{k_{f_h}}{\bar{J}_h} & 0 & 0 & 0 & -\frac{l_t \cdot k_{F_h} \cdot \sin(\bar{\alpha}_v) + k_{vh} \cdot \sin(\bar{\alpha}_v) \cdot \bar{u}_v}{\bar{J}_h} \\ 0 & 1 & 0 & 0 & 0 & 0 \\ 0 & 0 & 0 & -\frac{1}{k_{H_v} \cdot I_v} & 0 & 0 \\ 0 & 0 & 0 & \frac{l_m \cdot k_{F_v}}{J_v} & -\frac{k_{f_v}}{J_v} & -\frac{R_v + 2k_5 \cos(2\bar{\alpha}_v)}{J_v} \\ 0 & 0 & 0 & 0 & 1 & 0 \end{pmatrix} x(t) + \begin{pmatrix} \frac{1}{I_h} & 0 \\ 0 & -\frac{k_{vh} \cdot \cos(\bar{\alpha}_v)}{\bar{J}_h} \\ 0 & 0 \\ 0 & \frac{1}{I_v} \\ \frac{k_{hv}}{J_v} & 0 \\ 0 & 0 \end{pmatrix} u(t); \quad (34a)$$

$$y(t) = \begin{pmatrix} 0 & 0 & 0 & 0 & 1 & 0 \\ 0 & 0 & 0 & 0 & 0 & 1 \end{pmatrix} x(t) + \begin{pmatrix} 0 & 0 \\ 0 & 0 \end{pmatrix} u(t). \quad (34b)$$

The singular values of the twin rotor aerodynamic system plant having the parameters presented in Table 1, before augmentation, are presented in Figure 3.

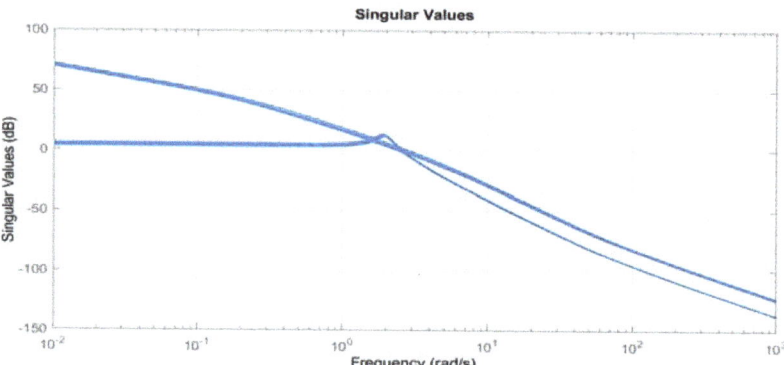

Figure 3. Singular value plot of the twin rotor aerodynamic system.

4. Numerical Results

The controller design procedure proposed in this paper will be applied on a twin rotor aerodynamic system (TRAS). The physical stand from INTECO [27] is presented in Figure 4. The numerical values of the parameters described in Section 3 are presented in Table 1, along with their nominal values and tolerances.

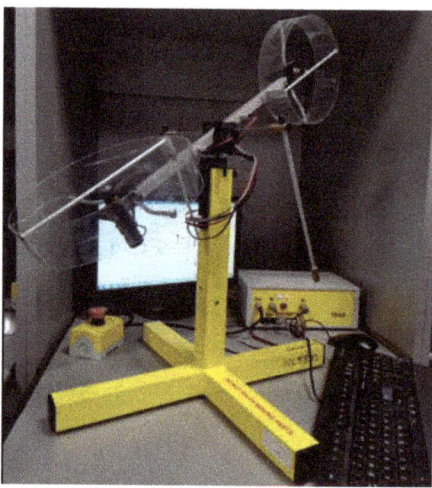

Figure 4. Twin rotor aerodynamic system used for practical experiments.

In order to illustrate the power of the proposed method, a comparison between the numerical simulations for the linearized system using MATLAB and the experimental results on the physical stand has been performed. For the numerical results, the block diagram is presented in Figure 5, where the reference signals $\mathbf{w}_1 \equiv \mathbf{r} = \begin{pmatrix} \alpha_h^\star & \alpha_v^\star \end{pmatrix}^\top$ are considered the inputs of the linearized system, while the performance output vector is:

$$\mathbf{z} = \begin{pmatrix} z_{S,\alpha_h} & z_{S,\alpha_v} & z_{T,\alpha_h} & z_{T,\alpha_v} & z_{KS,\alpha_h} & z_{KS,\alpha_v} \end{pmatrix}. \tag{35}$$

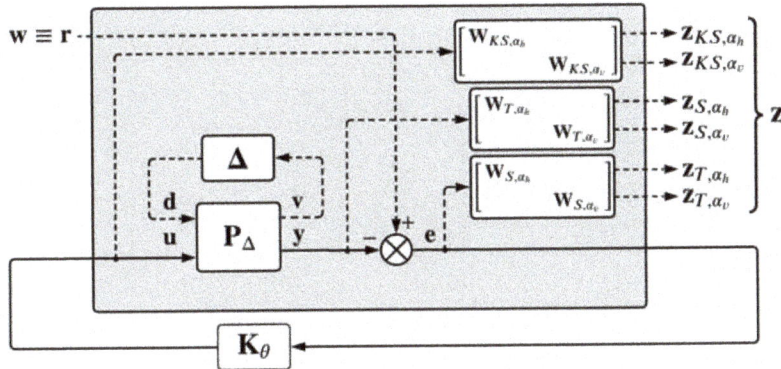

Figure 5. The block diagram of the proposed experiment containing the augmented plant which contains as inputs the reference signals only.

For the numerical simulation part, the plant augmentation has been done with the following weighting functions parameters: $\omega_{B,\alpha_h} = 0.2$ [rad/s], $\omega_{B,\alpha_v} = 0.05$ [rad/s], $A_{S,\alpha_v} = A_{S,\alpha_h} = 1 \times 10^{-2}$, $M_{S,\alpha_v} = M_{S,\alpha_h} = 2$, $n_{S,\alpha_v} = n_{S,\alpha_h} = 1$ (the reference is considered to be a unity step signal), $\omega_{BT,\alpha_h} = 20$ [rad/s], $\omega_{B,\alpha_v} = 5$ [rad/s], $A_{T,\alpha_v} = A_{T,\alpha_h} = 1 \times 10^{-2}$, $M_{T,\alpha_v} = M_{T,\alpha_h} = 2$, $n_{T,\alpha_v} = n_{T,\alpha_h} = 1$, while the DC component of the control effort weighting functions is 1, being the maximum value of the command signal, and the maximum value at high-frequency is of magnitude 5. The weighting functions result as follows:

$$W_S(s) = \begin{pmatrix} W_{S,\alpha_h}(s) & 0 \\ 0 & W_{S,\alpha_v}(s) \end{pmatrix}, \text{ where } W_{S,\alpha_h}(s) = \frac{0.5s + 0.2}{s + 2 \times 10^{-3}}, W_{S,\alpha_v}(s) = \frac{0.5s + 0.05}{s + 5 \times 10^{-4}}, \quad (36)$$

$$W_T(s) = \begin{pmatrix} W_{T,\alpha_h}(s) & 0 \\ 0 & W_{T,\alpha_v}(s) \end{pmatrix}, \text{ where } W_{T,\alpha_h}(s) = \frac{s + 20}{0.01s + 40}, W_{T,\alpha_v}(s) = \frac{s + 5}{0.01s + 10}, \quad (37)$$

$$W_{KS}(s) = \begin{pmatrix} W_{KS,\alpha_h}(s) & 0 \\ 0 & W_{KS,\alpha_v}(s) \end{pmatrix}, \text{ where } W_{KS,\alpha_h}(s) = W_{KS,\alpha_v}(s) = \frac{0.2s + 0.8532}{s + 0.8532}. \quad (38)$$

As noted in Figure 3, the frequency range is between $\omega_l = 1 \times 10^{-2}$ [rad/s] and $\omega_u = 1 \times 10^3$ [rad/s], which will be also used for ORA, along with the order of approximation $N = 5$. Using the augmented plant presented in Figure 5, the fixed-structure mixed-sensitivity loop shaping μ-synthesis problem (25) is solved using the musyn command from MATLAB with the following specifications: the maximum number of D–K iterations is 10, the threshold for the upper bound of the $\mu_\Delta(\text{LLFT}(P_{\text{aug}}, K))$ is 1, and the maximum number of iterations for asserting the lack of progress is 4.

The fixed-structured μ-synthesis control problem was solved using three D–K iterations, having the upper bound of the structured singular value $\mu_\Delta(\text{LLFT}(P_{\text{aug}}, K)) \leq 0.9902 < 1$, which means that the resulting FO-PID controller manages to fulfill both robust stability and robust performance. The resulting FO-PID controller is:

$$K^{\theta^*}_{FO-PID}(s) = \begin{pmatrix} 0.1149 + 0.0603 \cdot s^{-1.267} + 0.0909 \cdot \frac{s^{1.1442}}{0.1154s+1} & -7.1329 + 5.0864 \times 10^3 \frac{s^{1.0001}}{712.97s+1} \\ 0.0315 - 0.0832 \frac{s^{1.2251}}{27.3377s+1} & -0.0297 + 0.1013 \cdot s^{-1.0001} + 0.0232 \cdot \frac{s^{1.2851}}{0.149s+1} \end{pmatrix}, \quad (39)$$

where the low-pass component needs to be added in order to implement the derivative element of order greater than 1 having one extra degree of freedom for each such element. The results obtained after each step are summarized in Table 2, where after x steps the controller design problem has been successfully solved. The upper bound of the structural singular value is presented in Figure 6.

Table 2. The evolution of the structural singular value in the D–K iteration procedure used to solve the mixed-sensitivity fixed structure μ-synthesis problem for the case study—FO-PID structure.

D–K Iteration Number	1	2	3	4
Peak Value of μ (FO-PID)	2.657	1.066	1.007	**0.9902**

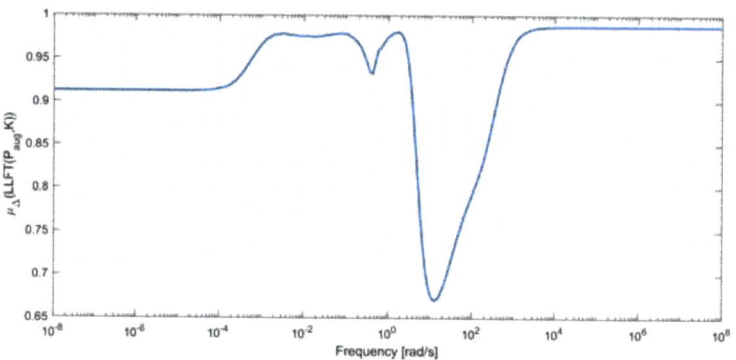

Figure 6. Upper bound of the structural singular value $\mu_\Delta(\text{LLFT}(P_{aug}, K)(j\omega))$ for the frequency range used for solving the Parrot problems.

In order to illustrate the frequency-domain performance, the sensitivity function, complementary sensitivity function and control effort are presented in Figure 7. The nominal plant has been analyzed along with 100 Monte Carlo simulations for the given uncertainty range. Also, in order to underline that the control problem has been successfully solved, the weighting functions are also depicted and it can be noticed that all the simulated functions are under the imposed thresholds. Additionally, the Bode magnitude characteristics of the resulting controller are provided in Figure 8.

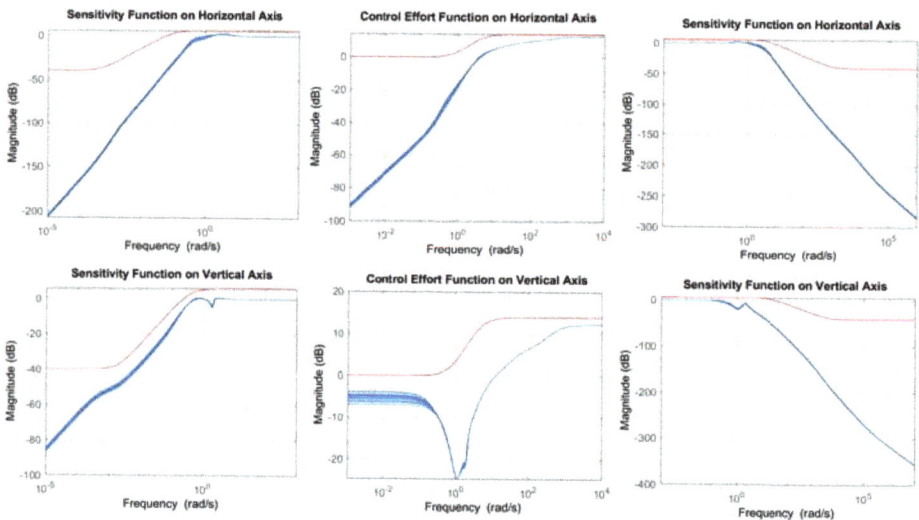

Figure 7. Sensitivity, control effort and complementary sensitivity functions for the TRAS design phase: specified and synthesized.

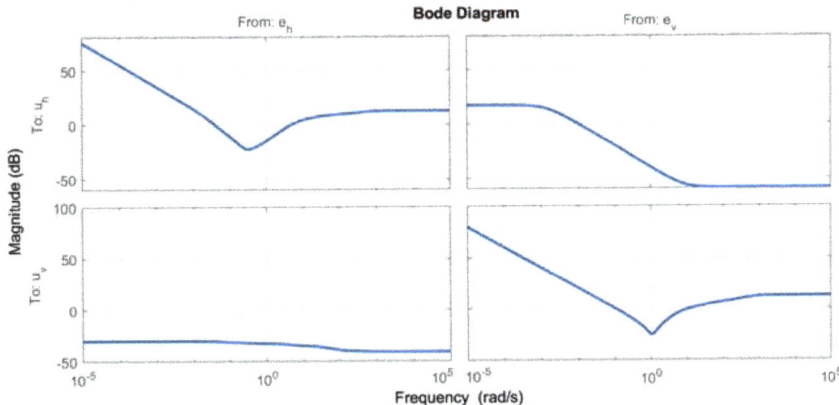

Figure 8. Bode magnitude characteristic of the resulting controller (39).

The time-domain performance of the lower linear fractional transform between the linearized plant and the controller are presented using a step response in Figure 9. In a similar manner, the nominal plant is illustrated along with 100 Monte Carlo simulations. The rise time for the azimuth position varies between 0.796 [s] and 1.05 [s], having a settling time between 14.8 [s] and 16.1 [s] and an overshoot between 16.9 [%] and 24.2 [%], with no steady-state error. Similarly, the rise time for the pitch position is between 6.79 [s] and 11.7 [s], having a settling time between 10.7 [s] and 19 [s], with no overshoot and no steady-state error.

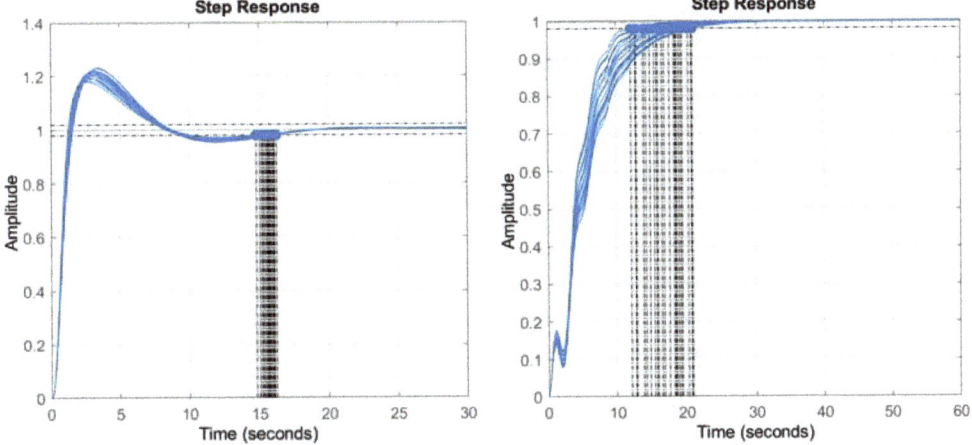

Figure 9. Closed-loop simulated step responses for azimuth and pitch positions, respectively.

Numerical results will further be compared with the experimental results. The first set of experiments, shown in Figure 10, have been made for a square reference with an amplitude of ±0.1 [rad] and a period of 100 [s] for both axes. The initial conditions were varied in practice in order to illustrate the capability of the method. It can be noticed that for the azimuth position, the practical overshoot is a bit higher than in the linear case, with a comparable settling time and near-zero steady-state error due to the quantization effects. The pitch position presents overshoot for the initial step, while for the second step the behavior is similar to that of the linear system, with no overshoot, no steady-state error and comparable settling time.

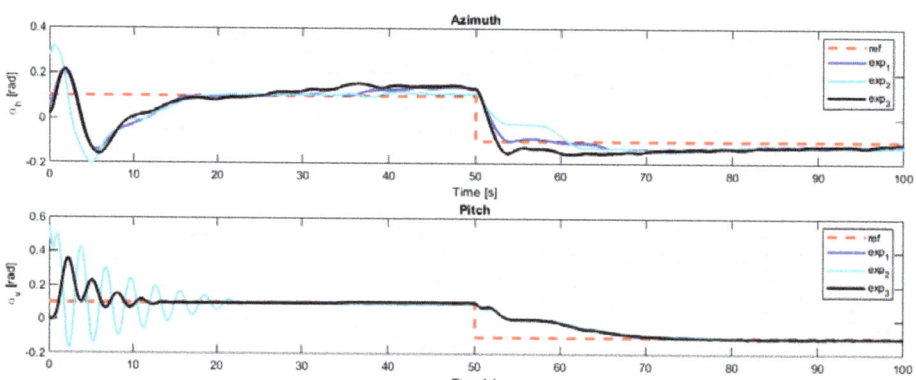

Figure 10. Practical experimental results for reference tracking using various initial conditions.

The second set of experiments are made for the stabilization problem, where the reference for both axes is $\alpha_h^\star = \alpha_v^\star = 0.1$ [rad]. It can be noticed that the azimuth position presents an initial overshoot comparable to that obtained for the linear system, while the second part of the oscillation is more aggressive, underling the influence of the nonlinear components. In a similar manner, the pitch position presents an overshoot along with several oscillations, while the settling time is comparable with the linear case's. The experimental results are shown in Figure 11. Moreover, three different disturbances have been applied after 50 [s]: a perturbation on the vertical axis which leads the pitch position at the maximum value (blue), a perturbation on the horizontal axis with the same characteristics (cyan), and another small perturbation on the horizontal axis (black). It can be noticed that all disturbances have been successfully rejected.

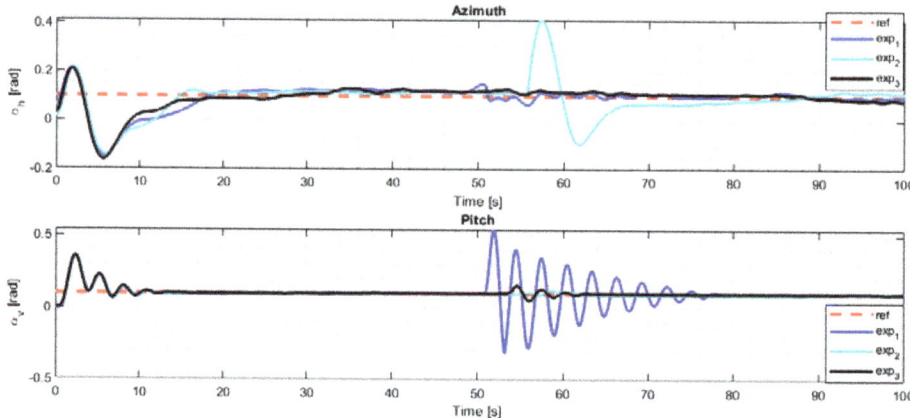

Figure 11. Practical experimental results for disturbance rejection, by alternatively perturbing both the azimuth and pitch axes alike.

Finally, the third set of experiments, depicted in Figure 12, illustrates the behaviour of the proposed method for an operating point far from the forced equilibrium point used in the linearization procedure and controller synthesis. As such, a step of $\alpha_h^\star = \alpha_v^\star = 0.7$ [rad] has been initially applied, with a different pair $\alpha_h^\star = \alpha_v^\star = -0.3$ [rad] applied at the moment $t_1 = 50$ [s]. For the horizontal axis, the overshoot is a bit higher than in the linear case, but with similar settling time and no steady-state error. Also, for the vertical axis, the overshoot is negligible for the first step and zero for the second step, having comparable settling times and no steady-state error. Therefore, the controller can be used for other operating points.

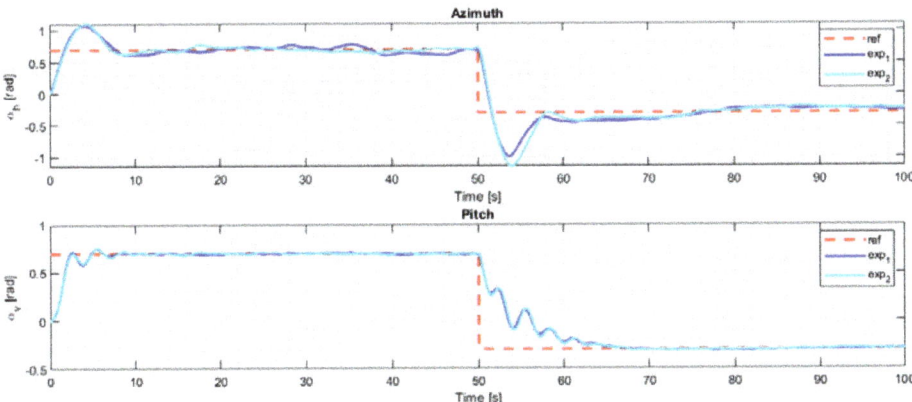

Figure 12. Practical experimental results for reference tracking using high-valued reference signals and, as such, validating the controller for variable plant operating points.

5. Discussion

In order to compare the current iteration of FO-PID μ-synthesis with previous methods, the fixed-structure part of the μ-synthesis mixed-sensitivity loop shaping control problem (25) is solved using the artificial bee colony (ABC) approach presented in [17]. The hyperparameters used for this experiment are: the swarm dimension $N = 50$, the maximum number of ABC cycles 50, the maximum number of cycles with no improvements 10, the limit for the abandonment counter 10, the maximum number of D–K iterations 10, the maximum window length for assessing lack of progress 4, while the parameters for the cost functions are $\alpha = 1$ and $\beta = 10^5$. Using this setup, the fixed-structure mixed-sensitivity loop shaping μ-synthesis problem (25) is solved using five D—K iterations. The resulting controller is:

$$K^{\theta^*}_{FO-PID,ABC}(s) = \begin{pmatrix} 0.1642 + 0.0892 \cdot s^{-1.1834} + 0.0913 \cdot \frac{s^{1.1209}}{0.1173s+1} & -0.0016 + 0.8355 \frac{s^{1.0001}}{104.8s+1} \\ 0.0106 - 6.769 \times 10^{-4} \frac{s^{1.4938}}{100s+1} & -0.0741 + 0.0981 \cdot s^{-1.185} + 0.001 \cdot \frac{s^{1.157}}{0.0735s+1} \end{pmatrix}, \quad (40)$$

Additionally, an experiment with unstructured μ-synthesis has also been performed, leading to an upper bound of the structured singular value of 99.86 and a controller of 34th order, which means that robust stability and robust performance are not guaranteed, with the controller additionally of high order. The optimization algorithm has been stopped after three iterations because the diverging stopping criteria has been reached. A summary of the obtained results with the proposed method, the ABC method [17] and the unstructured μ-synthesis is presented in Table 3.

Table 3. A comparison between the evolution of the structural singular values in the D–K iteration procedure used to solve the mixed-sensitivity fixed structure μ-synthesis problem for the case study.

D–K Iteration Number	1	2	3	4	5
Peak Value of μ (FO-PID)	2.657	1.066	1.007	0.9902	-
Peak Value of μ (unstructured)	100	99.8	99.4748	-	-
Peak Value of μ (ABC approach [17])	105.7741	2.4095	1.2452	1.1255	0.9989

As such, the unstructured version of the μ-synthesis control problem could not be solved, resulting a high-order controller which does not guarantee robust stability and performance. On the other hand, both remaining methods managed to solve the control problem described in (25). The new method introduced in this paper manages to solve the problem faster due to the advantages of the nonsmooth optimization techniques implemented in MATLAB.

As a future iteration, we propose to find a decentralized controller having a nonlinear component and a linear and time-invariant (LTI) robust component. The nonlinear component needs to be designed such that the lower linear fractional transform interconnection between the plant and such a component is asymptotically stable, as in [28], where the passivity-based control framework has been extended for quasi-linear input-affine systems. Additionally, the LTI robust controller can be designed using the proposed method, the decentralized controller managing to ensure robust stability and performance for the nonlinear system. Moreover, the fractional-order element can be approximated using the presented ORA method only for s^λ, with $\lambda \in (0,1)$. As such, another research direction is to find a method to integrate all positive values of $\lambda \in \mathbb{R}_+$. On the other hand, the presented methods were considered in the continuous-time domain, although, for practical implementation, the controller must be discretized and also quantized. A starting point for this aspect could be the work presented in [24].

6. Conclusions

The current paper presents an algorithm which manages to integrate the MIMO fractional-order PID (FO-PID) controller in the fixed-structure mixed-sensitivity loop shaping μ-synthesis control problem by constructing an element isomorphic with the fractional order. In order to expose the method capacity and potential, a twin rotor aerodynamic system experimental stand has been utilized. After the simplified nonlinear and linearized models were presented, the linear system has been augmented with weighting functions which managed to impose the desired performance. The fixed-structure μ-synthesis control problem has been successfully solved using four D–K iterations, resulting a controller which manages to ensure both robust stability and robust performance. A comparative analysis between the results obtained with the designed controller used for the linearized plant and for the practical experimental stand has also been performed.

As future work, the proposed design method will be added into a next iteration of the toolbox initially proposed in [9] in order to automatically perform the fractional-order fixed-structure μ-synthesis. Also, the proposed method can be integrated into a control scheme with a decentralized controller having an extra nonlinear component which ensures that the robust stability and robust performance are also guaranteed for the nonlinear system.

Author Contributions: Conceptualization, V.M. and M.Ş.; methodology, V.M.; software, V.M.; validation, M.Ş. and E.H.D.; formal analysis, V.M. and E.H.D.; investigation, V.M.; resources, V.M. and M.Ş.; data curation, V.M. and M.Ş.; writing—original draft preparation, V.M., M.Ş. and E.H.D.; writing—review and editing, V.M., M.Ş. and E.H.D.; visualization, M.Ş.; supervision, E.H.D.; project administration, V.M.; funding acquisition, E.H.D. All authors have read and agreed to the published version of the manuscript.

Funding: This paper was financially supported by the Project "Entrepreneurial competences and excellence research in doctoral and postdoctoral programs—ANTREDOC", project co-funded by the European Social Fund financing agreement no. 56437/24.07.2019.

Institutional Review Board Statement: Not applicable.

Informed Consent Statement: Not applicable.

Conflicts of Interest: The authors declare no conflict of interest.

Abbreviations

The following abbreviations are used in this manuscript:

CACSD	Computer-Aided Control System Design
DOF	Degrees of Freedom
FO-PID	Fractional-Order PID

LLFT	Lower Linear Fractional Transform
LMI	Linear Matrix Inequality
LTI	Linear and Time-Invariant
MIMO	Multiple-Input Multiple-Output
NP	Non-Deterministic Polynomial-Time
PID	Proportional-Integral-Derivative
TRAS	Twin Rotor Aerodynamic System

References

1. Doyle, J.; Glover, K.; Khargonekar, P.; Francis, B.A. State-space solutions to standard $\mathcal{H}_2$ and $\mathcal{H}_\infty$ control problems. *IEEE Trans. Autom. Control.* **1989**, *34*, 831–847. [CrossRef]
2. Ionescu, V.; Oară, C.; Weiss, M. *Generalized Riccati Theory and Robust Control—A Popov Function Approach*; John Wiley & Sons: Chichester, UK, 1999.
3. Gahinet, P.; Apkarian, P. A linear matrix inequality approach to $\mathcal{H}_\infty$ control. *Int. J. Robust Nonlinear Control.* **1994**, *4*, 421–448. [CrossRef]
4. Șușcă, M. Solving Algebraic Riccati Equations Using Proper Deflating Subspaces for $\mathcal{H}_2/\mathcal{H}_\infty$ Synthesis. Master's Thesis, Technical University of Cluj-Napoca, Cluj-Napoca, Romania, 2019.
5. Mihaly, V. General Purpose Linear Matrix Inequality Solver with Applications in Robust and Nonlinear Control. Master's Thesis, Technical University of Cluj-Napoca, Cluj-Napoca, Romania, 2020.
6. Packard, A.; Doyle, J.; Balas, G. Linear Multivariable Robust Control with a μ Perspective. *J. Dyn. Syst. Meas. Control.* **1993**, *115*, 426–438. [CrossRef]
7. Apkarian, P., Noll, D. Nonsmooth $\mathcal{H}_\infty$ Synthesis. *IEEE Trans. Autom. Control.* **2006**, *41*, 71–86. [CrossRef]
8. Apkarian, P. Nonsmooth μ synthesis. *Int. J. Robust Nonlinear Control.* **2011**, *21*, 1493–1508. [CrossRef]
9. Șușcă, M.; Mihaly, V.; Stănese, M.; Morar, D.; Dobra, P. Unified CACSD Toolbox for Hybrid Simulation and Robust Controller Synthesis with Applications in DC-to-DC Power Converter Control. *Mathematics* **2021**, *9*, 731. [CrossRef]
10. Apkarian, P.; Bompart, V.; Noll D. Nonsmooth Structured Control Design with Application to PID Loop-Shaping of a Process. *Int. J. Robust Nonlinear Control.* **2007**, *17*, 1320–1342. [CrossRef]
11. Dulf, E.H.; Șușcă, M.; Kovacs, L. Novel Optimum Magnitude Based Fractional Order Controller Design Method. *IFAC-PapersOnLine* **2018**, *51*, 912–917. [CrossRef]
12. Dulf, E.H. Simplified Fractional Order Controller Design Algorithm. *Mathematics* **2019**, *7*, 1166. [CrossRef]
13. Muresan, C.I.; Birs, I.R.; Dulf, E.H. Event-Based Implementation of Fractional Order IMC Controllers for Simple FOPDT Processes. *Mathematics* **2020**, *8*, 1378. [CrossRef]
14. Vinagre, B.M.; Monje, C.A.; Calderón, A.J.; Chen, Y.Q.; Feliu, V. The fractional integrator as a reference function. In Proceedings of the 1st IFAC Workshop on Fractional Differentiation and its Application, Bordeaux, France, 19–21 July 2004.
15. Mihaly, V.; Dulf, E. Novel fractional order controller design for first order systems with time delay. In Proceedings of the 2020 IEEE International Conference on Automation, Quality and Testing, Robotics (AQTR), Cluj-Napoca, Romania, 21–23 May 2020; pp. 1–4.
16. Sabatier, J.; Lanusse, P.; Melchior, P.; Outaloup, A. *Fractional Orde Differentiation and Robust Control Design*; Springer: Dordrecht, The Netherlands, 2015; Volume 77.
17. Mihaly, V.; Șușcă, M.; Morar, D.; Stănese, M.; Dobra, P. μ-Synthesis for Fractional-Order Robust Controllers. *Mathematics* **2021**, *9*, 911. [CrossRef]
18. Petkov, P.H.; Christov, N.D.; Konstantinov, M.M. Robust Real-Time Control of a Two-Rotor Aerodynamic System. In Proceedings of the 17th World Congress, Seoul, Korea, 6–11 July 2008; pp. 6422–6427.
19. Ahmad, M.; Ali, A.; Chou, M.A. Fixed-Structure $\mathcal{H}_\infty$ Controller Design for Two-Rotor Aerodynamical System (TRAS). *Arab. J. Sci. Eng.* **2016**, *41*, 3619–3630. [CrossRef]
20. Khalid, M.U.; Saleem, F.; Shaikh, I.U.H.; Ali, A. Decentralized 2 degree of freedom loop shaping $\mathcal{H}_\infty$ controller for twin rotor aerodynamic system. In Proceedings of the 13th International Conference on Emerging Technologies (ICET), Islamabad, Pakistan, 27–28 December 2017. [CrossRef]
21. Saleem, F.; Ali, A.; Shaikh, I.H.; Wasim, M. A Hybrid $\mathcal{H}_\infty$ Control Based ILC Design Approach for Trajectory Tracking of a Twin Rotor Aerodynamic System. *Mehran Univ. Res. J. Eng. Technol.* **2021**, *40*, 169–179. [CrossRef]
22. Al-Mahturi, A.; Wahid, H. Optimal Tuning of Linear Quadratic Regulator Controller Using a Particle Swarm Optimization for Two-Rotor Aerodynamical System. *Open Sci. Index Electron. Commun. Eng.* **2017**, *11*, 196–202.
23. Faisal, R.F.; Abdulwahhab, O.W. Design of a PID-Lead Compensator for a Twin Rotor Aerodynamic System(TRAS). *J. Eng.* **2021**, *27*, 79–88. [CrossRef]
24. Kim, B.M.; Yoo, S.J. Approximation-Based Quantized State Feedback Tracking of Uncertain Input-Saturated MIMO Nonlinear Systems with Application to 2-DOF Helicopter. *Mathematics* **2021**, *9*, 1062. [CrossRef]
25. Balas, G.; Chiang, R.; Packard, A.; Safonov, M. *Robust Control Toolbox—User's Guide*; The MathWorks: Natick, MA, USA, 2020.
26. Caputo, M. Linear model of dissipation whose Q is almost frequency independent-II. *Geophys. J. Int.* **1967**, *13*, 529–539. [CrossRef]

27. *Two Rotor Aero-Dynamical System—User's Manual*; INTECO: Moscow, Russia, 2009.
28. Mihaly, V.; Şuşcă, M.; Dobra, P. Krasovskii Passivity and μ-Synthesis Controller Design for Quasi-Linear Affine Systems. *Energies* **2021**, *14*, 5571. [CrossRef]

Article

Adaptive Control of CO_2 Production during Milk Fermentation in a Batch Bioreactor

Jožef Ritonja

Faculty of Electrical Engineering and Computer Science, University of Maribor, Koroška cesta 46, 2000 Maribor, Slovenia; jozef.ritonja@um.si; Tel.: +386-2-220-7074

Abstract: The basic characteristic of batch bioreactors is their inability to inflow or outflow the substances during the fermentation process. This follows in the simple construction and maintenance, which is the significant advantage of batch bioreactors. Unfortunately, this characteristic also results in the inability of the current industrial and laboratory batch bioreactors to control fermentation production during the process duration. In some recent studies, it was shown that changing the temperature could influence the execution of the fermentation process. The presented paper shows that this phenomenon could be used to develop the closed-loop control system for the fermentation production control in batch bioreactors. First, based on theoretical work, experiments, and numerical methods, the appropriate structure of the mathematical model was determined and parameters were identified. Next, the closed-loop control system structure for batch bioreactor was proposed, and the linear and adaptive control system based on this structure and the derived and identified model were developed. Both modeling and adaptive control system design are new and represent original contributions. As expected, due to the non-linearity of the controlled plant, the adaptive control represents a more successful approach. The simulation and experimental results were used to confirm the applicability of the proposed solution.

Keywords: biotechnology; fermentation process; batch bioreactors; modeling; control system design and synthesis; linear control; adaptive control; model reference adaptive control; control system realization

1. Introduction

1.1. Basic Facts about Fermentation Process and Batch Bioreactors

The fermentation process represents a planned use of microorganisms (bacteria, yeasts, molds, or algae) or cells (animal or plant cells) to make products advantageous to humans. In the food industry, fermentation refers to bioprocesses where microorganisms' activity creates a desirable change in food and beverages to improve flavor, provide health benefits, or preserve foodstuffs.

Fermentation processes are carried out in bioreactors. With regard to the type of fermentation process, bioreactors are divided into three groups: batch bioreactors, fed-batch bioreactors, and continuous bioreactors. The main difference between the individual types of bioreactors is in their ability to supply and discharge substances during the fermentation process. Batch bioreactors are the simplest and do not allow the input and output of substances during the fermentation process. This means that the time course of the fermentation process quantities depends entirely on the initial concentrations of bioreactor substances. During operation, the bioreactor is closed, and we do not have the ability to control the fermentation process. From an operational standpoint, this type of bioreactor is the least capable. However, due to their uncomplicated construction, these bioreactors are the cheapest to purchase and, at the same time, very easy to maintain. Fed-batch bioreactors allow the introduction of substances during the fermentation process but do not allow the removal of substances. All fermentation products remain in the bioreactor until the end of fermentation. The possibility of adding substances during operation makes it possible to influence the fermentation process in fed-batch bioreactors during operation. Unlike batch bioreactors, fed-batch bioreactors enable a relatively simple and efficient implementation of a closed-loop control system, ensuring the desired dynamics of the fermentation process. Continuous bioreactors are the most capable in terms of adding and removing substances. They allow the inflow and outflow of substances into/from the bioreactor continuously throughout the operation as a flowing stream. Although fed-batch and continuous bioreactors allow greater flexibility during operation, batch bioreactors are still used widely in the industrial environment. Based on data from manufactures and traders, industrial bioreactors are still made primarily for batch processing (some reports even 90% presence in certain areas) [1].

The goal of fermentation is to produce a lot of high-quality fermentation product in the shortest possible time. This goal is achieved when the time course of fermentation quantity follows the prescribed reference course. Therefore, the control of the fermentation process is extremely important.

While the control of the fermentation processes in fed-batch and continuous bioreactors is relatively easy to implement, the control of the batch bioreactors is very difficult to perform. The reason is simple: batch bioreactors do not have an input substance that could be changed through inflow or outflow during the fermentation process and used to control it. The fact that there are extremely rare examples in commercial offers or in academic publications that show the control system for the production control during fermentation processes in batch bioreactors posed a challenge for this study. This paper has focused on developing a control system for a batch bioreactor that utilizes **temperature changing** to control the growth of the fermentation product. The implementation of the adaptive control system represents an original approach that is not found in other publications.

1.2. Literature Review

The problem of the automatic control of the fermentation processes is very important, up to date, and attractive. The availability of non-expensive equipment for the development and manufacture of control systems has caused great topicality in this area in the last three decades. Therefore, in recent years, we have seen an enormous effort from academic institutions and industrial providers to find new control systems for bioreactors.

The initial phase of any control system research represents the determination of the mathematical model of the controlled plant. We can trace the intensive work and new publications in mathematical modeling of the fermentation processes in bioreactors. Still, the progress in the field of mathematical modeling does not reach the development in the field of the control of the fermentation process. In control studies in the last two decades, the fundamental kinetic mathematical model of the fermentation in the bioreactors has still been used commonly for quantitative simulations or theoretical analysis [2–5]. Unfortunately, in many cases, this model is not the most suitable for the design and synthesis of bioreactor control systems [6]. The new bioreactors enable an easy, fast and wide range of changes in the mechanical (by mixing) and thermal (by heating and cooling) conditions of fermentation processes, also during operation. It turns out that the course of the fermentation process can be influenced by changes in these fermentation quantities (stirrer speed and temperature), and it is not always necessary to control the fermentation process by feeding substances into the bioreactor [7]. To develop the control system for the fermentation process, which would use stirrer speed or heater temperature as an input quantity, we need a mathematical model that describes the influence of these quantities on the fermentation process. References [6,8] are some of the publications where the influence of stirrer speed and heater temperature on the fermentation process is analyzed, and appropriate mathematical models are also determined.

More publications are in the field of the control of bioreactors. The bioreactor fermentation process is a very suitable and attractive process for developing and testing conventional and advanced control theories. The presented review is focused on works dealing with the control of the time profile of the fermentation product.

As expected, most publications are in the field of control systems developed for continuous bioreactors, where control is possible through changes in inflow and outflow during the fermentation. Reference [8] shows the utilization of robust control for continuous bioreactors. The implementation of the sliding mode theory is presented in [9], the use of output linearization in [10], the application of output linearization taking into account the constraints of the input signals is studied in [11], the appropriateness of model predictive control (MPC) is shown in [12,13]. A multitude of new publications testifies to the topicality of the problem and intensive work in this area.

The intensive development of the control systems is also seen in the field of fed-batch bioreactors. Reference [14] shows the use of robust control for the fermentation process in fed-batch bioreactors. The use of an iterative learning controller is presented in [15]. The use of model-based optimization for a fed-batch bioreactor was studied in [16]. References [17,18] discuss the applicability of MPC for fed-batch bioreactor control. Reference [19] shows the implementation of the sliding mode control for the photobioreactor (which works initially in the fed-batch mode and then in continuous operation), but the reference deals only with the fed-batch stage. All publications demonstrate the advantage of advanced control concepts over the conventional closed-loop control of fed-batch bioreactors.

As opposed to continuous- and fed-batch bioreactors, relatively few publications have been observed that address the closed-loop control of the time profile of the fermentation product quantity during the fermentation process in batch bioreactors. Most batch bioreactors still operate autonomously, without closed-loop control, which would control the fermentation process. Publications in the field of batch bioreactor control are mainly limited to the control of bioreactors' subsystems. Many works show different control theories or different realizations for temperature regulation, pH regulation, oxygen control, and stirrer speed control. The most considered is temperature regulation. Reference [20] comparatively shows the use of MPC and sliding mode control for temperature regulation in a batch bioreactor. Reference [21] shows temperature control of fermentation bioreactor for ethanol production using internal model control (IMC) based PID controller. Modified fractional-order IMC-PID for ethanol production is proposed in [22]. Non-linear autoregressive moving average neuro controller for temperature control in bioreactors is shown in [23]. The temperature control of an alcoholic fermentation process through Takagi–Sugeno modeling is presented in [24]. A fuzzy–split range control system applied to the fermentation process is shown in [25].

Because of the importance of dissolved CO_2 for the fermentation process, it is also possible to find frequent publications considering CO_2 monitoring. Reference [26] discusses the importance of real-time CO_2 monitoring for the proper execution of the fermentation process. Reference [27] describes sensors for real-time dissolved CO_2 monitoring and control. A noninvasive approach for monitoring dissolved CO_2 in cell culture using a silicon sampling loop is presented in [28].

However, very few publications deal with the control of the growth of microorganisms in batch bioreactors. The growth is visible in the time course of the generation of the end-product quantity during the fermentation. Only a few references in this field were found. The gain scheduling control was used in [29]. Reference [30] shows the implementation of the PI-controller. Reference [31] studies the application of model reference adaptive control. The absence of publications studying the control of the yield of the fermentation product in batch bioreactors was also an additional motivation to work even more in-depth and intensively in this area.

1.3. Contributions and Novelties

There are two major contributions of this article.

- The first contribution of the presented study is the discovery of the solution for the closed-loop control of the growth of microorganisms (and, thus, control the time course of the fermentation product quantity), which will be valid for **batch bioreac-**

tors. The controlled operation mode has so far been reserved only for the fed-batch and continuous bioreactors, which are much more expensive to purchase and more difficult to maintain than batch bioreactors. All today's industrial or laboratory batch bioreactors operate without a closed-loop control system in an autonomous mode. The time course of the fermentation product quantity depends only on the initial concentrations of substances introduced into the bioreactor before the start of the fermentation process. The presented solution is based on the finding that changing the bioreactor's temperature could be used for the closed-loop control of the fermentation product profile. This discovery was obtained from the analysis of previous studies, from simulation calculations based on the derived mathematical model, and from the laboratory experiments. Based on this finding, the structure of the closed-loop control system was defined. This structure allows the use of different control approaches.

- The second contribution of the article is the finding that adaptive control is very convenient for the control of the time course of the fermentation product in batch bioreactors. A study of various adaptive theories was made. Model reference adaptive control based on almost strictly positive real theory proved to be convenient for the implementation of the founded control structure. This control approach assures stability, easy realization, and an undemanding choice of adaptation coefficients. The proposed adaptive control system was compared with the conventional linear control system. The advantages of the developed adaptive control system are to ensure the desired course of the fermentation process even when the parameters of the mathematical model of the fermentation process are unknown. An additional and important advantage of the presented adaptive control system is that it ensures the same performance even in the case of significantly different fermentation processes.

In such a way, the batch bioreactors, thanks to the advanced control theory, easily and cheaply acquire the possibility to significantly improve their performance. The implementation of the developed adaptive control system does not the require major modification of batch bioreactors, and all basic advantages of these reactors are retained. The adaptive control system is easy to start and does not require time-consuming bioreactor identification and the controller's parameter setting.

The shorter fermentation time and higher quality of the obtained fermentation products are guaranteed, which means greater efficiency of operation. In addition to simulations, the efficiency and stability of the proposed adaptive control system have also been proven by experiments on a laboratory bioreactor. Although the applicability of the adaptive control system is confirmed in the case of CO_2 production during milk fermentation, the proposed control system is universal and is suitable for controlling various fermentation processes in batch bioreactors.

The originally presented novelties in this article are:

- A new derived non-linear mathematical model which describes the impact of temperature on the fermentation process substances in batch bioreactors;
- The use of an optimization technique for mathematical model parameters estimation;
- The definition of the fundamental control structure for control of time courses of fermentation products in batch bioreactors using temperature changes;
- Implementation of conventional linear and adaptive control theories for control of the fermentation product response;
- Comparison of the efficiency of both control approaches and simulation-based numerical evaluation of both control approaches;
- Experimental confirmation of the proposed adaptive control system.

2. Materials and Methods

2.1. Studied Fermentation Process

The presented study focused on the production of CO_2 during milk fermentation with kefir grains. Traditionally, kefir is produced by inoculating kefir grains, which are a mixture of proteins, polysaccharides, mesophilic, homofermentative, and heterofermentative lactic acid streptococci, thermophilic and mesophilic lactobacilli, acetic acid bacteria, and yeast. The fermentation of milk by the inoculum proceeds for ca. 24 h, during which time, homofermentative lactic acid streptococci grow rapidly, initially causing a reduction in pH. This low pH favors the growth of lactobacilli but causes the streptococci number to decline. The presence of yeasts in the mixture, together with fermentation temperature, encourages the growth of aroma-producing heterofermentative streptococci. As fermentation proceeds, the growth of lactic acid bacteria is favored.

For the original fermentation, before the fermentation, kefir grains (40 g) were activated for 5 successive days so that they were washed daily with cold water and put into 500 mL of fresh pasteurized whole-fat milk at room temperature. To start the fermentation, 500 mL of fresh pasteurized whole-fat milk was preheated in the fermenter to the desired temperature and then inoculated with 40 g of active kefir grains. For the original fermentation, the desired starting milk temperature was 22 °C, and fully activated (5 days activation) kefir grains were used. Different modified fermentation processes were obtained by means of differently activated kefir grains.

During the fermentation, carbon dioxide, acetic acid, ethyl alcohol, and several other substances are formed, and these give the products their characteristic aroma. Milk fermentation with kefir grains propagation is an inherently very complex process because of the specific nature of the microbial metabolism, as well as the non-linearity of its kinetics. Therefore, fermentation control is extremely important to obtain high-quality products.

2.2. Laboratory Equipment

2.2.1. Batch Bioreactor

Laboratory fermentations were performed in the reaction calorimeter RC1e from Mettler Toledo. It is a computer-controlled benchtop batch bioreactor with a working volume of 0.7 L. By using specific modifications in hardware and software, it was used as a bioreactor. A more detailed description of RC1e can be found in [31].

2.2.2. Heating/Cooling System

The tested batch bioreactor was factory equipped with the combined heating/cooling (H/C) system. The silicone oil used as a heat transfer agent is pumped through the double jacket of the reactor in a closed circulation system. H/C system is equipped with an integrated closed-loop temperature control system with a proportional-integral (PI) controller. The H/C system enables the changing of the temperature of the bioreactor's contents in the range from 5 °C to 50 °C. The temperature control system enables operation without steady-state error for a constant reference temperature. The delay in the temperature control system is very short compared to the dynamics of the fermentation process. The H/C system was identified and modelled. The 1st-order differential equation with unit gain and estimated time constant $T_{\vartheta cs}$ = 0.1 h represents a satisfactory description of its dynamics.

2.2.3. Dissolved CO_2 Measurement

The selection of the output quantity that could be used as the feedback variable in the control system is crucial for the implementation of the control theory and the realization of the theoretical approach. It is necessary that the measured quantity contains as much information on the fermentation process as possible. At the same time, it is also important that the measurement should be accurate and could be performed on-line.

In the fermentation processes, dissolved oxygen and cell culture measurement are essential for ensuring optimal conditions for cell growth. The oxygen levels in bioreactors can have an impact on the growth rate, nutrients' uptake, cellular morphology, and metabolite synthesis, leading to end-product quality. Accurate oxygen control is only possible if measurements from dissolved oxygen sensors installed in fermenters/bioreactors are reliable. Biomass concentration is another critically needed measurement in fermentation studies.

In performing laboratory tests, it is not always possible to measure these two biochemical quantities and, thus, conclude whether their trajectories are such as to ensure the desired course of the fermentation process with a high-quality end product. This is the reason that in the proposed study, the measurement of the CO_2 dissolved in the bioreactor's medium was introduced. CO_2 is a product of the cellular metabolism of microorganisms. Assuming the growth medium with a sufficient carbon source, the measured CO_2 concentration profile could also be the indicator of the fermentation progress [32]. Accurate, real-time data on the concentration of CO_2 increases the understanding of the fermentation process and can get a better insight into cell metabolism, cell culture productivity, and other changes within bioreactors [33]. The distribution of the CO_2 in the bioreactor medium is very homogeneous. The sensors for the measurement of CO_2 concentration are reliable, accurate, maintenance free, have a long lifetime, and have a known measurement curve [33]. The duration of the measurement process is short; therefore, these sensors are convenient for implementation in real-time control systems.

For the measurement of the dissolved CO_2 in the laboratory bioreactor the ISE51B (Mettler Toledo) ion-selective electrode was used. A measuring system can be modelled with the 1st-order differential equation. The laboratory measuring system can be matched with the mathematical model with a gain k_{CO2ms} = 1 mmol/g and a time constant T_{CO2ms} < 0.01 h. The time constant of the measurement system is almost negligible compared to the inherently slow dynamics of the biotechnological systems.

2.2.4. Equipment for Data Acquisition and Control

For the connection of the dissolved CO_2 measurement sensor and signal adjustment, the SevenMulti (Mettler Toledo) basic device with an expansion module was used. The analogue 1st-order low-pass filter for the elimination of sensor noise is integrated into the expansion module.

For the transfer of measured signals from the SevenMulti basic unit to PC, the basic device was equipped with a digital output module (USB). For the comprehensive measurement of several quantities over a long time period and for the necessary signal processing, software LabX direct pH 2.3 was installed on the PC. This is professional equipment used for a data logger and a data analysis. The selected sampling time was 10 min. This sampling time was sufficient due to the slow dynamics of the fermentation process. During the performing the experiments, the sampling time was changed and adjusted to the dynamics of the measured signal. The measured data were saved into Microsoft Office 365 Excel documents, transferred into MathWorks MATLAB, and processed using MATLAB with its Optimization toolbox functions [7].

For the implementation of the control system, a dSpace 1103 PPC controller board was utilized. The controller is equipped with 16-bit A/D and D/A converters as well as serial and CAN interfaces [7]. An analogue output module of the basic device SevenMulti was used for the transfer of the measured signal of the CO_2 concentration from the bioreactor system to the controller's analogue input. The additional analogue 1st-order low-pass filter was used at the dSpace analogue input to eliminate the superposed noise signal. The analogue output signal from the controller is sent to the input of the heating/cooling system. To enable this connection, the heating/cooling system was equipped with an additional electronic interface.

2.2.5. Reference Profile Generation

The quantity and quality of the product in the batch bioreactor are decisively dependent on the trajectories of the biological quantities in the fermentation process, affecting the kinetics of the bioprocess. The developed control system makes the influencing of the time responses of the biological quantities possible. With the developed control system, we can change the time course of biological quantities in a batch bioreactor. In this way, we can influence the fermentation process and its result. The question arose about how to choose a reference trajectory. The reference trajectory selected should provide that the generated product will be high quality, abundant, and that the process will end in a shorter time, with as little energy and material resources as possible. There are many professional and scientific publications where the methods for determining the optimal trajectory for different fermentation processes for fed-batch bioreactors are discussed that could also be useful for batch bioreactors [3]. The presented article does not deal with the methods of determining the optimal trajectory. The primary purpose of this study is the development of a control system that will ensure that the fermentation process quantities will follow the previous set reference trajectory. Therefore, the reference trajectory used in this paper was determined by the empirically obtained expert knowledge of the consumers of the bioreactor's technology [33]. The reference trajectory of the dissolved CO_2 course was generated by means of dSpace 1103 PPC controller board.

2.3. Laboratory Set-Ups for Parameter Estimation and Control

To perform the parameter estimation and fermentation control, the laboratory batch bioreactor was supplemented with the controlled heating/cooling system, measuring system, PC for parameter estimation, and dSpace for control implementation. The laboratory set-up for the **measurement** of the time response of the dissolved CO_2 production on the temperature changes and **estimation** of model parameters is shown in Figure 1a. Laboratory set-up for the **control** of the dissolved CO_2 production is shown in Figure 1b.

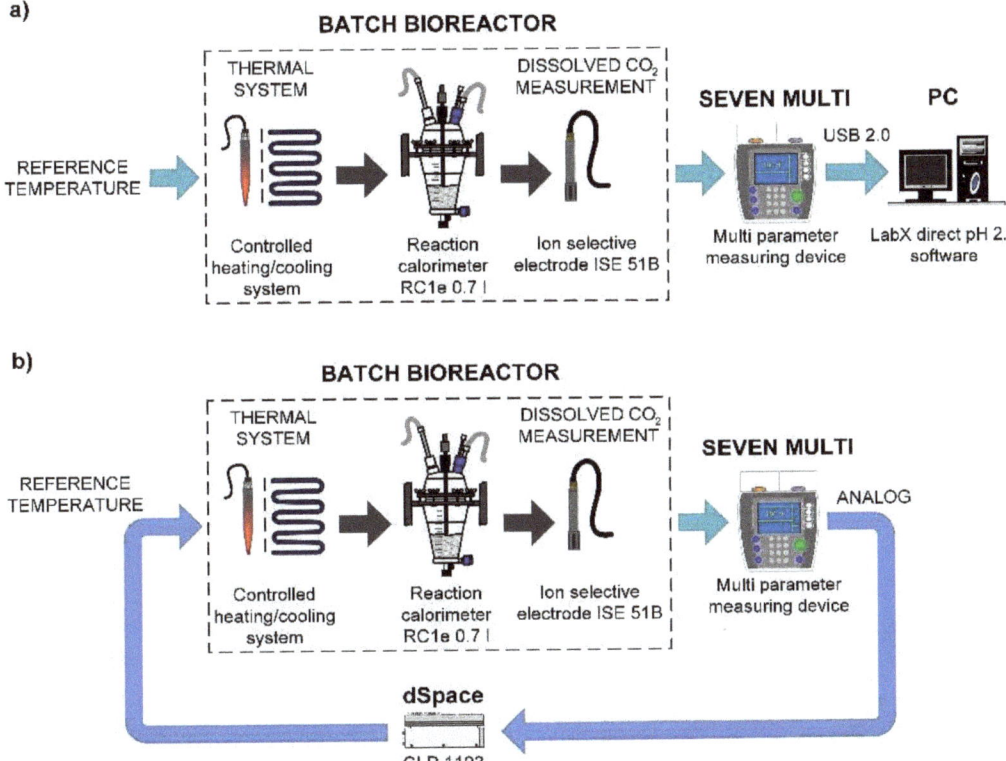

Figure 1. (a) Laboratory system for **measurement** of the time response of the dissolved CO_2 production on the temperature changes and **estimation** of model parameters. (b) Laboratory system for the **control** of dissolved CO_2 production.

2.4. Mathematical Model of the Fermentation Process

Fermentation is described as a process in which an agent causes an organic substance to break down into simpler substances. The agents are mainly microorganisms, the source substance is named a substrate, and the final substance is named a product [7].

Fermentation is a non-linear, time-dependent complex system with a poorly known structure and unknown parameters. There are many mathematical models of the fermentation process of different types and degrees of complexity. Almost all models are derived from the mass balance of microorganisms, substrate, and product [5].

The fundamental mathematical model of the fermentation process in batch bioreactor represents a state-space non-linear model of the 3rd order [2–5]. The state-space variables of this model are the concentrations of the microorganisms, substrate, and product. This model is autonomous—it has no input variable. This is expected because batch bioreactors do not have an input quantity to control the fermentation process. The transients of the model's variables are obtained as the response to the initial values of the substances. During the fermentation process, the quantity of the microorganisms and product increases, and the quantity of substrate decreases. All parameters of the fundamental kinetic model are supposed to be constant throughout the duration of the fermentation process.

The commonly accepted fundamental kinetic mathematical model of the fermentation process enables the simple and efficient simulation and analysis of the fermentation process in cases of different initial concentrations. In many studies, it has been proven that the profiles of the fermentation process substances can be influenced by changing the operating conditions during fermentation. The profiles are most easily influenced by the bioreactor's heating/cooling system and the stirrer system. In [6,30], an analysis of the influence of temperature change on the course of the fermentation process is made, and the influence of stirrer speed change is discussed in [6,7,31]. The fundamental mathematic model does not enable the evaluation of the impact of temperature changes on the time courses of concentrations of individual substances of the fermentation process. The knowledge of this dependence is essential for the design and synthesis of convenient control systems. This is the reason that in the article, a new model that involves the phenomenon of the impact of temperature on the fermentation process was derived and presented.

The derived model is the non-autonomous non-linear 4th-order state-space mathematical model, whose input is the reference temperature of the heating system, and the model's state variables are the concentrations of microorganisms, substrate, fermentation product, and bioreactor's temperature. The impact of temperature on the fermentation process is taken into account by assuming that temperature influences the fundamental model parameters. The derived model is presented with (1)–(4):

$$\dot{x}_1(t) = \frac{\mu_m (1 + k_{\mu_m}(x_4(t) - \vartheta_0))\left(1 - \frac{1}{P_i}x_3(t)\right)x_2(t)}{S_m + x_2(t) + \frac{1}{S_i}(x_2(t))^2} x_1(t) \quad (1)$$

$$\dot{x}_2(t) = -\frac{\mu_m (1 + k_{\mu_m}(x_4(t) - \vartheta_0))\left(1 - \frac{1}{P_i}x_3(t)\right)x_2(t)}{S_m + x_2(t) + \frac{1}{S_i}(x_2(t))^2} x_1(t) \quad (2)$$

$$\dot{x}_3(t) = \left(\alpha \frac{(1 + k_\alpha(x_4(t) - \vartheta_0))\mu_m(1 + k_{\mu_m}(x_4(t) - \vartheta_0))\left(1 - \frac{1}{P_i}x_3(t)\right)x_2(t)}{S_m + x_2(t) + \frac{1}{S_i}(x_2(t))^2} + \beta\right)x_1(t) \quad (3)$$

$$\dot{x}_4(t) = \frac{1}{T_{\vartheta cs}}(u(t) - x_4(t)) \quad (4)$$

where the input of the non-autonomous state-space model is:

$u(t)$—the reference temperature of the bioreactor's temperature control system (°C).

The state-space variables of the mathematical model denote the following biological and thermal quantities:

$x_1(t)$—the concentration of the microorganisms (g/L);

$x_2(t)$—the concentration of the substrate (g/L);
$x_3(t)$—the concentration of the product (g/L);
$x_4(t)$—the temperature in the bioreactor (°C).
Additionally, the parameters of the mathematical model are:
μ_m—the maximum microorganisms' growth rate (h^{-1});
P_i—the product inhibition constant (g/L);
S_m—the substrate saturation constant (g/L);
S_i—the substrate inhibition constant (g/L);
α—the parameter that describes the relation between product yield and microorganism growth;
β—the parameter that describes the product yield that is independent of the microorganism growth (h^{-1});
ϑ_0—the temperature of the bioreactor's contents at the beginning of the fermentation process (°C), normally ϑ_0 is equal to the outside temperature;
k_{μ_m}—the coefficient that describes the impact of the temperature changing on the maximum microorganisms' growth rate μ_m (°C);
k_α—the coefficient that describes the impact of temperature changing on the parameter that describes the relation between product yield and microorganism growth (°C);
$T_{\vartheta cs}$—time constant of the simple 1st-order model of the controlled heating system (h).

2.5. Conventional Control System with a Linear Controller

In order to improve the economy of the fermentation and the quality and quantity of the fermentation product, it is necessary to ensure that the actual time profile of the yield of the fermentation product is as close as possible to the reference one. The fermentation process is a non-linear process, but the deviation between the response of the non-controlled fermentation process and the reference trajectory is relatively small. Therefore, the control of the yield of the fermentation product is also possible with the conventional control system with a linear controller. By selecting the performance index and using the optimization method, we can ensure that the controller will provide optimum performance in the broadest possible operating range. The block diagram of the fermentation process control system with a linear controller is shown in Figure 2.

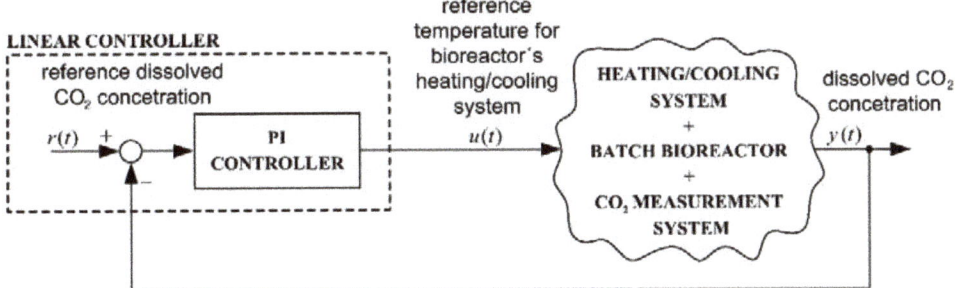

Figure 2. Block diagram of the conventional control system with linear controller.

The main disadvantage of this approach is the need for knowledge of the accurate mathematical model of the batch bioreactor's fermentation process. The structure and parameters of the mathematical model must be known to perform the tuning of the control system. Determining the appropriate mathematical model of a batch bioreactor is hugely time consuming. It is necessary to execute the whole fermentation process with a constant temperature and, after that, repeat the fermentation with the same charge but a changeable temperature. From the responses, all the parameters of the non-linear model (μ_m, P_i, S_m, S_i, α, β, k_{μ_m}, k_α, and $T_{\vartheta cs}$) must be estimated by means of the optimization technique. Finally, the control system tuning must be made by means of the estimated mathematical model. Due to the time-consuming and challenging identification procedure for determining the mathematical model, the use of the conventional linear control system proved to be less appropriate for industrial applications. It makes sense to use a control approach that will not require knowledge of the batch bioreactor's mathematical model.

2.6. Adaptive Control System

The adaptive theory represents an ideal tool for developing a control system for batch bioreactors. Adaptive control systems do not require accurate knowledge of the controlled plant's mathematical model and can adapt their parameters to the changing dynamics of the controlled plant. An additional reason that justifies the use of adaptive strategies for the control of batch bioreactors is that the fermentation processes are executed very slowly, allowing the unproblematic implementation of computationally complex adaptive algorithms.

There are two main approaches to the development of adaptive control systems. The first approach is called indirect adaptive control or self-tuning control (STC) [34]. The advantage of indirect control is its modularity, which allows a combination of different identification methods (least squares, maximum likelihood, instrumental variables, corrector least squares) and different tuning procedures (deterministic and stochastic).

The second approach is termed direct adaptive control because control input is, in general, calculated directly, without preliminary determination of the controlled plant mathematical model. Due to the mandatory reference model, this adaptive control is also called model reference adaptive control (MRAC). Almost all modern MRAC systems can be classified as evolving from one of the three following adaptive approaches:

- Adaptive control systems based on the full-state access method, which requires that all state variables of the controlled plant are measurable (MRAC-FSA) [35];
- Adaptive control systems based on the input–output description of a controlled plant, where an adaptive observer is incorporated into the controller to overcome the inability to access the entire state space vector (MRAC-AO) [36];
- Adaptive control systems, where the adaptive algorithm only requires that the controlled plant's outputs and the reference model states are available for measurement. The asymptotic stability of this adaptive approach is assured in the case when the plant is almost strictly positive real. This adaptive approach is called model reference adaptive control for almost strictly positive real plants (MRAC-ASPR) [37].

MRAC-ASPR is not new, but is more recent than the previously mentioned adaptive approaches. This approach is an output feedback method that requires neither full state feedback nor adaptive observers. The essential improvement of the MRAC-ASPR concept related to the other STC and MRAC concepts is that the MRAC-ASPR theory is also applicable for non-linear controlled plants [38,39]. The other significant qualities of this class of algorithms are given as follows:

- They are applicable to non-minimum phase systems;
- The order of the controlled plant need not be known to select the reference model and the adaptation mechanism;
- The adaptation mechanism is computationally undemanding.

Because of all these advantages, the MRAC-ASPR theory was used to develop the adaptive control system to control the fermentation process in a batch bioreactor. Due to the simple realization of this type of adaptive control system, the name simple adaptive control instead of MRAC-ASPR will also be used.

The proposed MRAC-ASPR was revealed primarily to control the continuous linear systems subject to uncertainty in the parameters [37]. Such consideration coincides with the derived and verified linearized mathematical model of the fermentation process of a batch bioreactor obtained with the linearization of the non-linear model around the fermentation process's trajectory. In 2009, the extension of the MRAC-ASPR theory to minimum-phase nonstationary and non-linear systems was made [38]. Reference [39] shows a detailed and complete description of this theory, with added new results considering non-linear system stability analysis. The MRAC-ASPR concept has been used successfully in different engineering areas to control non-linear controlled plants (electrical drives, robotics, power systems, and chemistry). The MRAC-ASPR concept was also used to design an adaptive system that controls the fermentation process in bioreactors by varying the rotational speed of the mixing system [7]. However, there are no publications showing the use of MRAC-ASPR to control the fermentation process in batch bioreactors by changing the temperature of the heating system.

The following is a brief description of the controller's adaptive algorithm. The MRAC-ASPR will be presented for the control of the controlled plant, which is described by a state-space model [38]:

$$\dot{x}(t) = A(x,t)x(t) + b(x,t)u(t) \tag{5}$$

$$y(t) = c^T(x,t)x(t) \tag{6}$$

where:

$x(t)$, $u(t)$, and $y(t)$ are the state-space vector, input scalar, and output scalar of the mathematical model of the controlled plant;

$A(x,t)$, $b(x,t)$, and $c^T(x,t)$ are the non-linear functions of the mathematical model of the controlled plant.

The desired static and dynamic behaviour of the closed-loop controlled system are defined with the state-space reference model [39]:

$$\dot{x}_m(t) = A_m x_m(t) + b_m u_m(t) \tag{7}$$

$$y_m(t) = c_m^T x_m(t) \tag{8}$$

where:

$x_m(t)$, $u_m(t)$, and $y_m(t)$ are the state-space vector, input scalar, and output scalar of the reference model, and

A_m, b_m, and c_m^T are the system matrix, input vector, and output vector of the reference model.

The reference model is assumed to be bounded-input/bounded-state stable. The task of the reference model is only to represent the desired input–output behaviour. The number of state-space variables of the reference model can be significantly less than the number of state-space variables of the controlled plant, as described by the equation:

$$\dim[x_m(t)] \ll \dim[x(t)] \tag{9}$$

Since the order of the reference model is in general not the same as the order of the mathematical model of the controlled plant, it is not possible to require the state-space variables of the controlled plant to follow the state-space variables of the reference model. Instead, a request is made that the controlled plant output $y(t)$ follows the output of the reference model $y_m(t)$ asymptotically. The output tracking error $e_y(t)$ is defined with:

$$e_y(t) = y_m(t) - y(t) \tag{10}$$

The extension of the Lyapunov stability theory to non-linear non-autonomous systems was applied for the derivation of the adaptive control algorithm [37–39]. The main request by the derivation of the adaptive control algorithm was to ensure the stability of the entire control system. The final goal was to obtain an adaptive algorithm that will generate such controlled plant input signal $u(t)$ which will assure that the controlled plant output $y(t)$ will approximate "reasonably well" the output of the reference model $y_m(t)$ without explicit knowledge of the controlled plant functions $A(x,t)$, $b(x,t)$ and $c^T(x,t)$ [32]. The determined adaptive control algorithm is expressed with:

$$u_p(t) = K_e(t)e_y(t) + K_x(t)x_m(t) + K_u(t)u_m(t) \tag{11}$$

where scalar $K_e(t)$ is the stabilizing output feedback parameter, and matrix $K_x(t)$ and scalar $K_u(t)$ are control gains. Parameters $K_e(t)$, $K_x(t)$ and $K_u(t)$ can be united in a vector of adaptive gains $K(t)$, and the variables $e_y(t)$, $x_m(t)$ and $u_m(t)$ can be linked in a vector of control variables $r(t)$:

$$K(t) = \begin{bmatrix} K_e(t) & K_x(t) & K_u(t) \end{bmatrix} \tag{12}$$

$$r^T(t) = \begin{bmatrix} e_y(t) & x_m(t) & u_m(t) \end{bmatrix} \tag{13}$$

The adaptive gains $K(t)$ can be represented as the sum of the two terms: of the proportional term $K_p(t)$ and integral term $K_i(t)$, as written in (14) [37].

$$K(t) = K_p(t) + K_i(t) \tag{14}$$

The proportional- and integral terms can be calculated with the following non-linear equations [37]:

$$K_p(t) = e_y(t)\,r^T(t)\,T' \tag{15}$$

$$\dot{K}_i(t) = e_y(t)\,r^T(t)\,T \tag{16}$$

where T' is a positive semi-definite matrix and T is a positive definite matrix.

The proportional term $K_p(t)$ drives the system very quickly towards a small tracking error, and the integral term $K_i(t)$ guarantees convergence. In the MRAC-ASPR concept, we cannot talk about the optimal gain value that the adaptive controller wants to achieve. The gain varies during operation according to the error [7].

In order to improve the convergence of the adaptive system, the following modification of the integral term was proposed [37]:

$$\dot{K}_i(t) = e_y(t)\,r^T(t)\,T - \sigma K_i(t) \tag{17}$$

where the task of the σ-term is to protect the integral gains from divergence if there are disturbances.

The block diagram of the control system with simple adaptive controller is shown in Figure 3.

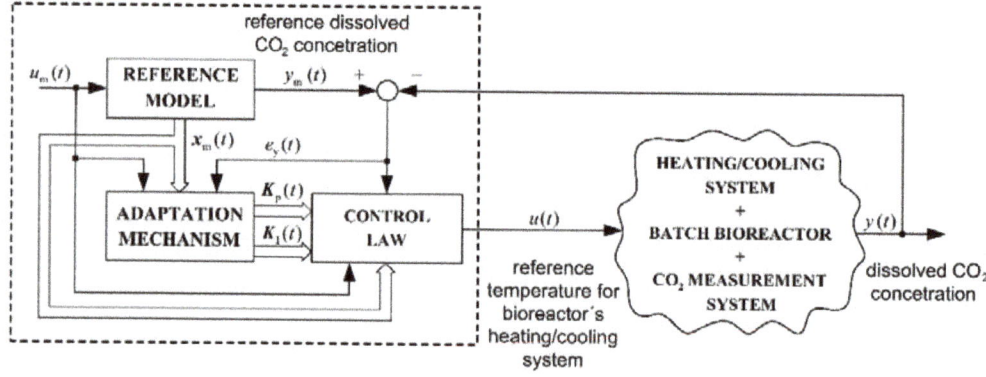

Figure 3. Block diagram of the control system with simple adaptive controller (MRAC-ASPR).

3. Results

3.1. Estimation of Model Parameters

The parameters μ_m, P_i, S_m, S_i, α, β, k_{μ_m}, k_α, and $T_{\vartheta cs}$ depend on the quality and quantity of the substances. They remain almost constant during the fermentation process. They can be calculated for the real bioreactors by different optimization methods from the measured trajectories of the bioreactors' substances. For the studied fermentation process in a laboratory bioreactor, the Particle Swarm algorithm was used to obtain the mathematical model's parameters. The integral absolute error between measured and model fermentation product variables was used as the optimization method's performance index.

For the estimation of the parameters, the studied fermentation process was executed two times in the laboratory bioreactor.

First, an appropriate amount of fully activated microorganisms (kefir grains) and substrate (milk) was introduced into the bioreactor. The dissolved CO_2 was selected as the fermentation product. The initial concentrations of microorganisms, substrate, and fermentation product were measured. The obtained values are given in Table 1. The fermentation process was then performed at a constant temperature. The time courses of concentrations of all substances were measured. From the obtained measurements, the parameter μ_m, P_i, S_m, S_i, α, and β were estimated by means of particle swarm optimization.

Table 1. Initial values of the fermentation process in the studied bioreactor.

Variable	Value
The initial value of the microorganisms' concentration	$x_1(0) = 0.26$ mg/L
The initial value of the substrate's concentration	$x_2(0) = 0.89$ mg/L
The initial value of the product's concentration	$x_3(0) = 0.02$ mg/L
The initial temperature of the bioreactor's contents	$x_4(0) = 22$ °C

Then, the fermentation process with the equal initial substances (fully activated kefir grains) was performed again. This time, during the operation, the reference temperature value of the heating system was changed from 22 °C to 29 °C. The step-change of the reference temperature (i.e., the input signal of the controlled plant) occurs in the fermentation's growing phase, 2 h after the beginning of the fermentation. The courses of the substances' concentrations were measured again, and the model parameters k_{μ_m}, k_α, and $T_{\vartheta cs}$ were estimated.

The parameters of the identified mathematical model of the milk fermentation process in the laboratory batch bioreactor where fully activated kefir grains were used are presented in Table 2.

Table 2. Parameters of the Mathematical Model for the Original Fermentation Process in the Studied Laboratory Batch Bioreactor where **fully activated kefir grains** were used.

Parameter	Value
The maximum microorganisms' growth rate	$\mu_m = 0.5 \, h^{-1}$
The product inhibition constant	$P_i = 7.0$ g/L
The substrate saturation constant	$S_m = 0.42$ g/L
The substrate inhibition constant	$S_i = 62.15$ g/L
The parameter of the product yield related to microorganisms' growth	$\alpha = 0.9 \, \frac{g/L}{g/L}$

Table 2. *Cont.*

Parameter	Value
The parameter of the product yield independent of the microorganisms' growth	$\beta = 0.001\ \text{h}^{-1}$
The coefficient of the impact of the temperature changing on the maximum microorganisms' growth rate μ_m	$k_{\mu_m} = 0.1\ (°C)^{-1}$
The coefficient of the impact of the temperature changing on the parameter α	$k_\alpha = 1.15\ (°C)^{-1}$
The temperature of the bioreactor's contents at the beginning of the fermentation process, normally this temperature is equal to the outside temperature	$\vartheta_0 = 22\ °C$
The time constant of the 1st-order model of the controlled heating system	$T_{\vartheta cs} = 0.1\ \text{h}$

The matching of the response of the measured CO_2 concentration in the laboratory bioreactor with the response of the CO_2 concentration calculated with the identified mathematical model is displayed in Figure 4. It shows the results of the fermentation with the changeable bioreactor's temperature. Note, in the experiment and in the simulations, the step increase in the reference temperature from 22 °C for 7 °C occurred at $t = 2$ h.

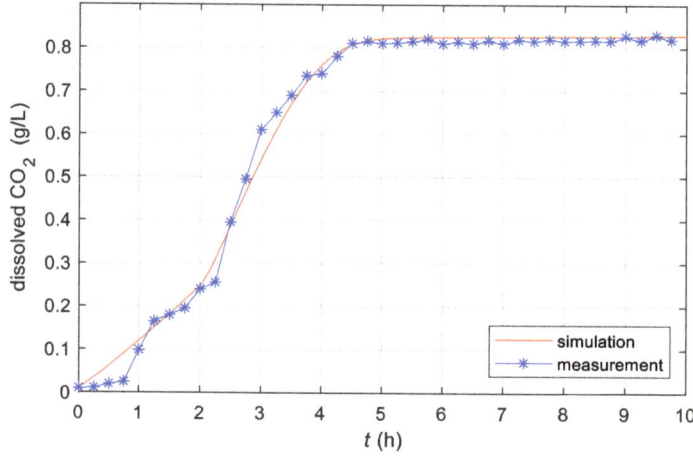

Figure 4. Measured and simulated time courses of dissolved CO_2 concentration during the fermentation process with changeable bioreactor temperature (step change of temperature from 22 °C to 29 °C occurred at time t = 2 h).

The simulation results show that the derived model can justifiably be used for bioprocess analysis and control system development. The calculation (with optimization techniques) of the model parameters is not complicated but can take a lot of time.

For the evaluation of the efficiency of the control system for different fermentations, the modified fermentation process was executed and identified. The difference between the original and modified fermentation process was in the kefir grains used for the fermentation. While kefir grains activated by washing with cold water and transferred into fresh milk for 5 successive days were used for the original fermentation, inactivated kefir grains were used in the modified process. This resulted in a slower fermentation and a lower final value of the fermentation product. The parameters of the mathematical model of the modified fermentation process are shown in Table 3.

Table 3. Parameters of the mathematical model for the modified fermentation process in the studied laboratory batch bioreactor where **inactivated kefir grains** were used.

Parameter	Value
The maximum microorganisms' growth rate	$\mu_m = 0.23\ h^{-1}$
The product inhibition constant	$P_i = 7.2\ g/L$
The substrate saturation constant	$S_m = 0.67\ g/L$
The substrate inhibition constant	$S_i = 74.58\ g/L$
The parameter of the product yield related to microorganisms' growth	$\alpha = 0.8\ \frac{g/L}{g/L}$
The parameter of the product yield independent of the microorganisms' growth	$\beta = 0\ h^{-1}$
The coefficient of the impact of the temperature on the maximum microorganisms' growth rate	$k_{\mu_m} = 0.12\ (°C)^{-1}$
The coefficient of the impact of the temperature on the temperature inhibition constant P_i	$k_\alpha = 0.04\ (°C)^{-1}$

3.2. Results Obtained with the Conventional Control System

The use of a linear controller makes sense since the analysis of a linearized mathematical model in the vicinity of the trajectory of the non-controlled fermentation process showed a relatively small range of variations in the model's parameters.

A simple PI-controller with transfer function $G_{PI}(s)$ (18) is used to demonstrate the efficiency of the conventional control system with linear controller for comparison with the advanced adaptive control system,

$$G_{PI}(s) = k_p \frac{sT_i + 1}{sT_i} \qquad (18)$$

where k_p is the gain and T_i is the time constant of the PI-controller.

The controller synthesis was done using the optimization method for the integral time square cost function J of the output error variable and input variable. The cost function J is presented with [30]:

$$J = \int_0^{t_f} \left\{ Q\left[r(t) - y(t)\right]^2 + R\left[u(t) - \vartheta_0\right]^2 \right\} dt \qquad (19)$$

where:

$u(t)$ is the input variable of the mathematical model of the controlled plant (i.e., the reference temperature of the bioreactor's temperature control system (°C));

$y(t)$ is the output variable of the mathematical model of the controlled plant (i.e., the output of the measurement system for the dissolved CO_2 concentration (mmol/L));

$r(t)$ is empirically determined reference trajectory for the dissolved CO_2 concentration;

ϑ_0 is the temperature of the bioreactor's mixture at the beginning of the fermentation process (°C), normally ϑ_0 is equal to the outside temperature;

Q, R—are the weighting parameters of the quadratic cost function;

t_f is the final time.

For the tested fermentation process in the laboratory bioreactor, the following reference trajectory $r(t)$ for the dissolved CO_2 was chosen:

$$r(t) = 0.8\left(1 - e^{-t/1.5}\right)\ mmol/L \qquad (20)$$

This reference trajectory was determined based on the dissolved CO_2 trajectory of the uncontrolled fermentation process in this bioreactor. The modification of the trajectory was made in such a way that the controlled fermentation process will finish in a shorter time and that the amount of the generated product will be higher. The biological limitations must be taken into consideration. It is necessary to ensure that the reference trajectory does not deviate excessively from the CO_2 concentration trajectory of the autonomous fermentation process.

The trajectory of the actual dissolved CO_2 of the batch bioreactor without a control system, and the reference (desirable) trajectory of the dissolved CO_2 of the batch bioreactor, are shown in Figure 5. The task of the developed control system is to ensure that the actual output value will follow the prescribed reference trajectory as closely as possible.

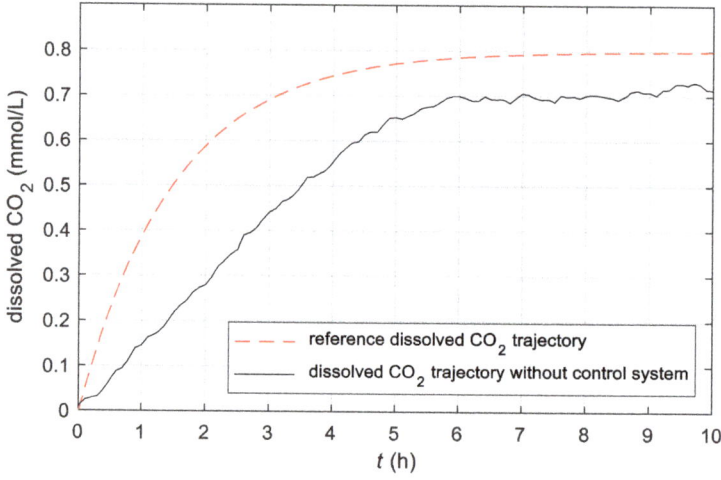

Figure 5. The trajectory of the dissolved CO_2 of the original fermentation process in batch bioreactor without a control system (solid line) and the reference trajectory of the dissolved CO_2 of the batch bioreactor (dashed line); data of the original fermentation process with fully activated kefir grains are shown in Table 2; the reference trajectory is described with Equation (20).

The cost function (19) was calculated by means of simulations of the closed-loop control system. Simulations were made with a non-linear model of the controlled plant (1)–(4) with the parameters from Table 2 and the initial concentration values given in Table 1.

In optimization calculations, the parameters k_p and T_i of the PI-controller were changed in order to obtain the minimum value of the cost function. Particle swarm optimization (PSO) was used for the calculation of the controller's parameters. PSO is a metaheuristic procedure that may provide a sufficiently effective solution to an optimization problem in cases where there are few, incomplete, imperfect or no assumptions about the problem being optimized. Functions from MathWorks MATLAB/Optimization Toolbox library were used for faster realization of the PSO for the calculation of the parameters of the mathematical model. Matlab function *particleswarm.m* is based on the algorithm described in [40], using the modifications suggested in [41,42]. The details of the PSO algorithm in the *particleswarm.m* function are written in [43].

The simulations were calculated for the time period from $t = 0$ h to $t = 10$ h. The weighting coefficients Q and R have been selected so that both terms of the cost function were proportionally weighted and that with the controller calculated reference temperature of the heating system remained within the realizable range. For the chosen cost function's parameters, Q, R, t_f, and the constant initial temperature of the bioreactor's filling, ϑ_0,

$$Q = 1 \ R = 0.1 \ t_f = 10 \ h \ \vartheta_0 = 292 \ °K \ ; \tag{21}$$

the following parameters of the PI-controller were calculated:

$$k_p = 22.0 \; T_i = 1.5 \text{ h} \tag{22}$$

The obtained control results are presented in Figures 6 and 7. The time response of the actual dissolved CO_2 concentration of a batch bioreactor controlled with a conventional PI-controller, together with the reference trajectory, is shown in Figure 6. The bioreactor's inner temperature $x_4(t)$, resultantly to the control of the heating/cooling system, is presented in Figure 7. No limiters or anti-windup were used. The controller's output stays in the feasible range, and it does not exceed the maximum or minimum limits.

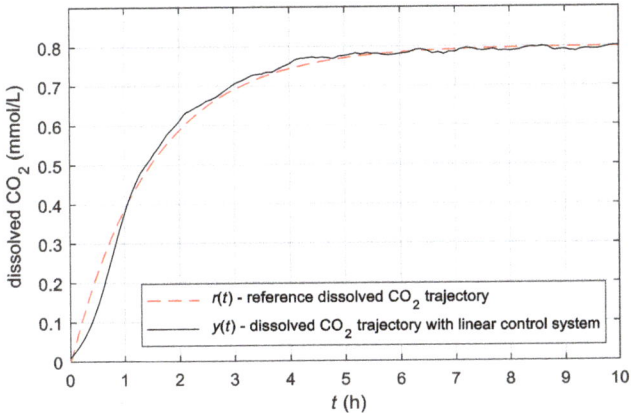

Figure 6. Time responses of the reference and actual dissolved CO_2 concentration of the original fermentation process in the batch bioreactor; the conventional control system with the linear controller with calculated parameters was used; data of the controlled original fermentation process with fully activated kefir grains are shown in Table 2; the controller parameters are shown in (22).

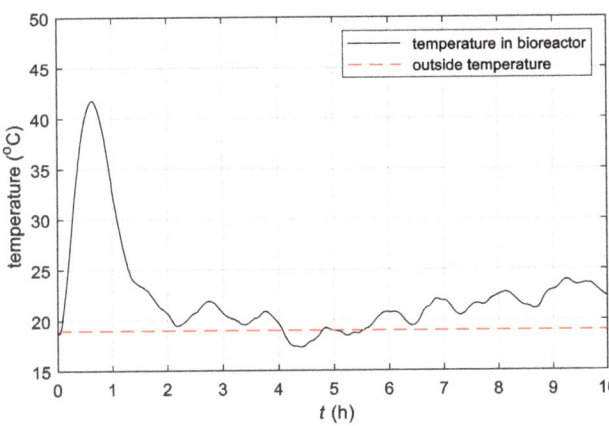

Figure 7. Time response of the temperature of the original fermentation process in the batch bioreactor and the constant outside temperature, corresponding to Figure 6.

It is expected that the controller provides good control of the fermentation process, for which its parameters have been optimized. Conversely, we cannot expect that the same controller will optimally control other fermentation processes.

The efficiency of the optimized control system was evaluated for the modified fermentation process. The parameters of the modified fermentation process are written in Table 3. Figure 8 shows the trajectory of the dissolved CO_2 of the modified fermentation process without a control system and the reference (desirable) trajectory of the dissolved CO_2.

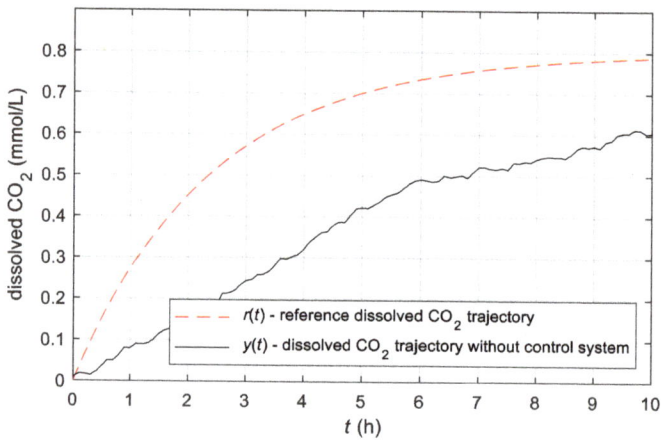

Figure 8. The trajectory of the dissolved CO_2 of the modified fermentation process in batch bioreactor without a control system (solid line) and the reference trajectory of the dissolved CO_2 of the batch bioreactor (dashed line); data of the modified fermentation process with inactivated kefir grains are shown in Table 3, and the reference trajectory is described in Equation (20).

Figures 5 and 8 seem similar, but there is a significant difference in the dynamics of the fermentation process. While the original fermentation process is completed in ca. 6 h, the modified fermentation process lasts more than 10 h.

The results of the control of the modified fermentation process with parameters in Table 3 with the PI-controller with original (non-modified) parameters (22) are shown in Figure 9.

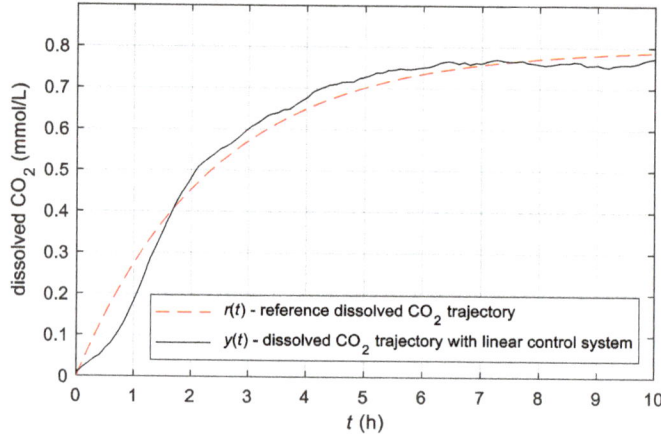

Figure 9. Time responses of the reference and actual dissolved CO_2 concentration of the modified fermentation process in the batch bioreactor; the conventional control system with the linear controller with calculated parameters was used; data of the controlled modified fermentation process with inactivated kefir grains are shown in Table 3; the controller parameters are shown in (22).

3.3. Results Obtained with the Simple Adaptive Control System

A simple adaptive control system based on the MRAC-ASPR control theory was used for the batch bioreactor's control implementation.

The presented adaptive control system assures that the batch bioreactor's output (i.e., the measured dissolved CO_2 concentration) follows the output of the reference model in the case of unknown and variable bioreactor's kinetics. In such a way, the adaptive controller enables that the bioreactor's dynamics stay the same during repetitions of the batch processes. The presented results were obtained for the adaptive control system with the reference model represented with the 1st-order term with gain $k_{rm} = 0.8$ and the time constant $T_{rm} = 1.5$ h. This reference model produces step response equal to the reference signal in (20). The parameters of the adaptation mechanism were obtained on the basis of numerical simulations with the non-linear model. No optimization technique was used to find the optimal values for the adaptation mechanism's parameters. Numerical simulations were used only to determine the approximate values of the parameters of the adaptation mechanism. The obtained values are convenient for different size batch bioreactors. An accurate setting of the parameters is not necessary. The following adaptation coefficient matrix based on a positive definite identity matrix:

$$T = 4000 \begin{bmatrix} 1 & 0 & 0 & 0 \\ 0 & 1 & 0 & 0 \\ 0 & 0 & 1 & 0 \\ 0 & 0 & 0 & 1 \end{bmatrix} \quad T' = 4000 \begin{bmatrix} 1 & 0 & 0 & 0 \\ 0 & 1 & 0 & 0 \\ 0 & 0 & 1 & 0 \\ 0 & 0 & 0 & 1 \end{bmatrix} \quad (23)$$

were used to carry out the adaptive control.

In addition, the σ-term wasused to avoid the divergence of integral gains in the presence of a disturbance.

$$\sigma = 0.95 \quad (24)$$

The results of the simple adaptive control technique for the fermentation process in the batch bioreactor are shown in Figures 10 and 11. Figure 10 shows the reference and the actual time response of the dissolved CO_2 concentration. The time response of the generated product of the fermentation process follows the reference variable despite the unknown parameters of the controlled plant and its structure uncertainties. The temperature of the bioreactor's filling, which was necessary to assure that actual dissolved CO_2 concentration follows the reference trajectory, is shown in Figure 11. The controller's output stays in the feasible range, and it does not exceed the maximum or minimum limits.

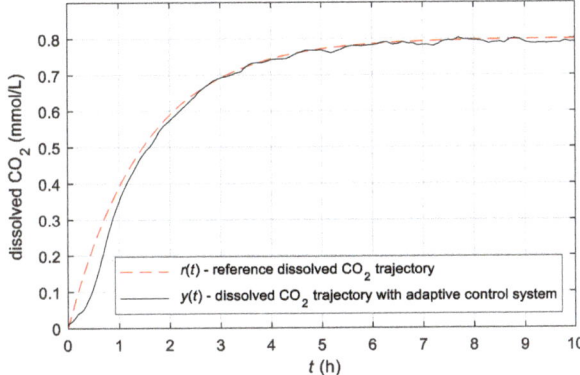

Figure 10. Time responses of the reference and actual dissolved CO_2 concentration of the original fermentation process in the batch bioreactor; simple adaptive control was used; data of the controlled original fermentation process with fully activated kefir grains are shown in Table 2, the adaptation mechanism parameters in (23) and (24).

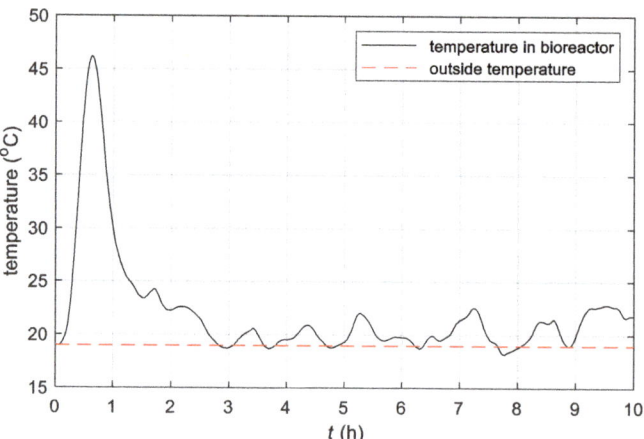

Figure 11. Time response of the temperature of the original fermentation process in the batch bioreactor and the constant outside temperature when the simple adaptive control was used, corresponding to Figure 10.

The results of the control of the modified fermentation process with parameters in Table 3 with the simple adaptive control (23) and (24) are shown in Figure 12. The time response of the dissolved CO_2 concentration of a batch bioreactor controlled with the simple adaptive controller, together with the reference trajectory, is shown.

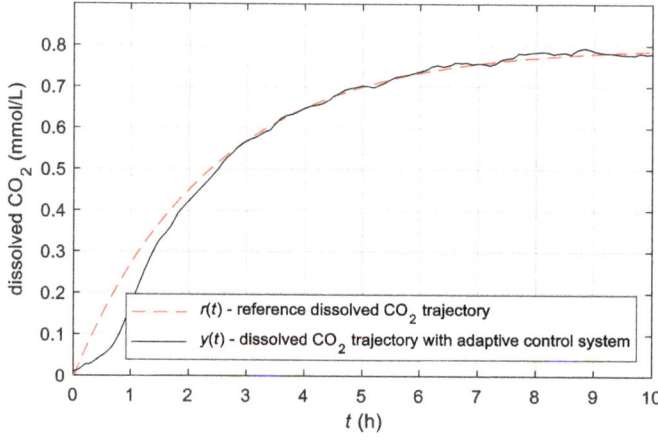

Figure 12. Time responses of the reference and actual dissolved CO_2 concentration of the modified fermentation process in the batch bioreactor; simple adaptive control was used; data of the controlled modified fermentation process with inactivated kefir grains are shown in Table 3, the adaptation mechanism parameters in (23) and (24).

3.4. Experimental Results Obtained with the Simple Adaptive Control System

Developed control systems were tested on the laboratory batch bioreactor. HW and SW equipment described in Sections 2.2.4 and 2.3 was used for the implementation of control systems. From the simulations was seen that the differences between responses are not significant. The results of laboratory tests were comparable. Due to additional reasons related to the realization and equipment used, it is difficult to objectively evaluate individual control concepts' effectiveness from the test results. Figure 13 shows the time responses of the reference and actual dissolved CO_2 concentration obtained with the experiment where the simple adaptive control was used for control of fermentation process of laboratory batch bioreactor with fully activated kefir grains. It can be seen from Figure 13 that the developed adaptive control system ensures the tracking of the actual dissolved CO_2 concentration to the reference trajectory. The deviation at the beginning of the transient is due to zero initial values in the integral elements of the adaptation mechanism. The reference trajectory tracking can be further improved by selecting the adaptation mechanism's initial values and weighting coefficients adjusted to the fermentation process. The response to the control of the fermentation process with inactivated kefir grains was very similar despite the fact that the controller's parameters stay unchanged.

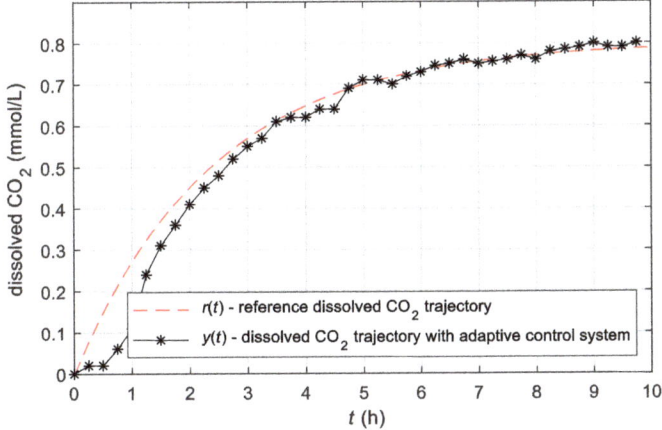

Figure 13. Time responses of the reference and actual dissolved CO_2 profiles obtained with the experiment where the simple adaptive control was used for control of fermentation process of laboratory batch bioreactor with fully activated kefir grains.

4. Discussion

At first sight, the results obtained with both presented control systems are excellent and very similar, especially for the fermentation process with parameters in Table 2. Almost identical dynamics of the fermentation process were obtained, as defined with the reference trajectory. To achieve these responses, acceptable changes were requested in the bioreactor's inner temperature.

Figures 5–7 show clearly that the developed linear controller provides very good tracking of the actual CO_2 concentration to the reference value for the original fermentation process for which controller's parameters were optimized. In this way, the fermentation process was made significantly more economical. The fermentation time was shortened. The duration of the non-controlled fermentation is about 6 h, and the duration of the controlled fermentation is approx. 5 h. An increase in concentration was also obtained of approx. 0.1 mmol/L. The efficiency of the same controller is lower if it is used to control modified fermentation processes, which is seen in Figures 8 and 9. Slight oscillations are visible from the results. The difference between the original and the modified fermentation process is that in the case of the original fermentation process, fully activated kefir grains were used. In the case of the modified fermentation process, the used kefir grains were inactivated. Original control parameters do not ensure optimal behavior in the case of the modified fermentation process. In this case, the fermentation process should be re-identified, and a new tuning of the controller parameters should be performed.

Figures 10 and 11 show the results obtained with the developed adaptive controller. Results are very similar to the results of the optimized linear controller. The advantage of the adaptive controller is visible when used to control fermentation processes that have different dynamics. The presented adaptive control system assures that the batch bioreactor's output (i.e., the measured dissolved CO_2 concentration) follows the output of the reference model in the case of unknown and variable bioreactor's kinetics. It can be seen from Figure 12 that the adaptive controller maintains the same time course of the output quantity, even in the case of significantly changed (and unknown) parameters of the fermentation process. In this case, the duration of the fermentation process was shortened from 10 h to 5 h, while an increase in concentration was also obtained of approx. 0.2 mmol/L.

To obtain better insight into the performances of the control systems, an evaluation was made based on the Performance Index. Since the purpose was to evaluate both control concepts as accurately as possible, these calculations were made on the basis of simulation results. The experimental results are affected by additional random external disturbances that obscure the comparison of control algorithms. The integral quadratic Performance Index, the same as the cost function shown in (19), was used for the comparative calculations. The same parameters of the Performance Index as for the PI-controller optimization were used to estimate the control quality (21). Disturbances and noise were added to the measured fermentation product variable to achieve the most realistic conditions, equal for both control systems. Disturbances and noise were estimated from the measured results. Band-limited white noise with the correlation time 0.1 h and the noise power 0.01 was used. The disturbances were generated by a PRBS signal with the amplitude 0.05, which was filtered through the transfer function $G(s) = \frac{s}{(5s^2+4s+1)}$. Results for both controllers for two fermentations processes are shown in Table 4.

Table 4. Performance indexes for the studied bioreactor's control systems.

Original Fermentation Process with the Data in Table 2 (Activated Kefir Grains)	
conventional control system with PI-controller	J = 4.2496
simple adaptive control system	J = 4.2864
Modified Fermentation Process with the Data in Table 3 (Inactivated Kefir Grains)	
conventional control system with PI-controller	J = 9.9223
simple adaptive control system	J = 9.0363

The difference between the calculated performance indexes for the original fermentation process (fermentation process data in Table 2) was minimal. It was expected that the conventional control system with the PI-controller would obtain good results because its parameters were optimized with the same cost function. On the other hand, it is encouraging that the simple adaptive control system led to almost the same results, even though the initial parameters of the control algorithm were zero, and those weighting coefficients of the adaptation mechanism were chosen very easily, without any optimization approach. The important advantage of the simple adaptive controller is visible in the control of the modified fermentation process (fermentation process data in Table 3). The results obtained with a linear regulator, which was not optimized for this process, were significantly worse (more than 10%) than the results of the adaptive controller, which itself adapted to the changing dynamics of the modified fermentation process.

Despite the similar performance indexes, the proposed adaptive control approach presents a much better choice for developing the control system for the batch bioreactor. The main advantage of the adaptive control system is that the detailed knowledge of the batch bioreactor and the substances used is not necessary. The simple adaptive controller adapts its operation automatically to the different dynamics of the fermentation processes. The pre-operation tuning is minimal. On the other hand, if we want to use a conventional linear controller with constant parameters, a preliminary determination of the mathematical model of the fermentation process is mandatory to ensure satisfactory control. Determining a mathematical model is time-consuming and involves determining a non-linear model and its linearization.

5. Conclusions

The article combines the fields of Control Engineering and Bioprocess Engineering. It shows the applicability of the conventional and advanced control approaches for the control of the fermentation processes in the batch bioreactors.

There are some important conclusions and contributions of this paper:

- The study confirmed that by changing the temperature in the batch bioreactor, the execution of the fermentation process could be controlled;
- The author showed that the proposed minor supplementation of the batch bioreactor with a controlled heating/cooling system and CO_2 measurement system enables the development of the closed-loop control system, which ensures that the time profiles of substantial biological quantities during the fermentation process will be the same as the reference profile;
- The author derived an original non-linear mathematical model of the fermentation process in the batch bioreactor, which describes the influence of temperature variations in the bioreactor on the courses of the fermentation quantities; the derived model was used for the design and synthesis of the closed-loop control systems;
- Based on the derived mathematical model, the author utilizes a conventional closed-loop control system with optimized parameters, which ensures that the course of trajectories of substantial biological quantities during the fermentation process will be the same as the course of the reference trajectories when the parameters of the mathematical model are known;
- The author reviewed various advanced control concepts and, on this basis, proposed and developed a control system based on the use of adaptive control theory. The proposed usage of the MRAC-ASPR theory for the development of the control of the growth in the fermentation process in batch bioreactor represents the main original contribution.
- It has been shown that the developed simple adaptive control system represents a very effective control for batch bioreactor operation. The advantage of the developed adaptive controller is significantly easier implementation while having the same or better performance as a conventional controller, which requires complicated and time-demanding tuning. The proposed adaptive control system was analysed theoretically and experimentally, and its advantages were confirmed with the results.

The similarity of the results obtained with the optimized conventional linear control system and advanced non-linear adaptive control system could lead to the opinion that both presented control concepts' efficiencies are very similar. However, there is a major difference between the tuning procedure and the related usability of both concepts. While the use of linear control with constant parameters requires knowledge of the exact mathematical model of each fermentation process, the adaptive control ensures the desired course of the fermentation process, even when the structure and parameters of the mathematical model are unknown.

Funding: This research received no external funding.

Institutional Review Board Statement: Not applicable.

Informed Consent Statement: Not applicable.

Data Availability Statement: Not applicable.

Conflicts of Interest: The authors declare no conflict of interest.

References

1. Biotechnology Market Growth & Trends. Available online: https://www.grandviewresearch.com/press-release/global-biotechnology-market (accessed on 2 April 2021).
2. Shuler, M.L.; Kargi, F. *Bioprocess Engineering: Basic Concepts*, 2nd ed.; Prentice Hall: Englewood Cliffs, NJ, USA, 2002.
3. Cinar, A.; Parulekar, S.J.; Undey, C.; Birol, G. *Batch Fermentation—Modelling, Monitoring and Control*; Marcel Dekker Inc.: New York, NY, USA, 2003.
4. Blanch, H.W.; Clark, D.S. *Biochemical Engineering*; Marcel Dekker, Inc.: New York, NY, USA, 1997.
5. Henson, M.A. Exploiting cellular biology to manufacture high-value products—Biochemical reactor modelling and control. *IEEE Control. Syst. Mag.* **2006**, *1066*, 54–62.
6. Goršek, A.; Ritonja, J.; Pečar, D. Mathematical model of CO_2 release during milk fermentation using natural kefir grains. *J. Sci. Food Agric.* **2018**, *98*, 4680–4684. [CrossRef] [PubMed]
7. Ritonja, J. Implementation of stir-speed adopted controllers onto a batch bioreactor for improved fermentation. *IEEE Access* **2021**, *9*, 16783–16806. [CrossRef]
8. Coutinho, D.; Vande Wouwer, A. A robust non-linear feedback control strategy for a class of bioprocesses. *IET Control. Theory Appl.* **2013**, *7*, 829–841. [CrossRef]
9. De Andrade, G.A.; Pagano, D.J.; Guzmán, J.L.; Berenguel, M.; Fernández, I.; Acién, F.G. Distributed sliding mode control of pH in tubular photobioreactors. *IEEE Trans. Control. Syst. Technol.* **2015**, *24*, 1160–1173. [CrossRef]
10. Battista, H.D.; Picó-Marco, E.; Santos-Navarro, F.N.; Picó, J. Output feedback linearization of turbidostats after time scaling. *IEEE Trans. Control. Syst. Technol.* **2018**, *27*, 1668–1676. [CrossRef]
11. Casenave, C.; Perez, M.; Dochain, D.; Harmand, J.; Rapaport, A.; Sablayrolles, J.-M. Antiwindup input–output linearization strategy for the control of a multistage continuous fermenter with input constraints. *IEEE Trans. Control. Syst. Technol.* **2020**, *28*, 766–775. [CrossRef]
12. Li, S.; Yueyang, L. Model predictive control of an intensified continuous reactor using a neural network Wiener model. *Neurocomputing* **2016**, *185*, 93–104. [CrossRef]
13. Vasičkaninová, A.; Bakošová, M.; Oravec, J.; Mészáros, A. Model predictive control of a tubular chemical reactor. In Proceedings of the 22nd International Conference on Process Control, Štrbské Pleso, Slovakia, 11–14 June 2019.
14. Rodriguez, A.E.; Munoz, J.A.T.; Luna, R.; Correa, J.R.P.; Bocanegra, A.R.D.; Ramirez, H.S.; Castro, R. Robust control for cultivation of microorganisms in a high density fed-batch bioreactor. *IEEE Lat. Am. Trans.* **2015**, *13*, 1927–1933. [CrossRef]
15. Estakhrouiyeh, M.R.; Vali, M.; Gharaveisi, A. Application of fractional order iterative learning controller for a type of batch bioreactor. *IET Control. Theory Appl.* **2016**, *10*, 1374–1383. [CrossRef]
16. Maria, G. Model-based optimization of a fed-batch bioreactor for mAb production using a hybridoma cell culture. *Molecules* **2020**, *25*, 5648. [CrossRef]
17. Grigs, O.; Galvanauskas, V.; Dubencovs, K.; Vanags, J.; Suleiko, A.; Berzins, T.; Kunga, L. Model predictive feeding rate control in conventional and single-use lab-scale bioreactors: A study on practical application. *Chem. Biochem. Eng. Q.* **2016**, *30*, 47–60. [CrossRef]
18. Chitra, M.; Pappa, N. Optimal tuning of model predictive controller tuning using chicken swarm optimization for real cultivation of Escherichia coli. In Proceedings of the IEEE Second International Conference on Control, Measurement and Instrumentation (CMI), Kolkata, West Bengal, India, 8–10 January 2021.
19. Rodriguez-Mata, A.E.; Luna, R.; Pérez-Correa, J.R.; Gonzalez-Huitrón, A.; Castro-Linares, R.; Duarte-Mermou, M.A. Fractional sliding mode nonlinear procedure for robust control of an eutrophying microalgae photobioreactor. *Algorithms* **2020**, *13*, 50. [CrossRef]
20. Gharagozloo, M.H.; Ghasemi, R.; Sedighi, M. *Nonlinear Model Predictive Versus Sliding Mode Controller Design for a Class of Nonlinear Non-Affine Chemical Batch Reactor Dynamics*; EasyChair Preprint: Budapest, Hungary, 2020.
21. Kumar, M.; Prasad, D.; Shekher Giri, B.; Sharan Singh, R. Temperature control of fermentation bioreactor for ethanol production using IMC-PID controller. *Biotechnol. Rep.* **2019**, *22*. [CrossRef] [PubMed]
22. Pachauri, N.; Rani, A.; Singh, V. Bioreactor temperature control using modified fractional order IMC-PID for ethanol production. *Chem. Eng. Res. Des.* **2017**, *122*, 97–112. [CrossRef]
23. Imtiaz, U.; Jamuar, S.; Sahu, J.N.; Ganesan, P.B. Bioreactor profile control by a non-linear auto regressive moving average neuro and two degree of freedom PID controllers. *J. Process. Control.* **2014**, *24*, 1761–1777. [CrossRef]
24. Flores-Hernández, A.A.; Reyes-Reyes, J.; Astorga-Zaragoza, C.M.; Osorio-Gordillo, G.L.; García-Beltrán, C.D. Temperature control of an alcoholic fermentation process through the Takagi–Sugeno modeling. *Chem. Eng. Res. Des.* **2018**, *140*, 320–330. [CrossRef]
25. Fonseca, R.R.; Schmitz, J.E.; Frattini, A.M.; Vasconcelos da Silva, F. A fuzzy–split range control system applied to a fermentation process. *Bioresour. Technol.* **2013**, *142*, 475–482. [CrossRef]
26. The Importance of Real-Time CO2 Monitoring in Cell Culture. Available online: https://https://bioprocessintl.com/sponsored-content/the-importance-of-real-time-co2-monitoring-in-cell-culture/ (accessed on 8 July 2021).
27. Chopda, V.R.; Holzberg, T.; Ge, X.; Folio, B.; Tolosa, M.; Kostov, Y.; Tolosa, L.; Rao, G. Real-time dissolved carbon dioxide monitoring I: Application of a novel in situ sensor for CO_2 monitoring and control. *Biotechnol. Bioeng.* **2020**, *117*, 981–991. [CrossRef]

28. Chatterjee, M.; Ge, X.; Uplekar, S.; Kostov, Y.; Croucher, L.; Pilli, M.; Rao, G. A unique noninvasive approach to monitoring dissolved O_2 and CO_2 in cell culture. *Biotechnol. Bioeng.* **2015**, *112*. [CrossRef]
29. Arévalo, H.; Snáchez, F.; Ruiz, F.; Guerrero, D.; Patino, D.; Alméciga-Díaz, C.; Rodríguez-López, A. Gain-scheduled oxygen concentration control system for a bioreactor. *IEEE Lat. Am. Trans.* **2018**, *16*, 2689–2697. [CrossRef]
30. Ritonja, J.; Goršek, A.; Pečar, D. Use of a heating system to control the probiotic beverage production in batch bioreactor. *Appl. Sci.* **2021**, *11*, 84. [CrossRef]
31. Ritonja, J.; Goršek, A.; Pečar, D. Model reference adaptive control for milk fermentation in batch bioreactor. *Appl. Sci.* **2020**, *10*, 9118. [CrossRef]
32. Goršek, A.; Tramšek, M. Kefir grains production—An approach for volume optimization of two-stage bioreactor system. *Biochem. Eng. J.* **2008**, *42*, 153–158. [CrossRef]
33. Zosel, J.; Oelßner, W.; Decker, M.; Gerlach, G.; Guth, U. Topical review: The measurement of dissolved and gaseous carbon dioxide concentration. *Meas. Sci. Technol.* **2011**, *22*, 072001. [CrossRef]
34. Isermann, R.; Lachmann, K.H.; Matko, D. *Adaptive Control Systems*; Prentice Hall International: New York, NY, USA, 1992.
35. Landau, Y.D. *Adaptive Control: The Model Reference Approach*; Marcel Dekker, Inc.: New York, NY, USA, 1979.
36. Narendra, K.S.; Annaswamy, A.M. *Stable Adaptive System*; Prentice Hall Inc.: Englewood Cliffs, NJ, USA, 1989.
37. Kaufman, H.; Bar-Khana, I.; Sobel, K. *Direct Adaptive Control. Algorithms*; Springer: New York, NY, USA, 1993.
38. Barkana, I. Output feedback stabilizability and passivity in nonstationary and non-linear systems. *Int. J. Adapt. Control. Signal. Process.* **2009**, *24*, 568–591.
39. Barkana, I. Adaptive control? But is so simple! A tribute to the efficiency, simplicity and beauty of adaptive control. *J. Intell. Robot. Syst.* **2016**, *83*, 3–34. [CrossRef]
40. Kennedy, J.; Eberhart, R.C. Particle swarm optimization. *Proc. IEEE Int. Conf. Neural Netw.* **1995**, *4*, 1942–1948.
41. Mezura-Montes, E.; Coello, C.A. Constraint-handling in nature-inspired numerical optimization: Past, present and future. *Swarm Evol. Comput.* **2011**, *1*, 173–194. [CrossRef]
42. Pedersen, M.E.H. Good parameters for particle swarm optimization. *Hvass Lab. Cph. Den. Tech. Rep. HL1001* **2010**, 1551–3203.
43. Particle Swarm Optimization Algorithm. Available online: https://mathworks.com/help/gads/particle-swarm-optimization-algorithm.html (accessed on 12 May 2021).

Article

μ-Synthesis for Fractional-Order Robust Controllers

Vlad Mihaly *, Mircea Șușcă *, Dora Morar [†], Mihai Stănese [†] and Petru Dobra [†]

Department of Automation, Technical University of Cluj-Napoca, Str. G. Barițiu nr. 26-28, 400027 Cluj-Napoca, Romania; Dora.Sabau@aut.utcluj.ro (D.M.); mihai.stanese@aut.utcluj.ro (M.S.); Petru.Dobra@aut.utcluj.ro (P.D.)
* Correspondence: vlad.mihaly@aut.utcluj.ro (V.M.); mircea.susca@aut.utcluj.ro (M.Ș.)
† These authors contributed equally to this work.

Abstract: The current article presents a design procedure for obtaining robust multiple-input and multiple-output (MIMO) fractional-order controllers using a μ-synthesis design procedure with D–K iteration. μ-synthesis uses the generalized Robust Control framework in order to find a controller which meets the stability and performance criteria for a family of plants. Because this control problem is NP-hard, it is usually solved using an approximation, the most common being the D–K iteration algorithm, but, this approximation leads to high-order controllers, which are not practically feasible. If a desired structure is imposed to the controller, the corresponding K step is a non-convex problem. The novelty of the paper consists in an artificial bee colony swarm optimization approach to compute the nearly optimal controller parameters. Further, a mixed-sensitivity μ-synthesis control problem is solved with the proposed approach for a two-axis Computer Numerical Control (CNC) machine benchmark problem. The resulting controller using the described algorithm manages to ensure, with mathematical guarantee, both robust stability and robust performance, while the high-order controller obtained with the classical μ-synthesis approach in MATLAB does not offer this.

Keywords: μ-synthesis; robust control; fractional-order control; swarm optimization; artificial bee colony optimization; CNC machine; mixed sensitivity; D–K iteration; Linear Matrix Inequality

Citation: Mihaly, V.; Șușcă, M.; Morar, D.; Stănese, M.; Dobra, P. μ-Synthesis for Fractional-Order Robust Controllers. *Mathematics* **2021**, 9, 911. https://doi.org/10.3390/math9080911

Academic Editor: António M. Lopes

Received: 28 February 2021
Accepted: 16 April 2021
Published: 20 April 2021

Publisher's Note: MDPI stays neutral with regard to jurisdictional claims in published maps and institutional affiliations.

Copyright: © 2021 by the authors. Licensee MDPI, Basel, Switzerland. This article is an open access article distributed under the terms and conditions of the Creative Commons Attribution (CC BY) license (https://creativecommons.org/licenses/by/4.0/).

1. Introduction

One of the active problems with major impact which have been studied for years in Control Theory refers to robustness. Robustness encompasses the sensitivity of a control system with respect to both internal and external disturbances. Several robust methods have been developed in order to achieve robust performance and stability in the presence of uncertainties. Robust control problems use $\mathcal{H}_2$ and $\mathcal{H}_\infty$ norms defined in frequency domain as a performance measure. To solve $\mathcal{H}_2/\mathcal{H}_\infty$ control problems, there are several approaches. One possible solution is presented in [1] and is based on Algebraic Riccati Equations (AREs). A more numerically stable approach to solve ARE was developed using Popov triplets in [2], approach recently implemented in an open-source manner in [3], with an iterative refinement method presented in [4], but an ARE-based solution presents a limitation due to the impossibility of solving singular problems. An alternative way which manages to solve such problems was introduced in [5], where AREs were replaced by Algebraic Riccati Inequalities (ARIs). ARIs are solved through Linear Matrix Inequalities (LMIs), while regular assumptions are no longer needed due to LMI system versatility. An open-source solver for Robust Control problems using LMIs is presented in [6].

The $\mathcal{H}_2/\mathcal{H}_\infty$ approach designs a suitable controller for the nominal plant, therefore only nominal stability and nominal performance are fulfilled. Additionally, generalized Robust Control framework allows to impose robust stability and robust performance, which cover the previous two aspects for an entire family of physical processes. As such, the μ-synthesis approach extends the $\mathcal{H}_\infty$ optimization in order to obtain a robust controller for the uncertain plant which includes parametric and dynamic uncertainties [7,8]. μ-synthesis is based on using structured singular values to quantify robustness margins and also on using linear fractional descriptions of the control problem containing the nominal plant model and uncertainty weighting functions. In [9], the authors present the μ-synthesis problem used with the so-called D–K iteration, which, in essence, provides two steps iteratively repeated until the robust performance stops improving: designing a $\mathcal{H}_\infty$ control law and μ analysis on closed-loop system.

μ-synthesis is, however, a non-convex problem and the D–K iteration represents only an approximation, without any convergence guarantees. Another significant concern of D–K iteration is that the method generates high order controllers. In order to solve this issue, various approaches based on fixed structure controller are proposed in different papers [10–12]. The method presented in [10] uses nonsmooth techniques for $\mathcal{H}_\infty$ synthesis. Then, using the same technique, the μ-synthesis was solved using D–K iteration and the result are presented in [13].

The main issue which appears when controller structure constraints are imposed is that the optimization problem in no longer convex and also, μ-synthesis is in general, considered nondeterministic polynomial time hard (NP-hard). A possible solution to that are swarm optimization algorithms. There are different approaches presented in papers [14–16] based on Genetic Algorithms (GA) and Particle Swarm Optimization (PSO). A solution for imposing a fixed structure controller, such as low-order or decentralized, was proposed in paper [14], which splits the problem in two parts: the convex part, solved using the classical ARE approach, and the non-convex part, solved using GA. The same authors proposed in [16] a new technique based on an evolutionary D–KD_0 iteration method, which combines the classical D-step with a KD_0 algorithm based on also a GA. Authors of the paper [15] propose an evolutionary approach to solve the μ-synthesis problem without order reduction by using an improved PSO.

Artificial Bee Colony (ABC) can also be used to solve complex optimization problems with constraints and could possibly outperform the other approaches and return the best solution in shorter execution time. The initial idea was presented in [17] as an extension of another metaheuristic algorithm, namely Honey Bee Swarm (HBS). The efficiency of the algorithm for several state-of-the-art optimization problems, along with an improvement for the stopping criterion, were underlined in [18].

One possible fixed structure controller is fractional-order proportional-integral-derivative (FO-PID). FO-PID is one of the most remarkable fractional order techniques with great interest in research [19]. It is used to generalize the classical PID control by adding extra degrees of freedom [20]. Compared to the integer PID controller, FO-PID brings the advantage of improving the robustness and providing better performance. In [21], the authors propose a FO-PID controller for a fractional-order plant model, presenting an analysis in both frequency and time domains, proving that achieving better control performance is one of the advantages of this approach. The FO-PID was used on a benchmark problem, i.e., the speed control of a DC motor [22], obtaining good results in terms of performance and robustness. Two generalized versions of Kessler's magnitude method with fractional order controllers were developed in [23,24]. A detailed comparison between classical modelling approaches and a fractional integrator approximation as a control baseline model is presented to the servo problem in [25]. A graphical method was developed in [26], while in [27] a fractional order internal model controller with event-based implementation was developed. Other applications comprise in an optimal FO-PID controller for a PMSM speed control, presented in [28], while a robust controller for a steam turbine was developed in [29].

In this paper, we present a new technique to design FO-PID robust controllers using μ-synthesis. The novelty of the current approach consists in implementing an algorithm able to find the nearly optimal values of the controller parameters using an artificial bee colony optimization. The cost function to be minimized is the $\mathcal{H}_\infty$ norm of the closed-loop system, when stable, and, otherwise, a large value affinely dependent on the largest real part of the eigenvalues of the closed-loop state matrix. Therefore, the non-convex part of the NP-hard μ-synthesis problem is solved using such a swarm optimization, while the D step is solved using the classical LMI technique. More than that, the `realp` object from MATLAB's Robust Control Toolbox embodies a limitation because it cannot be used as an exponent, necessary in approximating the fractional element with an integer-order system. Our approach manages to deal with this limitation in order to obtain the controller parameters, because the only information necessary in our approach is the range of the controller's parameters stored in such variables, which can be replaced with a simpler software object. Additionally, using our approach, a numerical example illustrates that the resulting controller manages to fulfill the robust stability and performance, while the controller obtained using unstructured μ-synthesis does not rigorously guarantee these specifications. Therefore, our method proposes a general framework able to synthesize arbitrary fixed structure fractional order controllers by optimizing their parameters in terms of robustness and performance, surpassing the well-established approach of manually tuning them for a desired problem, harnessing the Robust Control framework's design strongness.

We illustrate our proposed method on a benchmark problem: obtaining a controller for a two axis computer numerical control (CNC) system by solving a mixed-sensitivity μ-synthesis problem. Several control methods using the state space model of the machine are presented in literature. A comparison between the classical pole-placement method and Linear Quadratic Regulator (LQR) method is presented in [30]. For the LQR problem, an energy-based minimization algorithm was proposed, which proves to be suitable in terms of stability and robustness. As presented in [31], a different approach is recommended, in order to obtain a PI controller for each axis, using the state-feedback control algorithm. In this case, the state space model is augmented with an extra state which represents the integral of the position. The PI regulator parameters are obtained from the state-feedback gains.

The paper is organized in four sections. Section 2 introduces several ideas relevant to the proposed method, such as a mathematical foundation of FO-PID, continuing with the fundamental robust control problem based on μ-synthesis, and, finally, the ABC optimization algorithm. After that, the last subsection focuses on solving the non-convex problem of computing fixed structure controllers. Section 3 illustrates an application of the proposed method for position control of a two axis CNC machine, along with numerical results. In Section 4, the previously mentioned results are compared with those obtained using the well-established algorithms from MATLAB. Finally, conclusions are presented in Section 5.

2. Materials and Methods

In this section we present a controller synthesis procedure which manages to find a fixed structure fractional-order controller using the μ-synthesis technique from the Robust Control framework, where the non-convex subproblem involved in the classical D–K iteration is replaced by a swarm optimization algorithm. First, the mathematical background comprised in fractional-order control, robust control and artificial bee colony optimization is underlined, while the fourth subsection presents the proposed design procedure using all the mechanisms briefly described.

2.1. Fractional-Order Controller

The classical integer-order calculus was extended by Riemann and Liouville to fractional-order calculus by introducing the fractional integral operator [32]:

$$\mathcal{I}_a^\alpha \{f(t)\} = \frac{1}{\Gamma(\alpha)} \int_a^t f(\tau)(t-\tau)^{\alpha-1} d\tau, \tag{1}$$

where $\Gamma(\alpha) : \mathbb{C}_+ \to \mathbb{C}$ is the Euler Gamma function and the order of the integral operator is the complex parameter $\alpha \in \mathbb{C}_+$. This extension develops a new area of research in the

Control System domain. A commonly used definition of this operator was introduced in [33] as:

$$\mathcal{J}_C^\alpha\{f(t)\} = \frac{1}{\Gamma(\alpha)} \int_0^t (t-\tau)^{\alpha-1} f(\tau) d\tau = \mathcal{I}_0^\alpha\{f(t)\}, \qquad (2)$$

with $t > 0$ and $\alpha \in \mathbb{R}_+$, having the Laplace transform [34]:

$$\mathcal{L}\{\mathcal{J}_c^\alpha\{f(t)\}\}(s) = s^{-\alpha} F(s), \qquad (3)$$

where $F(s) = \mathcal{L}\{f(t)\}(s)$. One of the most common controller structures used in practice is the proportional-integral-derivative (PID) controller, having three degrees of freedom. Using fractional-order calculus, two new degrees of freedom can be added to a PID, having the notation $PI^\lambda D^\mu$. As such, the fractional order PID (FO-PID) has, as extra degrees of freedom, the order $\lambda \in \mathbb{R}_+$ of the integrator and the order $\mu \in \mathbb{R}_+$ of the differentiator, with the resulting transfer function:

$$H_c(s) = K_P + \frac{K_I}{s^\lambda} + K_D s^\mu. \qquad (4)$$

The time domain expression of the command signal $c(t)$ can be expressed using the error signal $\varepsilon(t)$ as:

$$c(t) = K_p \cdot \varepsilon(t) + K_I \cdot \mathcal{J}_c^\lambda\{\varepsilon(t)\} + K_D \cdot \mathcal{J}_c^{-\mu}\{\varepsilon(t)\}. \qquad (5)$$

One of the major issues of such a fractional element, i.e., $\mathcal{J}_c^\lambda$ or $\mathcal{J}_c^{-\mu}$, is its implementation. In order to solve this problem, the Oustaloup recursive approximation (ORA) was introduced [35], and allows the approximation of the fractional-order element with an LTI system of pre-specified order N:

$$s^\lambda = \prod_{k=1}^N \frac{1 + s/\omega_{z,k}}{1 + s/\omega_{p,k}}, \qquad (6)$$

where the frequency values of the singularities are obtained based on the desired fractional order $\lambda \in (0,1)$, the integer order of the approximation N, along with the frequency range in which the approximation is valid $[\omega_l, \omega_u]$. Using the following two coefficients:

$$\varepsilon = \left(\frac{\omega_u}{\omega_l}\right)^{\frac{\lambda}{N}} \quad \text{and} \quad \eta = \left(\frac{\omega_u}{\omega_l}\right)^{\frac{1-\lambda}{N}}, \qquad (7)$$

the above mentioned frequencies can be computed using:

$$\omega_{z,1} = \omega_l \sqrt{\eta}, \qquad (8)$$
$$\omega_{p,n} = \omega_{z,n} \cdot \varepsilon, \quad n = \overline{1,N}, \qquad (9)$$
$$\omega_{z,n+1} = \omega_{p,n} \cdot \eta, \quad n = \overline{1,N-1}. \qquad (10)$$

For the rest of the possible real values of λ, the approximation can be easily extended as: for $\lambda \in (-1,0)$ by inverting the relation (6), while for $|\lambda| \geq 1$, the components could be the integer part $[\lambda]$ and the fractional part $\{\lambda\}$, with the fractional part only approximated using (6).

2.2. Robust Control

The Robust Control framework assumes to minimize the $\mathcal{H}_2/\mathcal{H}_\infty$ norm of the lower linear fractional transformation (LLFT) of a plant P and a controller K:

$$P_o = \text{LLFT}(P,K) = P_{11} + P_{12}K(I - P_{22}K)^{-1}P_{21}, \qquad (11)$$

where the plant P can be written, in general form, as:

$$P : \begin{pmatrix} P_{11} & P_{12} \\ P_{21} & P_{22} \end{pmatrix} = \left(\begin{array}{c|cc} A & B_w & B_u \\ \hline C_z & D_{zw} & D_{zu} \\ C_y & D_{yw} & D_{yu} \end{array} \right), \quad (12)$$

where the signals involved in the above relation will be further detailed in a more general context. One approach to solving $\mathcal{H}_2/\mathcal{H}_\infty$ problems is using AREs, which presents several limitations that are removed by the LMI approach. However, this framework can be used to ensure nominal stability and nominal performance only, but, the plant P must also contain an augmented model of the real process in which uncertainties are also present. There are two classic uncertainties types: *parametric*, represented by δI, where δ is the maximum bound of the parameter for a physical variable, and *unstructured*, represented by a full block $\Delta \in \mathbb{R}^{m \times m}$. The latter illustrates neglected or unknown dynamics uncertainties. In the mixed-scenario case, the following set is considered:

$$\mathbf{\Delta} = \left\{ \Delta = \mathrm{diag}\left(\delta_1 I_{n_1}, \ldots, \delta_s I_{n_s}, \Delta_1, \ldots, \Delta_f \right) | \delta_k \in \mathbb{R}, \Delta_j \in \mathbb{R}^{m_j \times m_j}, k = \overline{1,s}, j = \overline{1,f} \right\}. \quad (13)$$

In the Robust Control field, one of the main tools used for robustness analysis is the structured singular value, defined as follows.

Definition 1. *For a square matrix $M \in \mathbb{C}^{N \times N}$ the structured singular value with respect to the set $\mathbf{\Delta}$ is:*

$$\mu_\Delta(M) = \frac{1}{\min_{\Delta \in \mathbf{\Delta}} \{ \overline{\sigma}(\Delta) | \det(I - M\Delta) = 0 \}}, \quad (14)$$

if there exists $\Delta \in \mathbf{\Delta}$ such that the matrix $I - M\Delta$ is rank deficient, otherwise 0.

For an LTI system described by the transfer matrix $M(s)$ and an upper linear fractional transformation (ULFT) connection shown in Figure 1 (left), the structured singular value $\mu_\Delta(M)$ can be defined as:

$$\mu_\Delta(M(s)) = \sup_{\omega \in \mathbb{R}_+} \mu_\Delta(M(j\omega)). \quad (15)$$

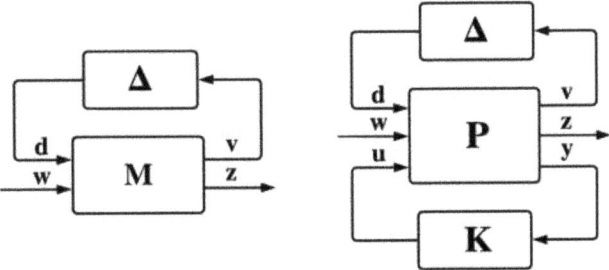

Figure 1. (**Left**) The generalized M-Δ structure containing the plant and the uncertainty block Δ. (**Right**) The closed-loop P-Δ-K structure containing the plant, controller and uncertainty block Δ.

Now, considering $M(s) = \mathrm{LLFT}(P, K)(s)$ as being the lower linear fractional transformation (LLFT) between plant P and controller K, the connection illustrated in Figure 1 (right) results. The generalized plant structure is:

$$P_\Delta(s) = \begin{pmatrix} P_{vd}(s) & P_{vw}(s) & P_{vu}(s) \\ P_{zd}(s) & P_{zw}(s) & P_{zu}(s) \\ P_{yd}(s) & P_{yw}(s) & P_{yu}(s) \end{pmatrix} \Leftrightarrow P_\Delta : \begin{pmatrix} \dot{\mathbf{x}}(t) \\ \mathbf{v}(t) \\ \mathbf{z}(t) \\ \mathbf{y}(t) \end{pmatrix} = \left(\begin{array}{c|ccc} A & B_d & B_w & B_u \\ \hline C_v & D_{vd} & D_{vw} & D_{vu} \\ C_z & D_{zd} & D_{zw} & D_{zu} \\ C_y & D_{yd} & D_{yw} & D_{yu} \end{array} \right) \begin{pmatrix} \mathbf{x}(t) \\ \mathbf{d}(t) \\ \mathbf{w}(t) \\ \mathbf{u}(t) \end{pmatrix}, \quad (16)$$

where three types of input signals are present—the command input $\mathbf{u} \in \mathbb{R}^{n_u}$, the performance input $\mathbf{w} \in \mathbb{R}^{n_w}$, and the disturbance input $\mathbf{d} \in \mathbb{R}^{n_d}$—and three types of outputs—the measurements vector $\mathbf{y} \in \mathbb{R}^{n_y}$, the performances vector $\mathbf{z} \in \mathbb{R}^{n_z}$, and the disturbance output $\mathbf{v} \in \mathbb{R}^{n_v}$.

Besides the well-known $\mathcal{H}_2/\mathcal{H}_\infty$ methods, a controller that meets robust stability and robust performance alike can be computed using the μ-synthesis framework. Robust stability implies that a specific controller manages to stabilize all the processes described by the upper linear fractional transformation (ULFT) presented in Figure 1 (left), while robust performance implies that the controller is able to impose the desired closed-loop performance in the worst case scenario. In order to have a mathematical guarantee that a controller K meets the robust stability and performance, the Main Loop theorem can be used. It implies that closed-loop system meets robust stability and performance if and only if the structural singular value of the LLFT of the plant and controller, with respect to Δ, fulfills:

$$\sup_{\omega \in \mathbb{R}_+} \mu_\Delta(\text{LLFT}(P,K)(j\omega)) < 1. \tag{17}$$

Therefore, the robust control problem can be written as:

$$\inf_{K \text{ stab.}} \sup_{\omega \in \mathbb{R}_+} \mu_\Delta(\text{LLFT}(P,K)(j\omega)), \tag{18}$$

which is not convex. More than that, the structural singular values are hard to be explicitly computed. In practice, there are various bounds which can be used to approximate the structural singular value. One of the most used approximations of the upper bound is in [9]:

$$\mu_\Delta(M) \leq \inf_{D \in \mathbf{D}} \overline{\sigma}(DMD^{-1}), \tag{19}$$

where $\overline{\sigma}$ denotes the largest singular value, and the set $\mathbf{D}$ is defined as:

$$\mathbf{D} = \left\{ \text{diag}\left(D_1, \ldots, D_s, d_1 I_{m_1}, \ldots, d_f I_{m_f}\right) | D_k = D_k^\top \in \mathbb{R}^{n_k \times n_k}, d_j > 0, k = \overline{1,s}, j = \overline{1,f} \right\}. \tag{20}$$

Based on this upper bound, a good approximation of the initial non-convex problem can be employed by solving the following quasi-convex problem:

$$\inf_{K \text{ stab.}} \sup_{\omega \in \mathbb{R}_+} \inf_{D \in \mathbf{D}} \overline{\sigma}\left(D(j\omega) \cdot \text{LLFT}(P,K)(j\omega) \cdot (D(j\omega))^{-1}\right). \tag{21}$$

If the scaling factor, represented by the system D, is fixed, then the problem (21) is nothing but a $\mathcal{H}_\infty$ optimization problem. On the other hand, fixing the controller K, the D scaling step can now be obtained by solving a Parrot problem for a desired frequency set $\Omega = \{\omega_1, \ldots, \omega_N\}$ using the following generalized eigenvalue problem:

$$\begin{aligned} &\min \gamma, \\ &\text{s.t.} \quad (\text{LLFT}(P,K)(j\omega_i))^* \cdot X \cdot \text{LLFT}(P,K)(j\omega_i) \leq \gamma^2 X, \end{aligned} \tag{22}$$

where from the solution $X = (D(j\omega_i))^* \cdot D(j\omega_i)$, the matrix $D(j\omega_i)$ can be extracted using a singular value decomposition. After all Parrot problems are solved, a minimum phase system is found in order to approximate the analytical solution $D(s)$. In summary, an iterative algorithm which solves the μ-synthesis problem starts by setting $D = I$, with the following steps applied successively:

1: Fix D and solve the $\mathcal{H}_\infty$ optimal problem to find a controller K:

$$K = \arg \inf_{K \text{ stab.}} ||\text{LLFT}(P,K)||_\infty. \tag{23}$$

2: Fix the controller K and solve the set of convex problems:

$$D(j\omega) = \arg \inf_{D \in \mathbf{D}} \bar{\sigma}\Big(D \cdot \text{LLFT}(P,K)(j\omega) \cdot D^{-1}\Big), \qquad (24)$$

for a given frequency range Ω and, then, fit a stable minimum phase transfer matrix $D(s)$.

Steps **1** and **2** are executed in a loop sequence until the difference between two consecutive $\mathcal{H}_\infty$ norms is less than a prescribed tolerance or the maximum number of iterations is reached.

2.3. Artificial Bee Colony Optimization

The artificial bee colony (ABC) optimization is a nature-inspired algorithm used to minimize a cost function:

$$f: D \to \mathbb{R}, \text{ where } D = [lb_1, ub_1] \times [lb_2, ub_2] \times \cdots \times [lb_d, ub_d] \subset \mathbb{R}^d. \qquad (25)$$

The ABC algorithm mimics the behaviour of real honeybees, where each food source represents a possible solution of the optimization problem described above. The location and the amount of nectar correspond to the design variables and the cost function, respectively. The bees are divided in two main groups: employed and unemployed bees, while the unemployed bees could be of two types as well: onlooker and scout bees. The employed bees are the ones that investigate the food source and return to the hive to inform the others by performing the waggle dance; the onlooker bees are the ones that watch the dance and decide whether or not a food source is worthy of being searched or not; the scout bees are former employed bees that have abandoned their previous food source, due to lack of nectar, and which now search for a new one. The Best Solution is represented by the food source, and the quality (or cost) of the solution is represented by the amount of nectar.

The number of employed bees coincides with number of the onlooker bees and represents the dimension of the swarm problem, denoted by N. The employed bees start the foraging process by randomly searching an initial position $\mathbf{x}_i^{(0)}$ in the domain D:

$$\mathbf{x}_i^{(0)} = \begin{pmatrix} lb_1 + \phi_{i,1}^{(0)} \cdot (ub_1 - lb_1) \\ lb_2 + \phi_{i,2}^{(0)} \cdot (ub_2 - lb_2) \\ \vdots \\ lb_d + \phi_{i,d}^{(0)} \cdot (ub_d - lb_d) \end{pmatrix} \in D, \qquad (26)$$

where $\phi_{i,\overline{1,d}}^{(0)} \in [-1, 1]$ are random numbers. After this initialization step, the first Best Solution appears.

Each employed bee searches a new food source based on the location of the current food source $\mathbf{x}_i^{(k)}$ and another food source $\mathbf{x}_j^{(k)}$ randomly selected. The new possible position is:

$$\overline{\mathbf{x}}_i^{(k)} = \text{sat}\Big(\mathbf{x}_i^{(k)} + \phi \odot (\mathbf{x}_i^{(k)} - \mathbf{x}_j^{(k)})\Big), \qquad (27)$$

where $\phi \in [-1, 1]^d$ is an array of random numbers, $\odot$ is the element-wise multiplication, and sat is the saturation function that does not allow the position to be outside the searching domain D. Now, the position of the ith employed bee for the next iteration will be:

$$\mathbf{x}_i^{(k+1)} = \arg \min \{f(\mathbf{x}_i^{(k)}), f(\overline{\mathbf{x}}_i^{(k)})\}, \qquad (28)$$

which means that an employed bee will never choose a source with less nectar. If the position for the next iteration will not be changed, the abandonment counter of the ith employed bee increments.

The onlooker bees use the information shared by each employed bee and choose a location around the position of the employed bees. The fitness value of a solution $\mathbf{x}_i^{(k)}$ is given by:

$$\log W(i) = \frac{-f(\mathbf{x}_i^{(k)})}{\frac{1}{N}\sum_j f(\mathbf{x}_j^{(k)})}, \tag{29}$$

and the probability that ith bee's source will be selected by an onlooker bee is:

$$p_i = \frac{W(i)}{\sum_j W(j)}. \tag{30}$$

Now, using a roulette wheel selection method, the onlooker bee will choose a source i and, using the same searching technique as an employed bee, a new position is computed using (27). If the outlooker bee founds a better solution than the ith employed bee, they change their roles, otherwise the abandonment counter for the ith source increments. After this step, we have N employed bees and N unemployed bees. However, if the abandonment counter for the ith source exceeds a threshold, the ith employed bee becomes a scout and tries to find a new location using the relation (26).

After every loop corresponding to employed, outlooker and scout bees, it is checked if there is a food source with a solution better than the last one. The algorithm is over when the maximum number of cycles is reached or when there is no improvement of the Best Solution after a prescribed number of cycles.

2.4. Proposed Method

The solution of the problem described in (21) using the classical D–K iteration with $\mathcal{H}_2/\mathcal{H}_\infty$ framework leads to a high-order controller. As a solution to this issue, a fixed structure family of controllers $\mathcal{K}$ can be considered, and the problem (21) becomes:

$$\inf_{\substack{K \in \mathcal{K} \\ K \text{ stab}}} \sup_{\omega \in \mathbb{R}_+} \inf_{D \in \mathbf{D}} \overline{\sigma}\Big(D(j\omega) \cdot \text{LLFT}(P, K)(j\omega) \cdot (D(j\omega))^{-1}\Big). \tag{31}$$

The new problem has the disadvantage of being non-convex in terms of the parameters of the proposed controller structure. However, the problem described in (31) will also be solved using a D–K iterative procedure, but the non-convex part of the problem has fewer degrees of freedom and it will be managed with an ABC optimization.

Starting with the model of the process G, a plant P is obtained after the augmentation step:

$$P: \begin{pmatrix} \dot{\mathbf{x}}(t) \\ \mathbf{z}(t) \\ \mathbf{y}(t) \end{pmatrix} = \begin{pmatrix} A & B_w & B_u \\ C_z & D_{zw} & D_{zu} \\ C_y & D_{yw} & D_{yu} \end{pmatrix} \begin{pmatrix} \mathbf{x}(t) \\ \mathbf{w}(t) \\ \mathbf{u}(t) \end{pmatrix}, \tag{32}$$

where the meaning of the inputs and outputs remains the same as in Section 2.2. Considering the advantages of the fractional-order controllers, the proposed structure of the controller is:

$$K^\theta(s) = \begin{pmatrix} C_{FO}^{(1,1)}(s) & C_{FO}^{(1,2)}(s) & \cdots & C_{FO}^{(1,n_y)}(s) \\ C_{FO}^{(2,1)}(s) & C_{FO}^{(2,2)}(s) & \cdots & C_{FO}^{(2,n_y)}(s) \\ \vdots & \vdots & \ddots & \vdots \\ C_{FO}^{(n_u,1)}(s) & C_{FO}^{(n_u,2)}(s) & \cdots & C_{FO}^{(n_u,n_y)}(s) \end{pmatrix}, \tag{33}$$

where $C_{FO}^{(i,j)}$ is a fractional order controller from the ith input to the jth output and has the form:

$$C_{FO}^{(i,j)}(s) = K_P^{(i,j)} + \frac{K_I^{(i,j)}}{s^{\lambda^{(i,j)}}} + K_D^{(i,j)} s^{\mu^{(i,j)}}, \tag{34}$$

with the tunable parameters described by the vector:

$$\theta^{(i,j)} = \begin{pmatrix} K_P^{(i,j)} & K_i^{(i,j)} & \lambda^{(i,j)} & K_D^{(i,j)} & \mu^{(i,j)} \end{pmatrix}^\top \in D_{ABC}^{(i,j)} \subset \mathbb{R}^5. \tag{35}$$

Using these considerations, the desired family of the fixed structure controllers can be described as follows:

$$\mathcal{K} = \left\{ K^\theta(s) \middle| \theta \in D_{ABC} \equiv D_{ABC}^{(1,1)} \times D_{ABC}^{(1,2)} \times \cdots \times D_{ABC}^{(n_u, n_y)} \right\}, \tag{36}$$

where all parameters are stored in a single vector θ describing all degrees of freedom of the tunable controller. Using the ORA mechanism, each component $K^\theta(s) \in \mathcal{K}$ has a state-space representation:

$$K^\theta : \begin{pmatrix} \dot{\mathbf{x}}_K(t) \\ \mathbf{u}(t) \end{pmatrix} = \left(\begin{array}{c|c} A_K(\theta) & B_K(\theta) \\ \hline C_K(\theta) & D_K(\theta) \end{array} \right) \begin{pmatrix} \mathbf{x}_K(t) \\ \mathbf{y}(t) \end{pmatrix} \tag{37}$$

Next, we denote by P_o^θ the closed-loop system represented by the lower linear fractional transformation between the augmented plant P and controller K^θ, which can be represented as:

$$P_o^\theta = \text{LLFT}(P, K^\theta) : \begin{pmatrix} \dot{\mathbf{x}}_o(t) \\ \mathbf{z}_o(t) \end{pmatrix} = \left(\begin{array}{c|c} A_o(\theta) & B_o(\theta) \\ \hline C_o(\theta) & D_o(\theta) \end{array} \right) \begin{pmatrix} \mathbf{x}_o(t) \\ \mathbf{w}_o(t) \end{pmatrix}, \tag{38}$$

where state vector, input vector and output vector of the closed-loop system are:

$$\mathbf{x}_o = \begin{pmatrix} \mathbf{x} \\ \mathbf{x}_K \end{pmatrix}, \quad \mathbf{w}_o \equiv \mathbf{w} \quad \text{and} \quad \mathbf{z}_o \equiv \mathbf{z}. \tag{39}$$

In Algorithm 1 the main steps necessary to obtain the parameters of the controller having the structure (33) are presented. The inputs of the algorithm are the closed-loop plant P_o^θ, containing the tunable controller parameters θ, and the parameters α and β which describe the cost function that needs to be minimized using an ABC optimization, but, P_o^θ must also contain the varying parameters and unmodelled dynamics. Therefore, the plant P_Δ obtained after the augmentation step with uncertainties Δ has the form presented in (16). Thus, the first step made in Algorithm 1 consists in transforming the closed-loop system P_o^θ in the generalized closed-loop system $P_{o,\Delta}^\theta$ described as follows:

$$P_{o,\Delta}^\theta = \text{LLFT}(P_\Delta, K^\theta) : \begin{pmatrix} \dot{\mathbf{x}}_o(t) \\ \overline{\mathbf{z}}_o(t) \end{pmatrix} = \left(\begin{array}{c|c} \overline{A}_o(\theta) & \overline{B}_o(\theta) \\ \hline \overline{C}_o(\theta) & \overline{D}_o(\theta) \end{array} \right) \begin{pmatrix} \mathbf{x}_o(t) \\ \overline{\mathbf{w}}_o(t) \end{pmatrix}, \tag{40}$$

where the new extended performance inputs and outputs are:

$$\overline{\mathbf{w}}_o = \begin{pmatrix} \mathbf{d} \\ \mathbf{w}_o \end{pmatrix} \quad \overline{\mathbf{z}}_o = \begin{pmatrix} \mathbf{v} \\ \mathbf{z}_o \end{pmatrix} \tag{41}$$

As mentioned in Section 2.2, the structured singular value can be bounded using two D-scaling factors, one for the left and one for the right scaling, denoted D_L and D_R, respectively. As it can be notice in Figure 2, a new closed-loop plant $\overline{P}_o^\theta$ is obtained:

$$\overline{P}_o^{(\theta)}(s) = D_L(s) \cdot P_{o,\Delta}^\theta(s) \cdot D_R^{-1}(s), \tag{42}$$

having the state-space representation:

$$\overline{P}_o^\theta : \begin{pmatrix} \dot{\hat{\mathbf{x}}}_o(t) \\ \hat{\mathbf{z}}_o(t) \end{pmatrix} = \left(\begin{array}{c|c} \hat{A}_o(\theta) & \hat{B}_o(\theta) \\ \hline \hat{C}_o(\theta) & \hat{D}_o(\theta) \end{array} \right) \begin{pmatrix} \hat{\mathbf{x}}_o(t) \\ \hat{\mathbf{w}}_o(t) \end{pmatrix}. \tag{43}$$

All the above mentioned plants are presented in Figure 2. First, the generalized plant P_Δ has a LLFT connection with the controller K^θ, resulting the closed-loop plant $P^\theta_{o,\Delta}$, having the input vector $\overline{w}_o$ and the output vector $\overline{z}_o$. After the D-scaling step, a new plant $\overline{P}^\theta_o$ is obtained, having the input vector $\hat{w}_o$ and the output vector $\hat{z}_o$.

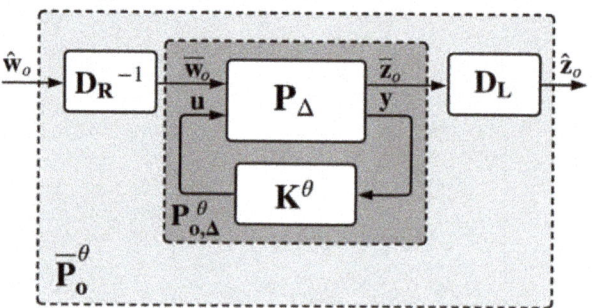

Figure 2. The augmented plant with uncertainties P_Δ in LLFT connection with controller K^θ forms the closed-loop plant $P^\theta_{o,\Delta}$. After each D-scale step, the plant used to find the controller parameters is $\overline{P}^\theta_o$.

Before starting the while loop of Algorithm 1, an initialization of the generalized closed-loop plant with D-scale $\overline{P}^\theta_o$ is performed with the initial scale factors $D_L = I_{n_w}$ and $D_R = I_{n_z}$, as seen in line 2. As noticed in line 4, the K step is performed using this generalized plant having as degrees of freedom the tunable parameters θ. In order to compute the controller parameters θ^*, the ABC optimization will be used. The cost function to be minimized is:

$$f : D_{ABC} \to \mathbb{R}_+, \quad f(\theta) = \begin{cases} ||\overline{P}^\theta_o||_\infty, & \text{if } \overline{P}^\theta_o \text{ is stable} \\ \alpha \lambda_{max}(\hat{A}_o) + \beta, & \text{if } \overline{P}^\theta_o \text{ is unstable} \end{cases}, \quad (44)$$

where the operator λ_{max} is defined by:

$$\lambda_{max} : \mathbb{R}^{M \times M} \to \mathbb{R}, \quad \lambda_{max}(A) = \max\{\text{Re}(\lambda) \mid \lambda \in \Lambda(A)\}. \quad (45)$$

Algorithm 1: Fixed Structure μ-Synthesis.

Input: P^θ_o, α, β
Output: K^{θ^*}
1 get uncertain closed-loop plant $P^\theta_{o,\Delta}$ as in (40)
2 set $D_L = I_{n_w}$ and $D_R = I_{n_z}$ and compute $\overline{P}^\theta_o = D_L \cdot P^\theta_{o,\Delta} \cdot D_R^{-1}$
3 **while** $N_{iter} \leq MAX\_ITER$ **and** *exists_improvement* **do**
4 $\quad$ $\overline{P}^{\theta^*}_o$ = computeKstep($\overline{P}^\theta_o$, α, β)
5 $\quad$ update the uncertain plant $P^{\theta^*}_{o,\Delta}$
6 $\quad$ $P^{\theta^*}_{o,nom}$ = getNominalPlant($P^{\theta^*}_{o,\Delta}$)
7 $\quad$ $[D_L, D_R]$ = computeDstep($P^{\theta^*}_{o,nom}$)
8 $\quad$ $\overline{P}^{\theta^*}_o = D_L \cdot P^{\theta^*}_{o,\Delta} \cdot D_R^{-1}$
9 $\quad$ check if improvement exists and increase N_{iter}

The procedure computeKstep used to obtain the controller parameters is briefly presented in Algorithm 2 and will be described below. The inputs of the algorithm are the generalized closed-loop plant with D-scaling step $\overline{P}^\theta_o$, along with the parameters α and β

which describe the cost function (44). The first step of this routine consists in computing the domain D_{ABC} of the cost function based on the constraints of the tunable parameters. The main limitation is represented by the fractional orders of the integral and derivative effects of the controller which must remain in $(0,1)$, according to ORA.

In the second step of routine Algorithm 2, an initial population is created. Let N be the dimension of the swarm problem. This parameter can be given as input, but as a good practice, this can be chosen 100 times larger than the number of tunable parameters. The initialization step consists in randomly generating the positions $\theta_1^{(0)}, \theta_2^{(0)}, \ldots, \theta_N^{(0)}$ of the food sources for each employed bee in the domain D_{ABC} using relation (26). After the initialization step, the first Best Solution (BS) is computed as:

$$\theta^\star = \arg\min\left\{f(\theta_i^{(k)}) \mid i = \overline{1,N}\right\} \Rightarrow BS = f(\theta^\star). \tag{46}$$

Using this initial population, the main while loop starts. In line 4, the employed bees step is performed. In this step, each employed bee searches a new position around using relation (27), resulting N new possible positions for the next step, denoted by $\overline{\theta}_1^{(k)}, \overline{\theta}_2^{(k)}, \ldots, \overline{\theta}_N^{(k)}$, and the proposed positions of the employed bees at step $k+1$ will be:

$$\hat{\theta}_i^{(k+1)} = \arg\min\{f(\theta_i^{(k)}), f(\overline{\theta}_i^{(k)})\}, \ i = \overline{1,N}. \tag{47}$$

If, for a specific food source i, the proposed position coincides with the previous position, the abandonment counter increments.

Using the proposed solutions of the above step, each onlooker bee selects a new possible solution based on the roulette wheel selection mechanism and relation (27), resulting a new set of proposed solutions: $\tilde{\theta}_1^{(k)}, \tilde{\theta}_1^{(k)}, \ldots, \tilde{\theta}_N^{(k)}$. If the ith onlooker bee has a better solution than ith employed bee, they exchange their roles, otherwise the abandonment counter for the ith food source increments. After this step, the new set of proposed positions for the employed bees are:

$$\hat{\theta}_i^{(k+1)} = \arg\min\{f(\tilde{\theta}_i^{(k)}), f(\hat{\theta}_i^{(k+1)})\}, \ i = \overline{1,N}. \tag{48}$$

Algorithm 2: Compute K Step.

Input: $\overline{P}_o^\theta, \alpha, \beta$
Output: $\overline{P}_o^{\theta^\star}$

1 compute the domain D of the optimization problem
2 create an initial population of employed bees using (26) and select the first Best Solution
3 **while** $N_{cycles} <$ MNC **and** exists_improvement **do**
4   perform the employed bees step
5   perform the outlooker bees step
6   perform the scout bees step
7   find the new Best Solution, check if improvements exist and increase N_{cycles}
8 **end**

After performing the outlooker bees step from line 5, the abandonment counter for each active food source will be interrogated. If the abandonment counter of the ith position exceeds a prescribed threshold, denoted by LIMIT, the employed bee becomes a scout bee and its new position is obtained using (26). As a good practice, this paper proposed an improved mechanism of converting employed bees in scout bees. If the value of the cost function at the proposed position $f(\hat{\theta}_i^{(k+1)})$ is over β, this means that, in the case of a good calibration of parameters α and β, the solution corresponds to an unstable closed-loop system and can be dropped. The positions of the employed bees for the next iteration become $\theta_1^{(k+1)}, \theta_2^{(k+1)}, \ldots, \theta_N^{(k+1)}$.

The last step of the main while loop, presented in line 7, consists in computing the Best Solution after this new iteration:

$$\theta^\star = \arg\min\left\{f(\theta_{old}^\star), f(\theta_1^{(k)}), f(\theta_2^{(k)}), \ldots, f(\theta_N^{(k)})\right\} \Rightarrow BS = f(\theta^\star), \tag{49}$$

and, then, checking if there exist any improvements after this step. In order to have a good trade-off between execution time and solution accuracy, it can be useful to establish a threshold for the number of steps when no improvement appears in Best Solution and mark if there exists such an improvement. Being a metaheuristic optimization algorithm, the runtime can be made deterministic and theoretically finite by imposing the variable *MNC*, without convergence guarantees. In practice, such methods have been successfully employed in various global optimization problems and, being that it is an offline optimization, it is not problematic for our approach.

Returning to Algorithm 1, after the *K* step is performed, a new parameter vector $\theta^\star$ results, which leads to a new generalized closed-loop plant $\overline{P}_o^{\theta^\star}$ and to a new uncertain plant $P_{o,\Delta}^{\theta^\star}$. The closed-loop plant $P_{o,\Delta}^{\theta^\star}$ is used to compute the next *D*-scale factors. From $P_{o,\Delta}^{\theta^\star}$ we extract a nominal plant $P_{o,nom}^{\theta^\star}$ by fixing the tunable parameters of the controller with the values determined in step 4.

The computeDstep routine from line 7 receives as input this nominal plant and returns the left and the right *D*-scale factors. Based on the poles and the transmission zeros of the nominal plant $P_{o,nom}^{\theta^\star}$, a set $\Omega = \{\omega_1, \omega_2, \ldots, \omega_F\}$ of frequencies is generated. Then, we need to get the frequency response data for each scaling factor by solving the following generalized eigenvalue problem:

$$\min \gamma,$$
$$\text{s.t. } \overline{\sigma}\left(D_L(j\omega_i) \cdot P_{o,nom}^{\theta^\star}(j\omega_i) \cdot D_R^{-1}(j\omega_i)\right) < \gamma, \tag{50}$$

for each $i = \overline{1, F}$, which is nothing but a Parrot problem which can be solved point by point using LMI techniques, as mentioned in Section 2.2. Once the frequency response data points are obtained for each value in Ω, we need to fit two minimum phase systems, one for each scaling factor, then perform the *D*-scaling step, giving a new generalized closed-loop system:

$$\overline{P}_o^{\theta^\star}(s) = D_L(s) \cdot P_{o,\Delta}^\star(s) \cdot D_R^{-1}(s). \tag{51}$$

In a similar manner with the computeKstep routine, there are possible stopping criteria which can be used. First, a threshold for the maximum number of *D–K* iterations can be imposed. Another important stopping condition appears if the upper bound of the structural singular value is less than 1, because this fact already guarantees that the controller ensures robust stability and robust performance. In accordance with the allowed maximum number of steps, a stopping criterion could be to check if there are any improvements after a certain number of steps.

3. Numerical Results

In this section we illustrate how the proposed method can be used on a benchmark problem. The process is represented by a Computer Numerical Control (CNC) machine with two orthogonal axis which are operated by two servo DC motors. A Trio Motion Coordinator family controller was used for the CNC motors. The programming language was Trio Basic, which provides various functions such as linear, circular and helical interpolation, variable speed and acceleration profile functions and control functions to ensure smooth and synchronized motions.

A mathematical model for each axis was determined on the basis of measured data: angular speeds ω_x and ω_y, and angular positions θ_x and θ_y. The state space mathematical model of the machine is described as follows:

$$G: \begin{pmatrix} \dot{\omega}_x \\ \dot{\theta}_x \\ \dot{\omega}_y \\ \dot{\theta}_y \\ \dot{\theta}_x \\ \dot{\theta}_y \end{pmatrix} = \left(\begin{array}{cccc|cc} -\frac{1}{T_{mx}} & 0 & 0 & 0 & \frac{K_{mx}}{T_{mx}} & K_{xy} \\ 1 & 0 & 0 & 0 & 0 & 0 \\ 0 & 0 & -\frac{1}{T_{my}} & 0 & K_{yx} & \frac{K_{my}}{T_{my}} \\ 0 & 0 & 1 & 0 & 0 & 0 \\ 0 & 1 & 0 & 0 & 0 & 0 \\ 0 & 0 & 0 & 1 & 0 & 0 \end{array} \right) \begin{pmatrix} \omega_x \\ \theta_x \\ \omega_y \\ \theta_y \\ u_x \\ u_y \end{pmatrix}, \quad (52)$$

where the state vector is $\mathbf{x} = \begin{pmatrix} \omega_x & \theta_x & \omega_y & \theta_y \end{pmatrix}^\top$, the input vector is $\mathbf{u} = \begin{pmatrix} u_x & u_y \end{pmatrix}^\top$ and the output vector is $\mathbf{y} = \begin{pmatrix} \theta_x & \theta_y \end{pmatrix}^\top$. The model parameters, along with their nominal values and uncertainty range are detailed in Table 1.

Table 1. Nominal values and uncertainty ranges of the CNC model parameters.

Parameter	Nominal Value	Uncertainty Range	Parameter	Nominal Value	Uncertainty Range
T_{mx}	0.02448	±10%	T_{my}	0.01139	±10%
K_{mx}	25.8017	±10%	K_{my}	25.1494	±10%
K_{xy}	26.65	±10%	K_{yx}	24.46	±10%

For this control problem a mixed-sensitivity loop shaping technique is used, which provides a good trade-off between performance and robustness. In order to use this technique, a new plant model will be obtained after the augmentation process, as in Figure 3. The performance inputs for the resulting augmented plant P are the references for both axis $\mathbf{w} \equiv \mathbf{r} = \begin{pmatrix} r_x & r_y \end{pmatrix}^\top$. For the augmentation procedure, the closed-loop transfer functions need to be weighted from the reference signals $\mathbf{r}$ to their corresponding error signals $\mathbf{e}$, output signals $\mathbf{y}$ and command signals $\mathbf{u}$, which are named: sensitivity function $S = (I + GK)^{-1}$, complementary sensitivity function $T = I - S$ and control effort KS, respectively. The performance output vector is composed from the weighted outputs of these three vector-valued functions:

$$\mathbf{z} = \begin{pmatrix} \mathbf{z}_S \\ \mathbf{z}_T \\ \mathbf{z}_{KS} \end{pmatrix}, \quad \text{where } \mathbf{z}_S = \begin{pmatrix} z_{S,x} \\ z_{S,y} \end{pmatrix}, \mathbf{z}_T = \begin{pmatrix} z_{T,x} \\ z_{T,y} \end{pmatrix} \text{ and } \mathbf{z}_{KS} = \begin{pmatrix} z_{KS,x} \\ z_{KS,y} \end{pmatrix}. \quad (53)$$

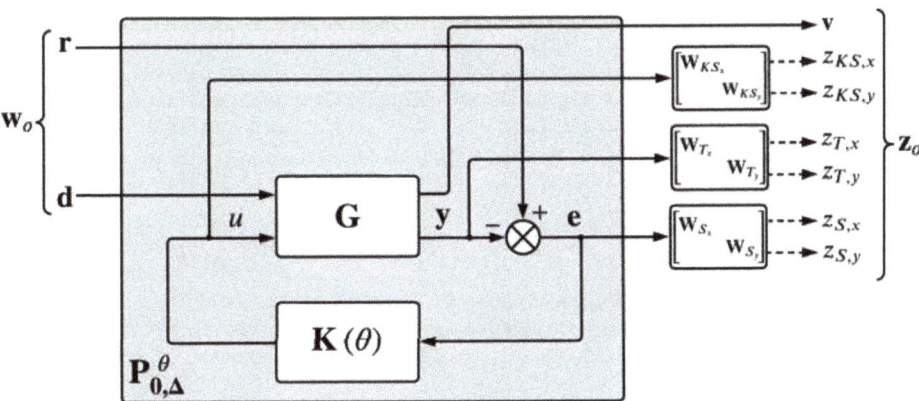

Figure 3. Closed-loop augmented plant.

In order to ensure good disturbance rejection, good reference tracking and stabilization of an unstable plant, the sensitivity function must have small magnitude, which means that the magnitude of the open loop must be large. On the other hand, in order to ensure mitigation of measurement noise and robust stability, the complementary sensitivity function must have small magnitudes, which implies that the magnitude of the open loop system must be small. More than that, the command signal must have small magnitude, which implies that the control effort transfer function must also have small magnitude. However, although these requirements seem to be conflicting, the frequency range where the magnitude of the open loop must be small and high are mostly disjunctive: the magnitude should be high for low frequencies and should be low for high frequencies. In order to ensure the desired shape of these three functions, three weighting functions must be designed, respectively.

The sensitivity function is a very good indicator of the closed-loop system tracking performance and has the advantage of being sufficient to consider only the magnitude. Typical specifications for sensitivity weighting functions are: minimum bandwidth frequency $\omega_B^\star$, maximum steady-state error A and maximum peak magnitude M, imposed by the following model [36]:

$$W_S(s) = \frac{\frac{1}{M}s + \omega_B^\star}{s + \omega_B^\star A}. \tag{54}$$

In a similar manner, the weighting function for the complementary sensitivity must be designed using the following specifications: the maximum peak amplitude M_T, the maximum value for high frequencies A_T, the minimum bandwidth $\omega_{BT}^\star$ and the roll-off n, formulated as:

$$W_T(s) = \frac{(s + \omega_{BT}^\star)^n}{\left(A_T^{1/n} s + \omega_{BT}^\star M_T^{1/n}\right)^n}. \tag{55}$$

For the control effort weighting function, the main performance specifications are the maximum value of the magnitude at low and high frequencies, denoted by DC and HF, respectively, and an intermediate point of interest. Sometimes, the main goal is simply to maintain the command signal under a prescribed value due to physical limitations of the system or other causes.

The major advantage of this approach consists in sculpting the relevant loop functions to impose performances implying good tracking and dynamic behaviour. Of great use are the rise time limitation through ω_B, steady-state error through A, while the roll-off slope of the closed-loop system imposed using n is directly coupled with sensor noise characteristics. These performances are specified for different frequency ranges, using the adequately selected weighting functions presented above.

Being a MIMO system with two inputs and two outputs, all weighting functions must be described by 2×2 transfer matrices, but, following a standard decoupling procedure, the weighting functions will be 2×2 diagonal transfer matrices. For the sensitivity, we consider two nearly similar weighting transfer functions, one for each axis, having the maximum bandwidth $\omega_{B,x}^\star = 3$ [rad/s] and $\omega_{B,y}^\star = 5$ [rad/s], the maximum steady-state error $A_x = A_y = 10^{-2}$ and the desired maximum sensitivity peak $M_x = M_y = 1.5$, resulting in:

$$W_S(s) = \begin{pmatrix} W_{S,x}(s) & 0 \\ 0 & W_{S,y}(s) \end{pmatrix}, \text{ where } W_{S,x}(s) = \frac{0.6667s + 3}{s + 0.03} \text{ and } W_{S,y}(s) = \frac{0.6667s + 5}{s + 0.05}. \tag{56}$$

For the complementary sensitivity weighting function, the same 2×2 diagonal structure approach will be used, imposing the same parameters for both axis: maximum complementary bandwidth $\omega_{BT,x}^\star = 50$ [rad/s], maximum peak magnitude $M_{T,x} = M_{T,y} = 1.5$, maximum magnitude at high frequencies $A_{T,x} = A_{T,y} = 10^{-2}$ and roll-off $n_x = n_y = 1$, resulting in:

$$W_T(s) = \begin{pmatrix} W_{T,x}(s) & 0 \\ 0 & W_{T,y}(s) \end{pmatrix}, \quad \text{where} \quad W_{T,x}(s) = W_{T,y}(s) = \frac{s+50}{0.01s+75}, \tag{57}$$

while the control effort weighting function being designed to encompass only the physical limitation of the command signal (between -1 and 1), resulting the transfer matrix:

$$W_{KS}(s) = \begin{pmatrix} W_{KS,x}(s) & 0 \\ 0 & W_{KS,y}(s) \end{pmatrix}, \quad \text{where} \quad W_{KS,x}(s) = W_{KS,y}(s) = 1. \tag{58}$$

The proposed structure of the controller is a decentralized one with two $PI^\lambda D^\mu$ controllers:

$$\mathcal{K} = \left\{ K^\theta(s) = \begin{pmatrix} K_{P,x} + K_{I,x}s^{-\lambda_x} + K_{D,x}s^{\mu_x} & 0 \\ 0 & K_{P,y} + K_{I,y}s^{-\lambda_y} + K_{D,y}s^{\mu_y} \end{pmatrix} \middle| \theta \in D \subset \mathbb{R}^{10} \right\}, \tag{59}$$

where the tunable parameters are:

$$\theta = \begin{pmatrix} K_{P,x} & K_{I,x} & \lambda_x & K_{D,x} & \mu_x & K_{P,y} & K_{I,y} & \lambda_y & K_{D,y} & \mu_y \end{pmatrix}^\top. \tag{60}$$

The problem to be solved with the proposed method is the mixed-sensitivity fixed structure μ-synthesis one, described as:

$$\min_{\substack{K \in \mathcal{K} \\ K \text{ stab}}} \sup_{\omega \in \mathbb{R}_+} \inf_{D \in \mathcal{D}} \overline{\sigma} \Big(D(j\omega) \cdot \text{LLFT}(P, K)(j\omega) \cdot (D(j\omega))^{-1} \Big),$$

$$\text{s.t.} \quad \left\| \begin{matrix} W_S S \\ W_T T \\ W_{KS} KS \end{matrix} \right\|_\infty < 1. \tag{61}$$

The settings for computeKstep used in the experiments are: the swarm dimension $N = 1000$, the maximum number of cycles $MNC = 50$, the maximum number of cycles with no improvements $NOIMP = 10$, the limit for the abandonment counter $LIMIT = 10$. The parameters necessary to describe the cost function (44) are $\alpha = 1$ and $\beta = 10^5$. The maximum number of D–K iterations is $MAX\_ITER = 10$ and the maximum window length for assessing lack of progress is 4. Using this setup, the mixed-sensitivity fixed structure μ-synthesis problem (61) is solved using four D–K iterations, as noticed in Table 2. The fractional order controller has been approximated using ORA with the following parameters: the frequency range is $[\omega_l, \omega_u] = [1 \times 10^{-4}, 1 \times 10^3]$ [rad/s] and the order of the approximation is $N = 3$. Given that, the resulting controller is:

$$K^{\theta^\star}(s) = \begin{pmatrix} 0.01 + 7.1116 \cdot s^{-0.7648} + 0.1133 \cdot s^{0.0909} & 0 \\ 0 & 0.1464 + 10 \cdot s^{-0.9926} + 0.1344 \cdot s^{0.0549} \end{pmatrix}. \tag{62}$$

Table 2. The evolution of the structural singular value in the D–K iteration procedure used to solve the mixed-sensitivity fixed structure μ-synthesis problem for the case study—FO-PID structure.

D–K Iteration Number	1	2	3	4
Number of ABC Iterations	50	39	50	45
Peak Value of μ	1.8956	1.0185	1.0021	0.9822

The imposed shape of the sensitivity, complementary sensitivity and control effort functions for both axis are depicted in Figure 4, along with the obtained shapes of those functions with the resulting controller for 100 Monte Carlo simulations. It can be noticed that all resulting Bode diagrams are under the prescribed shapes, which is guaranteed by the fact that the upper bound of the structured singular value $\mu_\Delta \left(P_{0,\Delta}^{\theta^\star} \right)$ is less than 1. The time-domain performances of the systems $r_x \to \theta_x$ and $r_y \to \theta_y$ are depicted in Figure 5, which are correlated with the frequency-domain performances. As such, the minimum values of the bandwidths for both x and y axis are over $\omega_{B,x}^\star = 3$ [rad/s] and $\omega_{B,y}^\star = 5$ [rad/s], which means that the rise time is less than 0.33 [s] and 0.2 [s], respectively. For the system $r_x \to \theta_x$, the rise time of the nominal system is 0.248 [s] and varies from 0.227 [s] to 0.281 [s], while the settling time is 0.558 [s] and varies from 0.496 [s] to 0.664 [s]. On the other hand, for the system $r_y \to \theta_y$, the rise time of the nominal system is 0.211 [s] and varies from 0.19 [s] to 0.235 [s], having the settling time for the nominal system 0.405 [s] and varying from 0.363 [s] to 0.454 [s]. According to the shape of the actual obtained sensitivity functions, there is no overshoot and no steady-state error for neither of the experiments presented using Monte Carlo simulations. Moreover, the reciprocal axis influence is small, as resulted from numerical simulations, where the peak amplitude from r_y to θ_x varies from 0.01 to 0.0143, and the peak amplitude from r_x to θ_y varies from 3.55×10^{-3} to 4.97×10^{-3}, respectively.

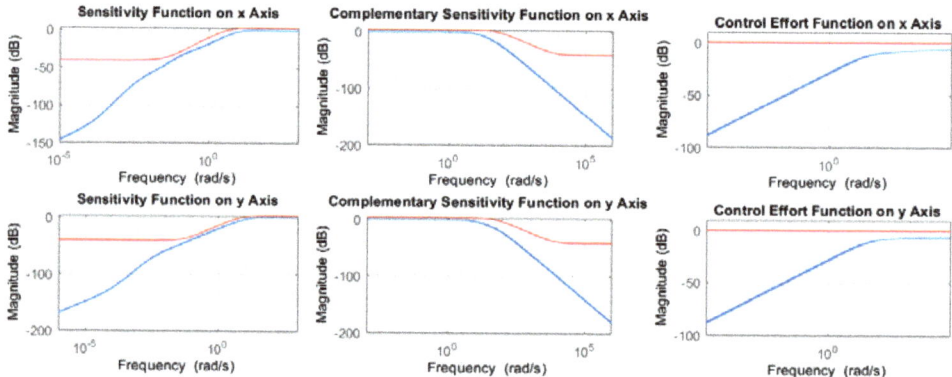

Figure 4. Imposed shapes of the sensitivity, complementary sensitivity and control effort frequency responses for both axis, along with 100 Monte Carlo simulations with *FO-PID* closed-loop systems.

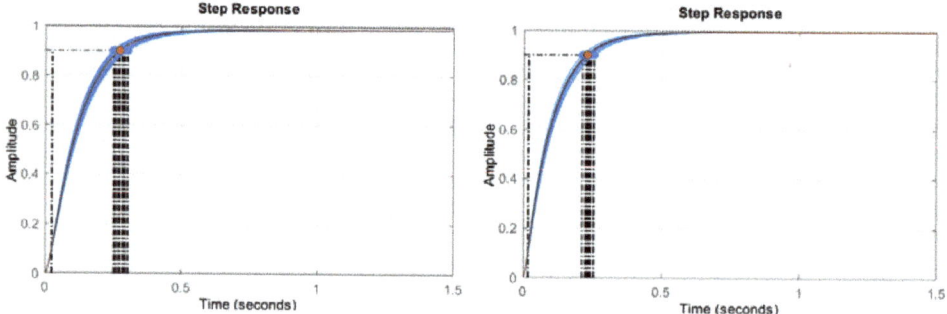

Figure 5. Time-domain closed-loop *FO-PID* performances for $r_x \to \theta_x$ and $r_y \to \theta_y$ systems.

4. Discussion

As noticed, the resulting controller contains a small fractional D term, which may be negligible. As such, the controller may be resynthesized with an imposed diagonal PI^λ structure:

$$\mathcal{K} = \left\{ K^\theta(s) = \begin{pmatrix} K_{P,x} + K_{I,x} s^{-\lambda_x} & 0 \\ 0 & K_{P,y} + K_{I,y} s^{-\lambda_y} \end{pmatrix} \middle| \theta \in D \subset \mathbb{R}^6 \right\}, \qquad (63)$$

where the tunable parameters are:

$$\theta = \begin{pmatrix} K_{P,x} & K_{I,x} & \lambda_x & K_{P,y} & K_{I,y} & \lambda_y \end{pmatrix}^\top. \tag{64}$$

The control problem remains the same as in (61), maintaining the constraints as in (56)–(58). The settings for this experiment were kept the same as for the previous case: $N = 1000$, $MNC = 50$, $NOIMP = 10$, $LIMIT = 10$, $\alpha = 1$, $\beta = 10^5$, $MAX\_ITER = 10$, $[\omega_l, \omega_u] = $ [1e-4, 1e3] [rad/s] and $N = 3$. Using this setup, the mixed-sensitivity fixed structure μ-synthesis problem (61) is solved using four D–K iterations, as noticed in Table 3. The resulting controller is:

$$K_{FO-PI}^{\theta^*}(s) = \begin{pmatrix} 0.2229 + 5.9392 \cdot s^{-0.7792} & 0 \\ 0 & 0.3949 + 10 \cdot s^{-0.9733} \end{pmatrix}. \tag{65}$$

Table 3. The evolution of the structural singular value in the D–K iteration procedure used to solve the mixed-sensitivity fixed structure μ-synthesis problem for the case study—FO-PI structure.

D–K Iteration Number	1	2	3	4
Number of ABC Iterations	30	30	24	30
Peak Value of μ	3.0676	1.1961	1.0026	0.9959

The imposed shape of the sensitivity, complementary sensitivity and control effort functions for both axis are depicted in Figure 6, along with the obtained shapes of those functions with the resulting controller for 100 Monte Carlo simulations. It can be noticed that all resulting Bode diagrams are under the prescribed shapes, which is guaranteed by the fact that the upper bound of the structured singular value $\mu_\Delta \left(P_{o,\Delta}^{\theta^*} \right)$ is less than 1.

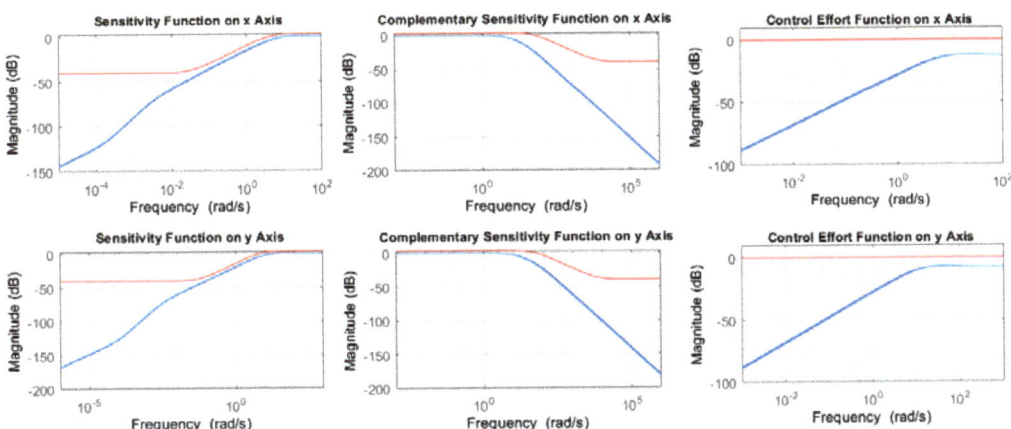

Figure 6. Imposed shapes of the sensitivity, complementary sensitivity and control effort frequency responses for both axis, along with 100 Monte Carlo simulations with actual FO-PI closed-loop systems.

The proposed method results are compared with those obtained using the musyn routine from MATLAB [37]. The musyn routine manages to solve both fixed structure and free-structure μ-synthesis control problems. The first solution of the problem (61) is obtained using the proposed method with results already presented in Section 3. Starting from the same control problem, the structured μ-synthesis version of the musyn routine with the same stopping criteria was used. The resulting controller after 2 iterations achieved its best performance $\mu_\Delta(P_{o,\Delta}^{\theta^*}) = 0.9842$, which means that there is a mathematical guarantee for robust stability and robust performance. The transfer matrix of this controller is:

$$K^{\theta^*}(s) = \begin{pmatrix} 0.0101 + 7.4689 \cdot s^{-0.7860} + 0.1729 \cdot s^{0.0542} & 0 \\ 0 & 0.0881 + 5.4840 \cdot s^{-0.9749} + 0.2031 \cdot s^{0.0348} \end{pmatrix}. \tag{66}$$

In opposition with the above approaches, the controller obtained with μ-synthesis procedure from MATLAB without imposing a fixed structure is of order 74 and, after 10 iterations, the best achieved performance is $\mu_\Delta(P_{o,\Delta}^{\theta^*}) = 1.003$. The frequency-domain data for structured singular values corresponding to these three numerical simulations are presented in Figure 7. As mentioned, the peak value of the μ values for the unstructured problem is over the critical value 1, while the FO-PID controller manages to fulfill all requirements. A comparison between the frequency responses of the controllers is shown in Figure 8.

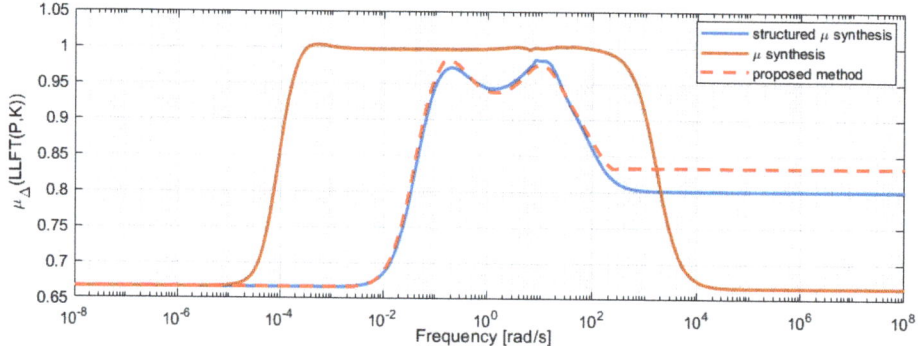

Figure 7. Comparison between structured singular values' frequency response obtained with the structured μ-synthesis from MATLAB (blue), with the unstructured μ-synthesis from MATLAB (orange) and with the proposed method (red), respectively.

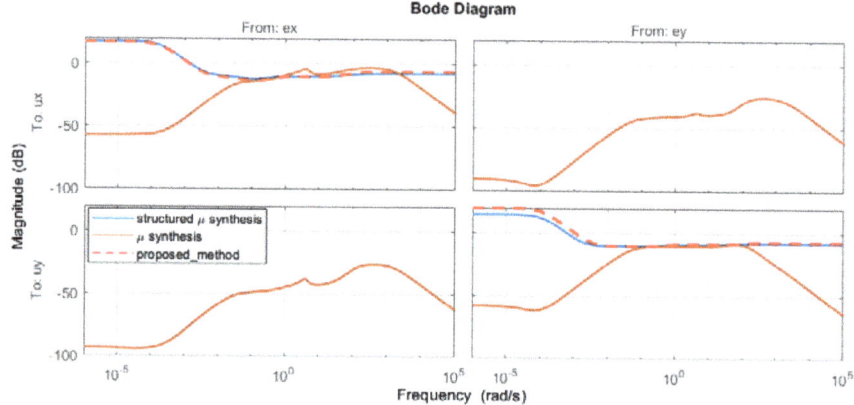

Figure 8. Obtained controllers' frequency responses with the structured μ-synthesis from MATLAB (blue), unstructured (orange) μ-synthesis from MATLAB, and with the proposed method (red).

The integer-order approximation of a fractional element using ORA contains two exponential terms, as seen in (7), which presently cannot be treated using the `realp` objects in MATLAB. The actual solution used for this paper is to approximate the exponential function using Taylor series truncation. As stated in the Introduction, our approach requires only the range of the controller parameters and, thus, another object could be used instead of the entire structure of `realp` necessary in nonsmooth optimization-based algorithms used in MATLAB `hinfstruct` and `systune` routines. Based on this limitation, the toolbox presented in [38] will be extended in order to address the previously mentioned mathematical issues, due to the fact that only the fixed-structure $\mathcal{H}_\infty$ synthesis algorithm uses the `realp` object, which can be replaced by our approach, where the NP-hard non-convex problem is solved using an ABC swarm optimization.

Compared to other approaches previously described in the introduction, where LQR methods were considered for CNC machines, the advantages of the proposed method consist in the generality of the method and the flexibility of the robustness, with the possibility to compensate measurement noise, unmodelled dynamics and input-output disturbances. Therefore, while LQR requires the complete state vector to be measured at runtime, the proposed method requires only the provided measurements, modelled by the actual outputs of the system. Although the LQR method can be augmented with a state estimator in order to obtain output feedback, the main limitation of this approach is that the model of the plant must be accurate, while in the Robust Control framework, utilized in our method, an entire family of uncertain plants can be taken into consideration at the design phase. Moreover, it is difficult to impose exact limitations on the maximum allowed command signals, using the energy-based approach, which generally is an intrinsic limitation of the execution element.

5. Conclusions and Future Work

The current paper presents a new design method for fixed structure fractional-order controllers using the Robust Control framework. The proposed method manages to return nearly optimal parameters of a MIMO FO-PID as a solution for a mixed-sensitivity fixed structure μ-synthesis control problem. Although the μ-synthesis control problem is NP-hard, the D–K iteration algorithm represents a good approximation which allows to convert it into a P-hard problem. However, the returned controller is of high order, which means that an order reduction must be performed in order to implement the control law. Therefore, the imposed structure is an increasingly explored approach, although such a problem presents a non-convex component for the K step. Our approach consists of an artificial bee colony swarm optimization as a solution to this non-convex fixed structure $\mathcal{H}_\infty$ subproblem. This solution requires only the range of the controller parameters, as opposed to the nonsmooth optimization-based approach from MATLAB's Robust Control Toolbox, where the parameters can be used only in polynomial structured expressions, which is an inherent limitation when fractional-order controllers are desired. Further, the case study of mixed-sensitivity μ-synthesis position control problem for both axis of a CNC machine manages to underline the strong points of the method and of the imposed structure of the controller: it provides a mathematical guarantee for robust stability and robust performance, while the unstructured version of μ-synthesis from MATLAB does not manage to offer it.

As future work, we want to include our method in a toolbox, as stated in the Discussion section, which starts from an initial process model and a set of desired performances, and manages to automatically obtain the augmented plant, the controller decoupling transfer matrix, the optimal values of the controller parameters, followed by closed-loop simulations and validations. Additionally, we propose to extend this technique for certain types of nonlinear systems.

Author Contributions: Conceptualization, V.M. and M.Ş.; methodology, V.M.; software, V.M.; validation, M.Ş. and M.S.; formal analysis, V.M. and P.D.; investigation, V.M. and D.M.; resources, V.M.; data curation, D.M. and M.S.; writing—original draft preparation, M.V., D.M. and M.S.; writing—review and editing, M.Ş., D.M., M.S. and P.D.; visualization, M.Ş.; supervision, P.D.; project administration, V.M.; funding acquisition, D.M. All authors have read and agreed to the published version of the manuscript.

Funding: This paper was supported by the Project "Entrepreneurial competences and excellence research in doctoral and postdoctoral programs—ANTREDOC", project co-funded by the European Social Fund.

Institutional Review Board Statement: Not applicable.

Informed Consent Statement: Not applicable.

Data Availability Statement: Not applicable.

Conflicts of Interest: The authors declare no conflict of interest.

Abbreviations

The following abbreviations are used in this manuscript:

ABC	Artificial Bee Colony
CNC	Computer Numerical Control
FO-PID	Fractional-Order PID
LMI	Linear Matrix Inequality
LLFT	Lower Linear Fractional Transform
LTI	Linear Time-Invariant
MIMO	Multiple-Input Multiple-Output
NP	Nondeterministic Polynomial Time
PID	Proportional-Integral-Derivative
ULFT	Upper Linear Fractional Transform

References

1. Doyle, J.; Glover, K.; Khargonekar, P.; Francis, B.A. State-space solutions to standard $\mathcal{H}_2$ and $\mathcal{H}_\infty$ control problems. *IEEE Trans. Autom. Control* **1989**, *34*, 831–847. [CrossRef]
2. Ionescu, V.; Oară, C.; Weiss, M. *Generalized Riccati Theory and Robust Control—A Popov Function Approach*; John Wiley & Sons: Chichester, UK, 1999.
3. Șușcă, M. Solving Algebraic Riccati Equations Using Proper Deflating Subspaces for $\mathcal{H}_2/\mathcal{H}_\infty$ Synthesis. Master's Thesis, Technical University of Cluj-Napoca, Cluj-Napoca, Romania, 2019.
4. Șușcă, M.; Mihaly, V.; Stănese, M.; Dobra, P. Iterative Refinement Procedure for Solutions to Algebraic Riccati Equations. In Proceedings of the 2020 IEEE International Conference on Automation, Quality and Testing, Robotics (AQTR), Cluj-Napoca, Romania, 21–23 May 2020.
5. Gahinet, P.; Apkarian, P. A linear matrix inequality approach to $\mathcal{H}_\infty$ control. *Int. J. Robust Nonlinear Control* **1994**, *4*, 421–448. [CrossRef]
6. Mihaly, V. General Purpose Linear Matrix Inequality Solver with Applications in Robust and Nonlinear Control. Master's Thesis, Technical University of Cluj-Napoca, Cluj-Napoca, Romania, 2020.
7. Liu, K.Z.; Yao, Y. *Robust Control—Theory and Applications*; John Wiley & Sons: Singapore, 2016.
8. Gu, D.-W.; Petkov, P.H.; Konstantinov, M.M. *Robust Control Design with MATLAB*; Springer London Limited: London, UK, 2005.
9. Packard, A.; Doyle, J.; Balas, G. Linear Multivariable Robust Control with a μ Perspective. *J. Dyn. Syst. Meas. Control* **1993**, *115*, 426–438. [CrossRef]
10. Apkarian, P.; Noll, D. Nonsmooth $\mathcal{H}_\infty$ Synthesis. *IEEE Trans. Autom. Control* **2006**, *41*, 71–86. [CrossRef]
11. Bompart, V.; Noll, D.; Apkarian, P. Second-order nonsmooth optimization for $\mathcal{H}_\infty$ synthesis. *Numer. Math.* **2007**, *107*, 433–454. [CrossRef]
12. Apkarian, P.; Noll, D. The $\mathcal{H}_\infty$ Control Problem is Solved. *Aerosp. Lab* **2017**, 1–11. [CrossRef]
13. Apkarian, P. Nonsmooth μ synthesis. *Int. J. Robust Nonlinear Control* **2011**, *21*, 1493–1508. [CrossRef]
14. Farag, A.; Werner, H. A Riccati-Genetic Algorithms Approach To Fixed-Structure Controller Synthesis. In Proceeding of the 2004 American Control Conference, Boston, MA, USA, 30 June–2 July 2004; pp. 2799–2804.
15. Lari, A.; Khosravi, A. An Evolutionary Approach to Design Practical μ Synthesis Controllers. *Int. J. Control Autom. Syst.* **2013**, *11*, 167–174. [CrossRef]
16. Farag, A.; Werner, H. Fixed-Structure μ-Synthesis—An Evolutionary Approach. In Proceeding of the 2006 American Control Conference, Minneapolis, MN, USA, 14–16 June 2006; pp. 4332–4337.
17. Dervis, K. *An Idea Based on Honey Bee Swarm For Numerical Optimization*; Technical Report—TR06; Erciyes University: Kayseri, Egypt, 2005.
18. Mihaly, V.; Covaci, R.; Andrei, S. Artificial Bee Colony Optimization. In Proceedings of the 21th International Conference on Automation, Quality and Testing, Robotics (AQTR), THETA, Student Forum, Cluj-Napoca, Romania, 24–26 May 2018.
19. Petráš, I. *Fractional Order Systems*; Printed Edition of the Special Issue Published in *Mathematics*; MDPI: Basel, Switzerland, 2019.
20. Monje, C.A.; Chen, Y.; Vinagre, B.M.; Xue, D.; Feliu, V. *Fractional-Order Systems and Controls: Fundamentals and Applications*; Springer Science and Business Media: London, UK, 2010.
21. Zhao, C.; Xue, D.; Chen, Y.Q. A fractional order PID tuning algorithm for a class of fractional order plants. In Proceedings of the IEEE International Conference Mechatronics and Automation, Niagara Falls, ON, Canada, 29 July–1 August 2005; pp. 216–221.
22. Mihaly, V.; Dulf, E. Novel fractional order controller design for first order systems with time delay. In Proceedings of the 2020 IEEE International Conference on Automation, Quality and Testing, Robotics (AQTR), Cluj-Napoca, Romania, 21–23 May 2020; pp. 1–4.

23. Dulf, E.H.; Șușcă, M.; Kovacs, L. Novel Optimum Magnitude Based Fractional Order Controller Design Method. *IFAC-PapersOnLine* **2018**, *51*, 912–917. [CrossRef]
24. Dulf, E.H. Simplified Fractional Order Controller Design Algorithm. *Mathematics* **2019**, *7*, 1166. [CrossRef]
25. Vinagre, B.M.; Monje, C.A.; Calderón, A.J.; Chen, Y.Q.; Feliu, V. The fractional integrator as a reference function. In Proceedings of the 1st IFAC Workshop on Fractional Differentiation and Its Application, Bordeaux, France, 19–21 July 2004.
26. Garrido, S.; Monje, C.A.; Martin, F.; Moreno, L. Design of Fractional Order Controllers Using the PM Diagram. *Mathematics* **2020**, *8*, 2022. [CrossRef]
27. Muresan, C.I.; Birs, I.R.; Dulf, E.H. Event-Based Implementation of Fractional Order IMC Controllers for Simple FOPDT Processes. *Mathematics* **2020**, *8*, 1378. [CrossRef]
28. Zheng, W.; Luo, Y.; Chen Y.Q.; Wang, X. A Simplified Fractional Order PID Controller's Optimal Tuning: A Case Study on a PMSM Speed Servo. *Entropy* **2021**, *23*, 130. [CrossRef] [PubMed]
29. Iannino, V.; Colla, V.; Innocenti, M.; Signorini, A. Design of a $\mathcal{H}_\infty$ Robust Controller with μ-Analysis for Steam Turbine Power Generation Applications. *Energies* **2017**, *10*, 1026. [CrossRef]
30. Sabău, D.; Dobra, P. State-feedback control algorithms for a CNC machine. In Proceedings of the 2018 22nd International Conference on System Theory, Control and Computing (ICSTCC), Sinaia, Romania, 10–12 October 2018; pp. 615–620.
31. Sabău, D.; Dobra, P. A PI controller based on state-feedback algorithm for an XY positioning system. In Proceedings of the 2019 22nd International Conference on Control Systems and Computer Science (CSCS), Bucharest, Romania, 28–30 May 2019; pp. 56–60.
32. Oldham, K.B.; Spanier, J. *The Fractional Calculus*; Academic Press: New York, NY, USA; London, UK, 1974.
33. Caputo, M. Linear model of dissipation whose Q is almost frequency independent-II. *Geophys. J. Int.* **1967**, *13*, 529–539. [CrossRef]
34. Podlubny, I. *Fractional Differential Equations*; Mathematics in Science and Engineering; Academic Press: San Diego, CA, USA, 1999; Volume 198.
35. Xue, D.; Chen, Y.Q.; Atherton, D.P. *Linear Feedback Control—Analysis and Design with MATLAB*; SIAM: Philadelphia, PA, USA, 2007.
36. Skogestad, S.; Postlethwaite, I. *Multivariable Feedback Control—Analysis and Design*, 2nd ed.; John Wiley & Sons: New York, NY, USA, 2005.
37. Balas, G.; Chiang, R.; Packard, A.; Safonov, M. *Robust Control Toolbox—User's Guide*; The MathWorks: Natick, MA, USA, 2020.
38. Șușcă, M.; Mihaly, V.; Stănese, M.; Morar, D.; Dobra, P. Unified CACSD Toolbox for Hybrid Simulation and Robust Controller Synthesis with Applications in DC-to-DC Power Converter Control. *Mathematics* **2021**, *9*, 731. [CrossRef]

Article

New Expressions to Apply the Variation Operation Strategy in Engineering Tools Using Pumps Working as Turbines

Frank A Plua [1], Francisco-Javier Sánchez-Romero [2], Victor Hidalgo [3], P. Amparo López-Jiménez [4] and Modesto Pérez-Sánchez [4,*]

1. Civil and Environmental Engineering Department, Escuela Politécnica Nacional, Quito 170525, Ecuador; frank.plua@epn.edu.ec
2. Rural and Agrifood Engineering Department, Universitat Politècnica de València, 46022 Valencia, Spain; fcosanro@agf.upv.es
3. Mechanical Engineering Department, Escuela Politécnica Nacional, Quito 170525, Ecuador; victor.hidalgo@epn.edu.ec
4. Hydraulic and Environmental Engineering Department, Universitat Politècnica de València, 46022 Valencia, Spain; palopez@upv.es
* Correspondence: mopesan1@upv.es

Citation: Plua, F.A; Sánchez-Romero, F.-J.; Hidalgo, V.; López-Jiménez, P.A.; Pérez-Sánchez, M. New Expressions to Apply the Variation Operation Strategy in Engineering Tools Using Pumps Working as Turbines. *Mathematics* **2021**, *9*, 860. https://doi.org/10.3390/math9080860

Academic Editors: Eva H. Dulf and Cristina I. Muresan

Received: 26 February 2021
Accepted: 12 April 2021
Published: 14 April 2021

Publisher's Note: MDPI stays neutral with regard to jurisdictional claims in published maps and institutional affiliations.

Copyright: © 2021 by the authors. Licensee MDPI, Basel, Switzerland. This article is an open access article distributed under the terms and conditions of the Creative Commons Attribution (CC BY) license (https://creativecommons.org/licenses/by/4.0/).

Abstract: The improvement in energy saving aspects in water systems is currently a topic of major interest. The utilization of pumps working as turbines is a relevant strategy in water distribution networks consisting of pressurized pipes, using these machines to recover energy, generate green energy and reduce leakages in water systems. The need to develop energy studies, prior to the installation of these facilities, requires the use of simulation tools. These tools should be able to define the operation curves of the machine as a function of the flow rate. This research proposes a new strategy to develop a mathematics model for pumps working as turbines (PATs), considering the modified affinity laws. This proposed model, which can be input into hydraulic simulation tools (e.g., Epanet, WaterGems), allows estimation of the head, efficiency, and power curves of the PATs when operating at different rotational speeds. The research used 87 different curves for 15 different machines to develop the new model. This model improves the results of the previously published models, reducing the error in the estimation of the height, efficiency, and power values. The proposed model reduced the errors by between 30 and 50% compared to the rest of the models.

Keywords: PAT model; modified affinity laws; hydraulic simulation tool

1. Introduction

Mathematical models have been a very useful tool to improve the management of water networks [1]. These models improved both pressurized systems [2], as well as free surface channels [3], improving their management and behavior in steady and unsteady flows. Some of these models were focused on the integration of the management into the new sustainability challenges of the infrastructures [4].

The improvement of the sustainability has been analyzed in water systems from different points of view, such as leakage reduction [5], minimizing consumed energy in pump systems [6], and quality parameters in the water supply [7], among others. One of these strategies has been the use of pumps working as turbines (PATs). These machines replace the pressure reduction valves, taking advantage of the excess of energy in the pressurized water systems [8]. A PAT is a pump which works in reverse mode and it is cheaper than classical turbines of the same small size [9]. When this machine operates in this mode, it generates energy. The efficiency of these machines is lower than traditional turbines and its hydraulic efficiency value is between 0.6 and 0.7 [10]. The global efficiency is between 0.5 and 0.6 when all the electromechanical equipment (electric and electronic devices) is considered. The traditional machines are classified as action (e.g., Pelton,

Turgo, among others) and reaction (Francis, Kaplan, among others), as described in [11]. In contrast, the PATs are pumps, and, therefore, their classification depends on the specific velocity (i.e., radial, mixed, or axial machines) [12,13].

Previously, different investigations were published in which the use and analysis of PATs focused on analyzing the theoretical energy recovery [11] as well as the duty point of these machines, when information about the manufacturer was not known [12]. When the curves are not known, the head, efficiency, and power curves (these curves are called characteristic curves of the PATs) should be estimated when the pump is used in turbine mode. These expressions are defined by the following equations:

$$H_0 = A + BQ_0 + CQ_0^2 \tag{1}$$

$$\eta_0 = E_4 Q_0^4 + E_3 Q_0^3 + E_2 Q_0^2 + E_1 Q_0 + E_0 \tag{2}$$

$$P_0 = P_4 Q_0^4 + P_3 Q_0^3 + P_2 Q_0^2 + P_1 Q_0 + P_5 \tag{3}$$

where H_0 is the recovered head in nominal rotational speed in m w.c. (water column); Q_0 is the flow rate in m^3/s; A, B, and C are the coefficients, which define the head curve of the PAT; η_0 is the efficiency of the machine for each flow (non-dimensional); E_4, E_3, E_2, E_1, and E_0 are the coefficients, which define the efficiency curve; P_0 is the generated power in kW; P_4, P_3, P_2, P_1, and P_5 are the coefficients, defining the power curve of the machine.

The head curve enables the determination of the recovered head as a function of the flow. The efficiency curve determines the efficiency of the machine according to the circulating flow; finally, the power curve establishes the generated power by the machine for each flow value. Previous references demonstrated the possibility to estimate these curves by use of non-dimensional parameters [14]. This estimation should be developed using non-dimensional parameters and they are head number (h), flow number (q), efficiency number (e), and torque number (b) [15]. The different non-dimensional parameters, which are used to regulate the machines by variation of the rotational speed, are the following:

$$q = \frac{Q_i}{Q_{BEP}} \tag{4}$$

$$h = \frac{H_i}{H_{BEP}} \tag{5}$$

$$e = \frac{\eta_i}{\eta_{BEP}} \tag{6}$$

$$p = \frac{P_i}{P_{BEP}} = qhe \tag{7}$$

where q, h, e, and p are the flow, head, efficiency, and power coefficients; Q_i is any flow value of the PAT in m^3/s; H_i is the head for Q_i according to the head curve in m w.c.; η_i is the efficiency of the machine when the flow is Q_i; P_i is the effective power for Q_i; Q_{BEP}, H_{BEP}, P_{BEP}, and η_{BEP} refer to the best efficiency point (BEP) of the machine, which define the best efficiency head (BEH) when the rotational speed is changed.

In line with this, the reduction of the uncertainties by estimating the characteristic curves with respect to their known behavior as pumps has been an objective of different studies [16]. Different semiempirical methods have been published, proposing polynomial expressions to estimate the PAT curves, when the machine operates with constant rotational speed [10,12,17,18]. The development of these mathematical expressions was crucial to improve the characterization of the PATs and the energy models to analyze the energy recovery.

However, the flow rate changes over time in the different pipes of the water networks due to the demands of the users. Therefore, the energy analyses are not maximized when they consider PATs, if they work under constant rotational speed. To increase energy recovery, different strategies have been published in which the energy maximization was reached when the machine operated at different rotational speeds, called the variable operation strategy (named VOS) [19]. The variation of the rotational speed is crucial to reach the best efficiency values in the water systems, and it is the focus of new challenges in hydropower systems also applied to Francis turbines [20]. Furthermore, when the rotation speed changes, it is necessary to introduce knowledge of PAT curves into mathematical models, which analyze energy recovery in water systems. The lack of mathematical expressions makes it difficult to improve energy estimates when applying the VOS strategy in the modeling of water systems [14].

In recent years, some researchers have published different methods which allow water managers and companies to estimate the characteristic curves of PATs, avoiding the experimental tests when developing preliminary energy studies. Efficiency and head curves operating without variation of rotational speed were described in [21,22]. The analysis of PAT curves was carried out using other methods, which proposed expressions considering specific speed as well as the best efficiency point [23–25]. These methods did not consider the variation in the rotational speed, which is of paramount importance to reach the maximization of the recovered energy [26].

A step forward was taken in 2014, when some researchers analyzed the variation in rotational speed through experimental tests to improve the maximization of energy recovery. Research described in [12] proposed empirical expressions using four different tested machines in 2016. These equations should only be considered when the specific speed is between 120 and 162 (m, kW). In 2018, two PATs were tested and they were used to define other expressions, which could estimate the characteristic curves when the best efficiency point was known [27]. Research published in [28] studied the efficiency, power, and head curves in one PAT, which was installed in water pressurized systems in 2020. All studies used between one and four machines [12,14,27,28]. The low number of machines reduces the applicability of the proposed expressions, when other machines are used. To solve this issue, the present research goes one step further, using 87 different tested characteristic curves (i.e., head, efficiency, and power) of the majority of hydraulic machines, which have been published in previous references.

New empirical expressions are here proposed. These expressions could be used by modelers, who could improve their energy analysis when they apply the VOS strategy in water systems. Previous research has conducted similar analyses to define the characteristic curves of the machine [12,14,27,28]. They used non-dimensional numbers (i.e., q, h, e, and p), which are calculated at the best efficiency point. These values were used to propose functions, which depended on the ratio of the rotational speed of the machine to modify the affinity laws. This proposal improves the use of PATs in the simulation tools. It will enable the reduction of the uncertainty in the previous energy analysis when the use of PATs is considered in a real case study. The proposed expressions reduced the error indexes when they were compared with the other published methods, as well as increasing the validity range. Furthermore, these expressions are based on fifteen different machines, which had 87 different curves, increasing the number of experimental curves.

2. Materials and Methods
2.1. Methodology

The methodology proposed herein is focused on obtaining some particular empiric expressions, which allow water managers to develop tools for modeling PATs in water systems, when they operate at variable rotational speed. The strategy is based on the knowledge of the operation curves (head, efficiency, and power) at nominal speed [29]. The proposed method is based on classical expressions of the hydraulic machines, proposing a strategy to modify them by the affinity laws.

The main objective of the strategy is to propose an empirical expression that allows water managers to introduce management tools to simulate the different scenarios under the VOS operation. Furthermore, the method will reduce the errors when the characteristic curves are estimated in variable velocity conditions. To achieve this, different steps were proposed to derive the new expressions considering the modification of the affinity laws of hydraulic machines. [30]. Finally, the method was validated with the different tested machines. Figure 1 shows the different proposed steps. These steps are the following:

1. Obtaining experimental characteristic curves of the PATs. The characteristic curves (i.e., head, efficiency, and power curve) were made available for the different machines using experimental data which were published by other researchers. Both the nominal curve and the curves for different rotational speeds were digitized using Equations (1)–(3).
2. Definition of the dimensionless values of the curve to apply the affinity laws. This is developed using the previously defined equations (Equations (4)–(8)). When the affinity laws are applied, the congruence parabola is defined by the following equation [29]:

$$H_{PC} = \frac{H_0}{Q_0^2} Q_j^2 = k_{AL} Q_j^2 \tag{8}$$

where Q_j is the new flow rate in m^3/s in which the machine has to operate. H_{PC} is a parabola, which has the same efficiency at each point. This consideration is theoretical, since (in practice) it is only acceptable for values around +/−20% of the best efficiency point of the machine [29]. This variation in the rotational speed of the machine is defined by the ratio between the rotational speed (n_j) of the machine to reach the value (Q_j) and the nominal rotational speed (n_0). This ratio between n_j and n_0 is called α.

The affinity laws are expressions which define points similar to each other under conditions of restricted similarity, neglecting the stresses due to viscosity. These expression are defined by the following expressions [29]:

$$\frac{Q_1}{Q_0} = \frac{n_1}{n_0} = \alpha \tag{9}$$

$$\frac{H_1}{H_0} = \left(\frac{n_1}{n_0}\right)^2 = \alpha^2 \tag{10}$$

$$\frac{P_1}{P_0} = \left(\frac{n_1}{n_0}\right)^3 = \alpha^3 \tag{11}$$

where Q_1 is the flow under the new conditions of rotational speed (n_1) in m^3/s; H_1 is the head under the new conditions in m w.c.; P_1 is the shaft power under the new conditions in kW.

When affinity laws are applied for different rotational speeds, the variable operation strategy (VOS) can be defined between ratios of α_{min} and α_{max}.

When affinity laws are applied, the dimensionless parameters are:

$$q = \alpha \tag{12}$$

$$h = \alpha^2 \tag{13}$$

$$e = 1 \tag{14}$$

$$p = \alpha^3 \tag{15}$$

Applying the affinity laws, $k_{AL, BEH}$ is defined by the following expression, considering that the ratio $\frac{h}{q^2} = 1$ (if the classical affinity laws is applied ideally):

$$k_{AL,BEH} = \frac{A}{Q_{BEP}^2} + \frac{B}{Q_{BEP}} + C \tag{16}$$

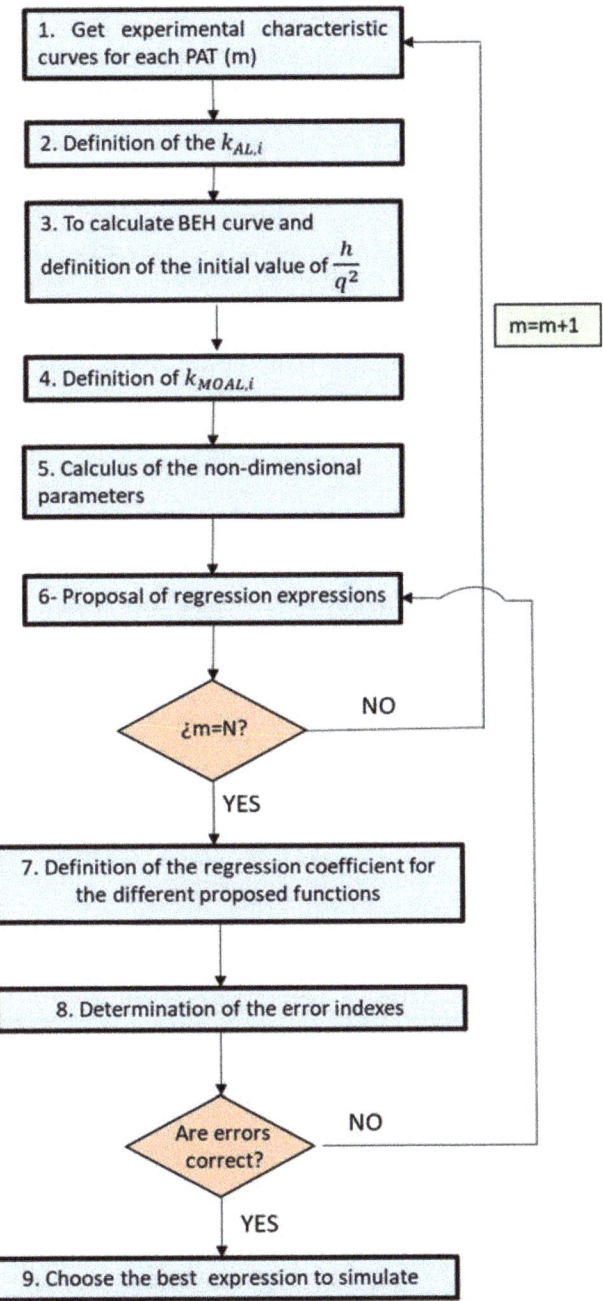

Figure 1. Methodology proposed to derive the expressions (m is the number of experimental machines, N is the maximum number of the tested machine).

3. Once the dimensionless parameters (q, h, e, and p) are defined, the best efficiency curve (BEH) of the machine is determined. BEH is the curve which establishes the

recovered head for each flow, maximizing efficiency and changing the rotation speed of the machine. This curve is defined in [27].

4. When the BEH is known for each machine, the ratio h/q^2 is defined for the different values, using the experimental data as well as the regression of the different head and efficiency curves. This parameter is defined for each rotational speed of the machine. The rotational speed varies between the α_{min} and α_{max} of the VOS. The operation area is defined by the maximum and minimum rotational speed, determined by the tested machine.

5. In [27,31], variations of the affinity laws are proposed, where the flow ratio (Q/Q_0) is a function that depends on α; taking into account this modification of the affinity laws, the corresponding parameter k_{AL} for the modified affinity laws (MOAL) can be defined when the affinity laws are modified by the following expression:

$$k_{MOAL,BEH} = \frac{h}{q^2}\left(\frac{A}{Q_{BEP}^2} + \frac{B}{Q_{BEP}} + C\right) \qquad (17)$$

6. The value of the $k_{MOAL,\,BEH}$ coefficient is defined for the different rotational speeds of the machine, determining the cut-off point with the hypothetical head surface and machine efficiency (Figure 2).

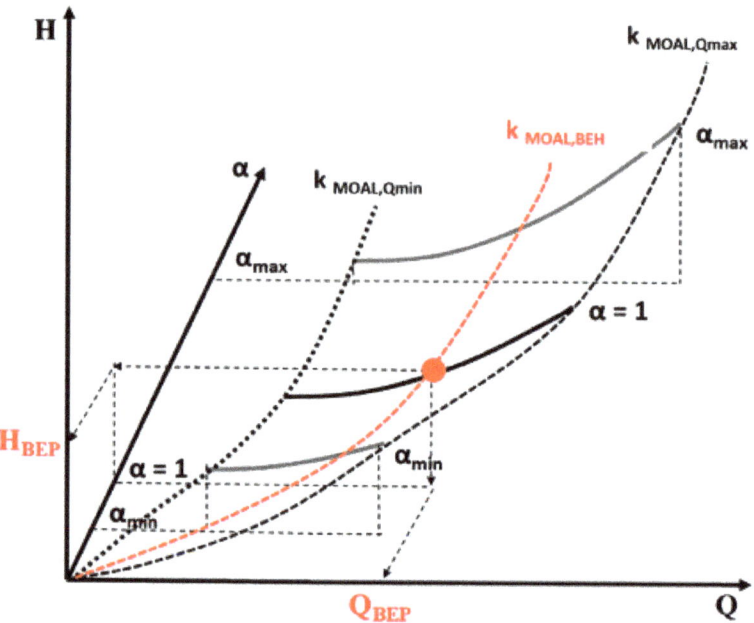

Figure 2. Congruence parabolas for the different values and rotational speeds when modified affinity laws (MOAL) is applied.

Once the $k_{MOAL,\,BEH}$ is defined using Q_{BEP} and H_{BEP}, k_{MOAL} is extended for different values of Q_0, defining the $k_{MOAL,i}$ for each rotational speed and the intersection points with head and efficiency areas are calculated. These points are Q_0, H_0, η_0, Q_{i,α_j}, H_{i,α_j} and η_{i,α_j} (Figure 3a,b). The values of these parameters enable definition of the new non-dimensional values, which will define the functions of the modified affinity laws. Each of these points is calculated considering the intersection point for each rotational speed curve. The new non-dimensional parameters are defined by the following expressions:

$$q_{i,j} = \frac{Q_{i,\alpha_j}}{Q_{i,0}} \tag{18}$$

$$h_{i,j} = \frac{H_{i,\alpha_j}}{H_{i,0}} \tag{19}$$

$$\eta_{i,j} = \frac{\eta_{i,\alpha_j}}{\eta_{i,0}} \tag{20}$$

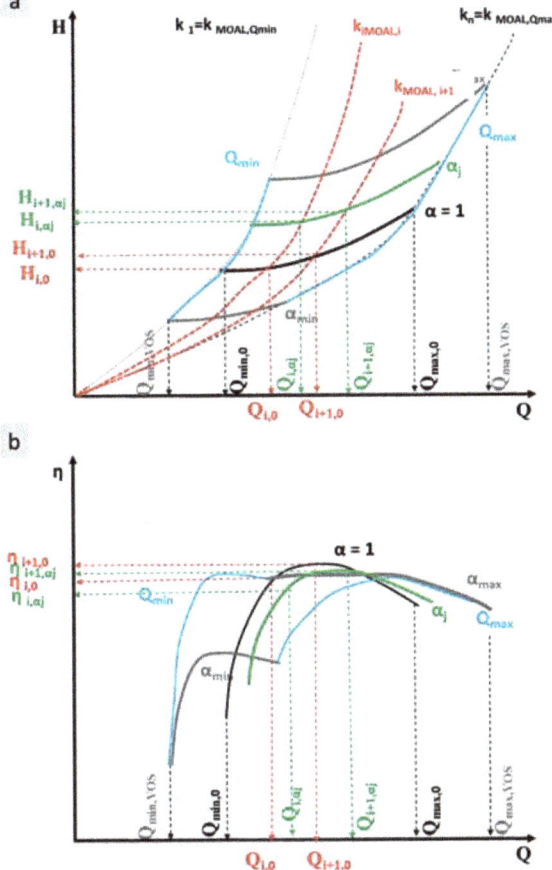

Figure 3. (a) Definition of head as a function of the flow for different rotational speeds; (b) Definition of the efficiency as a function of the flow for different rotational speeds considering the VOS area.

7. Once the non-dimensional parameters for the different rotational speeds are defined, the regression expressions are proposed. These functions depend on rotational speed (α), which is a significant variable [31] when the non-dimensional parameters are defined (i.e., h, q, and e). Moreover, different expressions are also proposed considering the ratio Q/Q_{BEP}. This parameter is considered since it measures the gap between the flow value and the flow for the best efficiency point. The incorporation of this param-

eter will improve the regression coefficient of the expressions, as well as reducing the errors. The modified affinity laws are then defined according to different expressions:

$$H = h\left(A + B\frac{Q}{q} + C\left(\frac{Q}{q}\right)^2\right) \qquad (21)$$

$$\eta = e\left(E_4\left(\frac{Q}{q}\right)^4 + E_3\left(\frac{Q}{q}\right)^3 + E_2\left(\frac{Q}{q}\right)^2 + E_1\left(\frac{Q}{q}\right) + E_0\right) \qquad (22)$$

$$P = p\left(P_4\left(\frac{Q}{q}\right)^4 + P_3\left(\frac{Q}{q}\right)^3 + P_2\left(\frac{Q}{q}\right)^2 + P_1\left(\frac{Q}{q}\right) + P_5\right) \qquad (23)$$

Ten different functions (F_i) were proposed, in order to be analyzed and obtain the best one to estimate the behavior of the machine when it operates at variable rotational speed. Table 1 shows the proposed functions in which the different coefficients (β_i) are calculated as a function on the analyzed F_i. This analysis proposes six polynomial functions and four exponential expressions.

Table 1. Proposed functions to be analyzed.

Function Model (FM)	Polynomial Function (from F_1 to F_6): $NP = \beta_1\left(\alpha\frac{Q}{Q_{BEP}}\right) + \beta_2\left(\frac{Q}{Q_{BEP}}\right)^2 + \beta_3\left(\frac{Q}{Q_{BEP}}\right) + \beta_4\alpha^2 + \beta_5\alpha + \beta_6$ Exponential Function (from F_7 to F_{10}): $NP = \left(\frac{Q}{Q_{BEP}}\right)^{\beta_3}\alpha^{\beta_5}\cdot exp^{\beta_6}$
F_1	$NP = \beta_4\alpha^2 + \beta_5\alpha$
F_2	$NP = \beta_4\alpha^2 + \beta_5\alpha + \beta_6$
F_3	$NP = \beta_2\left(\frac{Q}{Q_{BEP}}\right)^2 + \beta_4\alpha^2 + \beta_5\alpha$
F_4	$NP = \beta_2\left(\frac{Q}{Q_{BEP}}\right)^2 + \beta_4\alpha^2 + \beta_5\alpha + \beta_6$
F_5	$NP = \beta_1\left(\alpha\frac{Q}{Q_{BEP}}\right) + \beta_2\left(\frac{Q}{Q_{BEP}}\right)^2 + \beta_3\left(\frac{Q}{Q_{BEP}}\right) + \beta_4\alpha^2 + \beta_5\alpha$
F_6	$NP = \beta_1\left(\alpha\frac{Q}{Q_{BEP}}\right) + \beta_2\left(\frac{Q}{Q_{BEP}}\right)^2 + \beta_3\left(\frac{Q}{Q_{BEP}}\right) + \beta_4\alpha^2 + \beta_5\alpha + \beta_6$
F_7	$NP = \alpha^{\beta_5}$
F_8	$NP = \alpha^{\beta_5}\cdot exp^{\beta_6}$
F_9	$NP = \left(\frac{Q}{Q_{BEP}}\right)^{\beta_3}\alpha^{\beta_5}$
F_{10}	$NP = \left(\frac{Q}{Q_{BEP}}\right)^{\beta_3}\alpha^{\beta_5}\cdot exp^{\beta_6}$

* NP is the non-dimensional parameter. It can be h, q, e, $\frac{h}{q^2}$, $\frac{he}{q^2}$.

8. This step is related to the previous step and concerns the recalculation of the coefficients β_i considering the values of all the tested machines $q_{i,j,m}$, $h_{i,j,m}$, and $e_{i,j,m}$. The sub-index "m" refers to each tested machine.
9. Having the coefficients for the different functions (F_i) as well as the non-dimensional parameters (i.e., h, q, e, $\frac{h}{q^2}$, $\frac{he}{q^2}$) defined, the errors of the proposed functions by MOAL are calculated. The error indices considered were root mean square error (*RMSE*), mean absolute deviation (*MAD*), the mean relative deviation (*MRD*), and BIAS:

    (a) RMSE. This error index measures the error between the empirical expression and experimental values. When RMSE is zero, this value indicates a perfect fit. It is defined by (24):

$$RMSE = \sqrt{\frac{\sum_{i=1}^{x}[O_i - P_i]^2}{x}} \qquad (24)$$

where O_i are the estimated values; P_i the experimental values, and x is the number of observations.

(b) MAD. This index measures the average of the errors in the estimated values, using the absolute differences between estimated and experimental values. The perfect fit is defined when MAD is zero, and it is defined by the following expression (25):

$$MAD = \sum_{1}^{x} \frac{1}{x} |O_i - P_i| \qquad (25)$$

(c) MRD. This index considers the weight of the error to the variable value. If MRD is zero, this value indicates a perfect fit. Formally, it is defined as follows (26):

$$MRD = \sum_{1}^{x} \frac{|O_i - P_i|/P_i}{x} \qquad (26)$$

(d) BIAS. The index considers the variable tendency, analyzing whether the estimated values are greater (negative value) or smaller (positive value) than experimental values. It is defined by the following expression (27):

$$BIAS = \frac{\sum_{i=1}^{N}[O_i - P_i]}{x} \qquad (27)$$

If the error values are acceptable and the goodness of the expressions is correct, the best expression is chosen in order to be applied. The best expression should set low error values, and it should consider a smaller number of variables.

2.2. Materials

The proposed methodology was applied using different experimental machines. As indicated, 15 PATs were used in this research, as shown in Table 2. The experimental database was developed from different consulted studies. These PATs were tested considering different rotation speeds (Table 3), which allowed interpolation of the different experimental values among rotation speeds. The specific speed (n_{st}) of the used machines was between 5 and 50 rpm. n_{st} is defined as:

$$n_{st} = n_0 \frac{P_0^{\frac{1}{2}}}{H_0^{\frac{5}{4}}} \qquad (28)$$

Table 2. Characteristics of the used pumps working as turbines (PATs).

ID	Ref.	n_{st} (m, kW)	n_0 (rpm)	D (mm)	Q_{BEP} (l/s)	H_{BEP} (m w.c.)	η_{BEP}	RS	IP	AP
1	[32]	20.66	1020	139	3461	4144	0.615	4	766	2393
2		28.34	1200	200	24,460	12,437	0.596	7	621	3812
3	[33]	25.57	1100	225	22,295	11,941	0.714	7	851	5646
4		26.43	1100	250	23,731	11,910	0.766	7	766	5086
5		17.68	1200	210	16,755	18,126	0.718	6	846	4377
6	[34]	27.03	800	265	27,322	8305	0.800	5	680	2997
7		25.44	1200	255	28,392	15,859	0.715	6	580	3035
8	[35]	13.65	1200	139	4906	11,283	0.543	3	714	1937
9	[36]	5.67	1100	193	9762	51,267	0.703	6	680	3514
10		31.16	3000	127	17,985	30,288	0.695	6	802	4535
11	[37]	20.97	3000	158	17,975	51,355	0.727	6	777	4516
12		50.71	2700	127	36,909	22,207	0.705	7	609	4139
13	[38]	21.75	1000	419	95,591	34,428	0.795	4	745	2595
14	[39]	13.84	1250	175	8990	17,525	0.622	6	804	3020
15	[40]	33.1	2900	189	50,050	52,849	0.646	7	708	4848
							Total	87	10,949	56,450

RS, number of experimental curves, which were tested for different rotational speeds; IP, number of interpolated parabolas using the experimental curves for each rotational speed; AP, number of analyzed points.

Table 3. Values of the different β_i for each proposed function model and considering the different non-dimensional parameters.

q

FM	R^2	β_1	β_2	β_3	β_4	β_5	β_6
F_1	0.9945	–	–	–	−0.4603	1.4250	–
F_2	0.8630	–	–	–	−0.7538	2.0117	−0.2765
F_3	0.9955	–	0.1112	–	−0.4566	1.3510	–
F_4	0.8802	–	0.0925	–	−0.6867	1.8224	−0.2163
F_5	0.9956	−0.0109	0.2984	−0.2918	−0.5549	1.5705	–
F_6	0.8809	−0.1525	0.1958	−0.0118	−0.6429	1.8489	−0.2241
F_7	0.8310	–	–	–	–	0.7439	–
F_8	0.8243	–	–	–	–	0.6796	−0.0540
F_9	0.8949	–	–	0.1847	–	0.5541	–
F_{10}	0.8567	–	–	0.1675	–	0.5617	−0.0085

h

FM	R^2	β_1	β_2	β_3	β_4	β_5	β_6
F_1	0.9910	–	–	–	0.5072	0.4588	–
F_2	0.9474	–	–	–	0.0943	1.2844	−0.3890
F_3	0.9919	–	0.1161	–	0.5111	0.3816	–
F_4	0.9506	–	0.0874	–	0.1576	1.1055	−0.3321
F_5	0.9922	−0.0942	0.4740	−0.4828	0.3765	0.7450	–
F_6	0.9512	−0.3107	0.3172	−0.0546	0.2420	1.1708	−0.3426
F_7	0.9653	–	–	–	–	1.7017	–
F_8	0.9620	–	–	–	–	1.6646	−0.0312
F_9	0.9734	–	–	0.1392	–	1.5587	–
F_{10}	0.9684	–	–	0.1689	–	1.5457	0.0147

e

FM	R^2	β_1	β_2	β_3	β_4	β_5	β_6
F_1	0.9796	–	–	–	−1.2039	2.1823	–
F_2	0.2391	–	–	–	−0.8235	1.4219	0.3583
F_3	0.9798	–	0.0535	–	−1.2021	2.1467	–
F_4	0.2602	–	0.0896	–	−0.7586	1.2385	0.4167
F_5	0.9803	0.5100	−0.5435	0.4514	−1.2321	1.8052	–
F_6	0.2832	0.8271	−0.3187	−0.1758	−1.0350	1.1815	0.5019
F_7	0.0017	–	–	–	–	0.0306	–
F_8	0.0214	–	–	–	–	−0.1036	−0.1127
F_9	0.2677	–	–	0.3404	–	−0.3191	–
F_{10}	0.1019	–	–	0.2494	–	−0.2791	−0.0450

p

FM	R^2	β_1	β_2	β_3	β_4	β_5	β_6
F_1	0.9133	–	–	–	0.4447	0.4063	–
F_2	0.6836	–	–	–	−0.8423	2.9793	−1.2124
F_3	0.9244	–	0.3783	–	0.4574	0.1547	–
F_4	0.7109	–	0.2897	–	−0.6325	2.3865	−1.0239
F_5	0.9357	2.0926	0.4300	−2.2315	−1.0984	1.8245	–
F_6	0.7297	1.6724	0.1255	−1.4005	−1.3596	2.6509	−0.6651
F_7	0.8098	–	–	–	–	2.4762	–
F_8	0.8019	–	–	–	–	2.2406	−0.1978
F_9	0.8825	–	–	0.6644	–	1.7937	–
F_{10}	0.8374	–	–	0.5858	–	1.8282	−0.0388

h/q^2

FM	R^2	β_1	β_2	β_3	β_4	β_5	β_6
F_1	0.9781	–	–	–	−0.5118	1.6305	–
F_2	0.7375	–	–	–	1.0476	−1.4870	1.4690
F_3	0.9811	–	−0.2347	–	−0.5196	1.7866	–
F_4	0.7639	–	−0.1139	–	0.9651	−1.2538	1.3948
F_5	0.9862	−1.5078	−0.4706	1.9296	0.7143	0.3415	–
F_6	0.7780	−0.6902	0.1218	0.3127	1.2224	−1.2665	1.2940
F_7	0.2070	–	–	–	–	0.2140	–
F_8	0.4018	–	–	–	–	0.3055	0.0768
F_9	0.5060	–	–	−0.2302	–	0.4505	–
F_{10}	0.4786	–	–	−0.1660	–	0.4223	0.0317

he/q^2

FM	R^2	β_1	β_2	β_3	β_4	β_5	β_6
F_1	0.9792	–	–	–	−0.8255	1.8427	–
F_2	0.1508	–	–	–	−0.1706	0.5336	0.6169
F_3	0.9792	–	−0.0236	–	−0.8263	1.8584	–
F_4	0.1538	–	0.0316	–	−0.1478	0.4690	0.6374
F_5	0.9801	0.5319	−0.8280	0.7565	−0.7572	1.2873	–
F_6	0.1912	0.9930	−0.4939	−0.1555	−0.4706	0.3804	0.7298
F_7	0.1753	–	–	–	–	0.2447	–
F_8	0.1160	–	–	–	–	0.2019	−0.0359
F_9	0.2197	–	–	0.1102	–	0.1315	–
F_{10}	0.1288	–	–	0.0834	–	0.1432	−0.0132

Table 2 also shows the number of experimental curves (RS) which were tested considering different rotational speeds for each machine, the number of interpolated curves used, as well as the number of used points to develop the regression and database analysis. The analysis of the 87 tested curves for different PATs, which operated on different rotational speeds, enabled us to obtain 10,949 interpolated parabolas, as well as 56,450 work points, to develop the surface (Q, H, α; Figure 2).

3. Results

3.1. Proposed Function Models

Once the experimental data from the referred 15 tested PATs were analyzed, the β_i coefficients were determined for the different non-dimensional parameters (i.e., q, h, e, p, h/q^2, he/q^2). Table 3 shows the different values of coefficients β_i for each proposed function (F_i) to model the non-dimensional parameters (NP). Table 3 also shows the regression coefficient (R^2).

The goodness of these models was measured according to the different error indexes, which were described previously in the Methodology section by Equations (21)–(23). The different dimensionless parameters proposed for each machine and rotation speed were determined, defining the error rates for the ten different functions of the model. Table 4 shows the error values for each index, as well as its ranking compared among the ten functions. This table determines the average values of the error indexes, since these errors were calculated for each rotational speed in each tested machine (87 curves). BIAS shows the absolute value, in order to know the magnitude of this error in case of oversize or undersize of a variable (i.e., H, η, and P).

Table 4 shows the average error values for each FM. These errors values enable us to decide the best function model for each dimensionless parameter (i.e., h, q, e, and p). When the error analysis was developed, the best function model (FM) was F_6 for h and e dimensionless parameters. Although different FMs could be used, F_6 considered both rotational speed as well as the ratio $\frac{Q}{Q_{BEP}}$. The use of this ratio is interesting since it measures the distance between Q and Q_{BEP}. This is an important difference, since it allows water managers to fix the operation range of flow in order for the affinity laws to be applied [29].

Table 4. Average error indexes for the different characteristic curves using the defined $MOAL$.

	Expression (21) $H = h\left(A + B\frac{Q}{q} + C\left(\frac{Q}{q}\right)^2\right)$				Expression (22) $\eta = e\left(E_4\left(\frac{Q}{q}\right)^4 + E_3\left(\frac{Q}{q}\right)^3 + E_2\left(\frac{Q}{q}\right)^2 + E_1\left(\frac{Q}{q}\right) + E_0\right)$					Expression (23) $P = p\left(P_4\left(\frac{Q}{q}\right)^4 + P_3\left(\frac{Q}{q}\right)^3 + P_2\left(\frac{Q}{q}\right)^2 + P_1\left(\frac{Q}{q}\right) + P_5\right)$				
FM	RMSE	MAD	MRD	BIAS	FM	RMSE	MAD	MRD	BIAS	FM	RMSE	MAD	MRD	BIAS
F_1	0.6869 (6)	0.5733 (6)	0.0325 (8)	0.1695 (7)	F_1	0.0596 (8)	0.0486 (8)	0.1161 (8)	0.0198 (9)	F_1	0.2666 (8)	0.221 (9)	0.1467 (9)	0.0678 (8)
F_2	0.7234 (9)	0.6007 (9)	0.0296 (5)	0.1054 (6)	F_2	0.0656 (10)	0.0493 (10)	0.1185 (10)	0.0173 (7)	F_2	0.2391 (6)	0.1983 (7)	0.1313 (7)	0.049 (4)
F_3	0.6099 (3)	0.5109 (1)	0.0286 (3)	0.0272 (5)	F_3	0.0487 (4)	0.042 (6)	0.1139 (7)	0.004 (4)	F_3	0.3341 (10)	0.259 (10)	0.1626 (10)	0.1305 (10)
F_4	0.6535 (5)	0.5448 (5)	0.0274 (2)	0.0264 (4)	F_4	0.0469 (2)	0.0381 (3)	0.1015 (2)	0.0009 (1)	F_4	0.2732 (9)	0.213 (8)	0.1398 (8)	0.0222 (1)
F_5	0.6077 (2)	0.5113 (2)	0.0289 (4)	0.0026 (2)	F_5	0.0494 (5)	0.0424 (7)	0.1115 (6)	0.0052 (5)	F_5	0.2314 (5)	0.1775 (5)	0.1216 (5)	0.0646 (6)
F_6	0.6075 (1)	0.5154 (3)	0.0266 (1)	0.005 (3)	F_6	0.0397 (1)	0.0331 (1)	0.0883 (1)	0.0027 (3)	F_6	0.2472 (7)	0.194 (6)	0.1272 (6)	0.0654 (7)
F_7	0.6101 (4)	0.5186 (4)	0.0302 (6)	0.3739 (9)	F_7	0.054 (7)	0.0419 (5)	0.1033 (3)	0.0016 (2)	F_7	0.1169 (1)	0.1017 (1)	0.0858 (1)	0.023 (2)
F_8	0.6979 (7)	0.5911 (8)	0.0332 (9)	0.0021 (1)	F_8	0.0652 (9)	0.049 (9)	0.111 (5)	0.0222 (10)	F_8	0.174 (4)	0.1471 (4)	0.0886 (2)	0.1127 (9)
F_9	0.7697 (10)	0.6154 (10)	0.0349 (10)	0.4324 (10)	F_9	0.0506 (6)	0.0414 (4)	0.1174 (9)	0.0182 (8)	F_9	0.1486 (2)	0.13 (2)	0.0964 (4)	0.0364 (3)
F_{10}	0.7085 (8)	0.5847 (7)	0.0321 (7)	0.216 (8)	F_{10}	0.047 (3)	0.0379 (2)	0.1107 (4)	0.0064 (6)	F_{10}	0.1504 (3)	0.1345 (3)	0.0942 (3)	0.0629 (5)

The ranking of the F_i when the error indexes are compared from (1) to (9) as indicated.

When non-parameter p was analyzed, the F_7 function also was chosen since it only considered one variable (α), yielding good results in the estimation of the PATs curve. F_7 was used to determine the power curve directly by expression (23). However, when water managers wish to determine the power curve by the use of Q, H, and η, they should use the F_6 function.

3.2. Error Distribution Compared to Rotational Speed

Once F_6 was chosen, the error of the modified affinity laws was compared with all tested curves. All error indexes were calculated for head, efficiency, and power.

When head was analyzed, the MRD was smaller than 0.05, with a cumulated frequency equal to 91%. The maximum value was 0.089. In head values, RMSE was smaller than 0.6 in 57 compared curves and BIAS was smaller than 0.25 in 49 compared curves.

When efficiency was compared, RMSE was smaller than 0.035 in 58% of the comparisons and it was smaller than 0.07 in 88% of the comparisons. When MRD was checked, it was smaller than 0.15, showing a BIAS value smaller than 0.069 in 92% of the cases.

When the error values for the power curve using the F_6 function model were analyzed, RMSE was smaller than 0.2 (72% cumulated frequency). When MAD was analyzed, similar values were obtained. MAD was lower than 0.17, and the MRD was smaller than 0.2 in 90% of the samples.

However, when the errors of F_7 were analyzed for the power curve, they showed the best approach. Figure 4 shows the error values for the power curve using the F_7 expression. RMSE was analyzed (Figure 4a), and it was smaller than 0.18 (70% of cumulated frequency). This value was smaller than 0.09 in 51 cases. Moreover, when the α value was observed, smaller values were located between 0.8 and 1.2, reaching a minimum around 0.9.

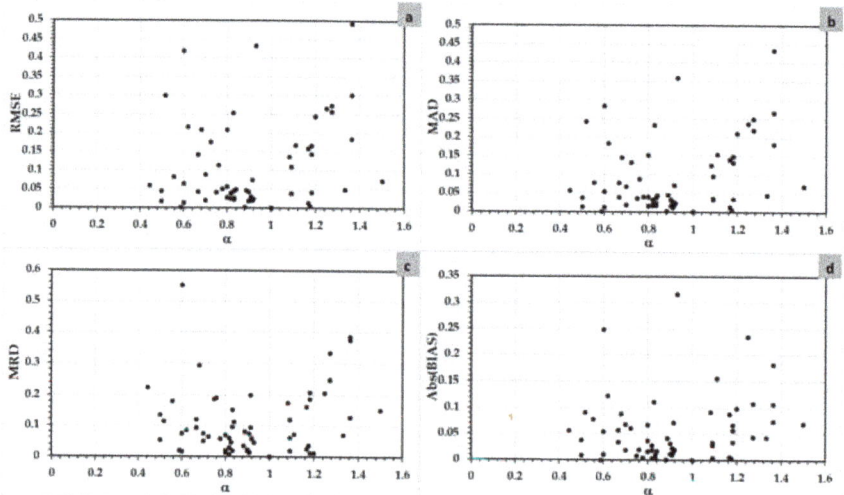

Figure 4. Error values when power is determined: (**a**) root mean square error (RMSE) (**b**) mean absolute deviation (MAD); (**c**) mean relative deviation (MRD), and (**d**) absolute value of BIAS.

When MAD was analyzed (Figure 4b), similar values were obtained. In this case, MAD values were smaller than 0.16 (92% of cumulated frequency). When MRD was analyzed (Figure 4c), this value was smaller than 0.2 in 94% of the samples. This value had a value of 65% of cumulated frequency for values lower than 0.07. Finally, BIAS had good accuracy, showing values lower than 0.1 in 85% of the sample. In all cases, the minimum errors were reached when the machine operated using α rates between 0.8 and 1.2, being the minimum for values near 0.9.

3.3. Proposed Functions vs. Other Published Functions

Once the relative errors of the selected function model (F_6) were compared for each rotational speed of the different tested machine in the different proposed functions of head, efficiency, and power, the proposed expressions were compared with other expressions which have already been published in the literature.

This research proposes the following particular functions to define the characteristic curves of the machine according to expressions (21)–(23). The model F_6 was chosen when head and efficiency curves should be estimated. F_6 showed the lowest errors compared to the rest of the models. Moreover, this model contained the variation of the rotational speed (α) as well as the use of the ratio Q/Q_{BEP}, enabling us to measure the closeness to BEP. To calculate the power, F_7 was chosen since it had the minimum error values, and it uses a simpler expression. The final expressions proposed herein are:

$$q = -0.1525\left(\alpha\frac{Q}{Q_{BEP}}\right) + 0.1958\left(\frac{Q}{Q_{BEP}}\right)^2 - 0.0118\left(\frac{Q}{Q_{BEP}}\right) - 0.6429\alpha^2 + 1.8489\alpha - 0.2241$$

$$h = -0.31070\left(\alpha\frac{Q}{Q_{BEP}}\right) + 0.3172\left(\frac{Q}{Q_{BEP}}\right)^2 - 0.0546\left(\frac{Q}{Q_{BEP}}\right) + 0.242\alpha^2 + 1.1708\alpha - 0.3426$$

$$e = 0.8271\left(\alpha\frac{Q}{Q_{BEP}}\right) - 0.3187\left(\frac{Q}{Q_{BEP}}\right)^2 - 0.1758\left(\frac{Q}{Q_{BEP}}\right) - 1.035\alpha^2 + 1.1815\alpha + 0.5019$$

$$p = \alpha^{2.4762};$$

$$q = \alpha^{0.7439}$$

The comparison concerns the model proposed in this research and four published proposals that are shown in Table 5.

Table 5. Methods used for the comparison.

Method	Reference	h	q	p	η
Carravetta et al. (2014)	[14]	$1.0253\alpha^{1.5615}$	$1.0323\alpha^{0.7977}$	$0.9741\alpha^{2.3207}$	$-0.4013\alpha^2 + 0.845\alpha + 0.5606$
Fecarotta et al. (2016)	[12]	$0.972\alpha^{1.603}$	$1.004\alpha^{0.825}$	–	$-0.317\alpha^2 + 0.587\alpha + 0.707$
Pérez-Sánchez et al. (2018)	[27]	$1.89\alpha^2 - 1.54\alpha + 0.74$	$1.08\alpha^{0.7}$	$4.59\alpha^2 - 6.33\alpha + 2.50$	$-0.36\alpha^2 - 0.69\alpha + 0.66$
Tahani et al. (2020)	[28]	$0.9962\alpha^{1.0851}$	$0.9974\alpha^{0.3651}$	$0.9767\alpha^{1.4888}$	$-4.3506\alpha^2 + 8.8879\alpha - 3.544$

Figure 5 shows the different values for error, when head, efficiency, and power were estimated using the proposed model (in black color, "this study") and the rest of the published models. In all cases, the present proposed model presented the best results.

When head curve was analyzed, the error indexes (RMSE, MAD, and MRD) were reduced between 20 and 45% compared to the second-best model (Carravetta et al.). The BIAS value for this characteristic curve was -0.005, compared to the second-best model (0.048). Similar values were shown when the efficiency curve was compared. When efficiency errors were compared, RMS, MAD, and MRD were reduced by 33% compared to the second-best model, while BIAS was ten times lower than the second-best model. Finally, when the power errors were checked using the F_7 model, the error indexes were reduced between 36 and 63% compared to the second-best model. Only when BIAS was checked, the second-best value was observed. Moreover, the F_6 model was also compared to the rest of the proposed models for the power curve. This model (F_6) showed good accuracy and the error indexes were 0.2209 (RMSE), 0.1884 (MAD), 0.0823 (MRD), and -0.097 (BIAS). All values were better than the second-best model, except for BIAS, which was the third-best value.

Finally, a visual comparison was carried out on the proposed model and the remaining models compared to an experimental PAT curve (Figure 6). To develop this comparison, the chosen PAT was a radial machine. The specific speed was 5.67 rpm (m, kW) and its nominal rotational speed was 1100 rpm. The best operation point of this machine was defined as 9.762 l/s and 51.267 m w.c., the efficiency being equal to 0.703 [24]

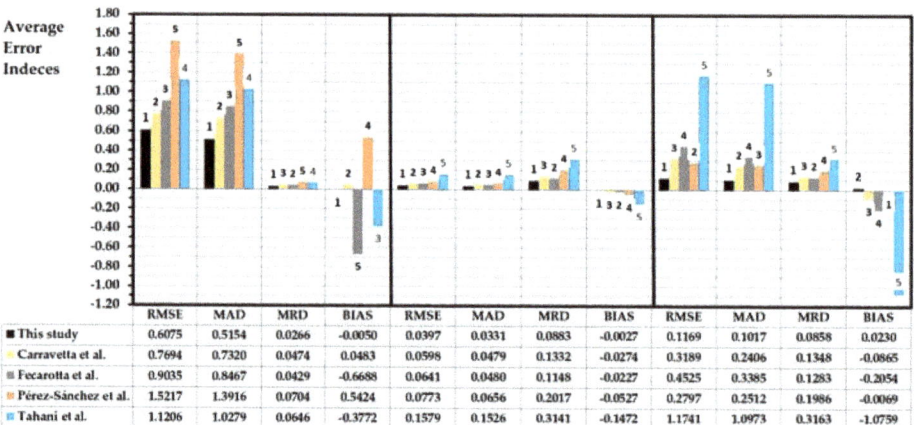

Figure 5. Error values for head, efficiency, and power, when the models are compared.

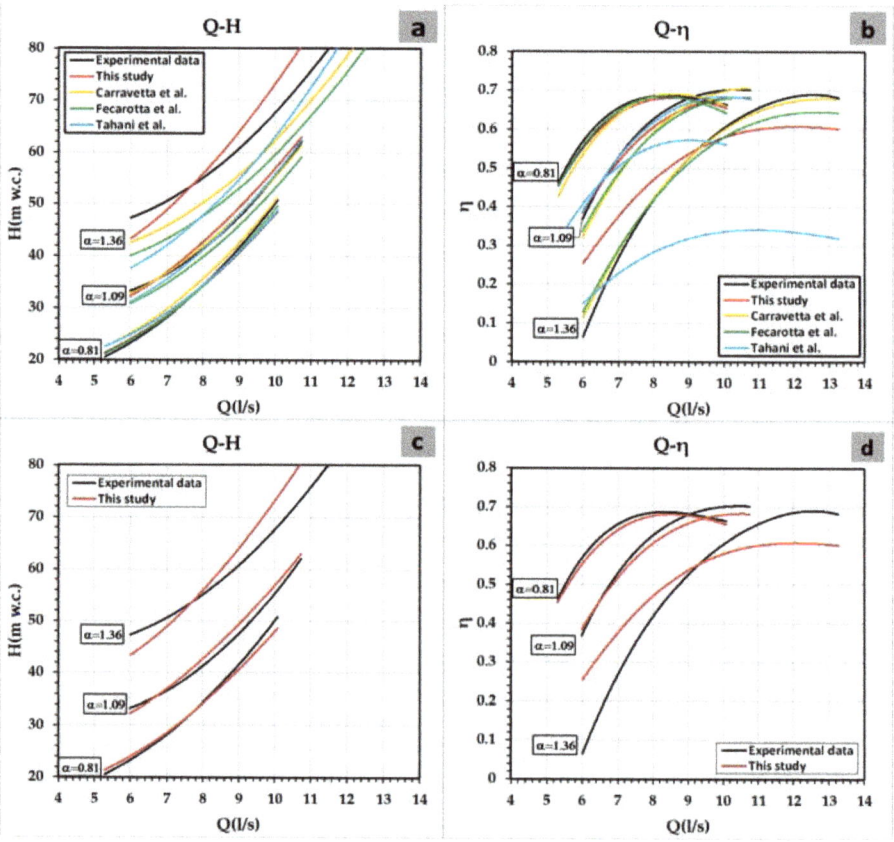

Figure 6. (a) Head curve comparison between proposed model, experimental data, and rest of published models; (b) Efficiency curve comparison between proposed model, experimental data, and rest of published models; (c) Head curve between proposed model and experimental curve; (d) Efficiency curve between proposed model and experimental curve.

Figure 6a,b show the good accuracy of the proposed model compared to the rest of the models. This accuracy can be observed for each α value. Figure 6a shows the accuracy of the proposed study, other published models, and experimental data. All models showed good accuracy when the head curve was compared with the experimental data. However, this accuracy decreased in the rest of the models when α was higher than one. The accuracy of the proposed expressions was much better when the efficiency curves were compared. This visual accuracy, which can only be observed, is supported by analysis of errors indexes shown in Figure 5. In this graph, the proposed expressions reduced over 20% of the error of the other published methods. The mean reduction in the error was 60%. To improve this perception, Figure 6c,d show the comparison between the proposed model and the experimental data. In all cases, the accuracy was good but, when the α was between 0.8 and 1.2, the estimation of the curves showed excellent accuracy.

4. Conclusions

This research proposed a modification of the affinity laws (MOAL) of the hydraulic machines that are used as pumps working as turbines. This modification was established according to a new methodology, which was defined in this research. The research proposed an analysis with ten general expressions (polynomial and exponential), considering the most significant variables (the ratio of the rotational speed, α, and the ratio of Q and Q_{BEP}). Finally, a polynomial model (namely F_6) depending on α and $\frac{Q}{Q_{BEP}}$ was selected, when head and efficiency were estimated, and a potential model (F_7) if the power is to be calculated directly. All proposed models exhibited good error indexes (RMSE, MAD, MRD, and BIAS) compared to the others, reducing the errors between 30 and 50% compared to the second-best model.

In addition, the proposed models were checked and compared to 15 different machines, which were tested by varying their rotational speed and its specific speed between 5 and 50 rpm (m, kW). The present model is based on 87 different curves and 56,450 operation points, using the largest database ever published.

The use of these models, which have excellent accuracy when α is between 0.8 and 1.2, is crucial to the development of mathematical models. These are of paramount importance to introduce the use of PATs when the manufacturer curve is not known. This is common when PATs are used, since the manufacturers do not publish these curves in their catalogue. Therefore, the inclusion of these equations will allow water managers to develop simulation tools, which can be introduced in the management of the water systems, improving the accuracy in their operation estimation. These models are expected to give a new impetus in the inclusion of the analysis tools when PATs operate at variable speed in water systems, and water modelers need mathematical expressions to develop simulations and operational limitations. Consequently, future works should be developed in which different procedures are proposed to establish the best variable operating strategy (VOS) in order to maximize the energy recovery using these expressions.

Author Contributions: Conceptualization, F.-J.S.-R. and M.P.-S., methodology, F.A.P.; software, V.H.; formal analysis, F.A.P.; writing—original draft preparation, F.A.P., F.-J.S.-R. and M.P.-S.; writing—review and editing, M.P.-S. and P.A.L.-J.; visualization, F.-J.S.-R. and M.P.-S.; supervision, P.A.L.-J. All authors have read and agreed to the published version of the manuscript.

Funding: This research received no external funding.

Institutional Review Board Statement: Not applicable.

Informed Consent Statement: Not applicable.

Data Availability Statement: Not applicable.

Conflicts of Interest: The authors declare no conflict of interest.

References

1. Mangalekar, R.D.; Gumaste, K.S. Residential water demand modelling and hydraulic reliability in design of building water supply systems: A review. *Water Supply* **2021**. [CrossRef]
2. Bach, P.M.; Rauch, W.; Mikkelsen, P.S.; McCarthy, D.T.; Deletic, A. A critical review of integrated urban water modelling–Urban drainage and beyond. *Environ. Model Softw.* **2014**, *54*, 88–107. [CrossRef]
3. Mousavi, S.N.; Bocchiola, D. A novel comparative statistical and experimental modeling of pressure field in free jumps along the apron of USBR Type I and II dissipation basins. *Mathematics* **2020**, *8*, 2155. [CrossRef]
4. Raboaca, M.S.; Bizon, N.; Trufin, C.; Enescu, F.M. Efficient and secure strategy for energy systems of interconnected farmers' associations to meet variable energy demand. *Mathematics* **2020**, *8*, 2182. [CrossRef]
5. Puust, R.; Kapelan, Z.; Savic, D.A.; Koppel, T. A review of methods for leakage management in pipe networks. *Urban Water J.* **2010**, *7*, 25–45. [CrossRef]
6. Vakilifard, N.; Anda, M.; Bahri, P.A.; Ho, G. The role of water-energy nexus in optimising water supply systems—Review of techniques and approaches. *Renew. Sustain. Energy Rev.* **2018**, *82*, 1424–1432. [CrossRef]
7. Orouji, H.; Bozorg Haddad, O.; Fallah-Mehdipour, E.; Mariño, M.A. Modeling of water quality parameters using data-driven models. *J. Environ. Eng.* **2013**, *139*, 947–957. [CrossRef]
8. Ramos, H.; Borga, A. Pumps as turbines: An unconventional solution to energy production. *Urban Water* **1999**, *1*, 261–263. [CrossRef]
9. García, I.F.; Novara, D.; Mc Nabola, A. A model for selecting the most cost-effective pressure control device for more sustainable water supply networks. *Water* **2019**, *11*, 1297. [CrossRef]
10. Pérez-Sánchez, M.; Sánchez-Romero, F.J.; Ramos, H.M.; López-Jiménez, P.A. Improved planning of energy recovery in water systems using a new analytic approach to PAT performance curves. *Water* **2020**, *12*, 468. [CrossRef]
11. Pérez-Sánchez, M.; Sánchez-Romero, F.J.; Ramos, H.; López-Jiménez, P. Energy recovery in existing water networks: Towards greater sustainability. *Water* **2017**, *9*, 97. [CrossRef]
12. Fecarotta, O.; Aricò, C.; Carravetta, A.; Martino, R.; Ramos, H.M. Hydropower Potential in Water Distribution Networks: Pressure Control by PATs. *Water Resour Manag.* **2015**, *29*, 699–714. [CrossRef]
13. Quaranta, E.; Bonjean, M.; Cuvato, D.; Nicolet, C.; Dreyer, M.; Gaspoz, A.; Rey-Mermet, S.; Boulicaut, B.; Pratalata, L.; Pinelli, M.; et al. Hydropower case study collection: Innovative low head and ecologically improved turbines, hydropower in existing infrastructures, hydropeaking reduction, digitalization and governing systems. *Sustainability* **2020**, *12*, 8873. [CrossRef]
14. Carravetta, A.; Conte, M.C.; Fecarotta, O.; Ramos, H.M. Evaluation of PAT performances by modified affinity law. *Procedia Eng.* **2014**, *89*, 581–587. [CrossRef]
15. Suter, P. Representation of pump characteristics for calculation of water hammer. *Sulzer Tech. Rev.* **1966**, *66*, 45–48.
16. Kougias, I.; Aggidis, G.; Avellan, F.; Deniz, S.; Lundin, U.; Moro, A.; Muntean, S.; Novara, D.; Perez-Diaz, J.I.; Quaranta, E.; et al. Analysis of emerging technologies in the hydropower sector. *Renew. Sustain. Energy Rev.* **2019**, *113*, 109257. [CrossRef]
17. Novara, D.; McNabola, A. A model for the extrapolation of the characteristic curves of Pumps as Turbines from a datum best efficiency point. *Energy Convers. Manag.* **2018**, *174*, 1–7. [CrossRef]
18. Shah, S.R.; Jain, S.V.; Patel, R.N.; Lakhera, V.J. CFD for centrifugal pumps: A review of the state-of-the-art. *Procedia Eng.* **2013**, *51*, 715–720. [CrossRef]
19. Carravetta, A.; Del Giudice, G.; Fecarotta, O.; Ramos, H. Pump as Turbine (PAT) design in water distribution network by system effectiveness. *Water* **2013**, *5*, 1211–1225. [CrossRef]
20. Iliev, I.; Trivedi, C.; Dahlhaug, O.G. Variable-speed operation of Francis turbines: A review of the perspectives and challenges. *Renew. Sustain. Energy Rev.* **2019**, *103*, 109–121. [CrossRef]
21. Singh, P. Optimization of the Internal Hydraulic and of System Design in Pumps as Turbines with Field Implementation and Evaluation. Ph.D. Thesis, Karlsruhe Institute of Technology, Karlsruhe, Germany, 3 June 2005.
22. Yang, S.S.; Derakhshan, S.; Kong, F.Y. Theoretical, numerical and experimental prediction of pump as turbine performance. *Renew. Energy* **2012**, *48*, 507–513. [CrossRef]
23. Rossi, M.; Nigro, A.; Renzi, M. Experimental and numerical assessment of a methodology for performance prediction of Pumps-as-Turbines (PaTs) operating in off-design conditions. *Appl. Energy* **2019**, *248*, 555–566. [CrossRef]
24. Novara, D.; Carravetta, A.; McNabola, A.; Ramos, H.M. Cost model for pumps as turbines in run-of-river and in-pipe microhydropower applications. *J. Water Resour. Plan. Manag.* **2019**, *145*, 04019012. [CrossRef]
25. Derakhshan, S.; Nourbakhsh, A. Experimental study of characteristic curves of centrifugal pumps working as turbines in different specific speeds. *Exp. Therm. Fluid Sci.* **2008**, *32*, 800–807. [CrossRef]
26. Carravetta, A.; del Giudice, G.; Fecarotta, O.; Ramos, H. PAT design strategy for energy recovery in water distribution networks by electrical regulation. *Energies* **2013**, *6*, 411–424. [CrossRef]
27. Pérez-Sánchez, M.; López-Jiménez, P.A.; Ramos, H.M. Modified affinity laws in hydraulic machines towards the best efficiency line. *Water Resour. Manag.* **2018**, *32*, 829–844. [CrossRef]
28. Tahani, M.; Kandi, A.; Moghimi, M.; Houreh, S.D. Rotational speed variation assessment of centrifugal pump-as-turbine as an energy utilization device under water distribution network condition. *Energy* **2020**, *213*, 118502. [CrossRef]
29. Mataix, C. *Turbomáquinas Hidráulicas*; Universidad Pontificia Comillas: Madrid, Spain, 2009.

30. Morrison, G.; Yin, W.; Agarwal, R.; Patil, A. Development of modified affinity law for centrifugal pump to predict the effect of viscosity. *J. Sol. Energy Eng. Trans. ASME.* **2018**, *140*. [CrossRef]
31. Fecarotta, O.; Carravetta, A.; Ramos, H.M.; Martino, R. An improved affinity model to enhance variable operating strategy for pumps used as turbines. *J. Hydraul. Res.* **2016**, *1686*, 1–10. [CrossRef]
32. KSB. PATs Curves. Catalogue. Available online: https://www.ksb.com/ksb-en/ (accessed on 5 November 2019).
33. Jain, S.V.; Swarnkar, A.; Motwani, K.H.; Patel, R.N. Effects of impeller diameter and rotational speed on performance of pump running in turbine mode. *Energy Convers. Manag.* **2015**, *89*, 808–824. [CrossRef]
34. Nygren, L. Hydraulic Energy Harvesting with Variable-Speed-Driven Centrifugal Pump as Turbine. Ph.D. Thesis, Lappeenrtanta University of Technology, Lappeenranta, Finland, 2017.
35. Abazariyan, S.; Rafee, R.; Derakhshan, S. Experimental study of viscosity effects on a pump as turbine performance. *Renew. Energy* **2018**, *127*, 539–547. [CrossRef]
36. Kramer, M.; Terheiden, K.; Wieprecht, S. Pumps as turbines for efficient energy recovery in water supply networks. *Renew. Energy* **2018**, *122*, 17–25. [CrossRef]
37. Delgado, J.; Ferreira, J.P.; Covas, D.I.C.; Avellan, F. Variable speed operation of centrifugal pumps running as turbines. Experimental investigation. *Renew. Energy* **2019**, *142*, 437–450. [CrossRef]
38. Stefanizzi, M.; Torresi, M.; Fortunato, B.; Camporeale, S.M. Experimental investigation and performance prediction modeling of a single stage centrifugal pump operating as turbine. *Energy Procedia* **2017**, *126*, 589–596. [CrossRef]
39. Postacchini, M.; Darvini, G.; Finizio, F.; Pelagalli, L.; Soldini, L.; Di Giuseppe, E. Hydropower generation through Pump as Turbine: Experimental study and potential application to small-scale WDN. *Water* **2020**, *12*, 958. [CrossRef]
40. Pugliese, F.; De Paola, F.; Fontana, N.; Giugni, M.; Marini, G. Experimental characterization of two Pumps as Turbines for hydropower generation. *Renew. Energy* **2016**, *99*, 180–187. [CrossRef]

Article

Statistical and Type II Error Assessment of a Runoff Predictive Model in Peninsula Malaysia

Lloyd Ling [1,*], Zulkifli Yusop [2] and Joan Lucille Ling [3]

1. Centre of Disaster Risk Reduction (CDRR), Civil Engineering Department, Lee Kong Chian Faculty of Engineering & Science, Universiti Tunku Abdul Rahman, Jalan Sungai Long, Kajang 43000, Malaysia
2. Centre for Environmental Sustainability and Water Security, Universiti Teknologi Malaysia, Skudai 81310, Malaysia; zulyusop@utm.my
3. Department of Liberal Arts and Languages, American Degree Programme, Taylor's University, Jalan Taylors, Subang Jaya 47500, Malaysia; linglucille@gmail.com
* Correspondence: linglloyd@utar.edu.my

Citation: Ling, L.; Yusop, Z.; Ling, J.L. Statistical and Type II Error Assessment of a Runoff Predictive Model in Peninsula Malaysia. *Mathematics* **2021**, *9*, 812. https://doi.org/10.3390/math9080812

Academic Editor: Eva H. Dulf

Received: 17 February 2021
Accepted: 29 March 2021
Published: 8 April 2021

Publisher's Note: MDPI stays neutral with regard to jurisdictional claims in published maps and institutional affiliations.

Copyright: © 2021 by the authors. Licensee MDPI, Basel, Switzerland. This article is an open access article distributed under the terms and conditions of the Creative Commons Attribution (CC BY) license (https://creativecommons.org/licenses/by/4.0/).

Abstract: Flood related disasters continue to threaten mankind despite preventative efforts in technological advancement. Since 1954, the Soil Conservation Services (SCS) Curve Number ($CN_{0.2}$) rainfall-runoff model has been widely used but reportedly produced inconsistent results in field studies worldwide. As such, this article presents methodology to reassess the validity of the model and perform model calibration with inferential statistics. A closed form equation was solved to narrow previous research gap with a derived 3D runoff difference model for type II error assessment. Under this study, the SCS runoff model is statistically insignificant (alpha = 0.01) without calibration. Curve Number $CN_{0.2} = 72.58$ for Peninsula Malaysia with a 99% confidence interval range of 67 to 76. Within these $CN_{0.2}$ areas, SCS model underpredicts runoff amounts when the rainfall depth of a storm is < 70 mm. Its overprediction tendency worsens in cases involving larger storm events. For areas of 1 km^2, it underpredicted runoff amount the most (2.4 million liters) at $CN_{0.2} = 67$ and the rainfall depth of 55 mm while it nearly overpredicted runoff amount by 25 million liters when the storm depth reached 430 mm in Peninsula Malaysia. The SCS model must be validated with rainfall-runoff datasets prior to its adoption for runoff prediction in any part of the world. SCS practitioners are encouraged to adopt the general formulae from this article to derive assessment models and equations for their studies.

Keywords: rainfall-runoff model; curve number; inferential statistics; 3D runoff difference model; model calibration

1. Introduction

Nearly 8.5 million casualties attributed to flood related disasters were reported between 1990 and 2020 all over the world, which is equivalent to one death every seven minutes. In the recent six decades, about 10,000 cases were reported with 1.3 million deaths and at least $3.3 trillion of financial losses. This financial loss is estimated to be an equivalent rate of almost USD$1800/s [1]. Floods are not only a nuisance to people but also impede the financial well-being, economic development, and natural and cultural heritage preservation efforts of a country. The impact is more profound amidst the COVID-19 pandemic. Uncertainties regarding different scenarios surrounding climate change also require us to safeguard agricultural production and manage water resources wisely to ensure sustainable development for the future. As such, there is an imminent need for hydrologists and modelers to reassess the rainfall-runoff model and improve the modelling approach for better applications in flood prediction.

In order to comply with the federal flood control program in 1954, the United States Department of Agriculture (USDA), Soil Conservation Services (SCS) developed a Curve Number (CN) runoff estimation procedure to implement across the nation. The hydrologic methods which were originally developed to address specific situations were adopted immediately without professional review and critics [2–5]. The work became the basic CN rainfall-runoff model:

$$Q = \frac{(P - I_a)^2}{P - I_a + S} \qquad (1)$$

Q = Amount of runoff depth (mm)
P = Depth of rainfall (mm)
S = Watershed maximum water retention potential (mm)
I_a = Rainfall initial abstraction amount (mm)

SCS also hypothesized that $I_a = \lambda S = 0.2S$ where λ is the initial abstraction ratio coefficient and fixed at $\lambda = 0.2$ as a constant. This equation was tenuously justified with daily rainfall and runoff data. The only official documentation source is the NRCS's National Engineering Handbook, Section 4 (NEH-4) [5]. Its substitution simplifies Equation (1) into the existing SCS CN model as:

$$Q = \frac{(P - 0.2S)^2}{P + 0.8S} \qquad (2)$$

if P < 0.2S, Q = 0.

The SCS CN methodology has been widely accepted since its inception in 1954. It has been incorporated in various types of software, adopted by many government agencies in design and even appears in every hydrology textbook. However, studies around the world from recent decades reported that Equation (2) inconsistently under and over-predicted runoff results. Curve Number (CN) selection from the SCS handbook for a watershed runoff prediction modelling were reported as subjective and often could not represent other watershed with similar land cover [2–4].

Despite that, many recent studies started to develop and propose extended applications with Equation (2). Some researchers even proposed a global gridded CN concept for runoff modelling [6,7] while others incorporated land-use information in their studies and the GIS modelling technique [8–12]. Contrarily, some reported that the usage of CN in representing a watershed is often contradictory in describing related land cover areas [13]. Some researchers still reported difficulty to calibrate the existing model [14,15] while other studies started to incorporate soil moisture and saturation-excess concepts in their modelling approach [16–19]. US researchers [2,20] were first to conduct large scale studies on the SCS CN model by analyzing more than half a million rainfall events across 24 states in the USA and reported an optimum $\lambda = 0.05$ to achieve better runoff modelling results than Equation (2) in USA. To date, SCS practitioners do not have a systematic approach to assess the SCS CN model framework and analyze the impact on runoff prediction when the model is not calibrated.

2. Data and Methods

The SCS CN model (Equation (2)) has been adopted in Malaysia for runoff prediction studies and design. However, no attempt has been made to validate previous study findings by performing hydrological characteristics calibration on the SCS CN model and to derive the λ value with inferential statistics for the entirety of Peninsula Malaysia. The impact of not calibrating the SCS CN model and the blind adoption of Equation (2) for runoff predictions in Peninsula Malaysia are unknown. Therefore, this study extended study results from US researchers [2,20] to develop assessment methods of the SCS CN model for SCS practitioners.

Slightly larger than England (130,395 km^2), the land area of Peninsula Malaysia is 132,265 km^2. It shares a land border with Thailand to the north and Singapore across the strait of Johor to its south. The formation of the Malaysian Department of Irrigation and Drainage (DID) in 1932 assumed all works in connection with drainage and irrigation from the Public Works Department. Flood mitigation and hydrology was made an additional responsibility of DID from 1972 onwards after the declaration of a national disaster due to severe floods in 1971. From 1986, coastal engineering has become an added function of the DID while river management became its official duty from 1990.

The Department has moved from the Ministry of Agriculture and Agro-based Industry (MOA) to Ministry of Natural Resource and Environment (NRE) on 27 March 2004. Over the years, DID took up new and expanded responsibilities. Today, the DID's duties encompass: River Basin Management and Coastal Zone, Water Resources Management and Hydrology, Flood Management and Eco-friendly Drainage projects in Malaysia.

The rainfall-runoff dataset from the DID, Hydrological Procedure no. 27 (DID HP 27) was used in this study. It is the latest official dataset published by this federal government agency that consists of 227 different storm events recorded between October 1970 to December 2000 from 41 different rural watersheds (Figure 1) across Peninsula Malaysia. The smallest storm event had a rainfall depth of 19 mm with a measurable runoff depth of 4.8 mm while the largest recorded storm event was 420 mm with 258 mm in runoff depth [21].

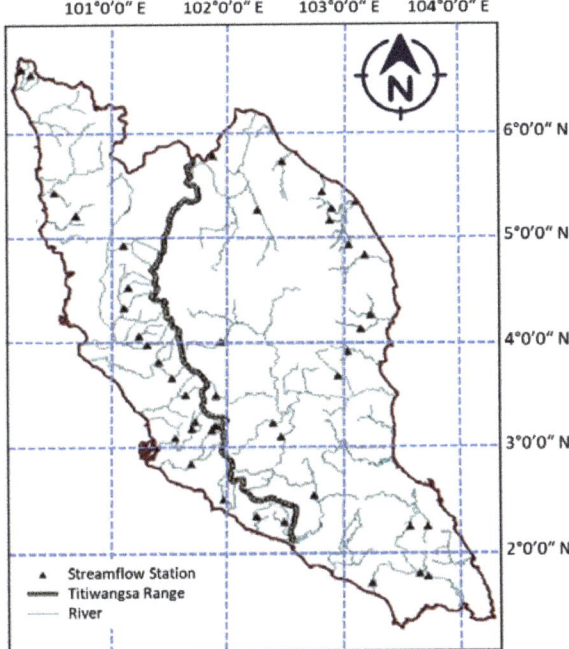

Figure 1. Locations of 41 streamflow stations with 227 rainfall-runoff (P-Q) data pairs used for λ derivation. Modified according to [21].

Objectives of this study are:

1. To assess the 1954 SCS assumption of: $I_a = 0.2S$ in $Q = \frac{(P-I_a)^2}{P-I_a+S}$ and determine its validity for runoff prediction use in Peninsula Malaysia according to the DID HP 27 dataset.
2. To solve the closed form mathematical equation of the "critical rainfall amount" and develop a statistically significant SCS CN model calibration methodology.
3. To assess the impact of not calibrating the existing SCS CN runoff predictive model (Equation (2)) for runoff prediction in Peninsula Malaysia with the official rainfall-runoff dataset from DID HP 27 [21].

2.1. The Reverse Derivation of λ and S Value

In hydrology, the difference between I_a and P is the effective rainfall depth (P_e) to initiate Q thus $P_e = P - I_a$. Substitute this relationship into SCS CN model (Equation (1)), it can be re-arranged to calculate the two key parameters of S and λ values according to the respective P-Q data pair [2,5,22]. Equation (1) can then be expressed as below after the substitution of $P_e = P - I_a$:

$$Q = \frac{(P_e)^2}{P_e + S} \tag{3}$$

rearrange Equation (3) to isolate S as:

$$S = \frac{(P_e)^2}{Q} - P_e \tag{4}$$

Equation (4) is subjected to the constraint where S must be a positive integer. SCS also proposed the correlation of $I_a = \lambda S$ thus λ can be calculated once I_a and S are known by rearranging the equation as:

$$\lambda = \frac{I_a}{S} \tag{5}$$

Equation (5) is subjected to the constraint defined by SCS that $S \geq I_a$ [5], and therefore the range of λ must be (0, 1). The upper limit for λ value is equal to 1 (where $I_a = S$) which is hardly realized in the real world as it implies the condition of a thick canopy interception. The infiltration during early parts of the storm and surface depression storage is equal to the maximum potential retention value (S) of a watershed [5].

Past studies reported different λ values in their work for model calibration. However, the statistical assurance of those new values was hardly mentioned [4]. Latest studies in this area started to report that the modelling approach with multiple CN and I_a values can reflect the heterogeneity of a watershed and the SCS CN model must be calibrated according to local rainfall-runoff data to improve the runoff prediction accuracy. Equation (2) may no longer be valid for runoff prediction modelling [23–25]. SCS defined $I_a = \lambda S$, the existence of multiple I_a values implied that multiple λ and S values can be found within a watershed. These latest study results [24,25] escalate the SCS CN model calibration difficulty to another level as SCS practitioners must identify a best collective representative I_a value to calibrate Equation (1). Therefore, this study proposed to use non-parametric inferential statistics as the guide to make a statistically significant selection of the two key parameters (S and λ values) to calibrate the fundamental SCS CN runoff framework (Equation (1)).

Under the SCS CN hydrological framework, the initial abstraction (I_a) amount must be less than the P value because I_a must first be fulfilled to initiate runoff. Therefore, a reasonable collective representative I_a value for runoff modelling must be less than the minimum P value from the entire P-Q dataset [5]. Given the P-Q dataset, an initial "I_a" value which was less than the minimum P value from the dataset was chosen as the first iterative value in order to calculate the corresponding S and λ values for each P-Q data pair according to Equations (4) and (5). In the event where either constraint in Equation (4) or (5) were to be violated, the "collective representative I_a" value must be reduced until every calculated λ and S values abide to their constraints for each P-Q data pair according to the SCS CN model framework [5].

The alpha value was set at a stringent level of 0.01 in this study to reduce the type I error in null assessment so that the SCS CN model will not be unnecessarily calibrated due to wrong null rejection under objective 1. It will also justify the urgent SCS CN model calibration need to the DID for runoff prediction work in Malaysia, review any past studies and projects that used Equation (2) when the null hypothesis is rejected. This study is only willing to accept 1% error chance because these DID processes are too costly to initiate by mistake.

According to the U.S. Geological Survey (USGS) Statistical methods in water resources guide, the minimum required sample size is 100 to be considered as a large dataset for water resources related study at the 0.01 alpha level [26]. As such, the DID HP 27 dataset will be sufficient for this study. Given the 227 rainfall-runoff (P-Q) data pairs from DID HP 27, corresponding λ and S values can be calculated. These 227 λ and S values will be bootstrapped independently with the Bias Corrected and Accelerated (BCa) procedure by using the IBM Predictive Analytics software (PASW) version 18.0 (commonly known as SPSS) [27]. The method neither assumes data normality nor has limitation to certain data distribution and performs random sampling with replacement in SPSS [27,28]. In this study, the Mersenne Twister seed number for random sampling generation was set at 2 million (by default) and 10 million to conduct 2000, 5000, and 10,000 sampling for the calculated λ and S dataset.

Consequently, the BCa option in SPSS was used to generate a sampling distribution and 99% confidence interval (CI) to optimize the parameter of interest such as S and λ. Additionally, it provides standard error statistics and CI for the median value, which are unavailable under most parametric tests in SPSS [27]. BCa procedure was chosen by this study for its ability to correct for skewness and bias in the bootstrap distribution [29]. When the dataset has a high positive skewness, BCa can also correct the issue that the bootstrap CI range might be too small [26]. BCa 99% CI has wider range than the 95% CI. Therefore, this study used BCa option in SPSS to generate 99% CI (instead of 95% CI) for both λ and S dataset so that the assessment of the initial claim from SCS that λ = 0.2 can be inferred from the wider BCa CI.

2.2. Supervised Numerical Optimization Analyses

Past researchers faced the dilemma of choosing between the mean and median of a dataset [2,30]. To address this issue, this study utilized an algorithm of numerical analysis guided by inferential statistics for decision making.

λ and S were optimized using Equation (1) with a supervised numerical analyses approach. To prevent the optimization algorithm from focusing on residual sum of squares (RSS) minimization only, the overall model bias (BIAS) will be minimized near to the value of zero concurrently during the parameter optimization process. This acts as a check with the BCa technique to ensure that the optimized λ and S value are not biased towards the dataset during the SCS model calibration. In the event of skewed data nature, the supervised numerical optimization would be conducted to search for an optimum value within the BCa median's confidence interval limits of the derived λ and S dataset, respectively. The optimized S value and its confidence interval range will lead to the calculation of CN value to represent the entire DID HP 27 dataset in Peninsula Malaysia (see Section 3.2).

2.3. Null Hypotheses Assessments with Inferential Statistics

A Null hypothesis was set up to assess the 1954 SCS proposal with inferential statistics as below:

H_0: $I_a = \lambda S$ where λ must be 0.2 in Equation (1) (as proposed by SCS) to model runoff conditions according to the DID HP 27 dataset in Peninsula Malaysia.

H_0 assesses the validity of Equation (2) for this study as pertained to the DID HP 27 dataset. The assessment of H_0 will be inferred from the BCa confidence interval of λ [28]. The rejection of H_0 indicates that the SCS CN model (Equation (2)) is invalid to model the dataset of this study. It requires the acceptance of H_0 to adopt Equation (2) for rainfall-runoff modelling while the rejection of H_0 will pave a way to derive a new λ value for the DID HP 27 dataset. The optimized λ and S values will be used to formulate a new calibrated runoff prediction model for Peninsula Malaysia. SCS practitioners are encouraged to validate the existing SCS CN model (Equation (2)) prior to runoff modelling adoption.

2.4. The S General Formula

Equation (1) was re-arranged into a general form of $S_\lambda = f(P, Q, \lambda)$ in a previous study [4]. When $\lambda = 0.2$, the corresponding $S_{0.2}$ value leads to the derivation of conventional CN values in use by SCS practitioners. Any other λ values will result in S_λ leading to the derivation of CN_λ values which are different from the SCS tabulated CN values. The general S_λ formula (see [4] for derivation steps) used by this study is:

$$S_\lambda = \frac{\left[P - \frac{(\lambda-1)Q}{2\lambda}\right] - \sqrt{PQ - P^2 + \left[P - \frac{(\lambda-1)Q}{2\lambda}\right]^2}}{\lambda} \tag{6}$$

S_λ = Total abstraction amount of any λ value (mm).

2.5. Correlation Between S_λ and $S_{0.2}$

According to previous researchers, when the optimum λ value is different from the conventional value where $\lambda = 0.2$, a correlation between the newfound λ value and 0.2 must be used in order to calculate the curve number again [2,3,20]. US researchers termed the batch of curve numbers derived from any λ value other than $\lambda = 0.2$ as "conjugate curve numbers" denoted by CN_λ which are different from the SCS tabulated curve numbers [2–4,20]. Given the P-Q dataset, S_λ and $S_{0.2}$ can be calculated using Equation (6). A correlation between the S_λ and $S_{0.2}$ dataset must be established before the calculation of conventional CN value (see Section 3.2). SCS practitioners must use the correlation equation between the S_λ and $S_{0.2}$ to calculate the conventional CN value to avoid the mistake of using conjugate curve number in their study.

2.6. The 3D Runoff Difference Model

Using P-Q datasets from multiple watersheds or from multiple locations within a watershed, a 3D runoff difference model can be created as a collective visual representation of multiple rainfall depths to compare with different $CN_{0.2}$ scenarios. If Equation (2) fails the Null assessment, this 3D model can reflect the runoff difference between it and the new calibrated runoff model for further analyses. The model will be a guide to visualize the runoff under and over prediction zones between two models. In 1954, SCS correlated S and CN. The SI unit version of the formula is:

$$S = \frac{25,400}{CN} - 254 \tag{7}$$

Equation (7) was derived from the SCS assumption where $\lambda = 0.2$, and therefore it will be more appropriate to denote CN as $CN_{0.2}$ and S with $S_{0.2}$. Substituting Equation (7) into Equation (2), the SCS model can be simplified to become: $Q_{0.2} = f(P, CN_{0.2})$ and represented in SI form of:

$$Q_{0.2} = \frac{\left[P - 50.8\left(\frac{100}{CN_{0.2}} - 1\right)\right]^2}{\left[P + 203.2\left(\frac{100}{CN_{0.2}} - 1\right)\right]} \tag{8}$$

$Q_{0.2}$ = Runoff depth (mm) of $\lambda = 0.2$
where $P > 0.2\, S_{0.2}$ else $Q_{0.2} = 0$.

The general form of Equation (1) after the substitution of $I_a = \lambda S$ for any λ value becomes:

$$Q_\lambda = \frac{(P - \lambda S_\lambda)^2}{P - \lambda S_\lambda + S_\lambda} \tag{9}$$

where $P > \lambda S_\lambda$, else $Q_\lambda = 0$. As such, the runoff difference between SCS model (uncalibrated) and the new calibrated runoff model (with new λ) can be quantified as the difference between Equations (8) and (9) as:

$$Q_v = \frac{\left[P - 50.8\left(\frac{100}{CN_{0.2}} - 1\right)\right]^2}{\left[P + 203.2\left(\frac{100}{CN_{0.2}} - 1\right)\right]} - \frac{(P - \lambda S_\lambda)^2}{P - \lambda S_\lambda + S_\lambda} \quad (10)$$

Q_v = Runoff depth prediction difference between 2 runoff models (mm)
$CN_{0.2}$ = the conventional curve number

As Equation (2) was widely adopted in many countries, it is important to assess the runoff prediction difference with Equation (10). It is a general equation that can be used by SCS practitioners to determine the impact of not calibrating Equation (2) for runoff predictions under their study.

In Equation (10), Q_v will be positive when the conventional SCS runoff model (Equation (2)) over-predicted runoff when compared to the calibrated new runoff equation and vice versa. If the newly derived $\lambda < 0.2$, Equation (10) is subject to the constraint where $P > \lambda S$. When the new derived $\lambda > 0.2$, Equation (10) will abide to the constraint of $P > 0.2 S_{0.2}$, else $Q_v = 0$ because there is no runoff difference as I_a of the lower λ value model is yet to be fulfilled to initiate the runoff process [2,5] and produce a runoff difference between two runoff models. All in all, the smaller λ runoff model will initiate runoff ahead of the larger λ runoff model [5].

2.7. Outer Boundary Equation

Equation (2) is subject to a constraint where $P > I_a$ or $P > \lambda S_\lambda$, else $Q_\lambda = 0$. The 3D runoff difference model captures the runoff difference of two different runoff models. When the I_a constraint of the lower λ value model has been fulfilled, runoff will be initiated. Base on this concept, the I_a constraint of the lower λ value model becomes the outer boundary of the 3D runoff difference model which also represents the runoff indifference boundary with the following general equation:

$$P = \lambda S_\lambda \quad (11)$$

2.8. Inner Boundary Equation

The second boundary is the "Inner Boundary" of the 3D runoff difference model. This boundary separates the runoff under-prediction zone from the over-prediction zone of the SCS runoff model. The runoff difference is equal to zero at the crossover boundary, which is also known as the runoff indifference boundary. Therefore, when $Q_v = 0$ (runoff indifference) in Equation (10), the form can be re-expressed as:

$$\frac{\left[P - 50.8\left(\frac{100}{CN_{0.2}} - 1\right)\right]^2}{\left[P + 203.2\left(\frac{100}{CN_{0.2}} - 1\right)\right]} = \frac{(P - \lambda S_\lambda)^2}{P - \lambda S_\lambda + S_\lambda} \quad (12)$$

Equations (11) and (12) are also general equations that can be used by SCS practitioners to analyze the 3D runoff difference model (created with Equation (10)) in their study.

2.9. Models Comparison

Runoff models are compared and benchmarked for their model predictive accuracy in this paper. Model's residual sum of squares (*RSS*), predictive model *BIAS* prediction and model efficiency index (*E*), also known as Nash–Sutcliffe index, were calculated with the following formulae to draw further comparison between them.

$$RSS = \sum_{i=1}^{n} \left(Q_{predicted} - Q_{observed}\right)^2 \quad (13)$$

$$E = 1 - \frac{RSS}{\sum_{i=1}^{n}\left(Q_{predicted} - Q_{mean}\right)^2} \qquad (14)$$

$$BIAS = \frac{\sum_{i=1}^{n}\left(Q_{predicted} - Q_{observed}\right)}{n} \qquad (15)$$

n = Total number of data pairs.

Lower RSS implies a better model. Index E lies on a spectrum of minus 1.0 to 1.0 whereby index value = 1.0 shows an ideal conjectured model. In the instance where $E < 0$, it is inferior to utilizing an average to predict the dataset. $BIAS$ is the overall model prediction error indicator. Zero $BIAS$ value indicates an error free model prediction while negative value indicates the overall predictive model's under-prediction tendency and vice versa.

2.10. Asymptotic Curve Number Fitting

Other than numerical optimization technique, many researchers [31–35] used asymptotic CN fitting method (AFM) to determine the best representative CN for the watershed of interest with P-Q dataset (λ value remains as 0.20 under this method). Therefore, AFM will be used to benchmark against the proposed method in this article. Under AFM, CN cannot be determined for the Complacent behavior watershed, but Standard behavior watershed follows the following formula [33]:

$$CN(P) = CN_\infty + (100 - CN_\infty)e^{(-\frac{P}{k})} \qquad (16)$$

$CN(P)$ = Fitted CN value of a specific rainfall depth
CN_∞ = CN of a watershed of interest
K = Fitting parameter

Violent behavior watershed follows the following formula [33]:

$$CN(P) = CN_\infty\left[1 - e^{-k(P-P_{th})}\right] \qquad (17)$$

P_{th} = Threshold Rainfall depth (mm).

2.11. Critical Rainfall Amount (P_{crit})

The concept of P_{crit} was initially suggested by US researchers [2,20,22] which can only be obtained through numerical analysis solving technique or by trial and error procedure. In their work, optimum λ was reported as 0.05 and the P_{crit} points were identified through the intersection of conjugate $CN_{0.05}$ and $CN_{0.2}$ curve on the graph in their study.

The concept of P_{crit} was built upon the runoff indifference between 2 runoff models. When $Q_v = 0$ (runoff indifference between two runoff models), Equation (10) becomes Equation (12). As such, this study introduces runoff difference curves which was created with numerical analysis technique as the visual presentation of Equation (12). Runoff difference curves can be plotted for specific $CN_{0.2}$ classes across multiple rainfall depth scenarios. Unlike previous research work, it combined two curves into a single curve and identify P_{crit} at where the curve crosses the x-axis.

2.12. The Closed Form Equation of Critical Rainfall Amount (P_{crit})

Through algebraic manipulation, this study successfully rearranged Equation (10) and solved the general closed-form equation of P_{crit} in terms of $CN_{0.2}$ when $Q_v = 0$. The breakthrough has also proven to be able to solve for P_{crit} value precisely of any pairing runoff models and replace the trial and error procedure used by previous researchers [2,20,22]. SCS practitioners can derive the P_{crit} equation for their study with proposed method in this article (see Section 3.10).

2.13. Critical Curve Number (CN_{crit})

With a similar concept (based upon Equation (12)) as the critical rainfall amount (P_{crit}), this study also introduces "critical curve number(s)" (CN_{crit}) to supplement the use of P_{crit}. Under a specific rainfall scenario, critical curve number value(s) can also be identified from the points where $Q_v = 0$ between 2 runoff models. Unlike the success of the P_{crit} closed-form equation derivation, the effort to realize the closed-form equation of CN_{crit} in term of P is still unfruitful to date. Therefore, the numerical analysis technique was applied to estimate CN_{crit} value(s) with visual aid from the runoff difference curves graph. Runoff difference curves methodology as Section 3.9 covered can be adopted to show that Equation (2) or Equation (8) will under-predict runoff amount in any curve number areas below the critical curve number value and vice versa.

2.14. Soft Computing and Data Mining of the 3D Model

In general, Equation (10) represents the runoff prediction errors of Equation (2) under multiple P and $CN_{0.2}$ scenarios but it is difficult to visualize the quantified effect by looking at Equation (10) and solve for the global maxima and minima in order to represent the worst under and over runoff prediction amounts between two runoff models.

Based on the rainfall depth range of the dataset [21], a numerical table can be compiled with Equation (10) through the substitution of different P, $CN_{0.2}$ scenarios and the λ value to quantify runoff depth prediction difference between two runoff models in a table. A 3D model can also be constructed with the collective information from the table (Section 3.7). With the visual aid of a 3D runoff difference model, it is possible to extract all minimum and maximum runoff prediction difference amount and represent them with statistically significant equations. The minimum under-prediction difference amount equation represents the worst under-design case incurred by Equation (2) and vice versa.

3. Results and Discussion

3.1. The Reverse Derivation of Optimum λ and S for Peninsula Malaysia

In all, 227 λ and S values were calculated according to corresponding rainfall-runoff (P-Q) data pairs. The calculated λ dataset was checked for normality in SPSS with Kolmogorov–Smirnov and Shapiro–Wilk test statistics, both tests concluded the λ dataset to be non-normal ($p < 0.001$). Nearly 95% (214 out of the 227) storm events calculated λ value below 0.2 while none was equal to 0.2 as proposed by SCS.

According to Section 2.1, as defined by the SCS [5], the "collective representative I_a" was reduced to 5.9 mm to fulfil both constraints of Equations (4) and (5) for the entire dataset of DID HP 27 [21]. 227 calculated λ and S values were independently used for 2000, 5000, and 10,000 random samplings prior to CI generations and cross checking (This study found that the CI upper and lower limits only differ at the fourth decimal places with 2000, 5000, and 10,000 random samplings while there were no difference between the use of 2 million (by default) and 10 million Mersenne Twister seed numbers for random sampling generation) in SPSS. The inferential statistics of the derived λ and S values are tabulated in Tables 1 and 2.

Table 1. Inferential Statistics of the derived λ dataset from Malaysian Department of Irrigation and Drainage (DID) Hydrological Procedure (HP) 27.

λ	Statistics	Bootstrap, BCa 99%			
		Bias	Std. Error	Confidence Interval	
				Lower	Upper
Skewness	5.125				
Kurtosis	36.456				
Mean	0.071	−0.00006	0.006	0.056	0.089
Median	0.042	0.00023	0.003	0.034	0.051

From Table 1, neither the mean nor the median BCa λ's 99% CI include the λ value of 0.2 (In comparison, the BCa 95% mean and the median CI for λ span across smaller range (0.036, 0.084)). Therefore, H_0 can be rejected at alpha = 0.01 level. As such, Equation (2) is statistically insignificant (not even significant at alpha = 0.05) and cannot be used to predict runoff conditions in this study. λ dataset is skewed (skewness of 5.125 in Table 1) thus the search of the optimum collective representative λ value via numerical optimization technique focusses on median λ's confidence interval [0.034, 0.051].

On the other hand, data distribution of the S dataset is somewhat skewed with a skewness of 1.624 (Table 2). The definition of skewness is non-uniform, some guidelines suggested skewness value less than 3.0 to be considered as normal while some set a more stringent limit at 1.0. To avoid the ambiguity of skewness determination, the search of the optimum S value was widened to include the lowest and the highest confidence interval limit of both mean and median values (118.125, 196.332) on S [2,30].

Table 2. Inferential Statistics of derived S dataset from DID HP 27.

S	Statistics	Bootstrap, BCa 99%			
		Bias	Std. Error	Confidence Interval	
				Lower	Upper
Skewness	1.624				
Kurtosis	4.392				
Mean	172.297	0.002	8.649	150.952	196.332
Median	141.54	−0.053	10.005	118.125	170.170

The optimum λ value was recognized as 0.051 (rounded) while 150.46 mm was the optimum S value in formulating the best runoff predictive model (based on Equation (1)) according to the entire dataset of DID HP 27 with an overall predictive model's *BIAS* near to zero. The collective representation of the I_a for the entire dataset was found from the product of the optimum λ and S and therefore, the best collective representative value of I_a to model the entire dataset in Peninsula Malaysia is 8.3 mm from this study.

As mentioned in Section 2.1 and 2.2, BCa technique produced confidence intervals (Tables 1 and 2) for the optimization of λ and S value to calibrate the SCS CN model. It also generated a range of λ and S value to enable the calculation of multiple I_a and CN values which is in line with the latest research development in this area [23–25]. Other than the best collective representative I_a value, SCS practitioners who use the proposed method in this article have an option to compare other possible I_a values with other research results in future.

3.2. The Correlation between S_λ and $S_{0.2}$ for Peninsula Malaysia

The derivation of S_λ formula (Equation (6)) proved mathematically that even with the same P-Q dataset, as λ varies, the corresponding total abstraction amount (S) varies as well and therefore, the corresponding CN value will change also. As such, it is more appropriate to re-represent Equation (7) in general form as:

$$CN_\lambda = \frac{25,400}{S_\lambda + 254} \qquad (18)$$

CN_λ = Curve number of any λ value (dimensionless)
S_λ = Total abstraction amount of any λ value (mm)

Given the P-Q dataset and λ value, the corresponding CN_λ can be derived from Equation (18). When λ = 0.2, its corresponding $S_{0.2}$ value gives rise to deriving the conventional curve number compiled by SCS. To differentiate the conventional SCS CN, the notation of "$CN_{0.2}$" is used in the remaining of this paper. When $\lambda \neq 0.2$, its corresponding S_λ value derives "Conjugate Curve Number" (CN_λ) [2,20,22]. As the optimum λ value = 0.051, the correlation between $S_{0.051}$ and $S_{0.2}$ was identified with SPSS for this study as:

$$S_{0.051} = 1.176 S_{0.2}^{1.063} \qquad (19)$$

$S_{0.051}$ = Total abstraction amount (mm) of λ = 0.051
$S_{0.2}$ = Total abstraction amount (mm) of λ = 0.2

Equation (19) has a R^2-adj of 0.946, standard error of 0.15 and $p < 0.001$. Equation (19) is also the key to convert $S_{0.051}$ back to its equivalent $S_{0.2}$ value for the calculation of $CN_{0.2}$ for SCS practitioners. The optimum $S_{0.051}$ is 150.46 mm (alpha = 0.01) from the range of 118.125 to 196.332 (Table 2) in Section 3.1. The equivalent $S_{0.2}$ value of $S_{0.051}$ = 150.46 mm is 95.97 mm (calculated from Equation (19)). By substituting $S_{0.2}$ = 95.97 mm into Equation (18), $CN_{0.2}$ = 72.58; thus, new λ of 0.051 derives an equivalent $CN_{0.2}$ value of 72.58 to model the entire DID HP 27 dataset. The 99% confidence interval of $S_{0.051}$ ranges from 118.125 to 196.332, those values can also be used to calculate its equivalent upper and lower $CN_{0.2}$ limits in the same manner through Equation (18) and therefore, for the DID HP 27 dataset [21], the best collective $CN_{0.2}$ = 72.58 (99% CI ranges from 67 to 76) for runoff predictions in Peninsula Malaysia.

3.3. Conjugate Curve Numbers (CN_λ) for Peninsula Malaysia

Given the P-Q data pairs from DID HP 27, conjugate curve number values (CN_λ) of each storm event can be calculated with aforementioned equations in the following steps:
Since the optimum λ value obtained was 0.051, Equation (18) becomes:

$$CN_{0.051} = \frac{25,400}{S_{0.051} + 254}$$

Substitute Equation (19) into Equation (18) will yield:

$$CN_{0.051} = \frac{25,400}{(1.176 S_{0.2}^{1.063}) + 254} \qquad (20)$$

where $S_{0.2}$ values can be calculated using Equation (6) (the S general formula) when P-Q data pairs are given. $CN_{0.051}$ is the conjugate curve number of $CN_{0.2}$. Equation (20) proves that conjugate curve number (CN_λ) is not the same as the conventional curve number $CN_{0.2}$ which was derived using Equation (7). Thus, it is inappropriate to use any conjugate curve number (CN_λ) with Equation (2) in any rainfall-runoff modelling work.

3.4. The 3D Runoff Difference Model for Peninsula Malaysia

According to the discussions from Sections 2.4 and 2.5, the S amount is specific to its corresponding λ value. The optimum λ value = 0.051 to model runoff conditions for the DID HP 27 dataset thus by substituting λ with 0.051 into Equation (9) yields a calibrated rainfall-runoff predictive model on Equation (1) in the form of:

$$Q_{0.051} = \frac{(P - 0.051 S_{0.051})^2}{P - 0.051 S_{0.051} + S_{0.051}}$$

The substitution of Equations (19) and (7) further simplifies it as:

$$Q_{0.051} = \frac{\left[P - 21.606 \left(\frac{100}{CN_{0.2}} - 1\right)^{1.063}\right]^2}{\left[P + 402.547 \left(\frac{100}{CN_{0.2}} - 1\right)^{1.063}\right]} \qquad (21)$$

Equation (21) re-expressed the runoff model in term of P and $CN_{0.2}$ and subjects to the constraint.

$P > 21.606 \left(\frac{100}{CN_{0.2}} - 1 \right)^{1.063}$ else $Q_v = 0$ on the 3D model

$CN_{0.2}$ = Conventional SCS tabulated curve number

$Q_{0.051}$ = Runoff depth (mm) of $\lambda = 0.051$

Equation (8) is the re-expression of Equation (2) in term of P and $CN_{0.2}$.

$$Q_{0.2} = \frac{\left[P - 50.8 \left(\frac{100}{CN_{0.2}} - 1 \right)\right]^2}{\left[P + 203.2 \left(\frac{100}{CN_{0.2}} - 1 \right)\right]} \qquad (22)$$

It subjects to the constraint $P > 50.8 \left(\frac{100}{CN_{0.2}} - 1 \right)$ else $Q_v = 0$.

Equation (8) or Equation (2) represents the un-calibrated SCS CN model. The runoff depth prediction differences between Equations (8) and (21) were collectively quantified by Equation (22) of which the 3D runoff difference model (Section 3.7 and Figure 2) was constructed with. Equation (22) also quantifies type II errors from Equation (2) (existing SCS model) if it is not calibrated for runoff prediction in Peninsula Malaysia.

3.5. Outer Boundary Equation

As per Section 2.7, the calibrated new λ value (0.051) is less than 0.2; thus, its model's constraint can be adopted to represent the runoff indifference boundary where runoff has not been initiated. Therefore, Equation (22) is also subject to the constraint, $P > 0.051 S_{0.051}$ or $P > 21.606 \left(\frac{100}{CN_{0.2}} - 1 \right)^{1.063}$ else $Q_v = 0$. Equation (19) can be substituted into 11 to preserve the conventional curve number ($CN_{0.2}$) through following the steps.

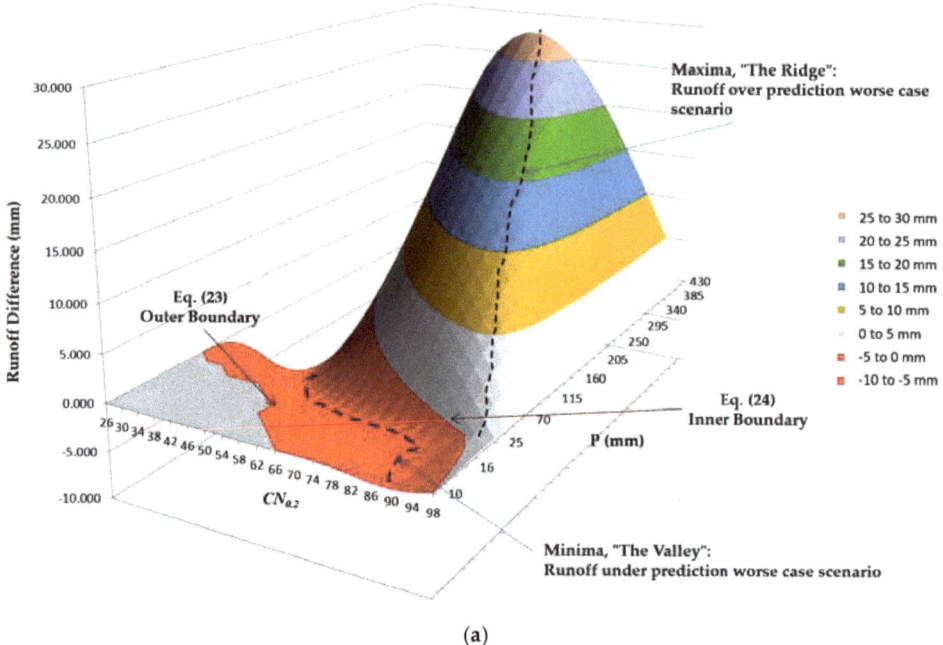

(a)

Figure 2. *Cont.*

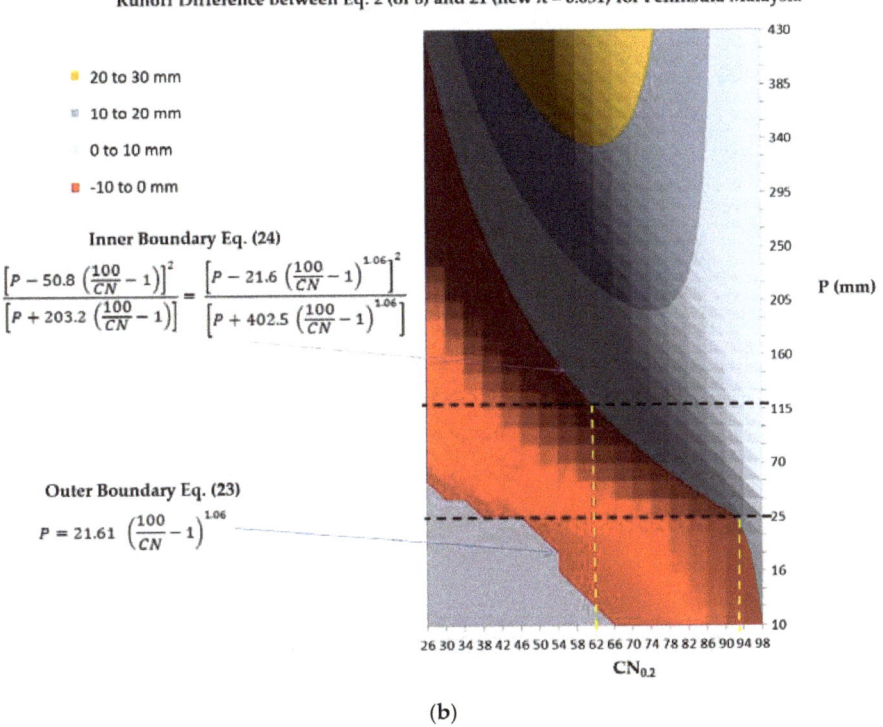

Figure 2. (**a**) The 3D runoff difference model (between Equations (2) and (21)) of Peninsula Malaysia with DID HP 27 dataset for Type II error assessment. (**b**) Top view of the 3D runoff difference model for Peninsula Malaysia with DID HP 27 dataset.

Substitute λ with 0.051, Equations (7) and (19) into Equation (11) yields:

$$P = 21.606 \left(\frac{100}{CN_{0.2}} - 1 \right)^{1.063} \qquad (23)$$

Equation (23) is the runoff indifference boundary equation between two runoff models. It is otherwise recognized as the "Outer Boundary" equation of the 3D runoff difference model (Figure 2a,b).

3.6. Inner Boundary Equation

When $Q_v = 0$ in Equation (22), the form can be expressed as:

$$\frac{\left[P - 50.8 \left(\frac{100}{CN_{0.2}} - 1 \right) \right]^2}{\left[P + 203.2 \left(\frac{100}{CN_{0.2}} - 1 \right) \right]} = \frac{\left[P - 21.606 \left(\frac{100}{CN_{0.2}} - 1 \right)^{1.063} \right]^2}{\left[P + 402.547 \left(\frac{100}{CN_{0.2}} - 1 \right)^{1.063} \right]} \qquad (24)$$

Equation (24) is also known as the "Inner Boundary" equation of the 3D runoff difference model for Peninsula Malaysia that demarcates the runoff under-prediction and over-prediction zones between two runoff models in this study.

3.7. The Construction of the 3D Runoff Difference Model

DID HP 27 dataset consist of 227 storm events ranging from 19 mm to 420 mm. In order to analyze and quantify the runoff prediction depth difference between Equation (2) (or Equation (8)) and 21 under multiple rainfall and $CN_{0.2}$ scenarios, rainfall depth (P) ranging from 10 mm to 430 mm across different $CN_{0.2}$ values (from 26 to 98) were entered into Equation (22) to calculate the runoff depth prediction difference that can be found in Figure 3. Those tabulated values are runoff prediction errors (or type II errors) from Equation (2) which are in line with previous studies that reported more profound error in forested watersheds represented by $CN_{0.2}$ values < 60 [2,20,22] Similarly, for Peninsula Malaysia, both runoff under and over prediction errors worsen when the value of CN0.2 reduces (Figure 3).

P (mm) \ CN	26	30	34	38	42	46	50	54	58	62	67	72	73	76	82	86	90	94	98
10												-0.028	-0.042	-0.105	-0.359	-0.620	-0.698	-0.488	0.001
12											-0.017	-0.105	-0.134	-0.246	-0.629	-0.808	-0.762	-0.429	0.068
14										0.005	0.072	-0.230	-0.275	-0.444	-0.859	-0.933	-0.773	-0.344	0.128
16									-0.002	0.039	-0.165	-0.401	-0.464	-0.696	-1.033	-1.007	-0.745	-0.244	0.181
18								-0.024	-0.104	-0.296	-0.615	-0.699	-0.952	-1.159	-1.039	-0.687	-0.134	0.228
20							0.009	-0.072	-0.198	-0.461	-0.872	-0.963	-1.165	-1.245	-1.036	-0.606	-0.021	0.269
25						0.027	-0.127	-0.302	-0.561	-1.025	-1.429	-1.474	-1.533	-1.313	-0.913	-0.340	0.264	0.354
40			-0.019	-0.151	-0.400	-0.766	-1.252	-1.816	-2.167	-2.275	-2.050	-1.968	-1.654	-0.757	-0.034	0.647	1.008	0.511
55		-0.003	-0.146	-0.475	-0.948	-1.600	-2.371	-3.029	-3.492	-2.422	-1.685	-1.509	-0.938	0.305	1.072	1.589	1.558	0.597
70	-0.015	-0.267	-0.781	-1.516	-2.456	-3.252	-3.609	-3.613	-3.328	-1.896	-0.788	-0.553	0.162	1.499	2.149	2.391	1.965	0.652
85	-0.285	-0.942	-1.894	-3.094	-3.951	-4.284	-4.187	-3.764	-3.087	-2.719	-0.957	0.387	0.654	1.423	2.680	3.127	3.060	2.275	0.689
100	-0.890	-2.011	-3.458	-4.533	-4.910	-4.772	-4.249	-3.440	-2.427	-1.281	0.231	1.698	1.972	2.728	3.793	3.996	3.619	2.517	0.716
115	-1.817	-3.457	-4.937	-5.497	-5.395	-4.819	-3.904	-2.757	-1.466	-0.110	1.561	3.060	3.324	4.019	4.821	4.763	4.090	2.711	0.737
130	-3.056	-5.048	-6.000	-6.053	-5.491	-4.506	-3.241	-1.805	-0.292	1.214	2.966	4.423	4.665	5.263	5.761	5.440	4.491	2.871	0.753
145	-4.596	-6.292	-6.697	-6.257	-5.260	-3.902	-2.328	-0.651	1.031	2.633	4.400	5.760	5.969	6.448	6.618	6.040	4.835	3.003	0.766
160	-6.106	-7.204	-7.071	-6.159	-4.756	-3.060	-1.219	0.652	2.455	4.106	5.834	7.054	7.224	7.566	7.398	6.573	5.134	3.115	0.777
175	-7.314	-7.817	-7.162	-5.800	-4.021	-2.024	0.042	2.063	3.945	5.603	7.249	8.296	8.422	8.618	8.110	7.049	5.395	3.211	0.786
190	-8.245	-8.162	-7.002	-5.216	-3.093	-0.833	1.420	3.552	5.473	7.104	8.632	9.482	9.561	9.604	8.760	7.476	5.626	3.295	0.793
205	-8.922	-8.264	-6.620	-4.438	-2.005	0.484	2.886	5.094	7.021	8.595	9.975	10.611	10.641	10.528	9.356	7.862	5.831	3.367	0.800
220	-9.366	-8.149	-6.043	-3.493	-0.781	1.902	4.419	6.669	8.572	10.064	11.274	11.683	11.663	11.395	9.902	8.211	6.014	3.431	0.806
235	-9.597	-7.837	-5.293	-2.404	0.555	3.399	6.000	8.263	10.116	11.505	12.527	12.701	12.631	12.207	10.405	8.529	6.178	3.488	0.811
250	-9.631	-7.348	-4.391	-1.192	1.984	4.959	7.614	9.863	11.644	12.913	13.733	13.667	13.546	12.968	10.870	8.819	6.327	3.539	0.815
265	-9.484	-6.699	-3.355	0.126	3.489	6.566	9.248	11.460	13.150	14.284	14.891	14.582	14.411	13.684	11.300	9.084	6.461	3.585	0.819
280	-9.173	-5.906	-2.201	1.535	5.056	8.208	10.893	13.047	14.629	15.617	16.003	15.451	15.231	14.356	11.698	9.329	6.584	3.626	0.823
295	-8.709	-4.983	-0.943	3.020	6.673	9.876	12.541	14.619	16.079	16.911	17.070	16.276	16.007	14.989	12.069	9.554	6.697	3.664	0.826
310	-8.105	-3.944	0.407	4.570	8.330	11.560	14.186	16.170	17.497	18.165	18.093	17.059	16.743	15.585	12.414	9.763	6.800	3.698	0.829
325	-7.374	-2.799	1.837	6.175	10.018	13.253	15.822	17.699	18.881	19.379	19.075	17.803	17.440	16.148	12.737	9.956	6.895	3.729	0.831
340	-6.524	-1.560	3.336	7.826	11.729	14.949	17.444	19.202	20.231	20.554	20.016	18.511	18.103	16.680	13.039	10.136	6.983	3.758	0.834
355	-5.567	-0.236	4.896	9.514	13.456	16.644	19.050	20.677	21.546	21.690	20.919	19.185	18.732	17.183	13.322	10.304	7.065	3.784	0.836
370	-4.510	1.164	6.510	11.233	15.194	18.332	20.637	22.123	22.826	22.790	21.786	19.827	19.331	17.659	13.588	10.461	7.141	3.809	0.838
385	-3.362	2.633	8.168	12.976	16.937	20.011	22.202	23.540	24.072	23.853	22.618	20.438	19.902	18.111	13.839	10.608	7.212	3.832	0.840
400	-2.131	4.163	9.866	14.739	18.683	21.678	23.744	24.927	25.284	24.881	23.417	21.022	20.445	18.540	14.075	10.746	7.278	3.854	0.842
415	-0.823	5.747	11.597	16.515	20.426	23.329	25.262	26.283	26.462	25.875	24.184	21.580	20.963	18.948	14.298	10.875	7.340	3.873	0.843
430	0.555	7.379	13.356	18.301	22.164	24.964	26.755	27.609	27.607	26.836	24.922	22.113	21.458	19.336	14.508	10.997	7.398	3.892	0.845

Figure 3. Runoff differences generated from Equation (22) for various rainfall (P) and Curve Number ($CN_{0.2}$) scenarios. Note: 1 mm = 1 million liters runoff volume in a 1 km^2 area.

Red zone cells in Figure 3 are where Equation (2) under-predicted runoff amount against Equation (21). On the other hand, the white zone cells are where Equation (2) over-predicted runoff amount. The empty cells on the upper left corner of the figure are where I_a has not been fulfilled yet to initiate any runoff amount. Collectively, Figure 3 can also be presented as a 3D model as seen in Figure 2a,b. Equations (23) and (24) represent boundary lines as indicated on the 3D model, respectively. SCS practitioners can refer to Figure 3 to perform runoff prediction correction on Equation (2).

For areas in Peninsula Malaysia with $CN_{0.2}$ value from 67 to 76 (marked by the dash line), the existing SCS model underpredicts runoff amount as indicated in red zone when rainfall depth of a storm is < 70 or 85 mm. SCS model tends to overpredict runoff amount after 85 mm and its overprediction tendency worsens toward larger storm events as indicated in white zone. Without model calibration, the SCS model worst runoff underprediction within these areas happens at $CN_{0.2} = 67$ area at rainfall depth of 55 mm, the model underpredicted runoff amount by 2.4 million liters in 1 km^2 area while it nearly overpredicted runoff amount by 25 million liters when the storm depth reaches 430 mm in Peninsula Malaysia. Blind adoption of the existing SCS CN model is likely to over-predict runoff amount when the rainfall depth of a storm event is larger than 85 mm in Peninsula Malaysia. As such, any past study or engineering projects based upon the return period concept of rainfall amount below 70 mm might be under-designed.

3.8. Soft Computing and Data Mining of the 3D Runoff Difference Model

Even though the 3D runoff difference model can be expressed using the closed form Equation (22), it is not easy to obtain the minimum (global minima) or maximum (global maxima) runoff depth difference equations. However, with the 3D runoff difference model as a visual aid accompanied by soft computing techniques, the data mining of this vital information becomes attainable.

The minimum and maximum runoff depth prediction errors across multiple P and $CN_{0.2}$ scenarios between the two runoff models can be extracted from Figure 3. The statistically significant equations can then be determined using the SPSS to formulate the worst under and over-estimated runoff prediction error equations from Equation (2) or Equation (8) against Equation (21).

The data mining process extracts all the minimum and maximum runoff prediction differences (bold numbers, highlighted in red and yellow color, respectively in Figure 4) according to each rainfall depth scenarios (in row).

Figure 4. Soft computing, data mining of minimum and maximum runoff depth difference of each rainfall class (in row). Note: 1 mm = 1 million liters runoff volume in a 1 km² area.

Two statistically significant and best correlation equations were identified through SPSS regression modelling as:

$$\text{Min } Q_v = 5.14 \times 10^{-5} \, P^2 - 0.052 \, P - 0.222 \tag{25}$$

$$\text{Max } Q_v = 5.14 \times 10^{-5} \, P^2 + 0.045 \, P - 0.734 \tag{26}$$

where Min Q_v represents worse under-predicted runoff scenarios while Max Q_v represents the maximum over-predicted runoff scenarios. Equation (25) has an R^2-adj of 0.999, standard error of 0.037 and $p < 0.001$ while Equation (26) has an R^2-adj of 0.999, standard error of 0.191 and $p < 0.001$. Given a specific rainfall depth, the worst under-estimated and over-estimated runoff prediction errors of Equation (2) or Equation (8) due to a specific rainfall depth can be estimated by Equations (25) and (26), respectively.

It is also possible to employ soft computing technique to derive similar runoff prediction error equations in term of curve number. From Figure 5, the minimum and maximum runoff prediction differences can be extracted as per their respective curve number (in column) which induced the runoff difference (bold numbers, highlighted in red and yellow color, respectively in Figure 5).

Two statistically significant and best correlation equations from SPSS regression modelling results are:

$$\text{Min } Q_v = 2.594 - (329.896/CN_{0.2}) \qquad (27)$$

$$\text{Max } Q_v = 2.2 \times 10^{-4} CN_{0.2}{}^3 - 0.061 CN_{0.2}{}^2 + 4.77 CN_{0.2} - 86.519 \qquad (28)$$

where Min Q_v, Max Q_v and $CN_{0.2}$ have been defined earlier. Equation (27) has an R^2-adj of 0.992, standard error of 0.242 and $p < 0.001$ while Equation (28) has an R^2-adj of 0.999, standard error of 0.255 and $p < 0.001$. Given a specific curve number, the worst underestimated and over-estimated runoff prediction errors of Equation (2) or Equation (8) due to a specific $CN_{0.2}$ area can be estimated with Equations (27) and (28), respectively.

P (mm) \ CN	26	30	34	38	42	46	50	54	58	62	67	72	73	76	82	86	90	94	98
10												-0.028	-0.042	-0.105	-0.359	-0.620	-0.698	-0.488	0.001
12											-0.017	-0.105	-0.134	-0.246	-0.629	-0.808	-0.762	-0.429	0.068
14											-0.072	-0.230	-0.275	-0.444	-0.859	-0.933	-0.773	-0.344	0.128
16									-0.005	-0.072	-0.165	-0.401	-0.464	-0.696	-1.033	-1.007	-0.745	-0.244	0.181
18								-0.002	-0.039	-0.165	-0.296	-0.615	-0.699	-0.952	-1.159	-1.039	-0.687	-0.134	0.228
20							-0.009	-0.024	-0.104	-0.198	-0.461	-0.872	-0.963	-1.165	-1.245	-1.036	-0.606	-0.021	0.269
25							-0.027	-0.127	-0.302	-0.561	-1.025	-1.429	-1.474	-1.533	-1.313	-0.913	-0.340	0.264	0.354
40				0.019	0.151	-0.400	-0.766	-1.252	-1.816	-2.167	-2.375	-2.050	-1.968	-1.654	-0.757	-0.034	0.647	1.008	0.511
55		-0.003	-0.146	-0.475	-0.970	-1.623	-2.371	-2.829	-2.982	-2.873	-2.422	-1.685	-1.509	-0.938	0.305	1.072	1.589	1.558	0.597
70	-0.015	-0.267	-0.781	-1.516	-2.456	-3.252	-3.609	-3.613	-3.328	-2.806	-1.896	-0.788	-0.553	0.162	1.499	2.149	2.391	1.965	0.652
85	-0.285	-0.942	-1.894	-3.094	-3.961	-4.284	-4.187	-3.764	-3.087	-2.219	-0.957	0.387	0.654	1.423	2.680	3.127	3.060	2.275	0.689
100	-0.890	-2.011	-3.458	-4.533	-4.910	-4.772	-4.249	-3.440	-2.427	-1.281	0.231	1.698	1.972	2.728	3.793	3.996	3.619	2.517	0.716
115	-1.817	-3.457	-4.937	-5.497	-5.395	-4.819	-3.904	-2.757	-1.466	-0.110	1.561	3.060	3.324	4.019	4.821	4.763	4.090	2.711	0.737
130	-3.056	-5.048	-6.000	-6.053	-5.491	-4.506	-3.241	-1.805	-0.292	1.214	2.966	4.423	4.665	5.263	5.761	5.440	4.491	2.871	0.753
145	-4.596	-6.292	-6.697	-6.257	-5.260	-3.902	-2.328	-0.651	1.031	2.633	4.400	5.760	5.969	6.448	6.618	6.040	4.835	3.003	0.766
160	-6.106	-7.204	-7.071	-6.159	-4.756	-3.060	-1.219	0.652	2.455	4.106	5.834	7.054	7.224	7.566	7.398	6.573	5.134	3.115	0.777
175	-7.314	-7.813	-7.162	-5.800	-4.021	-2.024	0.042	2.063	3.945	5.603	7.249	8.296	8.422	8.618	8.110	7.049	5.395	3.211	0.786
190	-8.245	-8.162	-7.002	-5.216	-3.093	-0.833	1.420	3.552	5.473	7.104	8.632	9.482	9.561	9.604	8.760	7.476	5.626	3.295	0.793
205	-8.922	-8.264	-6.620	-4.438	-2.005	0.484	2.886	5.094	7.021	8.595	9.975	10.611	10.641	10.528	9.356	7.862	5.831	3.367	0.800
220	-9.366	-8.149	-6.043	-3.493	-0.781	1.902	4.419	6.669	8.572	10.064	11.274	11.683	11.663	11.395	9.902	8.211	6.016	3.431	0.806
235	-9.597	-7.837	-5.293	-2.404	0.555	3.399	6.000	8.263	10.116	11.505	12.527	12.701	12.631	12.207	10.405	8.529	6.178	3.488	0.811
250	-9.631	-7.348	-4.391	-1.192	1.984	4.959	7.614	9.863	11.644	12.913	13.733	13.667	13.546	12.968	10.870	8.819	6.327	3.539	0.815
265	-9.484	-6.699	-3.355	0.126	3.489	6.566	9.248	11.460	13.150	14.284	14.891	14.582	14.411	13.684	11.300	9.084	6.461	3.585	0.819
280	-9.173	-5.906	-2.201	1.535	5.056	8.208	10.893	13.047	14.629	15.617	16.003	15.451	15.231	14.356	11.698	9.329	6.584	3.626	0.823
295	-8.709	-4.983	-0.943	3.020	6.673	9.876	12.541	14.619	16.079	16.911	17.070	16.276	16.007	14.989	12.069	9.554	6.697	3.664	0.826
310	-8.105	-3.944	0.407	4.570	8.330	11.560	14.186	16.170	17.497	18.165	18.093	17.059	16.743	15.585	12.414	9.763	6.800	3.698	0.829
325	-7.374	-2.799	1.837	6.175	10.018	13.253	15.822	17.699	18.881	19.379	19.075	17.803	17.440	16.148	12.737	9.956	6.895	3.729	0.831
340	-6.524	-1.560	3.336	7.826	11.729	14.949	17.444	19.202	20.231	20.554	20.016	18.511	18.103	16.680	13.039	10.136	6.983	3.758	0.834
355	-5.567	-0.236	4.896	9.514	13.456	16.644	19.050	20.677	21.546	21.690	20.919	19.185	18.732	17.183	13.322	10.304	7.065	3.785	0.836
370	-4.510	1.164	6.510	11.233	15.194	18.332	20.637	22.123	22.826	22.790	21.786	19.827	19.331	17.659	13.588	10.461	7.141	3.809	0.838
385	-3.362	2.633	8.168	12.976	16.937	20.011	22.202	23.540	24.072	23.853	22.618	20.438	19.902	18.111	13.839	10.608	7.212	3.832	0.840
400	-2.131	4.163	9.866	14.739	18.683	21.678	23.744	24.927	25.284	24.881	23.417	21.022	20.445	18.540	14.075	10.746	7.278	3.854	0.842
415	-0.823	5.747	11.597	16.515	20.426	23.329	25.262	26.283	26.462	25.875	24.184	21.580	20.963	18.948	14.298	10.875	7.340	3.873	0.843
430	0.555	7.379	13.356	18.301	22.164	24.964	26.755	27.609	27.607	26.836	24.922	22.113	21.458	19.336	14.508	10.997	7.398	3.892	0.845

Figure 5. Soft computing, data mining of minimum and maximum runoff depth difference of each $CN_{0.2}$ class (in column). Note: 1 mm = 1 million liters runoff volume in a 1 km² area.

The dash line on the 3D model in the valley of the red zone is described by Equations (25) and (27) while Equations (26) and (28) represent the dash line found on the ridge of the 3D runoff difference model (see Figure 2a). SCS practitioners can adopt Equations (25)–(28) to estimate the worst-case runoff prediction errors of Equation (2) when compared to the newly found λ (0.051) model in Peninsula Malaysia. On the other hand, regional or watershed specific equations can also be established by SCS practitioners for their study as proposed.

3.9. Runoff Difference Curves of the Critical Rainfall Amount

This study introduced runoff difference curves which were created with numerical analysis technique to visually present Equation (22) and to identify P_{crit}. Runoff difference curves graph combines two runoff curves (of conjugate curve numbers) into a single runoff difference curve to represent the concept of 2 previous studies [2,20,22] in another view. The graph can be plotted for specific $CN_{0.2}$ classes across multiple rainfall depth scenarios to show P_{crit} at where the curve crosses x-axis (Figure 6).

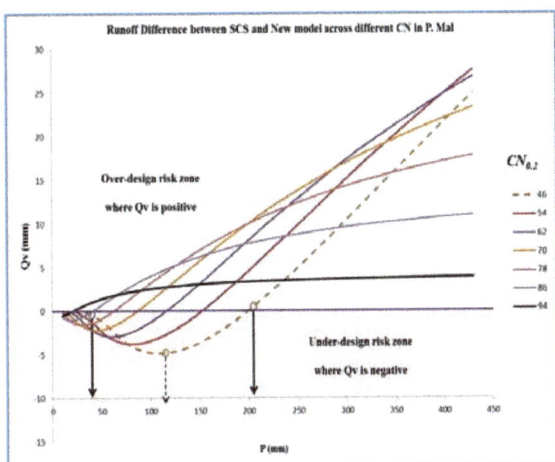

Figure 6. Runoff difference curve graph of Peninsula Malaysia. The graph was created to identify P_{crit} point(s) of different $CN_{0.2}$ classes. P_{crit} is/are the point(s) where the runoff difference curve crosses x-axis, marked by circle(s) with solid down arrow lines. The dotted down arrow line estimates the rainfall depth of maximum "under-design" risk for $CN_{0.2} = 46$. Note: When $CN_{0.2} = 46$ (dash line curve), Equation (29) solved $P_{crit} = 199.6$ mm (right bold down arrow). Equation (23) calculated the outer boundary is at $P = 25.6$ mm while the lower P_{crit} value $= 45.2$ mm (left bold down arrow). In conclusion, for $CN_{0.2} = 46$, Equation (2) under predicts runoff amount from any rainfall depth >25.6 mm until 199.6 mm (P_{crit}) and over predicts runoff amount for any rainfall depths >199.6 mm when compared to Equation (21).

Runoff difference curve can be used as a visual aid to identify the P_{crit} amount where the curve intersects the x-axis (when $Q_v = 0$). Possible true solution(s) as initial guess(es) of the trial and error process from the curve can be visually identified rather than guessing an arbitrary starting point for numerical solution as proposed by previous researchers [2,20]. Equation (22) is a quadratic model that yields two potential P_{crit} solutions.

Figure 6 illustrates the use of runoff difference curves to identify the "critical rainfall amount" (P_{crit}) of several $CN_{0.2}$ scenarios. For example, at $CN_{0.2} = 46$ (dash line curve), P_{crit} is approximately 40 mm and 205 mm (eyeballed from the graph, P_{crit} points are marked by solid downwards arrow where the curve intersects the x-axis, implying that Q_v is near to 0). However, the I_a amount has not been initiated for rainfall less than 40 mm according to Figure 3 and therefore, only 205 mm was used as the original trial and error estimate to satisfy Equation (22) and solve for the final solution of P_{crit} of $CN_{0.2} = 46$.

Runoff difference curve provides a brief overview and shows that Equation (2) will under-predict runoff amount at $CN_{0.2}$ area of 46 with any rainfall depths below the P_{crit} value (around 205 mm) and becomes an over-prediction thereafter. A non-linear under-design risk is therefore exhibited in the curve, with a peak of approximately 115 mm in rainfall depth (shown as dotted downwards arrow). Runoff difference curve provides additional insight of the worst under-estimated and over-estimated runoff prediction errors due to Equation (2) of specific rainfall depth which can be estimated with Equations (25) and (26), respectively.

3.10. The Critical Rainfall Amount (P_{crit}) Closed Form Equation

Through completing the square technique, this study has successfully used Equation (22) to obtain the closed form equation of P_{crit} in terms of $CN_{0.2}$. The closed form equation can be applied to solve for the P_{crit} in any pairing runoff models with any λ values. The equation can calculate the P_{crit} amount precisely and replace the trial and error procedure mentioned in Sections 2.11 and 2.12. SCS practitioners can refer to the proposed method in this article to derive the specific P_{crit} equation for their studies.

The derivation of the closed form equation of the critical rainfall depth (P_{crit}) from this study is shown below. From Equation (22),

$$Q_v = \frac{\left[P - 50.8\left(\frac{100}{CN_{0.2}} - 1\right)\right]^2}{\left[P + 203.2\left(\frac{100}{CN_{0.2}} - 1\right)\right]} - \frac{\left[P - 21.606\left(\frac{100}{CN_{0.2}} - 1\right)^{1.063}\right]^2}{\left[P + 402.547\left(\frac{100}{CN_{0.2}} - 1\right)^{1.063}\right]}$$

$$\text{Let}: \mathbf{A} = 21.606\left(\frac{100}{CN_{0.2}} - 1\right)^{1.063}$$

$$\text{Let}: \mathbf{B} = 50.8\left(\frac{100}{CN_{0.2}} - 1\right)$$

When $Q_v = 0$ (Runoff indifferent between 2 models), substitute A and B and solve for P (P_{crit}).

$$\frac{[P-B]^2}{[P+4B]} = \frac{[P-A]^2}{[P+18.631A]}$$

After grouping and simplifying, P ($\mathbf{P_{crit}}$) can be solved via quadratic form as below:

$$a = 4B - 2A + 2B - 18.631A$$

$$b = A^2 - 8AB - B^2 + 2(18.631)AB$$

$$c = 4BA^2 - 18.631AB^2$$

$$P_{crit} = \frac{-b \pm \sqrt{b^2 - 4ac}}{2a} \tag{29}$$

P_{crit} = Critical rainfall depth (mm)
$CN_{0.2}$ = Conventional curve number of a watershed

Equation (29) is a quadratic model that yields two potential P_{crit} solutions. The outer boundary (Equation (23)) can be used as checkpoint to determine if the lower P_{crit} value is a valid solution because any rainfall depths beyond the outer boundary will start to yield runoff difference between the two models after fulfilling the I_a requirement. The lower P_{crit} value is usually discarded due to its proximity to (or less than) the outer boundary.

If the P_{crit} value < the P value of Equation (23) (outer boundary equation), the I_a is yet to be fulfilled thus it is impossible to have any runoff or runoff difference amount. Runoff difference curves graph is also an effective visual aid to supplement the P_{crit} closed-form equation (refer to Figure 6 example).

Results from several derived formulae were compiled in Table 3 to provide another quick overview of the P_{crit} for Peninsula Malaysia across multiple $CN_{0.2}$ scenarios. According to the DID HP 27 dataset, the lowest calculated $CN_{0.2}$ is 48.8; hence, column A tabulates $CN_{0.2}$ range from 47 to 99 to cover the entire possible $CN_{0.2}$ scenario in Peninsula Malaysia. Column B and D were calculated using Equation (6), column C used Equation (20) and column E used Equation (29). Column F calculated $CN_{0.2}$ percentage change into $CN_{0.051}$.

Column A and E can be used to construct another P_{crit} overview curve across multiple $CN_{0.2}$ scenarios (Figure 7) with a statistically significant equation regressed via SPSS as:

$$P_{crit} = -245.4 \ln(CN_{0.2}) + 1132.6 \tag{30}$$

Equation (30) has an R^2-adj of 0.997, standard error of 3.047 and $p < 0.001$. Given $CN_{0.2}$ value of a watershed, the corresponding P_{crit} value can be estimated with Equation (30). Equation (2) under predicts runoff amount at any rainfall depths below the P_{crit} overview curve in Figure 7 and vice versa. Figure 7 is also in line with the research outcome reported by [2] that Equation (2) had the tendency to under-estimate runoff amount in rural and forested watersheds as $CN_{0.2}$ decreases.

Table 3. Conjugate $CN_{0.051}$ and P_{crit} for Peninsula Malaysia.

(A) $CN_{0.2}$	(B) $S_{0.2}$	(C) CN_λ (0.051)	(D) $S_{0.051}$	(E) P_{crit} (mm)	(F) %
99	2.57	98.76	3.20	7.38	0.2%
97	7.86	96.02	10.52	12.65	1.0%
95	13.37	93.20	18.52	17.83	1.9%
93	19.12	90.36	27.10	22.86	2.8%
91	25.12	87.52	36.23	27.85	3.8%
89	31.39	84.69	45.92	32.85	4.8%
87	37.95	81.89	56.18	37.91	5.9%
85	44.82	79.11	67.05	43.05	6.9%
83	52.02	76.38	78.57	48.31	8.0%
81	59.58	73.68	90.75	53.71	9.0%
79	67.52	71.02	103.67	59.26	10.1%
77	75.87	68.40	117.35	65.00	11.2%
75	84.67	65.83	131.88	70.94	12.2%
73	93.95	63.30	147.29	77.11	13.3%
71	103.75	60.81	163.68	83.53	14.4%
69	114.12	58.37	181.14	90.22	15.4%
67	125.10	55.98	199.74	97.22	16.4%
65	136.77	53.63	219.60	104.56	17.5%
63	149.18	51.33	240.84	112.26	18.5%
61	162.39	49.07	263.60	120.38	19.6%
59	176.51	46.86	288.03	128.94	20.6%
57	191.61	44.69	314.31	138.00	21.6%
55	207.82	42.57	342.65	147.61	22.6%
53	225.25	40.49	373.28	157.83	23.6%
51	244.04	38.46	406.49	168.74	24.6%
49	264.37	36.46	442.59	180.41	25.6%
47	286.43	34.51	481.96	192.95	26.6%

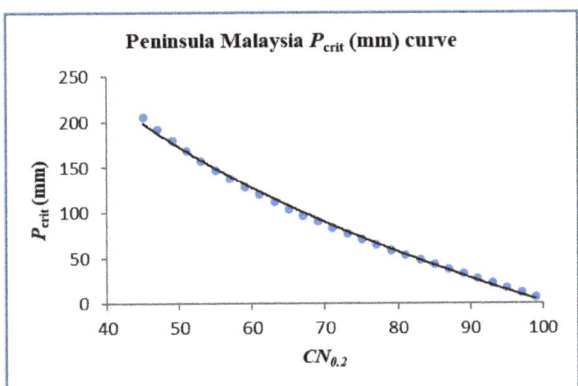

Figure 7. P_{crit} overview curve for Peninsula Malaysia. Equation (2) under predicts runoff amount for any rainfall depths below the curve at respective CN0.2 area. The underprediction tendency worsens as $CN_{0.2}$ value decreases.

Using the same concept as presented in ?? and Section 3.10, the closed form P_{crit} can also be derived to verify previous study results where the optimum λ value was identified as 0.05 in the USA. The correlation between S_λ and $S_{0.2}$ is best represented by $S_{0.05} = 1.33 S_{0.2}^{1.15}$ [2,20,22]. It is noteworthy to mention that US researchers used inches in their dataset; hence, Equation (18) (CN formula, SI version) needs to be converted and $CN_\lambda = \frac{1000}{S_\lambda + 10}$ should be used instead. The closed form P_{crit} equation can be derived with the same method as proposed in Section 3.10 to verify their published P_{crit} (inches) values (Table 4) in USA [2,22].

Table 4. The P_{crit} (inches) values with its corresponding $CN_{0.2}$ and $CN_{0.05}$ values for runoff prediction studies in USA (Modified from [2,22]).

Conjugate Curve Numbers and P_{crit} Values				
$CN_{0.2}$	$S_{0.2}$ (in)	$CN_{0.05}$	$S_{0.05}$ (in)	P_{crit} (in)
100	0	100	0	-
95	0.526	94.02	0.636	2.44
90	1.111	86.95	1.501	1.72
85	1.765	79.64	2.556	1.95
80	2.5	72.39	3.815	2.27
75	3.333	65.31	5.311	2.63
70	4.286	58.51	7.091	3.05
65*	5.385	52.03	9.219	3.52 (4.51)*
60	6.667	45.9	11.785	4.04
55	8.182	40.14	14.915	4.64
50**	10	34.74	18.787	5.33 (5.35)**
45	12.222	29.71	23.663	6.15
40	15	25.03	29.947	7.13
35	18.571	20.71	38.285	8.35

Note: (4.51)* old value for $CN_{0.2}$ = 65. (5.35)** old value for $CN_{0.2}$ = 50.

The closed form P_{crit} equation verified all P_{crit} values in Table 4 except for $CN_{0.2}$ = 50** and 65*. For $CN_{0.2}$ = 50**, the calculated P_{crit} using the closed form equation method is 5.33 inches (instead of 5.35 inches)**. The variance to the published value is about 0.5 mm. However, for $CN_{0.2}$ = 65*, the calculated P_{crit} is 3.52 inches (instead of 4.51 inches)*, which is much lower than the published value by about 25 mm.

Verification of Table 4 Pcrit values prove that the P_{crit} closed form equation can be used to calculate the exact P_{crit} value for any comparing SCS CN models for SCS practitioners. The success in the closed form equation derivation narrows the study gap from previous work. It can be adopted to replace the trial and error technique used by previous researchers [2,20,22].

3.11. Critical Curve Number (CN$_{crit}$)

Equation (29) will yield two possible CN_{crit} solutions (when Q_v = 0 in Equation (22)). Although it is possible for those CN_{crit} values to exist, all values must be verified. Potential CN_{crit} solution(s) as the initial guess(es) to the trial and error process to satisfy Equation (22) can be identified when visually aided by runoff difference curves.

For an example, when rainfall = 100 mm (dash line curve in Figure 8), potential CN_{crit} value is about 66 (marked by bold solid down arrows where the curve intersects with the x-axis or Q_v = 0). Other possible CN_{crit} value were discarded because the dash line curve intersects the x-axis at the left end at $CN_{0.2}$ around 22 and 99 on the right end, those values remain as a theoretical $CN_{0.2}$ value only.

3.12. Asymptotic Curve Number of Peninsula Malaysia

According to the AFM (Section 2.10), the DID HP 27 dataset resembles the standard behavior pattern (Figure 9) and thus Equation (16) was adopted to derive CN_∞ as the best representative $CN_{0.2}$ value for the dataset. Through least square fitting method under AFM, the fitting parameter k was identified to be 40.79 and CN_∞ = 67.77. When rounded to the closest positive integer, CN_∞ = $CN_{0.2}$ = 68.

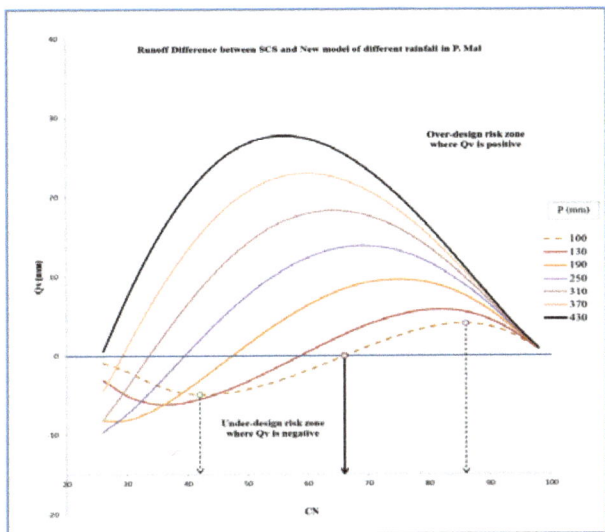

Figure 8. Runoff difference curves between Equation (2) or Equation (8) and (21). CN_{crit} is the point that the runoff difference curve intersects the x-axis, marked by circle with solid down arrows lines. The dotted down arrow lines estimate the rainfall depth of maximum "under and over-design" risk for P = 100 mm, respectively. Note: when rainfall = 100 mm (dash line curve), runoff difference curve also suggests that the return period design base on rainfall depth of 100 mm is likely to cause under-design risk (negative Q_v) in watersheds where $CN_{0.2}$ value(s) is (are) <66, meanwhile incurring over-design risk (positive Q_v) in $CN_{0.2}$ values >66. Estimated worst under-design risk (marked with dotted down arrows) occurs around $CN_{0.2}$ = 42 while the worst over-design risk at about 86. The worst under and over-estimated runoff prediction errors due to Equation (2) of those $CN_{0.2}$ area can be estimated with Equations (27) and (28), respectively.

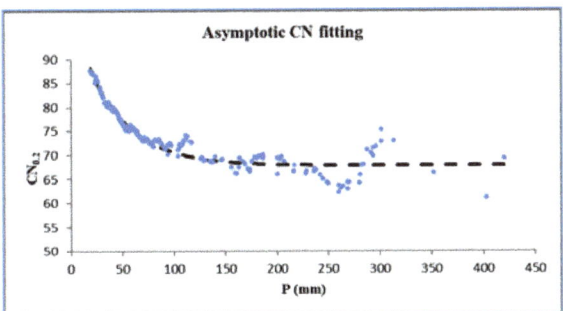

Figure 9. Asymptotic CN fitting of the dataset. For standard behavior pattern, CN_∞ is the point where a near to stable state of $CN_{0.2}$ fits to the higher rainfall depths.

The AFM $CN\infty$ result is in proximity to the equivalent $CN_{0.2}$ value of 72.58 which was derived in Section 3.2, whereas CN_∞ = 68 also falls within the 99% $CN_{0.2}$ confidence interval of this study. This proves that the proposed SCS CN model calibration methodology in this article is capable to produce results that are in line with other method introduced by previous study.

Using Equation (18), the calculated $S_{0.2}$ value of the AFM CN is 120.78 mm and I_a = 0.20 × 120.78 mm = 24.16 mm. These numbers are used in formulating the SCS runoff model with Equation (1) for benchmarking (Table 5).

Table 5. Asymptotic CN fitting method (AFM) and new λ runoff model's residual analyses comparison with descriptive and inferential statistics at alpha = 0.01 level.

	AFM Model	New λ Model
$\lambda\ value$	0.20	0.051
E	0.910	0.919
RSS	69,933	62,926
Residual Standard Deviation	17.083	16.556
Residual Standard Deviation: BCa 99% CI	[14.200, 19.552]	[13.875, 18.898]
Residual Skewness	0.401	−0.098
Mean Residual:	−4.188	−2.079
Mean Residual: BCa 99% CI	[−6.953, −1.035]	[−4.814, 0.920]
Residual: Range	96.89	101.45
Residual Variance	291.822	274.091
Residual Variance: BCa 99% CI	[201.207, 382.593]	[192.434, 358.014]

The newly calibrated λ model has lower RSS with higher E index compared to the runoff model formulated with the Asymptotic CN value. The models' residual skewness is near to zero, thus the mean residual value can act as an indicator for the predictive model's accuracy. The new λ model has lower mean residual with 99% confidence interval range which spans across zero, indicating its capability to achieve zero (residual) runoff prediction error. On the other hand, the AFM model tends to under-predict runoff volumes since their mean residual confidence interval range is within negative value range. The descriptive statistics indicates that the AFM model has a lower residual range. However, the standard deviation and variance in the model's residual are lower in the new λ model with smaller confidence interval ranges. Hence, the new λ model has higher stability and reliability for the dataset of this study.

AFM model faced another issue, whereby the calculated I_a value (24.16 mm) is larger than nearly 3.10% (seven recorded rainfall events) of the DID HP 27 dataset. According to the runoff constraint defined by SCS (as stated in Section 2.1) any rainfall depths < I_a value would not initiate any runoff; hence, AFM model failed to comply with the SCS constraint for those seven P-Q data pairs. On the other hand, New λ model does not have this issue.

4. Conclusions

This article presented the methodology to perform the SCS CN model calibration under the guide of inferential statistics with regional rainfall-runoff data. The study honed the runoff prediction accuracy of a popular rainfall-runoff model and based on its mathematical framework to develop engineering applications. Key highlights are as below:

1. The methodology to reassess the validity of a popular runoff model was presented. Under this study, the existing SCS runoff model is invalid for runoff modelling (alpha = 0.01), and therefore the model must be calibrated. λ = 0.051 (99% CI ranges from 0.034, 0.051) and $CN_{0.2}$ = 72.58 (99% CI ranges from 67 to 76) are the calibrated results for runoff prediction in Peninsula Malaysia according to the dataset of this study. Within these $CN_{0.2}$ areas, SCS model underpredicts runoff amount when rainfall depth of a storm is <70 to 85 mm and its overprediction tendency worsens toward larger storm events if it is not calibrated. The SCS CN model underpredicted runoff amount the most (2.4 million L/km² area) at $CN_{0.2}$ = 67 area and rainfall depth of 55 mm while it nearly overpredicted runoff amount by 25 million L/km² area when the storm depth reaches 430 mm in Peninsula Malaysia.
2. The closed form equation of the "Critical Rainfall Amount (P_{crit})" was solved (Sections 2.12 and 3.10) to narrow the research gap. Figure 6 example illustrated its use and past publication errors were detected (Table 4). The "Critical Curve Number (CN_{crit})" concept and the use of the runoff difference curves graph were also introduced in this article (Sections 2.11–2.13 and Sections 3.9–3.11) with demonstrated applications shown in Figures 6–8.

3. The 3D runoff difference model (Figure 2a,b) was created with Equation (22) to assess the runoff prediction results of the existing SCS CN model and its type II errors. Equations (25)–(28) to estimate the worst-case runoff prediction errors of the SCS CN model when it is not calibrated with λ = 0.051 for runoff predictions in Peninsula Malaysia. Any past study or engineering projects using this model and based upon the return period concept of rainfall amount below 70 mm might be under-designed while the model has over-design risk when a storm depth is larger than 85 mm. SCS practitioners are encouraged to refer to the general formulae (Equations (10)–(12)) and proposed methods in this article to derive the specific model and equations for their studies. Equation (2) must be validated with rainfall-runoff dataset prior to its adoption for runoff prediction in any part of the world.
4. Authors cautioned that there are several limitations of the proposed methodology. Minimum sample size should be at least 100 observations while the alpha level setting for Null assessment is pending upon research need. BCa should be used instead of bootstrapping and the choice of the statistical software must come with the option to provide confidence interval for median value to cater for model calibration need when the dataset is skewed. Runoff error analyses beyond the confidence interval or dataset limit may not be meaningful for interpretations.

Author Contributions: Conceptualization, L.L. and Z.Y.; methodology, L.L.; software, L.L. and J.L.L.; validation, L.L. and Z.Y.; formal analysis, L.L.; investigation, L.L. and Z.Y.; resources, L.L. and Z.Y.; data curation, L.L.; writing—original draft preparation, L.L. and J.L.L.; writing—review and editing, L.L., J.L.L., and Z.Y.; visualization, L.L. and J.L.L.; supervision, Z.Y.; project administration, L.L. and Z.Y.; funding acquisition, L.L. and Z.Y. All authors have read and agreed to the published version of the manuscript.

Funding: The authors would like to thank the Institute of Postgraduate Studies & Research (IPSR) of Universiti Tunku Abdul Rahman (UTAR) for financial support in this study (IPSR/RMC/UTARRF/2019-C2/L07). This study was also partly supported by the Brunsfield Engineering Sdn. Bhd., Malaysia (Brunsfield 8013/0002) and partly funded by FRGS (RJ130000.7809.4F208) from the Centre for Environmental Sustainability and Water Security of Universiti Teknologi Malaysia.

Acknowledgments: The authors appreciate the guidance from R. H. Hawkins at The University of Arizona, Tucson, AZ, USA and 3 anonymous reviewers who provided their feedback during the review process of this article.

Conflicts of Interest: The authors declare no conflict of interest.

References

1. EM-DAT, CRED/UCLouvain, Brussels, Belgium. International Disasters Database, 1900–2020 Hydrological & Meteorological Categories (Flood, Landslide & Storms). Available online: www.emdat.be (accessed on 16 December 2020).
2. Hawkins, R.H.; Ward, T.; Woodward, D.E.; Van Mullem, J. *Curve Number Hydrology: State of the Practice*; ASCE: Reston, VA, USA, 2009; p. 106.
3. Hawkins, R.H. Curve Number Method: Time to Think Anew? *J. Hydrol. Eng.* **2014**, *19*, 1059. [CrossRef]
4. Ling, L.; Yusop, Z.; Yap, W.-S.; Tan, W.L.; Chow, M.F.; Ling, J.L. A Calibrated, Watershed-Specific SCS-CN Method: Application to Wangjiaqiao Watershed in the Three Gorges Area, China. *Water* **2019**, *12*, 60. [CrossRef]
5. National Engineering Handbook; Part 630 Hydrology, Chapter 10, Figure 10.1: USDA, NRCS. 2004. Available online: https://directives.sc.egov.usda.gov/OpenNonWebContent.aspx?content=17752.wba (accessed on 30 March 2021).
6. Ross, C.W.; Prihodko, L.; Anchang, J.; Kumar, S.; Ji, W.J.; Hanan, N.P. HYSOGs250m, global gridded hydrologic soil groups for curve-number-based runoff modeling. *Sci. Data* **2018**, *5*, 180091. [CrossRef] [PubMed]
7. Jaafar, H.H.; Ahmad, F.A.; El Beyrouthy, N. GCN250, new global gridded curve numbers for hydrologic modeling and design. *Sci. Data* **2019**, *6*, 1–9. [CrossRef]
8. Chen, Y.; Wang, Y.; Zhang, Y.; Luan, Q.; Chen, X. Flash floods, land-use change, and risk dynamics in mountainous tourist areas: A case study of the Yesanpo Scenic Area, Beijing, China. *Int. J. Disaster Risk Reduct.* **2020**, *50*, 101873. [CrossRef]
9. Park, K.; Won, J.-H. Analysis on distribution characteristics of building use with risk zone classification based on urban flood risk assessment. *Int. J. Disaster Risk Reduct.* **2019**, *38*, 101192. [CrossRef]
10. Sun, R.; Gong, Z.; Gao, G.; Shah, A.A. Comparative analysis of Multi-Criteria Decision-Making methods for flood disaster risk in the Yangtze River Delta. *Int. J. Disaster Risk Reduct.* **2020**, *51*, 101768. [CrossRef]

11. Feng, B.; Wang, J.F.; Zhang, Y.; Hall, B.; Zeng, C.Q. Urban flood hazard mapping using a hydraulic-GIS combined model. *Nat. Hazards* **2020**, *100*, 1089–1104. [CrossRef]
12. Yalcin, E. Assessing the impact of topography and land cover data resolutions on two-dimensional HEC-RAS hydrodynamic model simulations for urban flood hazard analysis. *Nat. Hazards* **2020**, *101*, 995–1017. [CrossRef]
13. Zelelew, D.G. Spatial mapping and testing the applicability of the curve number method for ungauged catchments in Northern Ethiopia. *Int. Soil Water Conserv. Res.* **2017**, *5*, 293–301. [CrossRef]
14. Durán-Barroso, P.; González, J.; Valdés, J.B. Sources of uncertainty in the NRCS CN model: Recognition and solutions. *Hydrol. Process.* **2017**, *31*, 3898–3906. [CrossRef]
15. Lal, M.; Mishra, S.K.; Pandey, A.; Pandey, R.P.; Meena, P.K.; Chaudhary, A.; Jha, R.K.; Shreevastava, A.K.; Kumar, Y. Evaluation of the Soil Conservation Service curve number methodology using data from agricultural plots. *Hydrogeol. J.* **2017**, *25*, 151–167. [CrossRef]
16. Fidal, J.; Kjeldsen, T. Accounting for soil moisture in rainfall-runoff modelling of urban areas. *J. Hydrol.* **2020**, *589*, 125122. [CrossRef]
17. Sumargo, E.; McMillan, H.; Weihs, R.; Ellis, C.J.; Wilson, A.M.; Ralph, F.M. A soil moisture monitoring network to assess controls on runoff generation during atmospheric river events. *Hydrol. Process.* **2021**, *35*. [CrossRef]
18. Hoang, L.; Schneiderman, E.M.; Moore, K.E.B.; Mukundan, R.; Owens, E.M.; Steenhuis, T.S. Predicting saturation-excess runoff distribution with a lumped hillslope model: SWAT-HS. *Hydrol. Process.* **2017**, *31*, 2226–2243. [CrossRef]
19. Davidsen, S.; Löwe, R.; Ravn, N.H.; Jensen, L.N.; Arnbjerg-Nielsen, K. Initial conditions of urban permeable surfaces in rainfall-runoff models using Horton's infiltration. *Water Sci. Technol.* **2017**, *77*, 662–669. [CrossRef]
20. Jiang, R. *Investigation of Runoff Curve Number, Initial Abstraction Ratio*; University of Arizona: Tucson, AZ, USA, 2001.
21. DID, Hydrological Procedure No. 27. Design Flood Hydrograph Estimation for Rural Catchments in Peninsula Malaysia. JPS, DID, Kuala Lum-Pur. 2010. Available online: https://www.water.gov.my/jps/resources/PDF/Hydrology%20Publication/Hydrological_Procedure_No_27_(HP_27).pdf (accessed on 30 March 2021).
22. Woodward, D.E.; Hawkins, R.H.; Jiang, R.; Hjelmfelt, J.A.T.; Van Mullem, J.A.; Quan, Q.D. Runoff Curve Number Method: Examination of the Initial Abstraction Ratio. In Proceedings of the World Water & Environmental Resources Congress 2003; American Society of Civil Engineers (ASCE), Philadelphia, PA, USA, 23–26 June 2003; pp. 1–10.
23. ASCE-ASABE (American Society of Agricultural and Biological Engineers)-NRCS (Natural Resources Conservation Service) Task Group on Curve Number Hydrology. *Report of Task Group on Curve Number Hydrology, Chapters 8 (Land Use and Land Treatment Classes), 9 (Hydrologic Soil Cover Complexes), 10 (Estimation of Direct Runoff from Storm Rainfall), 12 (Hydrologic Effects of Land Use and Treatment)*; Hawkins, R.H., Ward, T.J., Woodward, D.E., Eds.; ASCE: Reston, VA, USA, 2017.
24. Santikari, V.P.; Murdoch, L.C. Including effects of watershed heterogeneity in the curve number method using variable initial abstraction. *Hydrol. Earth Syst. Sci.* **2018**, *22*, 4725–4743. [CrossRef]
25. Hawkins, R.H.; Theurer, F.D.; Rezaeianzadeh, M. Understanding the Basis of the Curve Number Method for Watershed Models and TMDLs. *J. Hydrol. Eng.* **2019**, *24*, 06019003. [CrossRef]
26. Helsel, D.R.; Hirsch, R.M.; Ryberg, K.R.; Archfield, S.A.; Gilroy, E.J. Statistical methods in water resources. In *Techniques and Methods*; US Geological Survey: Reston, VA, USA, 2020; p. 458.
27. IBM. *IBM, SPSS Bootstrapping 21 Guide*; IBM Press: Indianapolis, IN, USA, 2012.
28. Efron, B. *Large-Scale Inference: Empirical Bayes Methods for Estimation, Testing, and Prediction*; Cambridge University Press: London, UK, 2010.
29. Puth, M.-T.; Neuhäuser, M.; Ruxton, G.D. On the variety of methods for calculating confidence intervals by bootstrapping. *J. Anim. Ecol.* **2015**, *84*, 892–897. [CrossRef]
30. Schneider, L.; McCuen, R.H. Statistical Guidelines for Curve Number Generation. *J. Irrig. Drain. Eng. ASCE* **2005**, *131*, 282–290. [CrossRef]
31. Hjelmfelt, A.T. Curve-number procedure as infiltration method. *J. Irrig. Drain. Eng. ASCE* **1980**, *106*, 1107–1111.
32. Hjelmfelt, A.T. Empirical Investigation of Curve Number Technique. *J. Irrig. Drain. Eng. ASCE* **1980**, *106*, 1471–1476.
33. Hawkins, R.H. Asymptotic determination of runoff curve numbers from data. *J. Irrig. Drain. Eng. ASCE* **1993**, *119*, 334–345. [CrossRef]
34. González, Á.; Temimi, M.; Khanbilvardi, R. Adjustment to the curve number (NRCS-CN) to account for the vegetation effect on hydrological processes. *Hydrol. Sci. J.* **2015**, *60*, 591–605. [CrossRef]
35. Kowalik, T.; Walega, A. Estimation of CN Parameter for Small Agricultural Watersheds Using Asymptotic Functions. *Water* **2015**, *7*, 939–955. [CrossRef]

Article

Unified CACSD Toolbox for Hybrid Simulation and Robust Controller Synthesis with Applications in DC-to-DC Power Converter Control

Mircea Şuşcă *, Vlad Mihaly *, Mihai Stănese †, Dora Morar † and Petru Dobra †

Department of Automation, Technical University of Cluj-Napoca, Str. G. Bariţiu nr. 26-28, 400027 Cluj-Napoca, Romania; mihai.stanese@aut.utcluj.ro (M.S.); Dora.Sabau@aut.utcluj.ro (D.M.); Petru.Dobra@aut.utcluj.ro (P.D.)
* Correspondence: mircea.susca@aut.utcluj.ro (M.Ş.); vlad.mihaly@aut.utcluj.ro (V.M.)
† These authors contributed equally to this work.

Abstract: The current article presents the design, implementation, validation, and use of a Computer-Aided Control System Design (CACSD) toolbox for nonlinear and hybrid system uncertainty modeling, simulation, and control using μ synthesis. Remarkable features include generalization of classical system interconnection operations to nonlinear and hybrid systems, automatic computation of equilibrium points for nonlinear systems, and optimization of least conservative uncertainty bounds, with direct applicability for μ synthesis. A unified approach is presented for the step-down (buck), step-up (boost), and single-ended primary-inductor (SEPIC) converters to showcase the use and flexibility of the toolbox. Robust controllers were computed by minimization of the $\mathcal{H}_\infty$ norm of the augmented performance systems, encompassing a wide range of uncertainty types, and have been designed using the well-known mixed-sensitivity closed loop shaping μ synthesis method.

Keywords: CACSD toolbox; operating point linearization; automatic uncertainty bound computation; Model-in-the-Loop simulation; hybrid simulation; robust control; $\mathcal{H}_\infty$ control; μ synthesis; DC-to-DC power converters; buck; boost; SEPIC

Citation: Şuşcă, M.; Mihaly, V.; Stănese, M.; Morar, D.; Dobra, P. Unified CACSD Toolbox for Hybrid Simulation and Robust Controller Synthesis with Applications in DC-to-DC Power Converter Control. *Mathematics* **2021**, 9, 731. https://doi.org/10.3390/math9070731

Academic Editor: Aleksandr Rakhmangulov

Received: 28 February 2021
Accepted: 23 March 2021
Published: 28 March 2021

Publisher's Note: MDPI stays neutral with regard to jurisdictional claims in published maps and institutional affiliations.

Copyright: © 2021 by the authors. Licensee MDPI, Basel, Switzerland. This article is an open access article distributed under the terms and conditions of the Creative Commons Attribution (CC BY) license (https://creativecommons.org/licenses/by/4.0/).

1. Introduction

Robust Control represents a massive point of interest when it comes to Control Theory, which has been heavily studied over the past decades. However, albeit Robust Control brings many benefits, it is still an open field in research which gathers increasingly more approaches over time. Basically, the goal of a robust controller is to accomplish a specified set of performances for bounded model uncertainties which can occur in practice due to various reasons. In other words, closed loop stability and performance are maintained even for model parameter variations and unmodeled dynamics alike.

Over the years, multiple and various approaches for designing robust controllers have been presented, some of them being implemented into dedicated toolboxes, such as MATLAB's Robust Control Toolbox [1]. This toolbox gathers the most efficient ones based on $\mathcal{H}_2$, $\mathcal{H}_\infty$, and μ synthesis methods, and it is often considered a reference in research. However, while using these types of toolboxes leads to controllers which are optimal for their prescribed criterion, they are not necessarily best in terms of conventional performances. Additionally, of great use for defining and optimizing difficult robust control problems is the Global Optimization Toolbox from in [2], providing ready-to-use solvers using various state-of-the-art algorithms, such as Particle Swarm Optimization (PSO) and Genetic Algorithms (GA). An important work in this direction, for computing optimal weighting functions for the generalized plant model, is presented in [3].

Even though there is a large variety of CACSD toolboxes in the field, their number is still expanding due to the necessity of overcoming drawbacks that the already existing ones have. At this point, the purpose of new toolboxes is not only to determine robust controllers for a specific process class, but to use a unified approach that would make them work for more types of systems, even multiple interconnected systems, in various configurations. User experience is also more accentuated, which is why some of them incorporate graphical user interfaces (GUIs), for improved usability.

An example is Multivar, which is a MATLAB-based application used for multiple-input and multiple-output (MIMO) control design, presented in [4]. This toolbox supports two working modes. It allows the user to work both in function and GUI mode (which represents a configuration wizard for determining the controller). Multivar can be used for LTI systems with or without time delay and it allows creating a model; converting, approximating, and analyzing it; input–output pairing and decoupling; and controller design and evaluation. Besides this, the user is able to export the control design and compare it with other saved designs. Another GUI-based robust controller design tool, which was created in LabVIEW, is presented in [5], based on the $\mathcal{H}_\infty$ loop shaping method. However, the goal was to provide a simple, user-friendly interface to make it easier to use, especially for educational purposes. Therefore, as mentioned by the authors, it does not provide the same flexibility as other design tools on the market.

LCToolbox, as presented in [6], is another MATLAB software package which is used for robust controller design. One of the advantages of using this toolbox instead of classic MATLAB routines is the fact that it gathers all necessary steps for controller design in one place, while cutting the need of preprocessing steps such as separate construction of the plant, and postprocessing steps, such as closed loop simulation. LCToolbox can be used for both linear time-invariant (LTI) and linear parameter-varying (LPV) models, and it also incorporates system identification methods. The controller is obtained by using the $\mathcal{H}_\infty$ loop shaping method. Other $\mathcal{H}_\infty$-based CACSD toolboxes have been presented over the past years. One example is represented in [7], which is based on linearizing or convexifying the conventional non-convex constraints on the classical robustness margins of $\mathcal{H}_\infty$ constraints. The controller parameters are then computed by using an optimization solver. This toolbox was created for MATLAB, and some of its main features are represented by the large variety of control problems in which it could be used, such as multi-model systems; the toolbox is designed to work with the output data of MATLAB's System Identification Toolbox [8]. The output of the toolbox is represented by a PID controller, which can be easily implemented. Another example of a $\mathcal{H}_\infty$-based CACSD toolbox is shown in [9], in which the main advantage is the reduced conservatism of almost all types of model uncertainties which are defined.

Controller order is an important factor when implementing it on real systems. Therefore, this might be an issue in some cases. However, methods that are determining a fixed structure controller are already presented, such as in [10], which is based on the $\mathcal{H}_2$ controller design method, but can be cumbersome to compute. In order to deal with the high order controller problem, other toolboxes include controller simplification steps to avoid the necessity of postprocessing, as presented in [11].

Currently acknowledged problems in this domain regard closed loop simulation, where performance validation is generally treated ad hoc, from one control problem to another. Another difficulty encountered is when the test cases were done only on the linearized system for which the controller is designed, without checking if the initially proposed performance values are additionally verified for the nonlinear plant model, and, also, uncertainty modeling is a very cumbersome operation. The purpose of the paper is to provide means for treating the previously stated problems in a unified manner, such as implementing automated testing, performance validation, and report generation.

In this current iteration of the toolbox, robust controllers were designed using the well-established routines from in [1]. The interface is scalable and the control logic and validation can be replaced with other user-defined methods, or the current robust control approach can be replaced with open source alternatives for the $\mathcal{H}_2$ and $\mathcal{H}_\infty$ optimization problems, such as presented in the thesis [12], with the possibility to refine the necessary solutions of the Algebraic Riccati Equations (AREs) using the algorithms from [13], while for the μ synthesis problem, the thesis [14] provides a flexible, open-source implementation using linear matrix inequalities (LMIs). A clear advantage over the ARE approach is that LMIs are capable of solving singular and close to singular problems. Alternatively, a mixed $\mathcal{H}_2/\mathcal{H}_\infty$ approach for stabilization and optimization using fixed-order controllers can be found in [15]. As such, the current iteration of the proposed toolbox is MATLAB-dependent for certain key functionalities, especially with regards to numerical simulation, robust control, and optimization, although the exposed ideas and mathematical framework can be directly implemented in other software environments, such as Python, Scilab, or LabVIEW.

The remainder of the paper is structured in the following manner. Section 2 describes the software structure and features of the proposed toolbox; Section 3 describes a proposed end-to-end workflow exemplified using modeling, control, validation, and simulation of several DC-to-DC converter topologies; and Section 4 illustrates comparative discussions, proposed improvements, and completions for future work and conclusions.

2. Toolbox Structure and Functionalities

The proposed toolbox has been designed with the target of end-to-end design and implementation of closed loop control systems, starting from the definition of the uncertainty set of plants to be controlled, their required operating point, along with control performance specifications and controller synthesis, and ending with the controller validation for the initial desired plant set.

2.1. Toolbox Features

Proposed features and advantages over existing toolboxes available in the literature:

- specify finite-dimensional dynamical systems with the general framework from Equation (1) to be used with the MATLAB ode framework; ability to interconnect such systems in series, parallel, and linear-fractional transformations; this functionality is described in Sections 2.2.1 and 2.2.2;
- specify hybrid dynamical systems in the framework from in [16] as in Equation (4), with the ability to interconnect such systems in series, parallel, and linear-fractional transformations, upper and lower; this feature is described in Section 2.2.3;
- automatically compute equilibrium points numerically, with the possibility to impose certain states, inputs, and/or outputs, while the remaining ones are deduced through numerical optimization; this feature is presented in Section 2.4;
- automatically compute the uncertainty model as requested alongside a nominal plant: additive, inverse additive, input and output multiplicative, etc. using a global optimization algorithm, such as particle swarm optimization, to be directly used as necessary for robust synthesis methods; removes the burden for the control engineer to manually do this process for each plant; this feature is presented in Section 2.5;
- flexible and scalable, all features are implemented through MATLAB code and does not need the use of Simulink, which can become cumbersome when treating families of plants and not a single, specific, plant at a time; also, to account for the operating point in the case of linearized, nonlinear, and hybrid systems, alike, the same interface for Model-in-the-Loop simulation is provided in the toolbox, as shown in Sections 2.2 and 2.3;
- besides the automatic validation of the frequency response for the desired operating point of the linearized plant family, the toolbox runs tests accounting for the uncertainty behavior of the desired nonlinear plant, not only on the linearization which the controller has been designed for. Every specification imposed in the designed phase will be automatically tested for the entire nonlinear system family, as illustrated in the case studies from Section 3.

2.2. Systems Specification

The scope of software classes implemented and described in this section aims to provide a flexible framework for simulation by using the ordinary-differential equation ode solver exclusively, with the low-level requirement of integrating a differential state equation. As such, exogenous signals would be reference signals and disturbances, known a priori in a simulation context. The intrinsic signals, i.e., commands and corresponding measurements, are passed to their corresponding subsystems by means of ode. Figure 1 encompasses an overview of the toolbox classes described in Sections 2.2–2.4. When the relationship between two classes is of type *inheritance*, the inherited class will not redundantly recall all previous properties and methods from the base class in the diagram, unless they overload the methods and is explicitly noted.

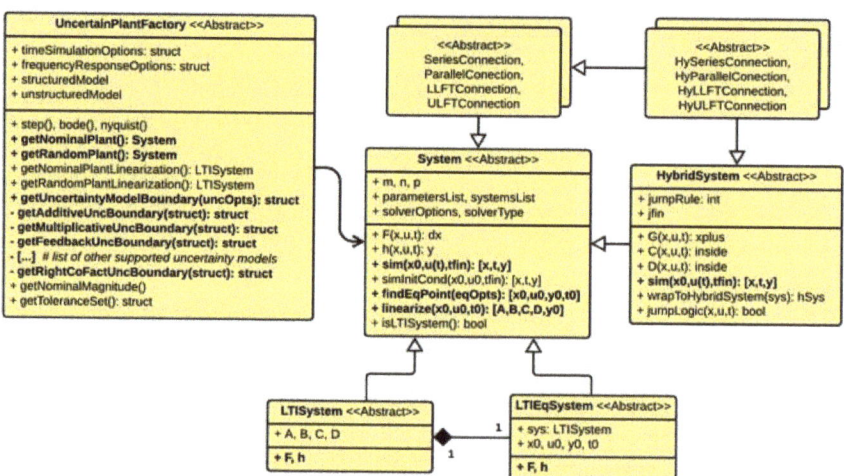

Figure 1. Class diagram for general-purpose nonlinear, LTI, linearized, and hybrid system implementations, along with the uncertain plant factory class, interconnections, and main functionalities.

2.2.1. Nonlinear Systems

For the purpose of this paper, we will focus on finite-dimensional systems: deterministic and stochastic. The so-called explicit or standard system form is obtained by writing the plant model in the following canonical form, using a set of differential equations and a set of output equations:

$$\begin{cases} \dot{\mathbf{x}}(t) = F(\mathbf{x}(t), \mathbf{u}(t), t); & (1a) \\ \mathbf{y}(t) = h(\mathbf{x}(t), \mathbf{u}(t), t), & (1b) \end{cases}$$

with the vector maps F and h being Lipschitz functions. The input signal $\mathbf{u}(t)$ has dimension m, state signal $\mathbf{x}(t)$ with dimension n, and output signal $\mathbf{y}(t)$ with dimension p, with $t \geq 0$. The initial conditions of the system are $\mathbf{x}(0) = \mathbf{x}_0 \in \mathbb{R}^n$.

Dynamical systems of the form (1) are implemented in class System. This will be the baseline interface for all systems the toolbox works with. Its most important methods are sim, findEqPoint, and linearize, which will be briefly described. The method sim simulates the dynamical system described by Equation (1a) from the initial condition x0, using the exogenous signal u(t), which is a predetermined anonymous function with at least the input argument time. tfin can be a scalar time value representing the final simulation time, a simulation interval, or a vector of predetermined time values. The solver options and type are based on MATLAB's ode framework options and are sent directly to it. The solver type can be selected from any of the supported functions: ode113, ode15s, ode15i, ode23, ode23t, ode23s, ode23tb, or ode45. After integrating the state equation, the output signal $\mathbf{y}(t)$ can be directly computed using the memoryless function h from (1b). A useful particularization is also the method `simInitCond`, with the only difference being that it replaces the time-varying input signal $\mathbf{u}(t)$ with a constant value $\mathbf{u}_0$, thus obtaining an impulse response. The method `findEqPoint` deduces an equilibrium point for the system given a set of specifications on the input, state, or output vectors and is described in detail in Section 2.4. After obtaining a valid equilibrium point, a linearized system can be obtained using the method `linearize`, also described there.

2.2.2. Linear Systems

Of particular interest for the framework and for control systems in general are linear time-invariant systems, which inherit the software interface from the `System` class, are implemented in the class `LTISystem` and are defined by

$$\begin{cases} \dot{\mathbf{x}}(t) = A\mathbf{x}(t) + B\mathbf{u}(t); & (2a) \\ \mathbf{y}(t) = C\mathbf{x}(t) + D\mathbf{u}(t). & (2b) \end{cases}$$

Separately, a nonlinear system can be linearized in the vicinity of an operating point, which is an equilibrium point for said system. The operating point $(\mathbf{u}_0, \mathbf{x}_0, \mathbf{y}_0, t_0)$ can be provided by the user or can be computed using the functionality from Section 2.4. The linearized system will work with variations of the initial variables and have the following model:

$$\begin{cases} \Delta\dot{\mathbf{x}}(t) = A \cdot \Delta\mathbf{x}(t) + B \cdot \Delta\mathbf{u}(t); \\ \Delta\mathbf{y}(t) = C \cdot \Delta\mathbf{x}(t) + D \cdot \Delta\mathbf{u}(t); \end{cases} \Leftrightarrow \begin{cases} \dot{\mathbf{x}}(t) = A(\mathbf{x}(t) - \mathbf{x}_0) + B(\mathbf{u}(t) - \mathbf{u}_0); \\ \mathbf{y}(t) = C(\mathbf{x}(t) - \mathbf{x}_0) + D(\mathbf{u}(t) - \mathbf{u}_0) + \mathbf{y}_0. \end{cases} \quad (3)$$

This latter structure is useful for MiL simulations and is implemented in the auxiliary class `LTIEqSystem`, seen as an affine nonlinear system. The great advantage of having the system from Equation (3) readily available is that it is interchangeable with the initial nonlinear interface in a closed loop context without making further adaptations in the source code and can be used to study the performance degradation obtained by replacing the controller from the linearized system to the nonlinear plant.

2.2.3. Hybrid Systems

A useful extension of framework (1) for hybrid systems, to account for system discontinuities, is with structures described in [16,17]:

$$\begin{cases} \dot{\mathbf{x}}(t) = F(\mathbf{x}(t), \mathbf{u}(t), t), & (\mathbf{x}, \mathbf{u}, t) \in \mathcal{C}; & (4a) \\ \mathbf{x}^+(t) = G(\mathbf{x}(t), \mathbf{u}(t), t), & (\mathbf{x}, \mathbf{u}, t) \in \mathcal{D}; & (4b) \\ \mathbf{y}(t) = h(\mathbf{x}(t), \mathbf{u}(t), t), & (4c) \end{cases}$$

with $F : \mathbb{R}^{n+m+1} \to \mathbb{R}^n$ as the flow function, $G : \mathbb{R}^{n+m+1} \to \mathbb{R}^n$ the jump function, and $h : \mathbb{R}^{n+m+1} \to \mathbb{R}^p$ the output function, while $\mathcal{C} \subset \mathbb{R}^{n+m+1}$ represents the flow set and $\mathcal{D} \subset \mathbb{R}^{n+m+1}$ is the jump set. When executing an ode simulation, a jump condition trigger is permanently verified and, based on the selected configuration, it allows prioritizing the flow logic, the jump logic, or a stochastic behavior which includes randomly selecting any of them. This jump condition will also be needed for hybrid system interconnections.

We propose a separate class in the toolbox, called `HybridSystem`, which inherits the previously described class `System`, includes the ode event-based mechanism from HyEQ Toolbox [16], and is extended to support time-varying differential equation systems and exogenous input signals. Besides the base interface from `System`, it also provides methods for functions G, $\mathcal{C}$, and $\mathcal{D}$. It also provides a wrapper function to promote any `System` object to the type `HybridSystem`, by adding dummy G, $\mathcal{C}$, and $\mathcal{D}$ methods, in order to be compatible for use in hybrid system interconnections. The flexibility added by this class in the toolbox allows model-in-the-Loop simulations using physical processes with hybrid dynamics, such as switching systems, i.e., electrical machines and power converters, or simulations of the closed loop control system, seen as hybrid system through the interconnection of a continuous-time process and a discrete-time controller, allowing the user to assess several performance analysis steps.

2.3. System Interconnections

After defining individual or *atomic* systems as in previous sections, the necessity for composing system interconnections readily appears. The classical interconnection operations are the series, parallel, lower, and upper linear-fractional transformations (LLFT and ULFT). Moreover, two separate approaches have been considered, i.e., to interconnect general-purpose nonlinear systems modeled by the class System and hybrid systems modeled by the class HybridSystem separately. The first case is useful for linearization near an operating point, studying its system theoretical properties, and designing control techniques, while the latter becomes useful in a model-in-the-Loop simulation context and for closed loop system property analysis. All provided system interconnections are implemented in classes which inherit the base class System.

The software classes presented in this section extend the series, parallel, feedback, and lft functions from MATLAB for nonlinear and hybrid systems, based on the interfaces from Equations (1) and (4). For hybrid system interconnections, the continuous and discrete dynamics sets $\mathcal{C}$ and $\mathcal{D}$, respectively, are obtained using union and intersection set operations.

Moreover, the next discrete state for each subsystem is triggered by its own logic, pre-defined in the jump function G and only when necessary; otherwise, it remains unchanged. For specifying this next discrete state $\mathbf{x}^+$ logic, as in the interface from Equation (4c), we will use the notation IF(CONDITION, THEN, ELSE), where CONDITION will be true when the point in the state-space is in the jump set, i.e., $(\mathbf{x}, \mathbf{u}, t) \in \mathcal{D}$ or $(\mathbf{x}, \mathbf{u}, t) \notin \mathcal{C}$; THEN gives the next state if a jump needs to be performed; and ELSE gives the next discrete state otherwise.

The state, output, and hybrid domain equations for nonlinear and hybrid system series connection, with the notations used in Figure 2, upper row, implemented in classes SeriesConnectionSystem and HybridSeriesConnectionSystem, are as follows:

$$
\begin{array}{c|c}
\begin{bmatrix} \dot{\mathbf{x}}_1 \\ \dot{\mathbf{x}}_2 \end{bmatrix} = \begin{bmatrix} F_1(\mathbf{x}_1, \mathbf{u}, t) \\ F_2(\mathbf{x}_2, h_1(\mathbf{x}_1, \mathbf{u}, t), t) \end{bmatrix}; &
\begin{bmatrix} \dot{\mathbf{x}}_1 \\ \dot{\mathbf{x}}_2 \end{bmatrix} = \begin{bmatrix} F_1(\mathbf{x}_1, \mathbf{u}, t) \\ F_2(\mathbf{x}_2, h_1(\mathbf{x}_1, \mathbf{u}, t), t) \end{bmatrix}; \\
\mathbf{y} = h_2(\mathbf{x}_2, h_1(\mathbf{x}_1, \mathbf{u}, t), t). &
\begin{bmatrix} \mathbf{x}_1^+ \\ \mathbf{x}_2^+ \end{bmatrix} = \begin{bmatrix} \text{if}(\text{jump}_1, G_1(\mathbf{x}_1, \mathbf{u}, t), \mathbf{x}_1) \\ \text{if}(\text{jump}_2, G_2(\mathbf{x}_2, h_1(\mathbf{x}_1, \mathbf{u}, t), t), \mathbf{x}_2) \end{bmatrix}; \\
 & \mathcal{C}(\mathbf{x}, \mathbf{u}, t) = \mathcal{C}_1(\mathbf{x}_1, \mathbf{u}, t) \cap \mathcal{C}_2(\mathbf{x}_2, h_1(\mathbf{x}_1, \mathbf{u}, t), t); \\
 & \mathcal{D}(\mathbf{x}, \mathbf{u}, t) = \mathcal{D}_1(\mathbf{x}_1, \mathbf{u}, t) \cup \mathcal{D}_2(\mathbf{x}_2, h_1(\mathbf{x}_1, \mathbf{u}, t), t); \\
 & \mathbf{y} = h_2(\mathbf{x}_2, h_1(\mathbf{x}_1, \mathbf{u}, t), t).
\end{array} \quad (5)
$$

Given two initial subsystems Sys1 and Sys2 with dimensions (m_1, n_1, p_1) and (m_2, n_2, p_2), respectively, the resulting series connection system will have dimensions $(m = m_1, n = n_1 + n_2, p = p_2)$.

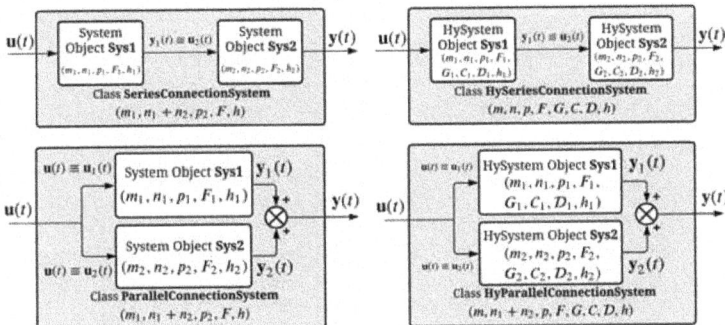

Figure 2. Series and parallel interconnections for general-purpose and hybrid systems.

The state, output, and hybrid domain equations for nonlinear and hybrid system parallel connection, with the notations used in Figure 2, bottom row, implemented in the classes named `ParallelConnectionSystem` and `HybridParallelConnectionSystem`, are as follows:

$$\begin{bmatrix}\dot{x}_1\\ \dot{x}_2\end{bmatrix} = \begin{bmatrix}F_1(x_1,u,t)\\ F_2(x_2,u,t)\end{bmatrix}; \quad \begin{bmatrix}\dot{x}_1\\ \dot{x}_2\end{bmatrix} = \begin{bmatrix}F_1(x_1,u,t)\\ F_2(x_2,u,t)\end{bmatrix};$$

$$\begin{bmatrix}x_1^+\\ x_2^+\end{bmatrix} = \begin{bmatrix}\text{if}(\text{jump}_1, G_1(x_1,u,t), x_1)\\ \text{if}(\text{jump}_2, G_2(x_2,u,t), x_2)\end{bmatrix}; \quad (6)$$

$$y = h_1(x_1,u,t) + h_2(x_2,u,t). \quad \mathcal{C}(x,u,t) = \mathcal{C}_1(x_1,u,t) \cap \mathcal{C}_2(x_2,u,t);$$

$$\mathcal{D}(x,u,t) = \mathcal{D}_1(x_1,u,t) \cup \mathcal{D}_2(x_2,u,t);$$

$$y = h_1(x_1,u,t) + h_2(x_2,u,t).$$

Given two initial subsystems `Sys1` and `Sys2` with dimensions (m_1, n_1, p_1) and (m_2, n_2, p_2), respectively, the resulting parallel connection system will have dimensions $(m = m_1 = m_2, n = n_1 + n_2, p = p_1 = p_2)$.

The state, output, and hybrid domain equations for nonlinear and hybrid system lower linear fractional trasformation (LLFT) connection, with the notations used in Figure 3, upper row, implemented in classes `LLFTConnectionSystem` and `HybridLLFTConnectionSystem`, are as follows:

$$\begin{bmatrix}\dot{x}_1\\ \dot{x}_2\end{bmatrix} = \begin{bmatrix}F_1(x_1,u_1^{\text{LLFT}},t)\\ F_2(x_2,u_2^{\text{LLFT}},t)\end{bmatrix};$$

$$\begin{bmatrix}\dot{x}_1\\ \dot{x}_2\end{bmatrix} = \begin{bmatrix}F_1(x_1,u_1^{\text{LLFT}},t)\\ F_2(x_2,u_2^{\text{LLFT}},t)\end{bmatrix}; \quad \begin{bmatrix}x_1^+\\ x_2^+\end{bmatrix} = \begin{bmatrix}\text{if}(\text{jump}_1, G_1(x_1,u_1^{\text{LLFT}},t), x_1)\\ \text{if}(\text{jump}_2, G_2(x_2,u_2^{\text{LLFT}},t), x_2)\end{bmatrix}; \quad (7)$$

$$y = \begin{bmatrix}h_1(x_1,u_1^{\text{LLFT}},t)\\ h_2(x_2,u_2^{\text{LLFT}},t)\end{bmatrix}. \quad \mathcal{C}(x,u,t) = \mathcal{C}_1(x_1,u_1^{\text{LLFT}},t) \cap \mathcal{C}_2(x_2,u_2^{\text{LLFT}},t);$$

$$\mathcal{D}(x,u,t) = \mathcal{D}_1(x_1,u_1^{\text{LLFT}},t) \cup \mathcal{D}_2(x_2,u_2^{\text{LLFT}},t);$$

$$y = \begin{bmatrix}h_1(x_1,u_1^{\text{LLFT}},t)\\ h_2(x_2,u_2^{\text{LLFT}},t)\end{bmatrix},$$

with the predefined notations:

$$u = \begin{bmatrix}u_1\\ u_2\end{bmatrix} = \begin{bmatrix}u_{11}\\ u_{12}^{\text{ref}}\\ u_{21}^{\text{ref}}\\ u_{22}\end{bmatrix}, \quad u_1^{\text{LLFT}} = \begin{bmatrix}u_{11}\\ u_{12}^{\text{ref}} + y_{21}^-\end{bmatrix} \equiv \begin{bmatrix}u_{11}\\ u_{12}\end{bmatrix}, \quad u_2^{\text{LLFT}} = \begin{bmatrix}u_{21}^{\text{ref}} + y_{12}^-\\ u_{22}\end{bmatrix} \equiv \begin{bmatrix}u_{21}\\ u_{22}\end{bmatrix}. \quad (8)$$

The common convention in the literature is to consider the last NCON values from the input vector $\mathbf{u}_1$, i.e., $\mathbf{u}_{12}$, as control input signals, while the last NMEAS values from the output vector $\mathbf{y}_1$, i.e., $\mathbf{y}_{12}$, as measurements signals. Only the vector $\mathbf{u}$ will be an exogenous signal, as the feedback components $\mathbf{y}_{12}^-$ and $\mathbf{y}_{21}^-$ are local and private feedback components computed implicitly at the previous time step, dictated by the selected ode solver. The exogenous signals $\mathbf{u}_{11}$ and $\mathbf{u}_{22}$ are seen as disturbance signals, while the signals $\mathbf{u}_{12}^{ref}$ and $\mathbf{u}_{21}^{ref}$ are seen as reference signals.

Given two initial subsystems Sys1 and Sys2 with dimensions (m_1, n_1, p_1) and (m_2, n_2, p_2), respectively, the resulting LLFT connection system will have dimensions $(m = m_1 + m_2, n = n_1 + n_2, p = p_1 + p_2)$. The subsystem Sys1 is usually seen as the controlled plant, while Sys2 is seen as the controller. In order to assure compatibility between the two, several assertions must be made: NMEAS = length($\mathbf{y}_{12}$) = length($\mathbf{u}_{21}$) and NCON = length($\mathbf{y}_{21}$) = length($\mathbf{u}_{12}$).

The state, output, and hybrid domain equations for nonlinear and hybrid system upper linear fractional trasformation (ULFT) connection, with the notations used in Figure 3, bottom row, implemented in classes ULFTConnectionSystem and HybridULFTConnectionSystem, are as follows:

$$\begin{bmatrix}\dot{\mathbf{x}}_1\\\dot{\mathbf{x}}_2\end{bmatrix} = \begin{bmatrix}F_1\left(\mathbf{x}_1,\mathbf{u}_1^{ULFT},t\right)\\F_2\left(\mathbf{x}_2,\mathbf{u}_2^{ULFT},t\right)\end{bmatrix}; \quad \left| \quad \begin{aligned}\begin{bmatrix}\dot{\mathbf{x}}_1\\\dot{\mathbf{x}}_2\end{bmatrix} &= \begin{bmatrix}F_1\left(\mathbf{x}_1,\mathbf{u}_1^{ULFT},t\right)\\F_2\left(\mathbf{x}_2,\mathbf{u}_2^{ULFT},t\right)\end{bmatrix};\\ \begin{bmatrix}\mathbf{x}_1^+\\\mathbf{x}_2^+\end{bmatrix} &= \begin{bmatrix}\text{if}\left(\text{jump}_1, G_1\left(\mathbf{x}_1,\mathbf{u}_1^{ULFT},t\right),\mathbf{x}_1\right)\\\text{if}\left(\text{jump}_2, G_2\left(\mathbf{x}_2,\mathbf{u}_2^{ULFT},t\right),\mathbf{x}_2\right)\end{bmatrix};\\ \mathcal{C}(\mathbf{x},\mathbf{u},t) &= \mathcal{C}_1\left(\mathbf{x}_1,\mathbf{u}_1^{ULFT},t\right) \cap \mathcal{C}_2\left(\mathbf{x}_2,\mathbf{u}_2^{ULFT},t\right);\\ \mathcal{D}(\mathbf{x},\mathbf{u},t) &= \mathcal{D}_1\left(\mathbf{x}_1,\mathbf{u}_1^{ULFT},t\right) \cup \mathcal{D}_2\left(\mathbf{x}_2,\mathbf{u}_2^{ULFT},t\right);\\ \mathbf{y} &= \begin{bmatrix}h_1\left(\mathbf{x}_1,\mathbf{u}_1^{ULFT},t\right)\\h_2\left(\mathbf{x}_2,\mathbf{u}_2^{ULFT},t\right)\end{bmatrix},\end{aligned}\end{aligned} \right. \qquad (9)$$
$$\mathbf{y} = \begin{bmatrix}h_1\left(\mathbf{x}_1,\mathbf{u}_1^{ULFT},t\right)\\h_2\left(\mathbf{x}_2,\mathbf{u}_2^{ULFT},t\right)\end{bmatrix}.$$

with the predefined notations:

$$\mathbf{u} = \mathbf{u}_{12}, \quad \mathbf{u}_1^{ULFT} = \begin{bmatrix}\mathbf{y}_2^-\\\mathbf{u}_{12}\end{bmatrix}, \quad \mathbf{u}_2^{ULFT} = \mathbf{y}_{11}^-. \qquad (10)$$

The common convention in the literature is to consider the first NU values from the input vector of the plant subsystem, i.e., $\mathbf{u}_{11}$, as input uncertainty signals, while the first NY values from the output vector of the plant, i.e., $\mathbf{y}_{11}$, as output uncertainty signals. Only the vector $\mathbf{u} \equiv \mathbf{u}_{12}$ will be an exogenous signal, as the feedback components $\mathbf{y}_{11}^-$ and $\mathbf{y}_2^-$ are local and private feedback components computed implicitly at the previous time step, dictated by the selected ode solver. The exogenous signal $\mathbf{u}_{12}$ is seen as set of performance and control signals for the plant, without any reference signals recalled explicitly compared to the LLFT case.

Given two initial subsystems Sys1 and Sys2 with dimensions (m_1, n_1, p_1) and (m_2, n_2, p_2), respectively, the resulting ULFT connection system will have dimensions $(m = m_1 + m_2, n = n_1 + n_2, p = p_1 + p_2)$. The subsystem Sys1 is usually seen as the augmented controlled plant, while Sys2 is seen as the unstructured uncertainty block. In order to assure compatibility between the two, several assertions must be made: NY = length($\mathbf{y}_{11}$) = length($\mathbf{u}_2$) and NU = length($\mathbf{y}_2$) = length($\mathbf{u}_{11}$).

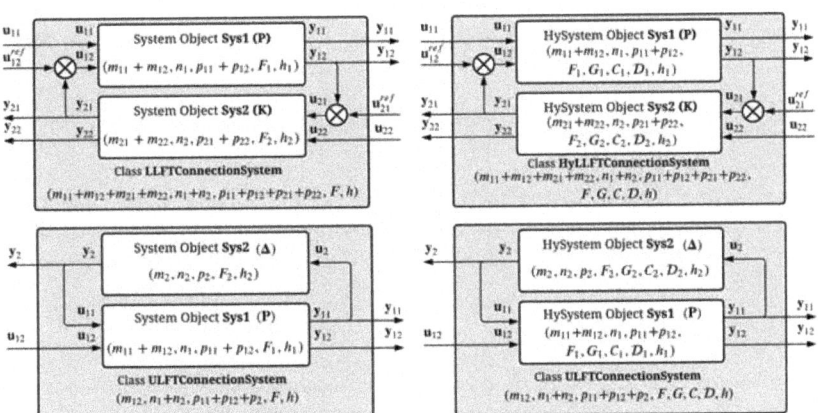

Figure 3. Upper (ULFT) and lower (LLFT) linear fractional transformation interconnections for general-purpose nonlinear and hybrid systems, with the ability to impose external reference signals.

2.4. Automatic Equilibrium Point Computation

Given a nonlinear system as in Equation (1), which may also include interconnections of systems, an operating point is desired with some of the input, state, and output variables imposed, such as a water level in a tank $y_h(t)$ controlled through two pumps $\mathbf{u}_{flow}(t)$, one with variable flow and one with a fixed flow, or a mechanical transportation system having a desired velocity $y_\omega(t)$ with respect to input forces and loads $\mathbf{u}(t)$. As such, a mechanism to automatically compute a partially imposed equilibrium point for an entire family of uncertain plants, relative to one which is considered nominal at the design phase, is proposed in this paragraph.

Starting from the system definition with dimensions m, n, and p, consider the sets of indexes, denoted by $\mathcal{I}$, and prescribed values, denoted by $\mathcal{V}$, for the input, state, and output variables, respectively:

$$\begin{cases} \mathcal{I}_\mathbf{u} := \{i_1^u, i_2^u, \ldots, i_{\overline{m}_u}^u\}, & \mathcal{V}_\mathbf{u} := \{\overline{\mathbf{u}}(i_1^u), \overline{\mathbf{u}}(i_2^u), \ldots, \overline{\mathbf{u}}(i_{\overline{m}_u}^u)\}, & 0 \leq \overline{m}_u \leq m; & (11a) \\ \mathcal{I}_\mathbf{x} := \{i_1^x, i_2^x, \ldots, i_{\overline{n}_x}^x\}, & \mathcal{V}_\mathbf{x} := \{\overline{\mathbf{x}}(i_1^x), \overline{\mathbf{x}}(i_2^x), \ldots, \overline{\mathbf{x}}(i_{\overline{n}_x}^x)\}, & 0 \leq \overline{n}_x \leq n; & (11b) \\ \mathcal{I}_\mathbf{y} := \{i_1^y, i_2^y, \ldots, i_{\overline{p}_y}^y\}, & \mathcal{V}_\mathbf{y} := \{\overline{\mathbf{y}}(i_1^y), \overline{\mathbf{y}}(i_2^y), \ldots, \overline{\mathbf{y}}(i_{\overline{p}_y}^y)\}, & 0 \leq \overline{p}_y \leq p, & (11c) \end{cases}$$

along with their complementary sets of values for the indexes, denoted by $\mathcal{UI}$, and the values, denoted $\mathcal{UV}$, to be computed through optimization by solving a system of equations:

$$\begin{cases} \mathcal{UI}_\mathbf{u} := \{i_1^u, i_2^u, \ldots, i_{\widetilde{m}_u}^u\}, & \mathcal{UV}_\mathbf{u} := \{\widetilde{\mathbf{u}}(i_1^u), \widetilde{\mathbf{u}}(i_2^u), \ldots, \widetilde{\mathbf{u}}(i_{\widetilde{m}_u}^u)\}, & 0 \leq \widetilde{m}_u \leq m; & (12a) \\ \mathcal{UI}_\mathbf{x} := \{i_1^x, i_2^x, \ldots, i_{\widetilde{n}_x}^x\}, & \mathcal{UV}_\mathbf{x} := \{\widetilde{\mathbf{x}}(i_1^x), \widetilde{\mathbf{x}}(i_2^x), \ldots, \widetilde{\mathbf{x}}(i_{\widetilde{n}_x}^x)\}, & 0 \leq \widetilde{n}_x \leq n; & (12b) \\ \mathcal{UI}_\mathbf{y} := \{i_1^y, i_2^y, \ldots, i_{\widetilde{p}_y}^y\}, & \mathcal{UV}_\mathbf{y} := \{\widetilde{\mathbf{y}}(i_1^y), \widetilde{\mathbf{y}}(i_2^y), \ldots, \widetilde{\mathbf{y}}(i_{\widetilde{p}_y}^y)\}, & 0 \leq \widetilde{p}_y \leq p, & (12c) \end{cases}$$

with

$$\begin{cases} \overline{m}_u + \widetilde{m}_u = m, & \mathcal{I}_u \cup \mathcal{UI}_u = \{1, 2, \ldots, m\}, & \mathcal{I}_u \cap \mathcal{UI}_u = \varnothing; & (13a) \\ \overline{n}_x + \widetilde{n}_x = n, & \mathcal{I}_x \cup \mathcal{UI}_x = \{1, 2, \ldots, n\}, & \mathcal{I}_x \cap \mathcal{UI}_x = \varnothing; & (13b) \\ \overline{p}_y + \widetilde{p}_y = p, & \mathcal{I}_y \cup \mathcal{UI}_y = \{1, 2, \ldots, p\}, & \mathcal{I}_y \cap \mathcal{UI}_y = \varnothing. & (13c) \end{cases}$$

a set of permutation matrices $P_\mathbf{u} \in \mathbb{R}^{m \times m}$, $P_\mathbf{x} \in \mathbb{R}^{n \times n}$, $P_\mathbf{y} \in \mathbb{R}^{p \times p}$ are obtained after sorting the indexes such as the following system of vector-valued equations needs to be solved:

$$\begin{cases} \mathbf{0} = F\left(P_\mathbf{x} \cdot \begin{bmatrix} \overline{\mathbf{x}} \\ \widetilde{\mathbf{x}} \end{bmatrix}, P_\mathbf{u} \cdot \begin{bmatrix} \overline{\mathbf{u}} \\ \widetilde{\mathbf{u}} \end{bmatrix}, \widetilde{t}\right); & (14a) \\ P_\mathbf{y} \cdot \begin{bmatrix} \overline{\mathbf{y}} \\ \widetilde{\mathbf{y}} \end{bmatrix} = h\left(P_\mathbf{x} \cdot \begin{bmatrix} \overline{\mathbf{x}} \\ \widetilde{\mathbf{x}} \end{bmatrix}, P_\mathbf{u} \cdot \begin{bmatrix} \overline{\mathbf{u}} \\ \widetilde{\mathbf{u}} \end{bmatrix}, \widetilde{t}\right). & (14b) \end{cases}$$

The system from Equation (14) becomes equivalent to directly solving a system of equations of the form $\mathbf{0} = \mathcal{F}(\mathbf{z})$ in the vector-valued unknown:

$$\mathbf{z} = \begin{bmatrix} \widetilde{\mathbf{x}}^T & \widetilde{\mathbf{u}}^T & \widetilde{\mathbf{y}}^T & \widetilde{t} \end{bmatrix}^T \in \mathbb{R}^{\widetilde{n}_x + \widetilde{m}_u + \widetilde{p}_y + 1}. \qquad (15)$$

If the dynamical system is time-invariant or if the required time value is known a priori, then the time variable can be removed from the solver or it can be imposed to a certain value $\widetilde{t}$ in the same manner as the for the other signals. Moreover, the method is flexible and allows imposing and solving only the subsystem (14a) if the output variables coincide with the states. The unknown variables from Equation (15) can be initialized to random values or a rough estimate for the entire family of uncertain plants can be obtained with the simulation to a step response of the nominal plant at the required amplitudes. All systems in the uncertain physical plant set will be found in the same mathematical vicinity. After solving the preferred algebraic system configuration from Equation (14), the desired equilibrium point $(\overline{\mathbf{u}}, \overline{\mathbf{x}}, \overline{\mathbf{y}}, \overline{t})$ can be reconstructed using the inverse permutation matrices $P_\mathbf{u}^{-1}, P_\mathbf{x}^{-1}, P_\mathbf{y}^{-1}$ and the notations from Equations (11) and (12). The method findEqPoint from class System forms and solves the system (14) and computes the desired equilibrium point for its predefined dynamical system based on the specifications from Equations (11) and (12) given in the structure eqOpts.

After acquiring the desired equilibrium point, the system linearization can be easily deduced through numeric differentiation methods. The most straightforward method is to compute the first-order Jacobian matrices of the functions F and h with respect to the state and input signals $\mathbf{x}$ and $\mathbf{u}$, respectively:

$$A = \left.\frac{\delta F}{\delta \mathbf{x}}\right|_{(\mathbf{x}_0, \mathbf{u}_0, t_0)}; \quad B = \left.\frac{\delta F}{\delta \mathbf{u}}\right|_{(\mathbf{x}_0, \mathbf{u}_0, t_0)}; \quad C = \left.\frac{\delta h}{\delta \mathbf{x}}\right|_{(\mathbf{x}_0, \mathbf{u}_0, t_0)}; \quad D = \left.\frac{\delta h}{\delta \mathbf{u}}\right|_{(\mathbf{x}_0, \mathbf{u}_0, t_0)}. \qquad (16)$$

The method linearize($\mathbf{x}_0, \mathbf{u}_0, t_0$) from class System of Section 1 computes the matrices from Equation (16) with a first-order derivative approximation and also the output equilibrium value $\mathbf{y}_0 = h(\mathbf{x}_0, \mathbf{u}_0, t_0)$:

$$A_{\overline{1,n},i} \approx \frac{F\left(\mathbf{x}_0 + \Delta \mathbf{x}_0^i, \mathbf{u}_0, t_0\right) - F(\mathbf{x}_0, \mathbf{u}_0, t_0)}{\Delta x}; \quad B_{\overline{1,m},j} \approx \frac{F\left(\mathbf{x}_0, \mathbf{u}_0 + \Delta \mathbf{u}_0^j, t_0\right) - F(\mathbf{x}_0, \mathbf{u}_0, t_0)}{\Delta u},$$

(17)

following that the output matrices C and D to be computed in a similar manner by replacing F with h in the above formulas. The shorthand notations are $\Delta \mathbf{x}_0^i = \begin{bmatrix} 0 & \cdots & 0 & \Delta x & 0 & \cdots & 0 \end{bmatrix}^T$ and $\Delta \mathbf{u}_0^j = \begin{bmatrix} 0 & \cdots & 0 & \Delta u & 0 & \cdots & 0 \end{bmatrix}^T$ for the disturbance vectors corresponding to the state with the index $i \in \overline{1,n}$ or input with index $j \in \overline{1,m}$, respectively, while the optimal [18] unit perturbations are, using double precision, $\Delta x = $ tol $\cdot (1 + ||\mathbf{x}_0||)$ and $\Delta u = $ tol $\cdot (1 + ||\mathbf{u}_0||)$, tol $= 10^{-5}$. Obviously, when linearizing the system, the static amplification of the initial nonlinear system is not accounted in the procedure, but will not be relevant in the actual control design process and implementation due to the consideration of only Lipschitz function-based systems and, as such, it will be correctly compensated. The correct simulation of the linearized system near the operating point is done using the class LTIEqSystem.

2.5. Automatic Least Conservative Uncertainty Bound Computation

Figure 4 encompasses an overview of the toolbox classes described in Sections 2.5 and 2.6, along with showing their relationship with the classes from the previous sections. Based on any desirable combination of the above system classes, i.e., System, LTISystem, LTIEqSystem, HybridSystem, and their interconnections, we propose a new functionality to aid in uncertain system modeling, ready for use in augmenting the plant for μ synthesis, implemented in the classes UncertainPlantFactory, seen in Figure 1, along with UncertaintyBoundOptimizationProblem, seen in Figure 4.

The common uncertainty model types considered in practice are gathered in Table 1. Besides the definition of the uncertain plant $G(s)$ starting from a nominal model $G_n(s)$ in relation to the uncertainty block $\Delta(s)$, the mathematical expression of $\Delta(s)$ is necessary to experimentally deduce its frequency response. Left and right coprime factor uncertainties are described by two blocks: Δ_M and Δ_N, and one of them can be selected as a free term, i.e., one degree of freedom (1-DOF). The class UncertainPlantFactory provides an interface to define a nominal plant and a random plant from a prespecified set. Besides the methods getNominalPlant and getRandomPlant, both returning a System object, it implements each uncertainty type $\Delta(s)$ from Table 1, obtained through Monte Carlo simulation using the magnitude characteristic in the frequency domain.

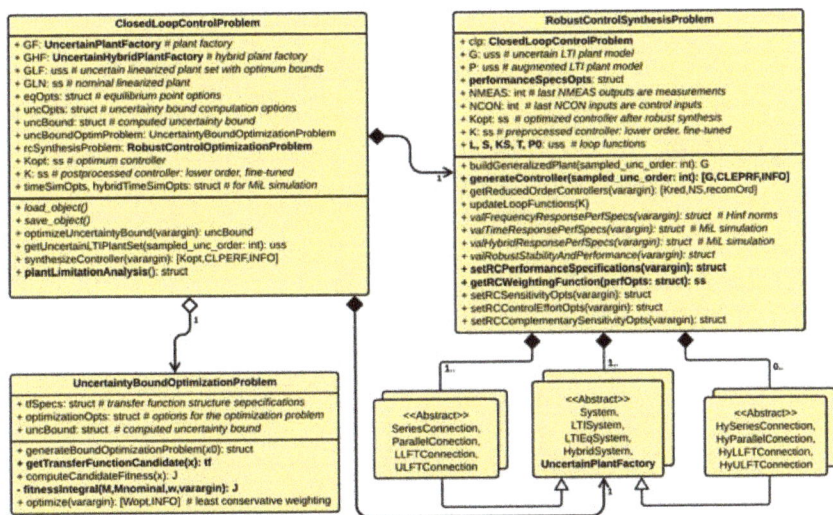

Figure 4. Closed loop control problem class which encompasses an uncertain plant set with operating point specification, options for automatically modeling the plant uncertainty, specifying robust control performances, synthesizing controller, and validating obtained results.

Table 1. Commonly used classes of perturbation and uncertainty models for multiple-input and multiple-output (MIMO) systems, implemented in class `UncertainPlantFactory`.

Uncertainty Type	Definition	Implementation
Additive	$G(s) = G_n(s) + \Delta(s)$	$\Delta(s) = G(s) - G_n(s)$
Inverse additive	$(G(s))^{-1} = (G_n(s))^{-1} + \Delta(s)$	$\Delta(s) = (G(s))^{-1} - (G_n(s))^{-1}$
Input multiplicative	$G(s) = G_n(s)[I + \Delta(s)]$	$\Delta(s) = (G_n(s))^{-1}(G(s) - G_n(s))$
Output multiplicative	$G(s) = [I + \Delta(s)]G_n(s)$	$\Delta(s) = (G(s) - G_n(s))(G_n(s))^{-1}$
Inverse input multiplicative	$(G(s))^{-1} = [I + \Delta(s)](G_n(s))^{-1}$	$\Delta(s) = (G(s))^{-1}(G_n(s)) - I$
Inverse output multiplicative	$(G(s))^{-1} = (G_n(s))^{-1}[I + \Delta(s)]$	$\Delta(s) = (G_n(s))(G(s))^{-1} - I$
Left coprime factor	$G(s) = (\tilde{M} + \Delta_{\tilde{M}})^{-1}(\tilde{N} + \Delta_{\tilde{N}})$	$\Delta_{\tilde{M}} = (\tilde{N} + \Delta_{\tilde{N}})(G(s))^{-1} - \tilde{M}$, **1-DOF**
Right coprime factor	$G(s) = (N + \Delta_N)(M + \Delta_M)^{-1}$	$\Delta_M = (G(s))^{-1}(N + \Delta_N) - M$, **1-DOF**

The frequency domain relevant for the studied plant is defined in logarithmic scale as

$$\Omega = \{\underline{\omega} = \omega_1 < \omega_2 < \ldots < \omega_{N-1} < \omega_N = \overline{\omega}\}, \tag{18}$$

along with the magnitude characteristic values sampled at points from Ω. By convention, $||\Delta(j\omega)||_\infty \leq 1$, $\forall \omega \geq 0$; thus, the worst-case uncertainty deduced experimentally through Monte Carlo simulation is written as $\Delta(s)W_{exp}(s)$. As an example for the additive uncertainty type, the uncertain plant family will be $G(s) = G_n(s) + \Delta(s) \cdot W_{exp}(s)$, with $||\Delta(j\omega)||_\infty \leq 1$, $\forall \omega \geq 0$ and the toolbox returns the worst-case experimental magnitude characteristic of $W_{exp}(s)$, sampled at points from $j\Omega$ through the high-level method `getUncertaintyModelBoundary` from class `UncertainPlantFactory`, which wraps over low-level methods such as `getAdditiveUncBoundary`, `getInverseAdditiveUncBoundary` etc.

Due to the fact that the sampled points from the previous paragraph cannot be directly accounted for in robust control synthesis and, moreover, they may not represent an actual transfer function, it appears the need to compute a least conservative low-order transfer function to model the desired uncertainty family. This problem has been solved by employing a global optimization algorithm, such as PSO, described in [19]. The PSO algorithm has been considered in favor of other global optimization algorithms, such as GA, due to its inherent structure of addressing semi-continuous functions, such as our strongly nonlinear semi-continuous function described in Equation (21) in the variable from Equation (19).

A particle $\mathbf{x} := \begin{bmatrix} x_1 & x_2 & \cdots & x_{n_4} \end{bmatrix}^T$ of the optimization problem is defined as the transfer function

$$W_x(s) = \frac{k}{s^p} \cdot \frac{\prod(\mathring{T}s + 1)\prod\left(\frac{s^2}{\mathring{\omega}_n^2} + \frac{2\mathring{\zeta}}{\mathring{\omega}_n}s + 1\right)}{\prod(\hat{T}s + 1)\prod\left(\frac{s^2}{\hat{\omega}_n^2} + \frac{2\hat{\zeta}}{\hat{\omega}_n}s + 1\right)} = \frac{x_1}{s^p} \cdot \frac{\prod_{1}^{n_1}(x_k s + 1)\prod_{n_1+1}^{n_2}\left(\frac{s^2}{x_k^2} + \frac{2x_{k+1}}{x_k}s + 1\right)}{\prod_{n_2+1}^{n_3}(x_k s + 1)\prod_{n_3+1}^{n_4}\left(\frac{s^2}{x_k^2} + \frac{2x_{k+1}}{x_k}s + 1\right)}, \tag{19}$$

with optimization variables

$$k > 0; \left\{\mathring{T}_i \in \left[\frac{1}{\overline{\omega}}, \frac{1}{\underline{\omega}}\right]\right\}, i \in \overline{2:n_1}; \quad \left\{\hat{T}_i \in \left[\frac{1}{\overline{\omega}}, \frac{1}{\underline{\omega}}\right]\right\}, i \in \overline{n_2+1:n_3}; \tag{20a}$$

$$\left\{\left(\mathring{\zeta}_i \in (0,1), \mathring{\omega}_{n,i+1} \in [\underline{\omega}, \overline{\omega}]\right)\right\}, i \in \overline{n_1+1:2:n_2}; \tag{20b}$$

$$\left\{\left(\hat{\zeta}_i \in (0,1), \hat{\omega}_{n,i+1} \in [\underline{\omega}, \overline{\omega}]\right)\right\}, i \in \overline{n_3+1:2:n_4}. \tag{20c}$$

The cost functional found to provide good results for most initial particle swarms is

$$J_{\Omega, W_{exp}}(W_x) = \sum_{k=1}^{N} \left| |W_{exp}(j\omega_k)|^{dB} - |W_x(j\omega_k)|^{dB} \right| \cdot \varphi_N(k) + \sum_{|W_x(j\omega_k)| < |W_{exp}(j\omega_k)|} \lambda, \quad (21)$$

where $\varphi_N : \{1, 2, \ldots, N\} \to \mathbb{R}^+$ from the first sum is a windowing function, based on the Gaussian window function, meant to amplify the penalization for low- and high-frequency components, while the second sum adds a high-cost term $\lambda = 10^9$ for each frequency point $j\omega_k$ for which the candidate function W_x is below the experimental reference uncertainty weight W_{exp}. The windowing function considered is defined by

$$\varphi_N(k) = \left(3 - 2 \cdot w_N \left(k - \frac{N+1}{2} \right) \right)^2, \quad (22)$$

with the Gaussian function $w_N : \mathbb{R} \to \mathbb{R}^+$ and value $\alpha = 2.5$:

$$w_N(x) = e^{-\frac{1}{2}\left(\alpha \frac{x}{(N-1)/2}\right)^2} = e^{-\frac{x^2}{2\sigma^2}} \Leftrightarrow \log w_N(x) = -\frac{1}{2}\left(\alpha \frac{x}{(N-1)/2}\right)^2 = -\frac{x^2}{2\sigma^2}. \quad (23)$$

The cost functional $J(W_x)$ is configurable, such that a different windowing function may be provided, or that the difference between W_{exp} and W_x in the integral may be considered in absolute magnitude values instead of decibels. An alternative formulation of this minimization problem would be to use a GA, where the optimization variable $x = W_x(s)$ could mutate in order to obtain different transfer function structures: add or remove real poles and zeros or, also, complex conjugate pole and zero pairs.

This functionality is implemented in the class `UncertaintyBoundOptimizationProblem`, using the methods `getTransferFunctionCandidate` for Equation (19), `computeCandidateFitness` and `fitnessIntegral` for Equation (21) based on the magnitudes of $|W_{exp}(j\omega)|$ and $|W_x(j\omega)|$ for $\omega \in \Omega$ from Equation (18) and `optimize` to use the PSO algorithm to compute the best candidate transfer function $W_{x,optim}(s)$. The plot function has been overloaded to facilitate seeing the fitness function values in real-time and, moreover, the Bode plot for the best candidate $W_x(s)$ compared to the experimental data $W_{exp}(s)$.

2.6. Robust Synthesis and Closed Loop Validation

Classical solutions to the $\mathcal{H}_2/\mathcal{H}_\infty$ problems are presented in [20–24] and others. The control problem is typically formulated for the nominal plant using the generalized framework depicted in Figure 5a.

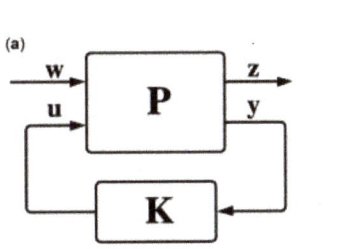
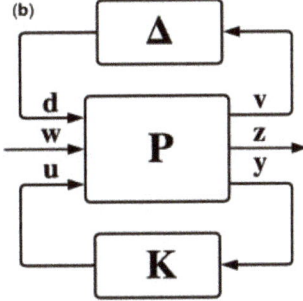

Figure 5. (a) Generalized plant framework; (b) generalized plant framework with uncertainties.

This generalized plant is obtained by augmenting the physical process model with a set of mathematical signals which aid the optimization procedure and has the following structure:

$$\begin{pmatrix} \mathbf{z}(t) \\ \mathbf{y}(t) \end{pmatrix} = \begin{pmatrix} P_{11} & P_{12} \\ P_{21} & P_{22} \end{pmatrix} \begin{pmatrix} \mathbf{w}(t) \\ \mathbf{u}(t) \end{pmatrix}, \text{ with } \begin{pmatrix} P_{11} & P_{12} \\ P_{21} & P_{22} \end{pmatrix} = \begin{pmatrix} A & B_1 & B_2 \\ \hline C_1 & D_{11} & D_{12} \\ C_2 & D_{21} & D_{22} \end{pmatrix}, \quad (24)$$

where $\mathbf{w} \in \mathbb{R}^{n_w}$ is the exogenous input vector, $\mathbf{u} \in \mathbb{R}^{n_u}$ is the control input vector, $\mathbf{z} \in \mathbb{R}^{n_z}$ is the error (output, performance) vector, and $\mathbf{y} \in \mathbb{R}^{n_y}$ is the measurement vector. The closed loop system is given by the LLFT interconnection of P and K:

$$P_{zw} = \text{LLFT}(P, K) = P_{11} + P_{12}K(I - P_{22}K)^{-1}P_{21}. \quad (25)$$

For the nominal plant, the target of the robust control problem is to minimize the $\mathcal{H}_\infty$ norm using a stabilizing controller K, which can be written as

$$\min_{K \text{ stab.}} \|P_{zw}\|_\infty = \min_{K \text{ stab.}} \sup_{\omega \in \mathbb{R}^+} \overline{\sigma}(P_{zw}(j\omega)), \quad (26)$$

obtaining a (sub)optimal value γ by iteration, which minimizes the effects of the input vector $\mathbf{w}(t)$ as seen through the performance output vector $\mathbf{z}(t)$.

However, this problem ensures only nominal stability and nominal performance. However, the plant is a model of a physical process, having uncertainties. There are two types of uncertainties: unstructured, which illustrates neglected and unmodelled dynamics and which are represented by a full block $\Delta \in \mathbb{R}^{m \times m}$, and parametric, which are represented by δI, where δ is the maximum bound of the variable parameter. In a mixed-scenario, the following set is considered:

$$\mathbf{\Delta} = \left\{ \Delta = \text{diag}\left(\delta_1 I_{n_1}, \ldots, \delta_s I_{n_s}, \Delta_1, \ldots, \Delta_f\right) | \delta_k \in \mathbb{R}, \Delta_j \in \mathbb{R}^{m_j \times m_j}, k = \overline{1,s}, j = \overline{1,f} \right\}. \quad (27)$$

In Figure 5b, the closed loop system containing a LLFT connection between plant P and controller K and an ULFT connection between plant P and uncertainty block Δ is presented. In this case, the generalized plant contains one extra input vector, i.e., disturbance inputs $\mathbf{d} \in \mathbb{R}^{n_d}$, and one extra output vector, i.e., disturbance outputs $\mathbf{v} \in \mathbb{R}^{n_v}$, giving the following structure:

$$P_\Delta(s) = \begin{pmatrix} P_{vd}(s) & P_{vw}(s) & P_{vu}(s) \\ P_{zd}(s) & P_{zw}(s) & P_{zu}(s) \\ P_{yd}(s) & P_{yw}(s) & P_{yu}(s) \end{pmatrix} \Leftrightarrow P_\Delta : \begin{pmatrix} \dot{x}(t) \\ \mathbf{v}(t) \\ \mathbf{z}(t) \\ \mathbf{y}(t) \end{pmatrix} = \begin{pmatrix} A & B_d & B_w & B_u \\ \hline C_v & D_{vd} & D_{vw} & D_{vu} \\ C_z & D_{zd} & D_{zw} & D_{zu} \\ C_y & D_{yd} & D_{yw} & D_{yu} \end{pmatrix} \begin{pmatrix} x(t) \\ \mathbf{d}(t) \\ \mathbf{w}(t) \\ \mathbf{u}(t) \end{pmatrix}. \quad (28)$$

A mathematical tool used for studying the robustness is the *structured singular value*, defined for a square matrix $M \in \mathbb{C}^{N \times N}$ with respect to the set $\mathbf{\Delta}$ as

$$\mu_\Delta(M) = \frac{1}{\min_{\Delta \in \mathbf{\Delta}}\{\overline{\sigma}(\Delta) | \det(I - M\Delta) = 0\}}, \quad (29)$$

if there exists $\Delta \in \mathbf{\Delta}$ such that the matrix $I - M\Delta$ is rank deficient; otherwise, it is 0. For the system presented in Figure 5b, the structured singular value of $\text{LLFT}(P, K)$, according to Δ, can be defined as

$$\mu_\Delta(\text{LLFT}(P, K)(s)) = \sup_{\omega \in \mathbb{R}_+} \mu_\Delta(\text{LLFT}(P, K)(j\omega)). \quad (30)$$

Besides the classical $\mathcal{H}_2/\mathcal{H}_\infty$ techniques, the μ synthesis framework manages to design a controller that meets the robust stability and robust performance specifications. The robust stability implies that a certain controller manages to stabilize all processes described by the upper linear fractional transformation between plant and uncertainty block, while the robust performance means that the controller is able to impose the desired closed loop performance in the worst-case scenario. Based on the main loop theorem, a controller K meets the robust stability and robust performance if and only if the structural singular value of the lower linear fractional transformation with respect to Δ is smaller than 1. Therefore, the minimization problem can be written as

$$\inf_{K \text{ stab.}} \sup_{\omega \in \mathbb{R}_+} \mu_\Delta(\text{LLFT}(P,K)(j\omega)), \tag{31}$$

which is not a convex problem. Additionally, the structural singular values are difficult to be explicitly computed. In order to solve this problem, the following upper bound is used [25]:

$$\mu_\Delta(\text{LLFT}(P,K)(j\omega)) \leq \inf_{D \in \mathcal{D}} \bar{\sigma}(D \cdot \text{LLFT}(P,K)(j\omega) \cdot D^{-1}), \tag{32}$$

where the set $\mathcal{D}$ is defined in relation to the uncertainty set Δ as

$$\mathcal{D} = \left\{ \text{diag}\left(D_1, \ldots, D_s, d_1 I_{m_1}, \ldots, d_f I_{m_f}\right) \middle| D_k = D_k^\top \in \mathbb{R}^{n_k \times n_k}, d_j > 0, k = \overline{1,s}, j = \overline{1,f} \right\}. \tag{33}$$

Now, using this bound, the solution of the initial non-convex problem can be practically approximated by solving the following quasi-convex problem:

$$\inf_{K \text{ stab.}} \sup_{\omega \in \mathbb{R}_+} \inf_{D \in \mathcal{D}} \bar{\sigma}\left(D(j\omega) \cdot \text{LLFT}(P,K)(j\omega) \cdot (D(j\omega))^{-1}\right). \tag{34}$$

Finally, if the system D is fixed, the problem (34) is nothing but a $\mathcal{H}_\infty$ control problem, in this case called the K step. Furthermore, for a fixed controller K, the D scale step can be obtained by solving a Parrot problem for a desired set of frequencies $\Omega = \{\omega_1, \ldots, \omega_N\}$ using a LMI and then obtain a minimum phase system after performing an identification step. Using these, an iterative algorithm, based on alternative D-K iterations, manages to solve the μ synthesis problem. This procedure starts with $D = I$ and successively applies a K step and a D scaling step until a stopping criterion is reached.

Of great use in the controller design phase are the sensitivity, complementary sensitivity, and control effort functions, respectively, defined by

$$S := (I + GK)^{-1}; \tag{35a}$$

$$T := GK(I + GK)^{-1}; \tag{35b}$$

$$R := K(I + GK)^{-1} = KS, \tag{35c}$$

where G is the open loop model. The great advantage of considering this approach is that it allows sculpting the relevant loop functions to impose steady-state and transitory regime performances, which are specified for different frequency ranges, using adequately selected weighting functions. Besides the minimization from Equation (34), different constraints can be added to the optimization problem to obtain a compromise between S, KS, and T at various frequencies:

$$\inf_{K \text{ stab.}} \sup_{\omega \in \mathbb{R}_+} \inf_{D \in \mathcal{D}} \bar{\sigma}\left(D(j\omega) \cdot \text{LLFT}(P,K)(j\omega) \cdot (D(j\omega))^{-1}\right), \tag{36a}$$

$$\text{such that } \left\| \begin{pmatrix} W_S S & W_T T & W_{KS} KS \end{pmatrix}^T \right\|_\infty < 1, \tag{36b}$$

also known as the mixed-sensitivity closed loop shaping μ synthesis method. The approach considered in this first iteration of the toolbox uses closed loop shaping with μ synthesis for the controller. Using the work in [26] as a starting point, different frequency response specifications, directly correlated with desired time-response performances, can be imposed in the weighting functions. The sensitivity weighting function W_S depends on four parameters as in

$$W_S(s) = \left(\frac{\frac{1}{M^{1/n}}s + \omega_B}{s + \omega_B A^{1/n}} \right)^n, \tag{37}$$

where ω_B represents the imposed bandwidth of the system; M imposes the $\mathcal{H}_\infty$ norm of the sensitivity function, in order to limit the overshoot of the system; n imposes the slope of the sensitivity function for low frequencies; and A imposes the maximum allowed steady-state error.

On the other hand, the complementary sensitivity weighting function W_T can be generally defined by the following structure, in a *symmetrical* manner compared to W_S:

$$W_T(s) = \left(\frac{s + \omega_{BT}}{A_T^{1/n}s + \omega_{BT} M_T^{1/n}} \right)^n, \tag{38}$$

with ω_{BT} being the imposed bandwidth of the system; M_T imposes the $\mathcal{H}_\infty$ norm of $T(s)$; n imposes the roll-off slope of the closed loop system, which should be directly coupled with sensor noise characteristics; and A_T imposes the least required attenuation for high frequencies. In practice, the complementary sensitivity bandwidth ω_{BT} can be adapted to the characteristics of the sensor in order to account for high-frequency noise.

Finally, the control effort weighting function with desired specifications $M_0 := |W_{KS}(0)|$, $M_\infty := |W_{KS}(\infty)|$ and $|W_{KS}(j \cdot \omega_d)| = M_d$, $M_0 < M_d < M_\infty$ can be synthesized by the following formula:

$$W_{KS}(s) = \frac{M_\infty s + M_0 \omega_d \sqrt{\frac{M_\infty^2 - M_d^2}{M_d^2 - M_0^2}}}{s + \omega_d \sqrt{\frac{M_\infty^2 - M_d^2}{M_d^2 - M_0^2}}}. \tag{39}$$

A higher-order counterpart can be generalized as for the previous cases, but was not found necessary for the proposed case studies and other tested benchmark plants.

Class `RobustControlSynthesisProblem` uses the nominal linearized plant around a required operating point using the system (14) and options (11)–(13), with a specified uncertainty type from Table 1, modeled through class `UncertaintyBoundOptimization-Problem`, allows imposing closed loop performance specifications with frequency weights (37)–(39), and synthesizes controller solutions for the problem (36) which cover the robust stability and robust performance problems. Additionally, the class also allows controller postprocessing, using order-reduction methods to compute easily implementable controllers. The aforementioned controller synthesis problem is illustrated in Figure 6, while the resulting LTI controller can be used in a MiL simulation context using the interface from class `LTIEqSystem`, encompassing system (3).

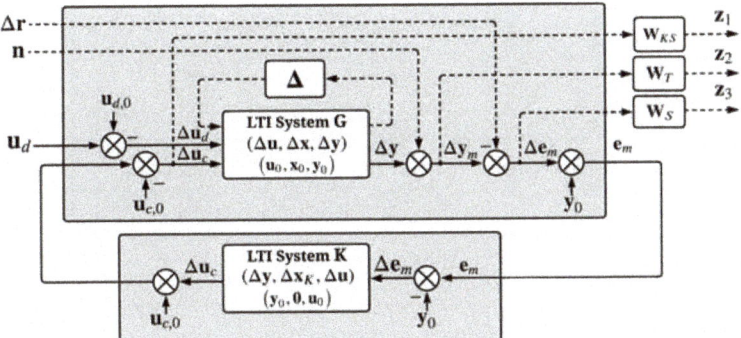

Figure 6. Robust controller synthesis diagram for the plant G, with the uncertainty set Δ determined a priori using the functionality from Section 2.5, linearized at the operating point $(\mathbf{u}_0, \mathbf{x}_0, \mathbf{y}_0)$; by convention, control inputs $\Delta \mathbf{u}_c$ of the linearized plant G are indexed after disturbance inputs $\Delta \mathbf{u}_d$.

Class `ClosedLoopControlProblem` gathers all data, options, computations, and results from the previously presented individual blocks. It is used to define the end-to-end control problem by aggregating the uncertain plant family with its corresponding least conservative bound optimization; the robust control synthesis procedure and metadata; and allows Model-in-the-Loop simulations with the nonlinear, linearized, and hybrid system models, with automatic validation of imposed performances for the nonlinear plant in the frequency and time-domain alike.

3. Results

To showcase the ease of use and functionalities of the proposed toolbox, a set of case studies will be illustrated for DC-to-DC power converter circuits in a unified manner to encompass modeling, control synthesis and performance validation, considering the topologies buck, boost, and single-ended primary-inductor converter (SEPIC). DC-to-DC converters have been considered as a case study due to their ubiquity in various practical domains and applications, as presented in [27], ranging from renewable energy, hybrid and electric vehicles, controlled power sources, and many more. They can be seen as a good benchmark for control systems, due to their switching behavior, nonlinear dynamics, and different tried and tested control methods, such robust techniques in [28], Lyapunov methods in [29], passivity theory in [30], or sliding mode control as in [31].

This section will be split in a subsection which presents the converter mathematical models, a subsection with a suggested workflow for a general purpose control problem, followed by a subsection with numerical results and simulations for each of the studied converter topologies. For the SEPIC converter, having the most highly nonlinear behavior, thus being more difficult to control, we will illustrate and detail all plots generated by the toolbox, while for the buck and boost circuits, for brevity, we will show only the relevant figures and maintain the mathematical results and discussions.

3.1. Mathematical Modeling

The nonideal step-down (buck), step-up (boost), and single-ended primary-inductor converter (SEPIC) circuits are presented in Figure 7, where each component is described as follows.

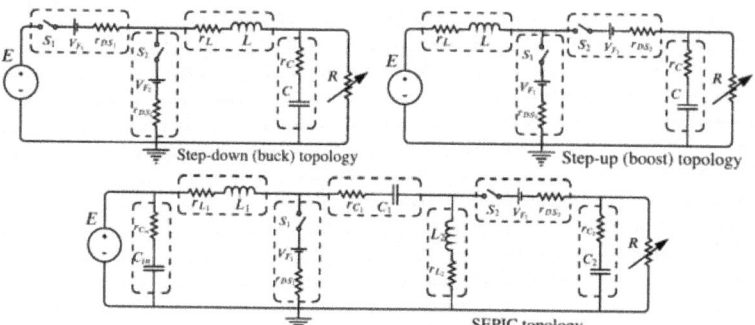

Figure 7. DC-to-DC power converter circuit topologies considered in the case studies: step-down (buck), step-up (boost), and single-ended primary-inductor (SEPIC).

- S_1: switching device, usually a transistor, and S_2: switching device, usually a diode or transistor;
- L, L_1, L_2: converter inductors;
- C, C_{in}, C_1, C_2: converter capacitors;
- R: (variable) output load resistance;
- E: external source voltage;
- r_L, r_{L_1}, r_{L_2}: resistances associated with the inductors;
- $r_C, r_{C_{in}}, r_{C_1}, r_{C_2}$: capacitors parasitic resistances;
- r_{DS_1}, r_{DS_2}: resistances associated with the ON state of the switching devices (usually drain source);
- V_{F_1}, V_{F_2}: constant voltage drops associated with the conducting phase of S_1 and S_2;
- $\mu \in [0, 1]$: normalized duty cycle applied to S_1; complementary to the PWM signal applied to S_2.

Each converter has one control switching device S_1, while the other one, S_2, will be complementary to the former. Although, typically S_2 is a diode, it is preferable to use two encapsulated transistors for S_1 and S_2. When working in continuous conduction mode (CCM), all converters will have an ON state, corresponding to S_1 being on and S_2 off, along with an OFF state, for its complementary behavior. The corresponding LTI models, for a constant load resistance R, for the ON and OFF states will be presented using the following structure with the external voltage seen as disturbance input $u(t) = E(t)$, the voltage drops V_{F_1} and V_{F_2} as constant DC inputs, states from the inductor currents and capacitor voltages, and the load resistor voltage as measured output:

$$\left(\begin{array}{c|c} A & \overline{B} \\ \hline C & \overline{D} \end{array}\right) = \left(\begin{array}{c|c:c} A & B & B_V \\ \hline C & D & D_V \end{array}\right), \tag{40}$$

where:

$$\begin{cases} \dot{x} = A_{ON}x + B_{ON}E + B_{V,ON}[V_{F_1}\,V_{F_2}]; \\ y = C_{ON}x + D_{ON}E + D_{V,ON}[V_{F_1}\,V_{F_2}], \end{cases} \begin{cases} \dot{x} = A_{OFF}x + B_{OFF}E + B_{V,OFF}[V_{F_1}\,V_{F_2}]; \\ y = C_{OFF}x + D_{OFF}E + D_{V,OFF}[V_{F_1}\,V_{F_2}]. \end{cases} \tag{41}$$

The control variable is the duty cycle of the switching devices $\mu(t) \in [0, 1]$. Using a convex combination of the ON and OFF equation systems from Equation (41), an averaged state-space nonlinear model of the process is obtained close to the hybrid model's behavior given a sufficiently high PWM frequency:

$$\dot{x}(t) = \mu(t) \cdot x_{ON}(t) + (1 - \mu(t)) \cdot x_{OFF}(t) \equiv F\Big(x(t), [E(t), R(t), \mu(t)]^T, t\Big). \tag{42}$$

As such, an affine nonlinear system, with respect to μ, with the state function F as above can be implemented by inheriting the class `System` from the toolbox. The disturbances which affect the system are the voltage source E and variable output load R, which are stochastic in nature, along with uncertainties of its components due to manufacturing tolerances, relevant on inductors and capacitors. As the toolbox easily allows using the output capacitor voltage or output load voltage with minor modifications, we used the resistor voltage as measurement variable due to its corresponding practical control use cases. By inheriting the class `UncertainPlantFactory`, a set of tolerances can be imposed on all relevant circuit parameters and, also, an LTI uncertain set can be automatically computed with the provided mechanisms. The equilibrium point will have only the steady-state values of $\overline{E}, \overline{R}$, and $\overline{u}_R$ imposed, with the following structure:

$$(\overline{\mathbf{u}}, \overline{\mathbf{x}}, \overline{\mathbf{y}}) = \left([\overline{E}, \overline{R}, \overline{\mu}], [\overline{i}_{L_1}, \overline{u}_{C_1}, \ldots], \overline{u}_R \right). \tag{43}$$

After synthesizing a robust controller, model-in-the-loop simulations would be desired for the averaged state-space models and, also, for a hybrid model description of the converters. For the hybrid approach, the class `UncertainPlantFactory` is inherited again, this time with individual plants of type `HybridSystem`. For the general approach for CCM, based on the structure from Equation (4), the input vector is comprised of $\mathbf{u} = [E, R, \mu]^T$, the state vector is extended to $\mathbf{x} := [\mathbf{z}^T, q, \tau]^T$, with $\mathbf{z} = [i_{L_1}, u_{C_1}, \ldots]^T$ being the physical continuous states; $q \in \{0,1\}$ the discrete state number, i.e., ON and OFF; and $\tau \in [0, T_{PWM})$ the time values for a single PWM period. After each PWM period completion, the auxiliary time state τ is reinitialized to 0. The model description becomes, for the state-space description

$$\begin{cases} \begin{bmatrix} \dot{\mathbf{z}}(t) \\ \dot{q}(t) \\ \dot{\tau}(t) \end{bmatrix} = \begin{bmatrix} (1-q) \cdot (A_{ON}\mathbf{z}(t) + \overline{B}_{ON}\mathbf{u}(t)) + q \cdot (A_{OFF}\mathbf{z}(t) + \overline{B}_{OFF}\mathbf{u}(t)) \\ 0 \\ 1 \end{bmatrix}, (\mathbf{x}, \mathbf{u}, t) \in \mathcal{C}; & (44a) \\ \begin{bmatrix} \mathbf{z}^+(t) \\ q^+(t) \\ \tau^+(t) \end{bmatrix} = \begin{bmatrix} \mathbf{z}(t) \\ \begin{cases} 1, \text{ if } q == 0 \\ 0, \text{ if } q == 1 \end{cases} \\ \begin{cases} \tau, \text{ if } q == 0 \\ 0, \text{ if } q == 1 \end{cases} \end{bmatrix}, (\mathbf{x}, \mathbf{u}, t) \in \mathcal{D}, & (44b) \end{cases}$$

while the flow, jump, and output functions are

$$\begin{cases} \mathcal{C}(\mathbf{x}, \mathbf{u}, t) = \{((q == 0) \land (\tau \leq \mu(t) \cdot T_{PWM})) \lor ((q == 1) \land (\tau > \mu(t) \cdot T_{PWM}))\}; & (45a) \\ \mathcal{D}(\mathbf{x}, \mathbf{u}, t) = \{((q == 0) \land (\tau > \mu(t) \cdot T_{PWM})) \lor ((q == 1) \land (\tau > T_{PWM}))\}; & (45b) \\ \mathbf{y}(t) = h(\mathbf{x}(t), \mathbf{u}(t), t). & (45c) \end{cases}$$

In order to encompass DCM regimes, the number of discrete states must be extended with new LTI blocks obtained by adding the mathematical constraint of canceling the diode S_2 voltage and, as such, the corresponding current signal for that branch, along with more sophisticated jump functions, flow and jump sets. For brevity, we will not insist on these extensions, although an example described for the boost converter can be found in [32].

3.2. Toolbox Workflow

A suggested end-to-end workflow for the toolbox can be summarized in the following steps, all of which should be run from intermediary methods of an instance of class `ClosedLoopControlProblem`:

- inherit class `System` to define the nonlinear model of the process as in Equation (1) and Figure 1;
- define equilibrium point specifications as in (11)–(13);

- inherit class `UncertainPlantFactory` and overload method `getRandomPlant`;
- using the desired operating point, linearize a set of systems using (16) and experimentally determine an uncertainty model W_{exp} based on Table 1;
- define uncertainty options, including transfer function structure for all particles, and execute the methods from class `UncertaintyBoundOptimizationProblem` in order to minimize the functional $J_{\Omega,W_{exp}}(W_x)$ from Equation (21), obtaining $W_{x,opt}$;
- run optimization to compute the uncertainty weight $W_{unc}(s)$ as in Table 1;
- define robust control specifications W_S, W_T, and W_{KS} as in Equations (37)–(39);
- synthesize robust μ controller based on Figure 6;
- apply order-reducing methods on the resulting controller;
- validate frequency and time-response performance specifications using the nonlinear system at operating point through Model-in-the-Loop simulations;
- optionally, inherit the classes `HybridSystem` and `UncertainPlantFactory`, respectively, to validate time-response performance specifications using the corresponding hybrid plant model at operating point; for DC-to-DC converter control, the last step should be adapted for CCM or DCM operation.

3.3. Numerical Results

We will briefly present the obtained results for the three converters using the approaches established in Sections 3.1 and 3.2.

3.3.1. SEPIC Converter

The SEPIC converter state-space model for the ON state of switch S_1, as structured in Equation (41), is

$$\left(\begin{array}{ccccc|ccc}
-\frac{1}{r_{C_{in}}C_{in}} & 0 & 0 & 0 & 0 & \frac{1}{r_{C_{in}}C_{in}} & 0 & 0 \\
\frac{1}{L_1} & -\frac{r_{C_{in}}+r_{L_1}+r_{DS_1}}{L_1} & 0 & \frac{r_{DS_2}}{L_1} & 0 & 0 & -\frac{1}{L_1} & 0 \\
0 & 0 & 0 & \frac{1}{C_1} & 0 & 0 & 0 & 0 \\
0 & -\frac{r_{DS_1}}{L_2} & -\frac{1}{L_2} & -\frac{r_{DS_1}+r_{C_1}+r_{L_2}}{L_2} & 0 & 0 & \frac{1}{L_2} & 0 \\
0 & 0 & 0 & 0 & -\frac{1}{(R+r_{C_2})C_2} & 0 & 0 & 0 \\
\hline
0 & 0 & 0 & 0 & \frac{R}{R+r_{C_2}} & 0 & 0 & 0
\end{array}\right), \quad (46)$$

while for the OFF state of the switch S_1 is

$$\left(\begin{array}{ccccc|ccc}
-\frac{1}{r_{C_{in}}C_{in}} & 0 & 0 & 0 & 0 & \frac{1}{r_{C_{in}}C_{in}} & 0 & 0 \\
\frac{1}{L_1} & -\frac{r_{aux}}{L_1} & -\frac{1}{L_1} & \frac{r_{DS_2}+r_{C_2}}{L_1} & -\frac{1}{L_1} & 0 & 0 & -\frac{1}{L_1} \\
0 & \frac{1}{C_1} & 0 & 0 & 0 & 0 & 0 & 0 \\
0 & \frac{r_{DS_2}+r_{C_2}}{L_2} & 0 & -\frac{r_{DS_2}+r_{C_2}+r_{L_2}}{L_2} & \frac{1}{L_2} & 0 & 0 & \frac{1}{L_2} \\
0 & \frac{R}{(R+r_{C_2})C_2} & 0 & -\frac{R}{(R+r_{C_2})C_2} & -\frac{1}{(R+r_{C_2})C_2} & 0 & 0 & 0 \\
\hline
0 & \frac{r_C R}{R+r_C} & 0 & -\frac{r_C R}{R+r_C} & \frac{R}{R+r_{C_2}} & 0 & 0 & 0
\end{array}\right), \quad (47)$$

with the auxiliary notation $r_{aux} = r_{L_1} + r_{C_1} + r_{DS_1} + r_{C_2} + r_{C_{in}}$.

The nominal SEPIC converter parameters and their tolerances are presented in Table 2.

Table 2. Single-ended primary-inductor converter (SEPIC) converter parameters, values, and corresponding tolerances.

Param.	Val.	Tol.	Param.	Val.	Tol.
L_1	2.57 [mH]	±20%	L_2	1.71 [mH]	±20%
r_{L_1}	130 [mΩ]	±10%	r_{L_2}	110 [mΩ]	±10%
r_{DS_1}	0.01 [Ω]	±10%	r_{DS_2}	80 [mΩ]	±10%
C_1	4.7 [µF]	±20%	C_2	3.57 [µF]	±20%
r_{C_1}	270 [mΩ]	±10%	r_{C_2}	350 [mΩ]	±10%
C_{in}	3.57 [µF]	±20%	$r_{C_{in}}$	270 [mΩ]	±10%
V_{F_1}	0.2 [V]	±10%	V_{F_2}	0.62 [V]	±10%

The desired operating point specifications are output signal $y(t) \equiv u_R(t)$ is at 400 [V], with nominal voltage source and load inputs $u_1(t) = E = 300$ [V], $u_2(t) = R = 80$ [Ω]. The initial guesses for the state equilibrium values where $\tilde{x} = [30, 0.5, 30, -0.5, 30]$ and $u_3(t) = \mu = 0.57$ for the duty cycle control input. After computation, the actual equilibrium point is $(\mathbf{u}, \mathbf{x}, \mathbf{y}) = ([300, 80, 0.5788], [300, 6.8711, 297.722, -5, 400], 400)$.

An input multiplicative uncertainty model, i.e., $G(s) = G_n(s)[1 + \Delta(s)W_{unc}(s)]$, $||\Delta||_\infty \leq 1$, as in Table 1 has been automatically computed from input $u_3(t)$ to output $y_1(t)$, with tolerances ±10 [V] and ±5 [Ω] for inputs $u_1(t) = E$ and $u_2(t) = R$, based on 1000 Monte Carlo simulations. The relevant frequencies vary in the interval $[\underline{\omega} = 10^{-2}, \overline{\omega} = 10^8]$, with 300 equally distributed samples in log domain. A successful set of hyperparameters for the particle swarm optimization algorithm is comprised of a swarm size of 1000, initial swarm span of 10^4, minimum neighbors fraction of 0.9, and inertia range of $[0.1, 1.1]$, for a transfer function structure as in Equation (19), with a complex pole pair and a complex zero pair, resulting in

$$W_{unc}(s) = \frac{0.67275(s^2 + 941.1s + 2.222 \times 10^5)}{s^2 + 147.3s + 5.422 \times 10^7}. \quad (48)$$

The linearized SEPIC plant family is comprised of fourth-order stable systems, in minimal form, with four zeros, three of which are of nonminimum phase. The nominal model is

$$G_n(s) = 4.1368 \frac{(s + 8.003\text{e}+5)(-s + 2.304\text{e}+4)(s^2 - 717.4s + 5.145\text{e}+7)}{(s^2 + 2673s + 3.749\text{e}+7)(s^2 + 1339s + 6.493\text{e}+7)}. \quad (49)$$

The entire uncertainty family has the same structure with poles, zeros, and equilibrium points in the vicinity of the nominal counterparts. The uncertain SEPIC plant family structure and behavior, along with the PSO conservative bound computation are illustrated in Figure 8. Figure 8-1 illustrates the pole-zero plot for the linearized uncertain SEPIC converter family, Figure 8-2 shows the best particle frequency-response fit $W_x(s)$ as in Equation (19) on the right y axis, and, also, the best functional fit on the left y axis, Figure 8-3 and 8-4 show the frequency response of the plant $G(s)$ family and uncertainty family $\Delta(s)W_{unc}(s)$, respectively, while Figure 8-5 illustrates the system states and outputs for a 2% step disturbance relative to the equilibrium input value $\mu_0 = 0.5788$.

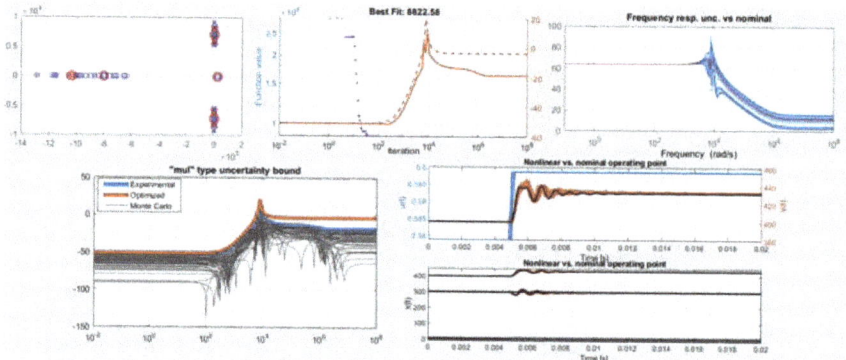

Figure 8. SEPIC multiplicative uncertainty set computation and open loop responses for the operating point with $E = 300[V]$, $R = 80[\Omega]$, $U_R = 400[V]$: step and frequency response, pole-zero placement; the step response is simulated for nonlinear plants sampled using `UncertainPlantFactory`.

For controller synthesis, the preferred loop shaping specifications were selected to highly penalize the control effort near the system resonance, as the obtained controllers would be difficult to implement in practice, needing high sampling frequencies, i.e., $f_e > 20$ [kHz]. As such, for the sensitivity function: $\omega_B = 200$ [rad/s], $A = 10^{-2}$, $M = 2$, $n = 1$; for the complementary sensitivity function: $\omega_{BT} = 2000$ [rad/s], $A_T = 10^{-4}$, $M_T = 2$, $n = 2$; and for the control effort $M_0 = 100$, $M_\infty = 10^5$, $|W_{KS}(j \cdot 200)| = 250$, resulting in

$$W_S(s) = \frac{0.5s + 200}{s + 2}, \quad W_T(s) = \frac{s^2 + 4000s + 4 \times 10^6}{1 \times 10^{-4}s^2 + 56.57s + 8 \times 10^6}, \quad W_{KS}(s) = \frac{10^5 s + 8.729 \times 10^6}{s + 8.729 \times 10^4}. \tag{50}$$

From the μ synthesis procedure, a controller of order 21 is obtained. After order reduction, the smallest controller which manages to assure all imposed specifications for the plant family, with a peak value $\mu_\Delta(\text{LLFT}(P, K)) \leq 0.8361 < 1$, is given by the third-order system

$$K_{red}^{SEPIC} = \left(\begin{array}{ccc|c} -1.997 & 3.056 & 3.227 & -0.5018 \\ -3.057 & -2197 & -6118 & -0.3838 \\ 3.225 & 6118 & -1.016 \times 10^4 & 0.4055 \\ \hline -0.5018 & 0.3838 & 0.4055 & 0 \end{array} \right). \tag{51}$$

The controller design phase, order reduction, and frequency response closed loop performance of the reduced-order one for the uncertain plant family are illustrated in Figures 9 and 10. In this case, the control system has very large stability margins, with a phase margin of $\approx 82.1[°]$ and gain margin in the interval $[19.4, 20.3]$ [dB]. Additionally, as specified by the $n = 2$ and $A_T = 10^{-4}$ parameters of the complementary sensitivity weighting function, W_T, the closed loop control system mitigates sensor noise signals with a considered spectrum starting from $\omega_{BT} > 2000$ [rad/s], using an initial roll-off of -40 [dB/dec], followed by an attenuation of at least four orders of magnitude. In the actual MiL simulations, the attenuation does not stop at the prescribed value, as the system manages to maintain at least a -20 [dB/dec] roll-off.

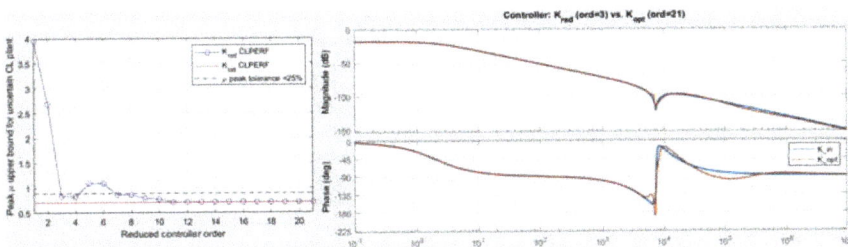

Figure 9. SEPIC-synthesized (order 21) vs. reduced (order 3) controllers; the peak μ value does not monotonically decrease with respect to controller order due to the influence of the $\approx$3750 [rad/s] notch over the converter resonance, which was not taken into account by the order reduction mechanism.

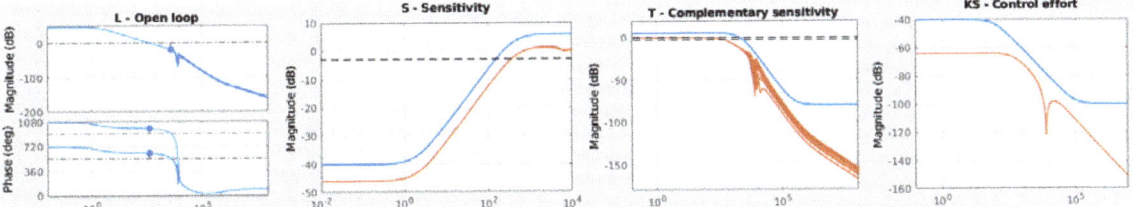

Figure 10. SEPIC open loop plant family with S, T, and KS functions using the reduced-order controller, which retains all imposed performance specifications for all test cases, provides high phase and gain margins, and guarantees closed loop response practically similar to that of a first-order low-pass filter; a relatively low bandwidth was imposed to compensate the SEPIC converter resonance and presence of multiple nonminimum phase zeros.

From Figure 11, the closed loop bandwidth can be observed as $\omega_B > 200$ [rad/s], equivalent to a rise time less than $\approx$5 [ms], a negligible steady-state error of $\approx 10^{-2} \times (y_{ss} - y_0) = 0.2$ [V], where y_{ss} represents the steady state value of the system, relative to the desired equilibrium value y_0. Moreover, the system has no overshoot, and it behaves like a first-order low-pass filter by design. The nonlinear MiL simulation options are $N = 50$ random plants from the uncertainty set, solver is `ode15i`, due to the difficulty of simulating the closed loop plant otherwise (it is numerically unstable), with a step on the reference signal of 5% from its initial equilibrium value of approximately 400 [V] and a simulation time of 0.05 [s]. Almost the same conditions apply to the hybrid MiL simulation, with the solver switched to `ode113`, as `ode15i` is not supported here, and a PWM period selected randomly for each experiment from a nominal value $T_{PWM} = 17.5$ [µs] with a $\pm 20\%$ fluctuation from one simulation to another. A comparison of the nonlinear and hybrid cases is presented also in Figure 11, where it can be seen that the transitory regime is practically identical, but for the hybrid case, a slightly lower command signal is necessary in steady state. Due to working with $u_R(t)$ instead of $u_C(t)$ only for the output signal, the current ripple is propagated into the measured voltage.

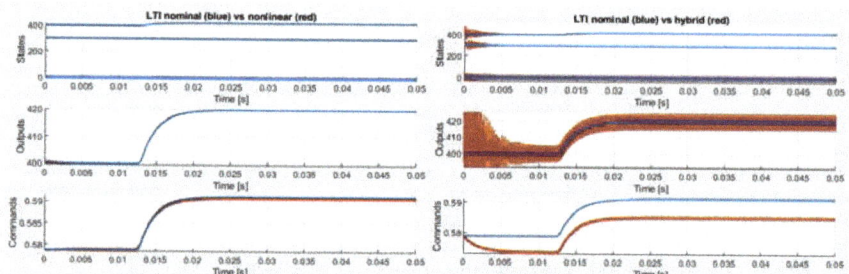

Figure 11. SEPIC-averaged state-space and hybrid model Monte Carlo closed loop simulations; due to a relatively low bandwidth imposed in order to obtain good stability margins and easily implementable controllers, all plants respond almost identically, although component tolerances reach values of ±20%; the initial transients exist due to starting the simulations from the nominal equilibrium point only.

3.3.2. Buck Converter

The buck converter state-space models for the *ON* and *OFF* states of switch S_1, respectively, as in Equation (41), are

$$\left(\begin{array}{ccc|ccc} -\frac{r_p+r_{DS_1}}{L} & -\frac{R}{(R+r_C)L} & \frac{1}{L} & -\frac{1}{L} & 0 \\ \frac{R}{(R+r_C)C} & -\frac{1}{(R+r_C)C} & 0 & 0 & 0 \\ \frac{r_C R}{R+r_C} & \frac{R}{R+r_C} & 0 & 0 & 0 \end{array}\right) ; \left(\begin{array}{ccc|ccc} -\frac{r_p-r_{DS_2}}{L} & -\frac{R}{(R+r_C)L} & 0 & 0 & -\frac{1}{L} \\ \frac{R}{(R+r_C)C} & -\frac{1}{(R+r_C)C} & 0 & 0 & 0 \\ \frac{r_C R}{R+r_C} & \frac{R}{R+r_C} & 0 & 0 & 0 \end{array}\right), \quad (52)$$

with the auxiliary notation $r_p = r_L + \frac{r_C R}{R+r_C} = r_L + r_C || R$.

The nominal buck converter parameters and their tolerances are presented in Table 3.

Table 3. Buck and boost converter parameters, values, and corresponding tolerances.

Param.	Val.	Tol.	Param.	Val.	Tol.
L	40 [µH]	±20%	r_L	10 [mΩ]	±10%
C	600 [µF]	±20%	r_C	0.2 [Ω]	±10%
r_{DS_1}	0.01 [Ω]	±10%	r_{DS_2}	0.01 [Ω]	±10%
V_{F_1}	0.2 [V]	±10%	V_{F_2}	0.2 [V]	±10%

The desired operating point specifications are as follows: the output signal $y(t) \equiv u_R(t)$ is at 5 [V], with nominal voltage source and load inputs $u_1(t) = E = 12$ [V], $u_2(t) = R = 15$ [Ω]. The initial guesses for the state equilibrium values where $\tilde{x} = [1.25, 5]$ and $u_3(t) = \mu = 0.5$ for the duty cycle control input. After computation, the actual equilibrium point is $(\mathbf{u}, \mathbf{x}, \mathbf{y}) = ([12, 15, 0.4335], [0.333, 4.999], 5)$.

An input multiplicative uncertainty model, i.e., $G(s) = G_n(s)[1 + \Delta(s)W_{unc}(s)]$, $||\Delta||_\infty \leq 1$, as in Table 1, has been automatically computed from input $u_3(t)$ to output $y_1(t)$, with tolerances ±1 [V] and ±1 [Ω] for inputs $u_1(t) = E$ and $u_2(t) = R$, respectively, based on 1000 Monte Carlo simulations. The relevant frequencies vary in the interval $[\underline{\omega} = 10^1, \overline{\omega} = 10^7]$, with 200 equally distributed samples in log domain. A successful set of hyperparameters for the particle swarm optimization algorithm is comprised of a swarm size of 1000, initial swarm span of 10^4, minimum neighbors fraction of 0.9, and inertia range of $[0.1, 1.1]$ for a transfer function structure as in Equation (19), with a real pole and a real zero, resulting in

$$W_{unc}(s) = \frac{0.51758(s+510.5)}{s+2906}. \quad (53)$$

The linearized buck plant family is comprised of second-order stable systems, in minimal form, with one zero. The nominal model is

$$G_n(s) = \frac{59178(s + 8333)}{s^2 + 5261s + 4.114 \times 10^7}. \tag{54}$$

The entire uncertainty family has the same structure with poles, zeros, and equilibrium points in the vicinity of the nominal counterparts. The uncertain buck plant family structure and behavior, along with the PSO conservative bound computation are illustrated in Figure 12.

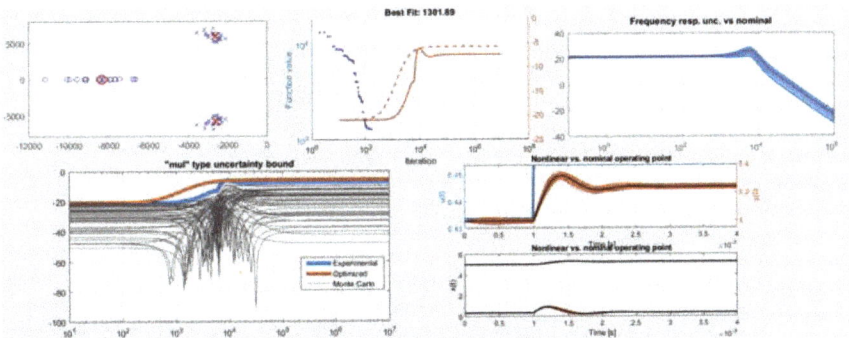

Figure 12. Buck multiplicative uncertainty set computation and open loop responses for the operating point with $E = 12[V]$, $R = 15[\Omega]$, $U_R = 5[V]$: step and frequency response, pole-zero placement; the step response is simulated for nonlinear plants sampled using UncertainPlantFactory.

For controller synthesis, the weighting function specifications were for the sensitivity function $\omega_B = 1200$ [rad/s], $A = 10^{-4}$, $M = 2$, $n = 1$; for the complementary sensitivity function $\omega_{BT} = 12 \times 10^3$ [rad/s], $A_T = 10^{-4}$, $M_T = 2$, $n = 2$; and for the control effort $M_0 = 0.1$, $M_\infty = 100$, $|W_{KS}(j \cdot 1200)| = 2$, resulting in

$$W_S(s) = \frac{0.5s + 1200}{s + 0.12}, \quad W_T(s) = \frac{s^2 + 24000s + 1.44 \times 10^8}{1 \times 10^{-4}s^2 + 339.4s + 2.88 \times 10^8}, \quad W_{KS}(s) = \frac{100s + 6006}{s + 6.006 \times 10^4}. \tag{55}$$

As specified by the $n = 2$ and $A_T = 10^{-4}$ parameters of the complementary sensitivity weighting function, W_T, the closed loop control system mitigates sensor noise signals with a considered spectrum starting from $\omega_{BT} > 12{,}000$ [rad/s], using an initial roll-off of -40 [dB/dec], followed by an attenuation of at least four orders of magnitude. From the μ synthesis, a controller of order 17 is obtained. After order reduction, the smallest controller which manages to assure all imposed specifications for the plant family, with a peak $\mu_\Delta(\text{LLFT}(P, K)) \leq 0.97 < 1$, is

$$K_{red}^{Buck} = \left(\begin{array}{ccc|c} -0.12 & -0.003479 & -0.4751 & 12.68 \\ -0.0001768 & -2.304 & -1.241 \times 10^4 & -0.1827 \\ -0.4611 & 1.241 \times 10^4 & -4.522 \times 10^4 & 25.09 \\ \hline 12.68 & 0.1838 & 25.09 & 0 \end{array} \right). \tag{56}$$

The nonlinear MiL simulation options are $N = 50$ random plants from the uncertainty set, with the ode23t solver, with a step on the reference signal of 5% from its initial equilibrium value of approximately 5 [V] and a simulation time of 0.02 [s]. The simulation conditions for the hybrid MiL case are identical, with an additional PWM period selected randomly for each experiment from a nominal value $T_{PWM} = 17.5$ [µs] with a $\pm 20\%$ fluctuation from one simulation to another. A comparison of the nonlinear and hybrid cases is presented in Figure 13, where it can be seen that the transitory regime is practically identical. Due to working with $u_R(t)$ instead of $u_C(t)$ only for the output signal, the current ripple is propagated into the measured voltage.

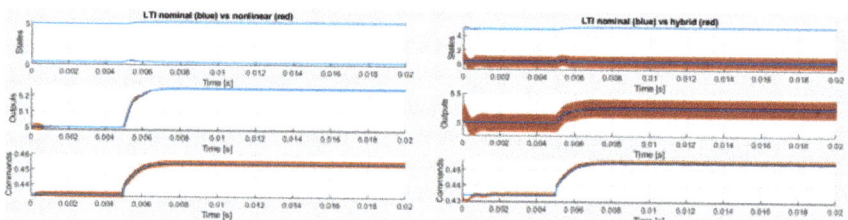

Figure 13. Buck averaged state-space and hybrid model Monte Carlo closed loop simulations with a closed loop bandwidth $\omega_B > 1200$ [rad/s], equivalent to a rise time of ≈ 0.83 [ms], negligible steady-state error, and no overshoot.

3.3.3. Boost Converter

The boost converter state-space models for the ON and OFF states of switch S_1, respectively, as in Equation (41), are

$$\left(\begin{array}{ccc|cc|c} -\frac{r_L+r_{DS_1}}{L} & 0 & 0 & \frac{1}{L} & -\frac{1}{L} & 0 \\ 0 & -\frac{1}{(R+r_C)C} & 0 & 0 & 0 & 0 \\ 0 & \frac{R}{R+r_C} & 0 & 0 & 0 & 0 \end{array} \right) ; \left(\begin{array}{ccc|cc|c} -\frac{r_{aux}}{L} & -\frac{R}{(R+r_C)L} & \frac{1}{L} & 0 & -\frac{1}{L} \\ \frac{R}{(R+r_C)C} & -\frac{1}{(R+r_C)C} & 0 & 0 & 0 \\ \frac{r_C R}{R+r_C} & \frac{R}{R+r_C} & 0 & 0 & 0 \end{array} \right) , \quad (57)$$

with the auxiliary notation $r_{aux} = r_L + r_{DS_2} + \frac{r_C R}{R+r_C} = r_L + r_{DS_2} + r_C || R$. The nominal boost converter parameters and their tolerances are presented in Table 3, as they correspond with the buck converter parameters.

The desired operating point specifications are as follows: the output signal $y(t) \equiv u_R(t)$ is at 24 [V], with nominal voltage source and load inputs $u_1(t) = E = 12$ [V], $u_2(t) = R = 15$ [Ω]. The initial guesses for the state equilibrium values where $\tilde{x} = [3, 24]$ and $u_3(t) = \mu = 0.5$ for the duty cycle control input. After computation, the actual equilibrium point is $(\mathbf{u}, \mathbf{x}, \mathbf{y}) = ([12, 15, 0.5179], [0.3189, 23.999], 24)$.

An input multiplicative uncertainty model, i.e., $G(s) = G_n(s)[1 + \Delta(s)W_{unc}(s)]$, $||\Delta||_\infty \leq 1$, as in Table 1, has been automatically computed from input $u_3(t)$ to output $y_1(t)$, with tolerances ± 1 [V] and ± 1 [Ω] for inputs $u_1(t) = E$ and $u_2(t) = R$, respectively, based on 1000 Monte Carlo simulations. The relevant frequencies vary in the interval $[\underline{\omega} = 10^1, \overline{\omega} = 10^7]$, with 200 equally distributed samples in log domain. A successful set of hyperparameters for the particle swarm optimization algorithm is comprised of a swarm size of 1500, initial swarm span of 10^4, minimum neighbors fraction of 0.9, and inertia range of $[0.1, 1.1]$ for a transfer function structure as in Equation (19), with two real poles and two real zeros, resulting in

$$W_{unc}(s) = \frac{0.26592(s+512.5)(s+3.535\times 10^4)}{(s+4016)(s+1.389\times 10^4)}. \quad (58)$$

The linearized boost plant family is comprised of second-order stable systems, in minimal form, with two zeros, one being of nonminimum phase. The nominal model is

$$G_n(s) = \frac{0.65505(-s+8.551\times 10^4)(s+8333)}{s^2+2988s+9.746\times 10^6}. \quad (59)$$

The entire uncertainty family has the same structure with poles, zeros, and equilibrium points in the vicinity of the nominal counterparts. The uncertain boost plant family structure and behavior, along with the PSO least conservative second-order bound computation are illustrated in Figure 14.

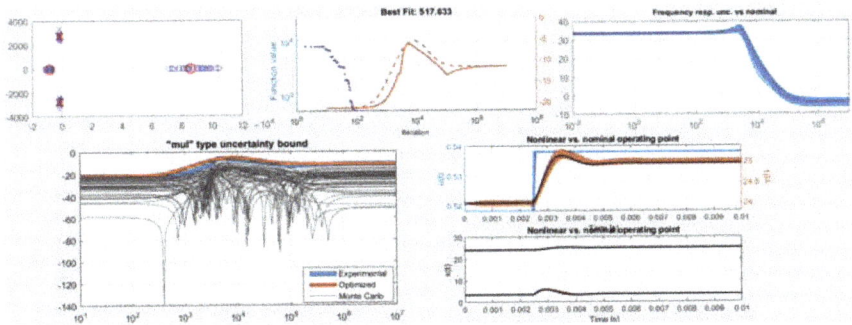

Figure 14. Boost multiplicative uncertainty set computation and open loop responses for the operating point with $E = 12[V]$, $R = 15[\Omega]$, $U_R = 24[V]$: step and frequency response, pole-zero placement; the step response is simulated for nonlinear plants sampled using `UncertainPlantFactory`.

For controller synthesis, the weighting function specifications were for the sensitivity function $\omega_B = 650$ [rad/s], $A = 10^{-4}$, $M = 2$, $n = 1$; for the complementary sensitivity function $\omega_{BT} = 3250$ [rad/s], $A_T = 10^{-4}$, $M_T = 2$, $n = 1$; and for the control effort $M_0 = 0.1$, $M_\infty = 100$, $|W_{KS}(j \cdot 650)| = 2$, resulting in

$$W_S(s) = \frac{0.5s + 650}{s + 0.065}, \quad W_T(s) = \frac{s + 3250}{0.0001s + 6500}, \quad W_{KS}(s) = \frac{100s + 3253}{s + 3.252 \times 10^4}. \tag{60}$$

An intrinsic limitation for the boost converter is that the bandwidth must not be imposed more than half of the value of the right-half plane non-minimum phase zero of the process, i.e., $\omega_B \leq \frac{z}{2} \approx 43{,}000$ [rad/s]. For this problem, this is a sufficiently high margin, as we also imposed a limitation for the command signal through W_{KS}. As specified by the $n = 2$ and $A_T = 10^{-4}$ parameters of the complementary sensitivity weighting function, W_T, the closed loop control system mitigates sensor noise signals with a considered spectrum starting from $\omega_{BT} > 3250$ [rad/s], using an initial roll-off of -40 [dB/dec], followed by an attenuation of at least four orders of magnitude. From the μ synthesis, a controller of order 17 is obtained. After order reduction, the smallest controller which retains all imposed specifications for the plant family, with a peak value $\mu_\Delta(LLFT(P, K)) \leq 0.9547 < 1$, is

$$K_{red}^{Boost} = \left(\begin{array}{ccc|c} -0.065 & 0.8025 & -0.002966 & 5.007 \\ 0.7116 & -9.715 \times 10^4 & 1.066 \times 10^4 & -30.91 \\ 0.002575 & -1.065 \times 10^4 & -1.422 & -0.113 \\ \hline 5.007 & -30.91 & 0.1142 & 0 \end{array}\right). \tag{61}$$

With this regulator, the boost converter control system with parameters from Table 3 has very large stability margins, with phase margins between $[81, 101]$ [°] and gain margins in the interval $[40, 46]$ [dB]. The obtained sensitivity bandwidths vary between $[828, 2510]$ [rad/s], all of them better than the prespecified value of 800.

Nonlinear MiL simulation options are $N = 50$ random plants from the uncertainty set, with the `ode15i` solver, with a step on the reference signal of 5% from its initial equilibrium value of approximately 5 [V] and a simulation time of 0.02 [s]. The simulation conditions for the hybrid MiL case are almost identical, with the use of the `ode113` solver instead, set to a relative tolerance of 10^{-8}, and with an additional PWM period selected randomly for each experiment from a nominal value $T_{PWM} = 17.5$ [µs] with a $\pm 20\%$ fluctuation from one simulation to another. A comparison of the nonlinear and hybrid cases is presented in Figure 15, where it can be seen that the transitory regime is practically identical. Due to working with $u_R(t)$ instead of $u_C(t)$ only for the output signal, the current ripple is propagated into the measured voltage.

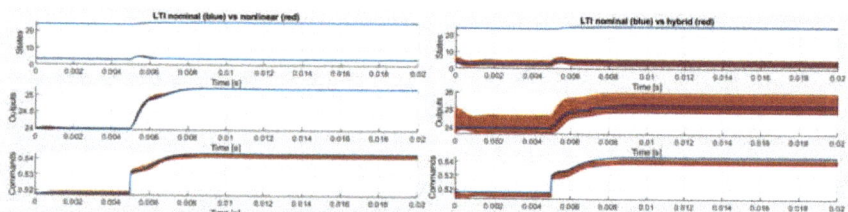

Figure 15. Boost averaged state-space and hybrid model Monte Carlo closed loop simulations with a closed loop bandwidth $\omega_B > 650$ [rad/s], equivalent to a rise time less than ≈ 1.53 [ms], negligible steady-state error and no overshoot.

4. Conclusions and Future Work

There are several aspects to be discussed with regards to the presented toolbox and, additionally, including future work to be implemented in upcoming iterations.

The main reason for which the hybrid system framework is considered for the proposed toolbox is that the continuous-time plant needs to be regulated with a numerical controller. The hybrid system context may include LTI or nonlinear systems, switching systems (which are hybrid by nature), and singular systems, represented using differential-algebraic equations (DAEs). As such, for a continuous-time plant, a continuous-time controller is designed using the robust control framework. The controller is required to be numerically implementable, therefore two interfaces are necessary, which lead to a hybrid system. The numerical implementation must be easily obtained and automatically validated using rapid control prototyping (RCP) techniques. However, the properties of the closed loop system need to be reanalyzed after the discretization of the controller. Differential-algebraic equations represent a useful framework for modeling dynamical systems in engineering with a network-based structure of components. They are used in various industry fields such as mechanics (e.g., multiple-link mobile manipulator model), chemical engineering (modeling of chemical reactions), electrical engineering, cyber-physical systems, etc. All categories of processes taken into consideration in the hybrid framework are described using an approximate model that incorporates their relevant behavior. However, these types of systems have model and structure uncertainties. Moreover, the nonlinear systems that are linearized around an equilibrium point introduce such uncertainties as well. Therefore, to consider these uncertainties for the controller design, the robust control framework is mandatory. The first main applicability of RCP was to derive the necessary C/C++ source code with drivers for a given target microprocessor, and to simulate in reproducible conditions the behavior of a complex system. For the latter case, the most relevant simulation types, given in increasing order of complexity and closeness to reality, are Model-in-the-Loop (MiL), Software-in-the-Loop (SiL), and Hardware-in-the-Loop (HiL).

Many modeling software programs return circuits or mechanical systems already in DAE form, and it would be difficult, or sometimes impossible, to reformulate them in an ODE form without changing variables and losing their intended physical significance. In the context of the robust control framework used for DAEs, a significant work is the monograph in [33]. Although the robust control theory was well formulated in the previous years, this is still an open domain for research and publication, as surveyed and described in [34]. The difficulty of using methods which work correctly for multiple operating points is mitigated by using adaptive methods, such as gain scheduling for tracking problems. Furthermore, other relevant problems are automatic C/C++ code generation for controller implementation and the commutation between the prescribed operating points. Considering use cases for modeling, simulation, computer-aided design, and RCP, a relevant set of examples in the domain of power electronics, hybrid vehicles, and renewable energy systems is given in [35,36], where, although the main limitation is that analysis, control, and implementation aspects must be performed individually for the presented applications, they can be generally included under the same software framework, with the important exception of the robust control design and verification, along with the quantization sensitivity analysis in a unitary manner.

The main hindrance is that the majority of applications involve DAE systems of index greater than 1, meaning they necessitate more than one derivation step in order to formulate the problem, so there is a great need of tools that can also deal with high order DAEs, i.e., index 2 or 3 [37–39]. Another reason for the proposal of considering DAEs model is that they are also explicitly related to control issues, regarding both physical and operational constraints. Two illustrative examples are the case of improper systems, such as an ideal PID controller, and the elements that realize the decoupling in the MIMO systems case and when the plant has impulsive dynamics. The robust control techniques' drawback is that the closed loop $\mathcal{H}_2/\mathcal{H}_\infty$ norm must be minimized using the prescribed weighting functions that penalize the exogenous outputs and these weighting functions need to be found ad hoc, which sometimes lead to some intermediary bad controllers and work overhead. Furthermore, after numerical implementation, the discrete controller loses a part of the imposed performances due to an inadequate sampling period or badly selected quantization levels. Although there are solutions in the literature, there is no unified approach to solve all these mentioned problems. As an extension to the framework and mindset given by the two previously mentioned RCP use cases, the current project proposed a highly automated toolbox for robust control design, which, in the current state-of-the-art is a highly iterative design process, when taking into consideration plant uncertainties, although the mathematical background for solving the optimization problems is well established. It proposes to eliminate design overhead when considering and modifying a specification set, manually redesigning the weighting functions, the optimization procedure, discretization of the regulators, quantization analysis, and closed loop analysis for the linearized and initial hybrid plant. Furthermore, in unison with high-performance numerical toolboxes, a justified report should automatically result after its use and explicitly state when unrealistic design specifications were considered.

Author Contributions: Conceptualization, M.Ş. and V.M.; methodology, M.Ş.; software, M.Ş.; validation, V.M. and P.D.; formal analysis, M.Ş. and P.D.; investigation, M.Ş. and M.S.; resources, M.Ş.; data curation, D.M.; writing—original draft preparation, M.Ş., M.S., and V.M.; writing—review and editing, V.M., D.M., and M.S.; visualization, M.Ş. and D.M.; supervision, P.D.; project administration, M.Ş.; funding acquisition, D.M. All authors have read and agreed to the published version of the manuscript.

Funding: This paper was supported by the Project "Entrepreneurial competences and excellence research in doctoral and postdoctoral programs—ANTREDOC", project co-funded by the European Social Fund.

Conflicts of Interest: The authors declare no conflicts of interest.

Sample Availability: Relevant source code implementation, including classes, objects, scripts, and data to reproduce the results presented in this paper are available from the GitHub repository: roconsys-toolbox-public.

Abbreviations

The following abbreviations are used in this manuscript:

CACSD	Computer-Aided Control System Design
CCM	Continuous Conduction Mode
DAE	Differential-Algebraic Equation
DC	Direct Current
DCM	Discontinuous Conduction Mode
DOF	Degree of Freedom
HyEQ	Hybrid Equations Toolbox
LLFT	Lower Linear Fractional Transformation

LMI	Linear Matrix Inequality
LTI	Linear Time-Invariant
MiL	Model-in-the-Loop
ODE	Ordinary Differential Equation
PSO	Particle Swarm Optimization
RCP	Rapid Control Prototyping
SEPIC	Single-ended primary-inductor converter
ULFT	Upper Linear Fractional Transformation

References

1. Balas, G.; Chiang, R.; Packard, A.; Safonov, M. *Robust Control Toolbox, Reference for MATLAB*; The MathWorks, Inc.: Natick, MA, USA, 2020.
2. *Global Optimization Toolbox, User's Guide for MATLAB*; The MathWorks, Inc.: Natick, MA, USA, 2020.
3. Feyel, P.; Duc, G.; Sandou, G. Optimal tuning of $\mathcal{H}_\infty$ fixed-structure robust controller against multiple high-level requirements using evolutionary computation. *Int. J. Robust Nonlinear Control* **2019**, *29*, 949–972. [CrossRef]
4. Blackwell, C.; Sastry, M.K.S. Multivar - A MATLAB based MIMO Control System Design Application. In Proceedings of the 8th International Conference on Computational Intelligence and Communication Networks, Tehri, India, 23–25 December 2016.
5. Keller, P. Robust and optimal $\mathcal{H}_\infty$ control in LabVIEW. *IEEE Conf. Control. Technol. Appl. (CCTA)* **2017**. [CrossRef]
6. Jacobs, L.; Verbandt, M.; De Preter, A.; Anthonis, J.; Swevers, J.; Pipeleers, G. *A Toolbox for Robust Control Design: An Illustrative Case Study*; IEEE: Tokyo, Japan, 2018.
7. Sadeghpour, M.; de Oliveira, V.; Karimi, A. A Toolbox for Robust PID Controller Tuning Using Convex Optimization. *IFAC Proc. Vol.* **2012**, *45*, 158–163. [CrossRef]
8. Ljung, L. *System Identification Toolbox, Reference for MATLAB*; The MathWorks, Inc.: Natick, MA, USA, 2020.
9. Karimi, A.S. Frequency-Domain Robust Control Toolbox. In Proceedings of the 52nd IEEE Conference on Decision and Control, Firenze, Italy, 10–13 December 2013. [CrossRef]
10. Sadeghzadeh, A.; Karimi, A.S. Fixed-structure $\mathcal{H}_2$ controller design for polytopic systems via LMIs. *Optim. Control Appl. Meth.* **2014**. [CrossRef]
11. Verbandt, M.; Swevers, J.; Pipeleers, G. An LTI control toolbox—Simplifying optimal feedback controller design. In Proceedings of the 2016 European Control Conference (ECC), Aalborg, Denmark, 29 June–1 July 2016. [CrossRef]
12. Şuşcă, M. Solving Algebraic Riccati Equations Using Proper Deflating Subspaces for $\mathcal{H}_2/\mathcal{H}_\infty$ Synthesis. Master's Thesis, Technical University of Cluj-Napoca, Cluj-Napoca, Romania, 2019. [CrossRef]
13. Şuşcă, M.; Mihaly, V; Stănese, M.; Dobra, P. Iterative Refinement Procedure for Solutions to Algebraic Riccati Equations. In Proceedings of the 2020 IEEE International Conference on Automation, Quality and Testing, Robotics (AQTR), Cluj-Napoca, Romania, 21–23 May 2020. [CrossRef]
14. Mihaly, V. General Purpose Linear Matrix Inequality Solver With Applications in Robust and Nonlinear Control. Master's Thesis, Technical University of Cluj-Napoca, Cluj-Napoca, Romania, 2020.
15. Gumussoy, S.; Henrion, D.; Millstone, M.; Overton, M.L. Multiobjective Robust Control with HIFOO 2.0. In Proceedings of the 6th IFAC Symposium on Robust Control Design, Haifa, Israel, 16–18 June 2009.
16. Sanfelice, R.G; Copp, D.A.; Nanez, P. *Hybrid Equations (HyEQ) Toolbox v2.04—A Toolbox for Simulating Hybrid Systems in MATLAB/Simulink®*; MathWorks®: Natick, MA, USA, 2017.
17. Goebel, R.; Sanfelice, R.G.; Teel A.R. *Hybrid Dynamical Systems: Modeling, Stability, and Robustness*; Princeton University Press: Princeton, NJ, USA, 2012.
18. Griewank, A.; Walther, A. *Evaluating Derivatives: Principles and Techniques of Algorithmic Differentiation*, 2nd ed.; SIAM: Philadelphia, PA, USA, 2008.
19. Kennedy, J.; Eberhart, R. Particle swarm optimization. In Proceedings of the ICNN'95—International Conference on Neural Networks, Perth, WA, Australia, 27 November–1 December 1995; Volume 4, pp. 1942–1948. [CrossRef]
20. Doyle, J.C.; Glover K.; Khargonekar, P.P.; Francis, B.A. State-Space Solutions to Standard $\mathcal{H}_2$ and $\mathcal{H}_\infty$ Control Problems. *IEEE Trans. Autom. Control.* **1989**, *34*, 831–847. [CrossRef]
21. Fortuna, L.; Frasca, M. *Optimal and Robust Control—Advanced Topics with MATLAB*; CRC Press: Boca Raton, FL, USA, 2012.
22. Gahinet, P.; Apkarian P. A Linear Matrix Inequality Approach to $\mathcal{H}_\infty$ Control. *Int. J. Robust Nonlinear Control.* **1994**, *4*, 421–448. [CrossRef]
23. Ionescu, V.; Oară, C.; Weiss, M. *Generalized Riccati Theory and Robust Control: A Popov Function Approach*; John Wiley & Sons Ltd.: Chichester, UK, 1999.
24. Zhou, K.; Doyle, J.C.; Glover, K. *Robust and Optimal Control*; Prentice Hall: Englewood Cliffs, NJ, USA, 1996.
25. Packard, A.; Doyle, J.; Balas, G. Linear, Multivariable Robust Control With a μ Perspective. *J. Dyn. Syst. Meas. Control.* **1993**, *115*, 426–438. [CrossRef]
26. Skogestad, S.; Postlethwaite, I. *Multivariable Feedback Control: Analysis and Design*, 2nd ed.; John Wiley & Sons Ltd.: Chichester, UK, 2005.

27. Raghavendra, K.V.G.; Zeb, K.; Muthusamy, A.; Krishna, T.N.V.; Kumar, S.V.S.; Kim, D.-H.; Kim, M.-S.; Cho, H.-G.; Kim, H.-J. A Comprehensive Review of DC–DC Converter Topologies and Modulation Strategies with Recent Advances in Solar Photovoltaic Systems. *Electronics* **2020**, *9*, 31. [CrossRef]
28. Chang, E.-C.; Cheng, C.-A.; Wu, R.-C. Robust Optimal Tracking Control of a Full-Bridge DC-AC. *Converter. Appl. Sci.* **2021**, *11*, 1211. [CrossRef]
29. Garcia, G.; Lopez Santos, O. A Unified Approach for the Control of Power Electronics Converters. Part I—Stabilization and Regulation. *Appl. Sci.* **2021**, *11*, 631. [CrossRef]
30. Mihaly, V.; Şuşcă, M.; Dobra, P. Passivity-Based Controller for Nonideal DC-to-DC Boost Converter. In Proceedings of the 2019 22nd International Conference on Control Systems and Computer Science (CSCS), Bucharest, Romania, 28–30 May 2019; pp. 30–35. [CrossRef]
31. Repecho, V.; Biel, D.; Olm, J.M.; Fossas, E. Robust sliding mode control of a DC/DC Boost converter with switching frequency regulation. *J. Frankl. Inst.* **2018**, *355*, 5367–5383. [CrossRef]
32. Lunze, J.; Lagarrigue, F.L. *Handbook of Hybrid Systems Control: Theory, Tools, Applications*; Cambridge University Press: Cambridge, UK, 2009.
33. Feng, Y.; Yagoubi M. *Robust Control of Linear Descriptor Systems*; Studies in Systems, Decision and Control 102; Springer: Singapore, 2017.
34. Liu, K.-Z.; Yao, Y. *ROBUST CONTROL—Theory and Applications*; John Wiley & Sons (Asia) Pte Ltd.: Singapore, 2016.
35. Zamboni, W.; Petrone, G. (Eds.) *ELECTRIMACS 2019, Selected Papers—Volume 1*; Springer International Publishing: Cham, Switzerland, 2020.
36. Zamboni, W.; Petrone, G. (Eds.) *ELECTRIMACS 2019, Selected Papers–Volume 2*; Springer International Publishing: Cham, Switzerland, 2020.
37. Ascher, U.M.; Petzold, L.R. *Computer Methods for Ordinary Differential Equations and Differential-Algebraic Equations*; SIAM: Philadelphia, PA, USA, 1998.
38. Brenan, K.E.; Campbell, S.L.; Petzold L.R. *Numerical Solution of Initial-Value Problems in Differential-Algebraic Equations*; SIAM Classics in Applied Mathematics: Philadelphia, PA, USA, 1996.
39. Milano, F.; Dassios, I.; Liu, M.; Tzounas, G. *Eigenvalue Problems in Power Systems*; CRC Press: Boca Raton, FL, USA, 2021.

Article

Robust Fractional-Order Control Using a Decoupled Pitch and Roll Actuation Strategy for the I-Support Soft Robot

Jorge Muñoz [1,*], Francesco Piqué [2,3], Concepción A. Monje [1] and Egidio Falotico [2,3]

1. Department of Systems Engineering and Automation, University Carlos III of Madrid, Avda. de la Universidad 30, 28911 Leganés, Madrid, Spain; cmonje@ing.uc3m.es
2. The BioRobotics Institute, Scuola Superiore Sant'Anna, Viale Rinaldo Piaggio 34, 56025 Pontedera, Pisa, Italy; Francesco.Pique@santannapisa.it (F.P.); egidio.falotico@santannapisa.it (E.F.)
3. Department of Excellence in Robotics and AI, Scuola Superiore Sant'Anna, 56025 Pontedera, Pisa, Italy
* Correspondence: jmyanezb@ing.uc3m.es

Abstract: Tip control is a current open issue in soft robotics; therefore, it has received a good amount of attention in recent years. The desirable soft characteristics of these robots turn a well-solved problem in classic robotics, like the end-effector kinematics and dynamics, into a challenging problem. The high redundancy condition of these robots hinders classical solutions, resulting in controllers with very high computational costs. In this paper, a simplification is proposed in the actuation setup of the I-Support soft robot, allowing the use of simple strategies for tip inclination control. In order to verify the proposed approach, inclination step input and trajectory-tracking experiments were performed on a single module of the I-Support robot, resulting in zero output error in all cases, including those where the system was exposed to disturbances. The comparative results of the proposed controllers, a proportional integral derivative (PID) and a fractional order robust (FOPI) controller, validate the feasibility of the proposed approach, showing a clear advantage in the use of the fractional robust controller for the tip inclination control of the I-Support robot compared to the integer order controller.

Keywords: soft robotics; robust control; fractional calculus

1. Introduction

Soft robotics is a growing research field which aims to incorporating softness in robotic bodies or in novel end effectors, enabling safe and adaptive interactions [1]. Soft robotics is bio-inspired, since it tries to reproduce the abilities of certain animals, such as worms, snakes or the octopus [2], to move without a rigid skeleton or exoskeleton, exploiting their softness in order to squeeze, and adapt to unstructured environments. The stiffness characteristics of traditional industrial robots were desirable because they enabled the fast, reliable and precise performance of tasks, such as those required in factory lines. Conversely, soft robotics finds application in tasks where safety and adaptability to unstructured environments is of paramount importance [1]. Such tasks include delicate food handling, medical procedures, and assistive tasks.

The compliance which characterizes soft robots, besides granting the desired properties, also introduces challenges from the perspective of modeling and control [3]. The hysteresis of the materials and their high redundancy, due to the virtually infinite number of degrees of freedom (DoF) of soft robots, makes them hard to model with high accuracy. Closed-form equations for describing the dynamics of soft robots are available [4], but are too computationally demanding for efficient use in control. The constant curvature (CC) or the piecewise constant curvature (PCC) approaches [5], which assume either all of the robot's body, or a number of robot sections, to be circular arcs, are computationally efficient, but tend to fail when the robot is highly nonlinear.

A different approach is to rely on neural networks [6] or reinforcement learning [7] for data-driven modeling of the soft robot. In [6], a dynamic model of a soft robot is learned through supervised learning using an auto-regressive network, and is employed for closed-loop control by model-based reinforcement learning. In [7], a multiagent reinforcement learning approach is used to learn the kinematic model of a robotic arm. A trajectory optimization method is also exploited for open-loop control of dynamic reaching tasks [8]. In [9], it was shown that data-driven models can exploit the retraining of their networks' weights to accommodate external disturbances. An extensive discussion of the challenges of such platforms can be found in [10]. While these data-driven approaches can accurately capture the nonlinearities of soft robot dynamics, their drawback is that neural networks are black box models which are unfit for the traditional controller design methods usually employed for state-space models.

Due to these limitations, many workspace control strategies applied to soft robots are based on nonlinear model-based controllers or linear model-free controller schemes. In the last case, as no model is available for the controller tuning, different alternatives must be used. For instance, an empirical estimation of the kinematic Jacobian matrix is proposed in [11], and later used in an optimal control scheme. Only the works in [12] propose workspace linear controllers, but use very complex control laws, involving the robot's Jacobian matrix and its derivative. See [3] for a complete survey on different soft robot control strategies.

Feedback control of nonlinear or time-varying systems has been a challenging problem not just for soft robotics, but since the early nonlinear control attempts at the beginning of the last century. Among the approaches proposed for dealing with nonlinearities, robust control has been extensively used for that purpose. This strategy aims to achieve constant system performance (in the sense of behavior), despite potential plant changes.

Some examples of robust control approaches can be found in [13], where a fractional controller is proposed in the robust control of a soft neck, or in [14], where a fuzzy approach is used to model a nonlinear plant (car steering), proposing an output feedback controller to obtain a robust behavior. Other, more advanced, control strategies have also been used, such as the sliding mode control of a wind turbine generator shown in [15], where a robust behavior is obtained in simulations under the conditions of variable wind-speed inputs and other parameter uncertainties. For a detailed discussion of nonlinear system control problems and possible solutions, see [16].

A desirable feature in robust systems consists of providing a constant overshoot despite changes in the plant parameters (usually the gain). This feature, often called iso-damping in the literature, provides a significant advantage in the control of time-varying or nonlinear systems. Often, this robustness specification is based on Bode's ideal function (see [17]), which features a flat phase diagram, and thus a constant damping. For instance, in [18], the tuning of a proportional integral derivative (PID) controller based on this flat-phase condition is proposed, showing the benefits of this robust specification in several case studies. A similar approach is found in [19], where a relay test is proposed to find the plant parameters, followed by the application of a tuning method based on the aforementioned condition.

Using that robust specification, a wide range of solutions are possible, from the use of a PID control, as described above, to more advanced strategies. A very interesting approach to the robust control problem is found using fractional calculus. Fractional order controllers (FOCs), based on non-integer-order derivative/integral operators, show greater flexibility in fulfilling the flat-phase condition compared to their integer-order alternatives, while keeping most of their benefits. An extensive review of fractional calculus applications in the field of robust control can be found in [20,21], including system modeling and controller design.

Although many fractional controller definitions have been proposed since the first works in [22], the non-integer-order generalization of the classic PID is generally preferred, probably due to its simple control law and strong similarities with the ubiquitous PID controller, allowing classic design tools to be adapted from integer to fractional exponents.

As described in [23], the fractional order PID (FOPID) controllers defined using Equation (1) are able to provide a robust performance despite plant parameter changes and nonlinearities

$$FOPID(s) = k_p + k_i \frac{1}{s^\lambda} + k_d s^\mu, \tag{1}$$

where k_p, k_i, k_d are the controller gains, and λ, μ are the fractional operator orders.

Given its benefits and convenience, the FOPID controllers have received particular attention in recent decades. Approaches using the definition in Equation (1) are found in many works. For instance, in [24], new tuning and auto-tuning methods are proposed for the controller parameters, showing excellent results in the control of real plants, like a water circuit or a servomotor. The same controller is used for the control of a DC motor model in [25], also proposing the possible electronic realization of the system. Again, in [26], an optimization method is proposed for the tuning of the same controller, showing excellent results in the control of a real servomotor system.

In this paper, a fractional-order robust control is proposed for I-Support, an assistive soft robot [27]. The particular cases of proportional integral derivative (PID) and fractional-order proportional integral controller (FOPI) are considered due to their plant and model characteristics. As a novelty, a dynamic model of the plant will be used for the controller tuning, achieving excellent results. This is an important contribution, as similar previous works are based on very complex control laws, while the proposed control scheme is based on simple PID or FOPI controllers.

In the following sections, the robotic platform hardware and the chosen model are described. In order to obtain a suitable model, the robot inputs are redefined, allowing for a direct relationship between the actuation variables and the work-space variables, such as orientation and inclination angles. Then, a plant model is obtained using a recursive least squares (RLS) parameter identification method, as described in [28]. Since the identification is done offline, other, simpler methods could be used, such as least-squares fit; however, given the tuning method proposed, the control strategy might be upgraded to an adaptive scheme, as in the case of [29]. Therefore, a recursive identification algorithm like RLS may have future advantages.

Once a plant model is available, it can be used for controller tuning. According to the iso-m procedure explained in [30], the magnitude, phase and slope of the plant are needed, which can be obtained from the RLS identification. In addition, the system's behavior must be defined using standard performance specifications, like the damping ratio (phase margin) and peak time (crossover frequency). The resulting controller parameters will be used in the robust control scheme proposed for the I-Support robot. See [30] for details on the method application.

It will be shown that the proposed controllers can track the robot's end effector configuration in termso f its orientation and inclination angles, and can effectively reject external disturbances, despite inaccuracies in the plant's model, thanks to the robust fractional order control.

2. Materials and Methods

A soft robotic manipulator for the assistance of elderly people, called I-Support, has been used in this work [27] (Figure 1). It belongs to the class of continuum manipulators that receive inspiration from biological models like elephant trunks or snakes. It is composed of three modules, each of them actuated by three coupled McKibben actuators and three tendon-driven actuators. In this work, the proximal module of the robot was selected and used independently of the others. McKibben actuators are artificial pneumatic muscles, based on an internal latex balloon surrounded by a bellow-shaped braid. The braided structure allows the McKibben actuator to perform uni-directional bending when inflated. The pneumatic actuators are placed within the module at 120° to enable the bending and elongation of the module in all directions. The cable-driven actuation, which was not used in this work for simplicity, allows for shortening and stiffness variation in the module. The McKibben actuators are controlled by Camozzi K8P pneumatic valves, which are controlled by an Arduino Due board. The Arduino is, in turn, controlled by a PC using a serial port within the Matlab environment. The module is kept together by plastic discs placed 10 mm from each other. An internal central channel is built to hold the hose, to provide water and/or soap.

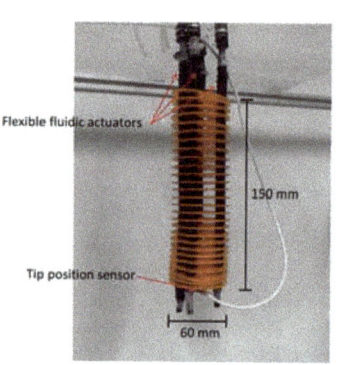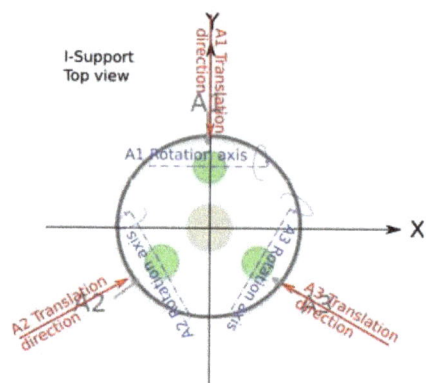

Figure 1. I-Support kinematic description.

2.1. Plant Model

According to [31,32], this robot is hyper-redundant, making the term degrees of freedom (DOF) not applicable in the classical sense. Nevertheless, in this specific case, the actuation parameters have a direct effect on measurable outputs like tip position and orientation. A correlation between the three available system inputs and the measurable outputs can be found and used to find a plant model.

The I-Support arm module is actuated through three evenly spaced, pressure-driven McKibben pneumatic actuators. As described in [33], the actuator elongation depends on the input pressure, which, in time, produces a change in the position and orientation of the end-effector according to its relative location within the robot. In this case, given the actuator disposition, the different input pressures result in a specific rotation and displacement, depending on the actuator used, as shown in Figure 1. Note that, as there is only one input variable per actuator, its resulting translations and rotations must be bounded.

The combined action of the three actuators produce the final end-effector's position and orientation in the workspace (Figure 2). As in a three-dimensional environment, the final orientation of the end effector can be defined using three Euler angles. More specifically, in our case, where the rotation in Z axis (yaw) cannot change, the final orientation can be described using two rotation angles in X and Y: pitch and roll. Therefore, the combination of the three angles produced by each actuator will result in a final rotation that can be defined or measured with two angles.

Given that translations and rotations are bound, either can be considered as an output. In this case, end-effector rotations will be considered as the system output. A deeper study of the robot geometry will show how the actuator pressure inputs are related to these final angle outputs.

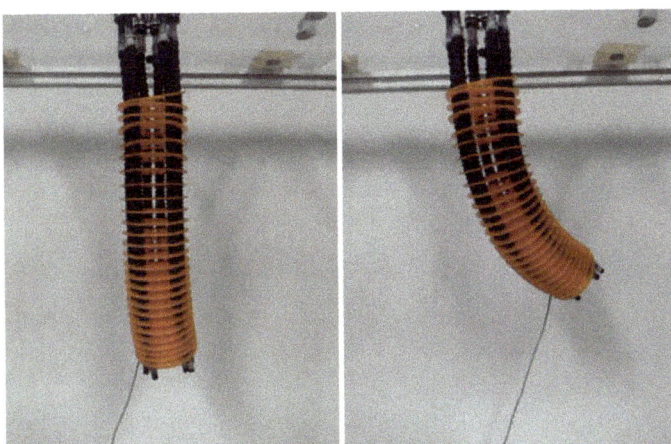

Figure 2. **Left**: elongated I-Support module, with three equally actuated chambers. **Right**: bended module with only one chamber inflated.

Starting with a single actuator ($A1$), and making its rotation axis parallel to the X axis in the frame of reference (see Figure 1), results in an output angle directly related to the input pressure of its chamber. Given that the angles can be negative, the pressures are also negative for the moment. Although the system does not allow this configuration, it can be solved later by adding an offset. Considering this actuator, with the index number 1, the equations describing angle α in X are as follows

$$\alpha_1 = f(P_1). \quad (2)$$

where α_1 is the angle contribution from the first actuator to the final X axis angle (α), P_1 is the actuator input pressure and f is a nonlinear function describing the relationship between them.

However, there are other two actuators with an effect on the final angle α. Given the proposed vertical robot setup, and using the same actuator type at all locations, we can assume that functions f relating the input pressure and actuator angle are also similar. Therefore, the same function applies, but including a projection factor that depends on the actuator relative angle (γ), resulting in

$$\alpha_2 = \cos(\gamma_2)f(P_2), \quad (3)$$
$$\alpha_3 = \cos(\gamma_3)f(P_3). \quad (4)$$

In fact, we can generalize the previous functions as follows

$$\alpha_i = \cos(\gamma_i)f(P_i). \quad (5)$$

Although the f functions are nonlinear, the resulting tip angles depend on the forces produced by the linear actuators; therefore, given the robot construction, the angles can be considered additive. The final angle in the X axis is then found by addition of the three actuator angles

$$\alpha = \alpha_1 + \alpha_2 + \alpha_3 = \cos(\gamma_{11})f(P_1) + \cos(\gamma_{12})f(P_2) + \cos(\gamma_{13})f(P_3). \quad (6)$$

Since the three actuators are symmetrically arranged, the angles are $\gamma_{11} = 0$ deg, $\gamma_{12} = 120$ deg and $\gamma_{13} = 240$ deg, and Equation (6) results in

$$\alpha = f(P_1) - 0.5f(P_2) - 0.5f(P_3) = f(P_1) - 0.5[f(P_2) + f(P_3)]. \quad (7)$$

This result shows how both actuators', A_2 and A_3, effects on the angle α are divided by two, with an opposite direction to actuator A_1. This leads to the first result of our approach. The α angle is defined by the pressure difference, which is positive when P_1 is larger than $0.5(P_2 + P_3)$, and negative otherwise. In the case of $P_1 = P_2 = P_3$, angle $\alpha = 0$, leading to different robot elongations depending on the pressure value, form zero ($P_1 = P_2 = P_3 = 0$) to full-length ($P_1 = P_2 = P_3 = P_{max}$).

Now, β angle is defined as the rotation around Y axis. Using the previous reasoning, but projecting in the Y axis (using $\sin(\gamma)$)

$$\beta = \beta_1 + \beta_2 + \beta_3 = \sin(\gamma_1)f(P_1) + \sin(\gamma_2)f(P_2) + \sin(\gamma_3)f(P_3). \tag{8}$$

In the case of $\gamma_1 = 0$ deg, $\gamma_2 = 120$ deg and $\gamma_3 = 240$ deg, Equation (8) results in

$$\beta = 0.866 f(P_2) - 0.866 f(P_3) = 0.866[f(P_2) - f(P_3)]. \tag{9}$$

Note that the value of β angle depends on the difference between P_2 and P_3, and the effects of the A_1 actuator cannot change it. Again, the angle depends on a pressure difference, and the elongation is a function of the minimum pressure values. For the case $P_1 = P_2 = P_3$, angle $\beta = 0$, leading to the previous result regarding robot elongation. As there are just two actuators involved in this case, the final elongation depends on the minimum values between those two pressures.

At this point, we can see that α and β angles depend on the pressure difference of actuators $A1$, A_2 and A_3, and the elongation depends on the minimum of these values. Based on that, we can define the new input variables β_i, α_i and l_i, as a linear combination of the pressure inputs without loss of generality.

Using the results from Equations (7) and (9), and considering the description for the elongation behavior of the robot, the following input redefinition is proposed

$$\alpha_i = P_1 - 0.5(P_2 + P_3), \tag{10}$$
$$\beta_i = 0.866(P_2 - P_3), \tag{11}$$
$$l_i = \min(P_1, P_2, P_3). \tag{12}$$

As β depends only on the input pressure difference of the actuators $A2$ and $A3$, the change in β_i will only lead to a change in β output angle. Likewise, α_i and l_i inputs will affect only the output values of α and l.

Based on thid, the I-Support can be modeled as three decoupled single-input, single-output (SISO) systems. The transfer functions G_α, G_β, and G_l will model the actual outputs (α, β, l) as a function of the new inputs (α_i, β_i, l_i), defined by Equations (10)–(12). Given the simplifications we have considered, the reality will be different in several aspects, such as the interference between actuators and the nonlinear plant behavior, as will be shown in the experimental sections. To deal with these problems, we propose use of a robust controller, since this will provide a constant behavior despite the plant parameter changes or nonlinearities, as discussed above in Section 1.

In order to find these models, recursive least squares (RLS) system identification is proposed. Based on the above discussion, redefined inputs (α_i, β_i and l_i) were considered instead of pressure inputs. Note that these are just the pressure input redefinition, and the output angles still depend on the system dynamics. Although f functions are unknown, they are considered within the resulting models, but the nonlinear part will be neglected due to the identification method. As a robust controller is proposed, the performance results will be constant in the entire operation range of the robot, despite these nonlinearities.

As the control system is now defined through angle and elongation inputs, the equivalence between these inputs and the pressure of each actuator is required in order to operate the robot. In that direction, Equations (10) and (11) can be used to solve P_1 and P_2.

$$P_2 = \beta_i + P_3, \tag{13}$$

$$P_1 = \alpha_i + 0.5(P_2 + P_3) = \alpha_i + \frac{\beta_i}{2} + P_3. \tag{14}$$

Note that values of P_1 and P_2 depend on α_i and β_i inputs, and also depend on P_3, according to Equations (13) and (14). Using these results in Equation (12) provides

$$l_i = \min(\alpha_i + \frac{\beta_i}{2} + P_3, \beta_i + P_3, P_3) = \min(\alpha_i + \frac{\beta_i}{2}, \beta_i, 0) + P_3, \tag{15}$$

where the min function properties are applied to obtain P_3 value out of the min function, leading to the definition of P_3 value based on the inputs α_i, β_i, and l_i detailed in the following equation

$$P_3 = l_i - \min(\alpha_i + \frac{\beta_i}{2}, \beta_i, 0). \tag{16}$$

This result means all the pressure results will be positive as long as l_i is greater than zero, which means that this input variable actually controls the robot elongation, as described before.

Once our system is defined, a model is needed for controller tuning and simulation. Given the complex behavior of the robot, system identification is the best option to obtain a linear model from captured data. This means that we neglect the possible nonlinear behavior, but, thanks to the proposed robust controller, a good performance will be obtained despite the model mismatch.

Using the described inputs and outputs definition, a set of experiments were carried out for different target inclinations in order to obtain a plant model. The experimental setup consists of different identification experiments where a changing target was set at one of the three inputs (for instance, α_i), while keeping the other two inputs fixed (for instance, β_i and l_i). A motion-capture system was used to record the real plant behavior, and later used to obtain the output angles (α and β) variation.

For example, Figure 3 shows the input and output captured data during two specific identification experiments.

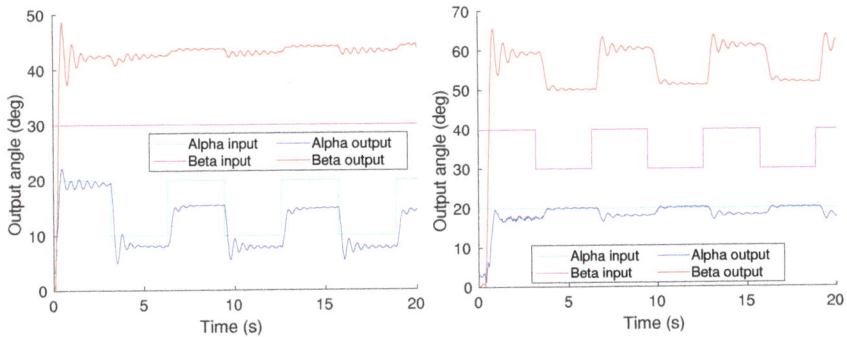

Figure 3. Two examples of identification experiments. The **left** figure shows the system response to variations in the input α_i, while the other inputs are kept constant ($\beta_i = 30$ and $l_i = 0$). **Right** figure shows the system response to variations in the input β_i, while the other inputs are kept constant ($\alpha_i = 20$ and $l_i = 0$).

Note that quite different behaviors can be observed for ascending and descending steps. This is probably due to the compressed air valve setting, which can result in plant differences when the air is pushed or released.

A relatively stable output is obtained for the fixed angle, despite the important variations in the changing angle, showing that the systems obtained are mainly decoupled, but a minimal influence still exists. Note that although the input values considered in the identification are α_i, β_i and l_i, the resulting model includes Equations (13), (14) and (16) dynamics (just the linear behavior, of course).

An appropriate number of experiments were performed in the I-Support, covering the entire robot workspace for different input combinations, resulting in a total of 62 separate datasets. Each set consists of the system input data (α_i, β_i) and the response obtained (α, β) over a period of 20 s (as shown in Figure 3). Then, RLS identification was applied to selected parts of the captured data, as shown in Figure 4. As expected, the system has an important variation in response over the range of possible inputs. The identification results show how the systems clearly split into two different classes, coincident with the two main observed behaviors. Figure 4 shows a validation example of the RLS results.

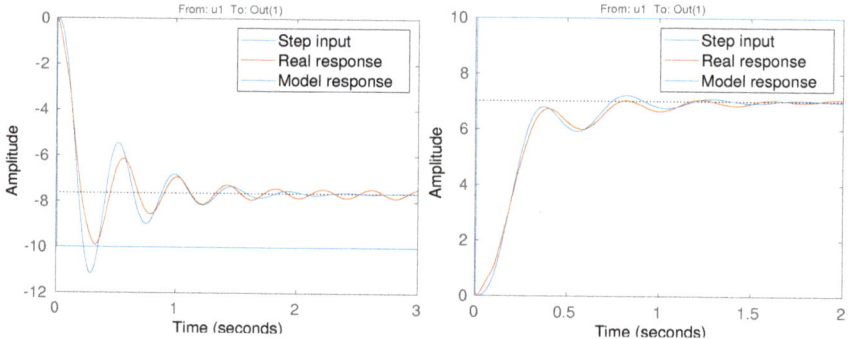

Figure 4. Validation example of the identified models showing two different behaviors. Data obtained from the case of variations in Alpha input ranging from 10 deg to 20 deg (left of Figure 3). Showing the Step input (Alpha input), and the Real response (Alpha output) used in the RLS identification. Resulting model time response is also shown for comparison.

Note that although the linear model captures the system behavior quite well, there are mismatches due to plant nonlinearity. In this case, the identification data were extracted from the capture data shown on the left side of Figure 3, but a different identification procedure was performed for every experiment.

Using RLS identification in every dataset will result in a different model for every single experiment. The frequency responses of these models are shown on the left side of Figure 5, using one color label for each identified model, showing experiment number, α_i, β_i, l. Note that two groups of frequency responses can be observed in the figure. One group shows a decayed resonance with low stationary gain values ($Mag < 0\,dB$ when $Freq \rightarrow 0\,rad/s$), and the other group shows a significant resonant peak and higher stationary gain values ($Mag > 0\,dB$ when $Freq \rightarrow 0\,rad/s$). These groups are highlighted on the right side of this figure, where only the systems with maximum and minimum gains are shown. In addition, an average model, obtained as the mean value of all resulting RLS parameters, is shown on the right side of Figure 5.

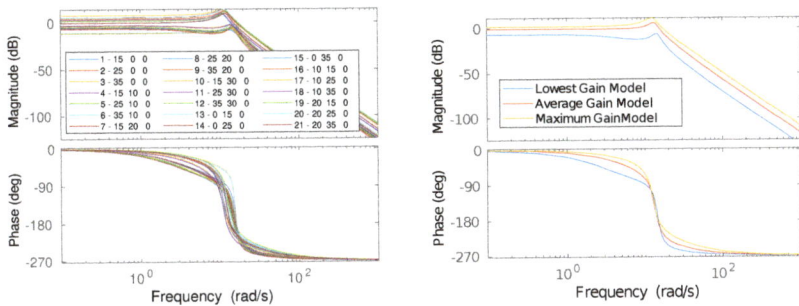

Figure 5. Frequency response of the models obtained using RLS, showing all the identification experiments (**left**) and the three most representative examples: Lowest, Average and Maximum Gains (**right**).

The transfer functions that describe these extreme systems are

$$G_{min} = \frac{311.86}{(s+3.154)(s^2+2.949s+210.9)}, \quad G_{max} = \frac{3294.8}{(s+17.29)(s^2+3.382s+155.5)}, \quad (17)$$

and the average system transfer function, with poles and gain found as the arithmetic mean of the poles and gain obtained from each dataset, is

$$G_{avg} = \frac{1403}{(s+8.665)(s^2+3.462s+176.7)}. \quad (18)$$

Therefore, two classes can be used to model the I-Support system behavior. One class is the low stationary gain case (G_{min}), consisting of a pair of complex conjugate poles shaped by the influence of a non-negligible real pole (three dominant poles). The other class shows a higher stationary gain, and is described by (G_{max}), with two complex dominant poles and one negligible real pole.

The unit input time response and *s* plane pole locations are shown in Figure 6 for the three described system models. An under-damped behavior is observed for the systems with negligible real poles (G_{max}, G_{avg}), while an oscillating over-damped response can be observed in the case with three dominant poles (G_{min}).

Note how the systems with less than 0 dB gain (G_{min}, G_{avg}) show stationary responses below the unit input value, while the other system (G_{max}) stationary response rises above this input level.

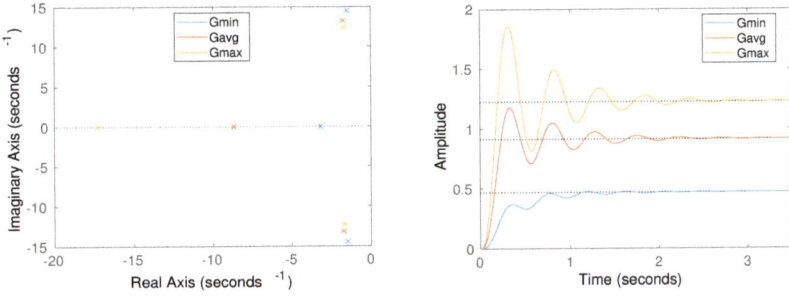

Figure 6. Zero-pole representation (**left**) and unit input time response (**right**) for the three most representative models obtained: Lowest Gain (G_{min}), Average Gain (G_{avg}) and Maximum Gain (G_{max}) models.

Given these results, a control scheme could be designed using two controllers, one for each system class (two in this case), with a switching supervisor applying the correct controller for each case. Nevertheless, the causes that affect the system behavior are not clear; therefore, the supervisor implementation is not possible in this case. That strategy could be considered in the future if the underlying reason leading to the differences in the plant parameters is found.

2.2. Control Strategy

Considering these conditions, a solution will be proposed using robust control techniques. As discussed before, robust controllers are able to show a constant performance despite plant parameter variations or nonlinearities. Therefore, the average plant parameters can be used for a robust controller tuning, in this way granting an invariant performance in the final system behavior despite changes in the plant parameters (usually gain) or neglected nonlinear plant dynamics.

As discussed in Section 1, the fractional order generalization of the integer order PID controller defined by Equation (1) is a convenient robust control approach, and it is suitable in this case. Given the plant characteristics, the derivative part of the controller is not needed and will only bring noise amplification. Therefore, the fractional order proportional integral (FOPI) variant of the controller, defined by Equation (19), will be used.

$$FOPI(s) = k_p + k_i \frac{1}{s^\lambda}. \qquad (19)$$

The three parameters (k_p, k_i, λ) must be tuned in order to achieve the desired system performance. Usual control specifications are stability and responsiveness, normally defined through frequency and damping ratio.

In order to provide a way to compare the robustness between the experiments, a small overshoot will be forced using a target damping ratio lower than 1. As described in [34], a phase margin of 70 deg will result in a damping ratio of 0.8, enough for a significant overshoot. This allows us to compare the overshoot between experiments, providing a measure of the system robustness by comparison. The design frequency must be low enough to avoid the resonance influence in the vicinity of 10 rad/s in order to enforce stability, with the fastest possible response. Based on this, the performance specifications are the following

- $\phi_m = 70$ deg
- $\omega_{gc} = 1.5$ rad/s

With the defined specifications, several tuning methods are available. The recently published iso-m method, described in [30], is straightforward and easy to apply. In order to tune a fractional order controller, a series of simple operations involving basic math and the use of a graph to find the fractional exponent are needed. Therefore, this method can be applied in the tuning of the controller described in Equation (19).

Using the average model defined in Equation (18) and the iso-m tuning method, the controller parameters shown in Table 1 were found.

Table 1. Fractional order controller parameters.

k_p	k_i	λ
0.1878	1.8279	1.19

Based on these parameters, the resulting controller is defined as follows

$$FOPI(s) = 0.1878 + 1.8279 \frac{1}{s^{1.19}}. \qquad (20)$$

An implementation of the fractional operator ($s^{1.19}$) is then needed in order to apply the previous controller in the feedback control scheme of the I-Support robot. One of the most common techniques is the equivalent pole-zero approximation described in [35], based on the operator frequency response (see, for example, [24] or [36]). Using that approximation, the $s^{1.19}$ operator implementation results in

$$s^{1.19} = \frac{0.6614s^3 + 1.763s^2 + 0.4491s + 0.01586}{s^4 + 1.589s^3 + 0.2861s^2 + 0.007414s + 2.53E - 06}. \tag{21}$$

The frequency response of the open-loop system cascading the controller and the average plant model ($FOPI(s) \cdot G_{avg}(s)$), and the closed-loop time response, are shown in Figure 7.

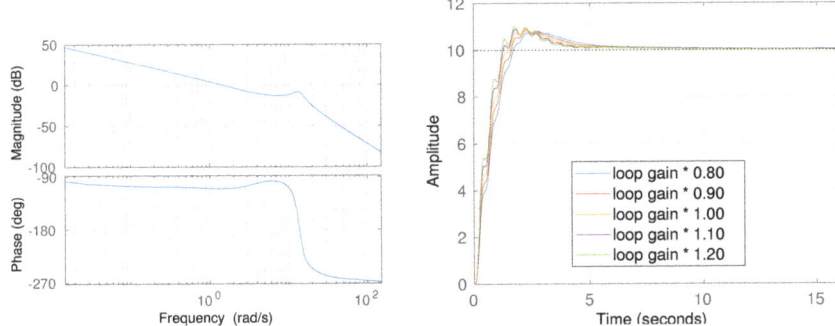

Figure 7. Frequency response (**left**) and time response (**right**) for the fractional order controller system.

Note that, in the left side of the figure, the phase is completely flat in the vicinity of the crossover frequency, leading to the desired iso-damping property and providing a constant overshoot in the expected step response, shown on the right side of Figure 7.

In a similar way, the same specifications and tuning method were used in an equivalent integer-order controller with the intention of a robustness comparison. The resulting parameters are shown in Table 2.

Table 2. Integer order controller parameters.

k_p	k_i	λ
0.0071	1.6402	1.00

With these parameters, the resulting controller is

$$IOPI(s) = 0.0071 + 1.6402 \frac{1}{s}. \tag{22}$$

Again, the frequency and time responses of the system with controller $IOPI(s)$ are shown in Figure 8.

See the significant phase slope around the crossover frequency, leading to an important difference in phase margin in the case of a gain change. Although the simulation predicts an underdamped step response, as shown on the right side of Figure 8, in the experimental section, how the overshoot variability is bigger in the case of the integer-order controller will be shown.

A set of experiments were performed for both fractional- and integer-order controllers. The results are shown and discussed in the following section.

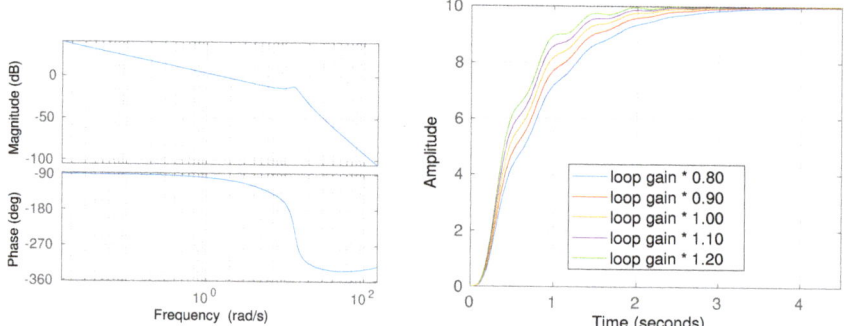

Figure 8. Frequency response (**left**) and time response (**right**) for the integer order controller system.

3. Results and Discussion

The experiments performed with the controllers defined in Equations (20) and (22) were designed to assess and compare their performance and robustness properties. As discussed earlier, the goal of robust control is to keep performance characteristics (like overshoot) invariant despite changes in plant parameters (like gain). In this case, we have seen that these parameters change for the different positions attained on the robot and, therefore, a robust system should provide a constant overshoot percentage despite the end effector position changes.

The first experiment consists of exciting the system with two-step input target angles α and β at the same time, showing the controller robustness by overshoot comparison. Tip orientation angles are recorded with an electromagnetic sensor (NDI Aurora®), as shown in Figure 9. Note that a robust system is expected to have the same performance despite plant parameter variations. Given the specifications defined, the difference in overshoot percent values will show the system robustness, with the results showing similar overshoot percentages in both output signals being more robust. An example of this first experiment for target angles $\alpha = 10$ and $\beta = 30$ is shown in Figure 10 for the FOPI and IOPI controllers.

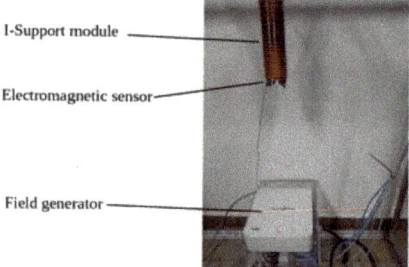

Figure 9. Experimental setup.

Since the plant parameters change with the inclination, introducing two different references for the target angles ($\alpha = 10, \beta = 30$) allows us to observe the dynamic behavior for two different parameter cases in a single experiment. Observe that, for the fractional-order controller system (left), the overshoot variation is much lower (from 11% to 16%) despite the difference in plant parameters compared to the integer controller (right), which shows a higher overshot difference (from 0% to 17%).

A video recording of this experiment is available at https://vimeo.com/517321273 (accessed on 26 February 2021) for the case of the robust controller, showing the overshoots during the tip positioning and the final controlled angles (Supplementary Material).

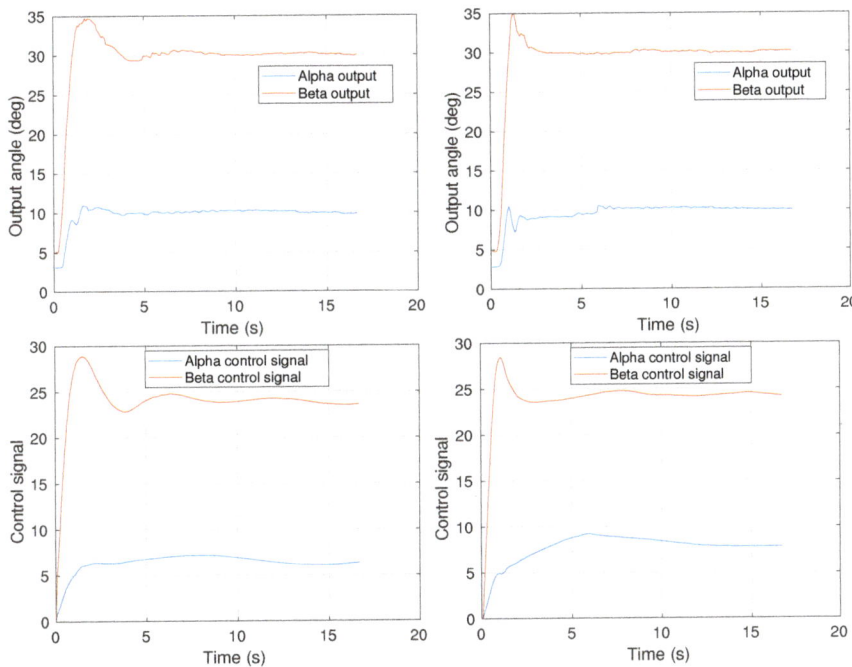

Figure 10. Experiment 1. Time response (**above**) and control signal (**below**) for the two controllers tested, fractional (**left**) and integer (**right**) orders.

The second experiment is a disturbance response test, showing how the control schemes respond to system disturbances. The targets from the first experiment are kept, but, in this case, a constant mass of 150 g is used as a disturbance during the experiment. The setup consists of a metal bar tied to the robot scaffolding in collision with the robot, which can be manually attached or released at any time. In this experiment, the mass was applied at $t = 5$ s, and removed at $t = 10$ s, both producing a sudden change in the feedback error, as shown in Figure 11. Disturbance rejection was correct for both controllers.

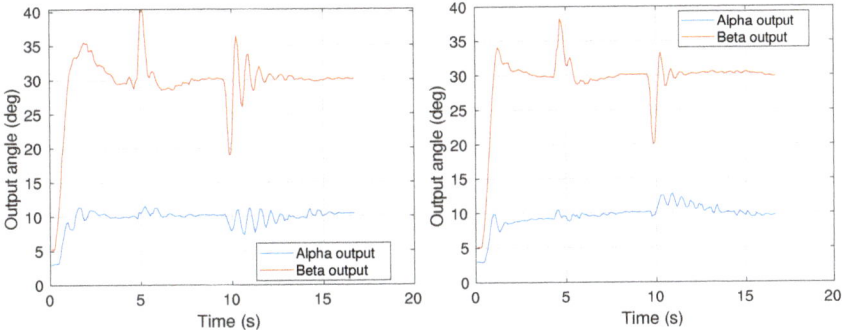

Figure 11. Experiment 2. Disturbance rejection. Time response for the two controllers tested, fractional (**left**) and integer (**right**) orders.

The third experiment is a trajectory in space describing a square of four targets. This experiment is the most demanding of all the experiments performed, presenting the most extreme parameter variations. In this experiment, a changing reference was programmed, following a trajectory of four positions. The references and points are shown in Table 3.

Table 3. Trajectory for experiment 3.

	α	β	l
Point 1	20	20	40
Point 2	−20	20	20
Point 3	−20	−20	40
Point 4	20	−20	20

The results are shown in Figure 12.

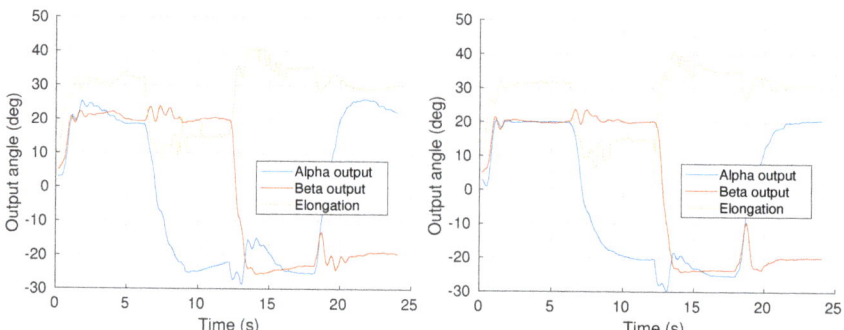

Figure 12. Experiment 3. Trajectory tracking. Time response for the two controllers tested, fractional (**left**) and integer (**right**) orders.

Note how the performances are varied in the case of the IOPI controller, ranging from under-damped to over-damped systems, which cannot be considered as robust in the sense described before. In contrast, the FOPI controller keeps a constant performance, showing similar overshoots and time responses during the whole trajectory, which is considered as a robust behavior, known in the literature as iso-damping.

Although elongation input was included to provide larger plant parameter variations, the feedback loop is only applied to orientation control (α and β). Therefore, as elongation positions are not feedback-controlled, their results show an important error. The position and elongation control of the I-Support robot will be addressed in future works.

4. Conclusions

A robust control for the I-Support soft robot tip orientation is proposed in this paper through the use of a FOPI controller, and compared to a similar PID controller in terms of performance and robustness.

Given the specific robot characteristics, a previous input variable transformation has been applied in order to split the MIMO system into three decoupled SISO systems. This new approach allows to define the model of each system independently and to apply a different feedback loop to each control variable.

With these decoupled SISO systems defined, a feedback control loop was designed and implemented in these systems, steering the robot tip orientation actuation (α and β angles). Given the simplifications made in the model, a robust controller is proposed to deal with the parameter variations and neglected dynamics.

The proposed robust control scheme is based on a fractional-order, proportional integral FOPI controller, tuned through a recent method (iso-m) that provides an easy and straightforward solution to the controller parameters. This is considered a major contribution of this paper, as the previous works using similar control strategies show higher control law complexity, resulting in much higher computational costs. This is probably the reason for the restriction of these control strategies to the simulation environment; therefore, none of these works provide experimental results.

Experimentation is then considered as another important contribution of this paper, as a thorough experimental comparison has been carried out between the two proposed controllers in the real I-Support soft robot platform.

The excellent results obtained for the I-Support tip angle control validate the application of this modeling and control scheme and open up the possibilities of position and elongation feedback control of the platform, which will be proposed in future works.

Besides, a further comparison with previous works based on open-loop configuration can be made in the future, to highlight the pros and cons of each control approach and show some hybrid (feedback-machine learning) control possibilities with that can be applied to the I-Support robot or similar platforms.

Supplementary Materials: The following are available online at https://www.mdpi.com/2227-7390/9/7/702/s1.

Author Contributions: conceptualization, J.M., F.P., C.A.M. and E.F.; methodology, J.M. and F.P.; software, J.M. and F.P.; validation, J.M. and F.P.; formal analysis, J.M.; investigation, J.M. and F.P.; resources, C.A.M. and E.F.; data curation, J.M.; writing—original draft preparation, J.M. and F.P.; writing—review and editing, J.M., F.P., C.A.M. and E.F.; visualization, J.M. and F.P.; supervision, C.A.M. and E.F.; project administration, C.A.M. and E.F.; funding acquisition, C.A.M. and E.F. All authors have read and agreed to the published version of the manuscript.

Funding: The research leading to these results has received funding from the project Desarrollo de articulaciones blandas para aplicaciones robóticas, with reference IND2020/IND-1739, funded by the Comunidad Autónoma de Madrid (CAM) (Department of Education and Research), from HUMASOFT project, with reference DPI2016-75330-P, funded by the Spanish Ministry of Economy and Competitiveness, and from RoboCity2030-DIH-CM, Madrid Robotics Digital Innovation Hub (Robótica aplicada a la mejora de la calidad de vida de los ciudadanos, FaseIV; S2018/NMT-4331), funded by "Programas de Actividades I+D en la Comunidad de Madrid" and cofunded by Structural Funds of the EU. This work was also funded by the European Union's Horizon 2020 research and innovation programme under grant agreement No. 863212 (PROBOSCIS) and No. 824074 (GROWBOT).

Conflicts of Interest: The authors declare no conflict of interest.

Abbreviations

The following abbreviations are used in this manuscript:

PID	Proportional integral derivative
FOPID	Fractional order proportional integral derivative
FOPI	Fractional order proportional integral

References

1. Kim, S.; Laschi, C.; Trimmer, B. Soft robotics: A bioinspired evolution in robotics. *Trends Biotechnol.* **2013**, *31*, 287–294. [CrossRef]
2. Laschi, C.; Cianchetti, M.; Mazzolai, B.; Margheri, L.; Follador, M.; Dario, P. Soft Robot Arm Inspired by the Octopus. *Adv. Robot.* **2012**, *26*, 709–727. [CrossRef]
3. George Thuruthel, T.; Ansari, Y.; Falotico, E.; Laschi, C. Control strategies for soft robotic manipulators: A survey. *Soft Robot.* **2018**, *5*, 149–163. [CrossRef]
4. Renda, F.; Giorelli, M.; Calisti, M.; Cianchetti, M.; Laschi, C. Dynamic model of a multibending soft robot arm driven by cables. *IEEE Trans. Robot.* **2014**, *30*, 1109–1122. [CrossRef]
5. Jones, B.A.; Walker, I.D. Kinematics for multisection continuum robots. *IEEE Trans. Robot.* **2006**, *22*, 43–55. [CrossRef]
6. Thuruthel, T.G.; Falotico, E.; Renda, F.; Laschi, C. Model-based reinforcement learning for closed-loop dynamic control of soft robotic manipulators. *IEEE Trans. Robot.* **2018**, *35*, 124–134. [CrossRef]
7. Ansari, Y.; Falotico, E.; Mollard, Y.; Busch, B.; Cianchetti, M.; Laschi, C. A Multiagent Reinforcement Learning approach for inverse kinematics of high dimensional manipulators with precision positioning. In Proceedings of the IEEE RAS and EMBS International Conference on Biomedical Robotics and Biomechatronics, Singapore, 26–29 June 2016; pp. 457–463.
8. Thuruthel, T.G.; Falotico, E.; Manti, M.; Laschi, C. Stable Open Loop Control of Soft Robotic Manipulators. *IEEE Robot. Autom. Lett.* **2018**, *3*, 1292–1298. [CrossRef]

9. Piqué, F.; Kalidindi, H.T.; Menciassi, A.; Laschi, C.; Falotico, E. A Learning-based Approach for Adaptive Closed-loop Control of a Soft Robotic Arm. In Proceedings of the I-RIM 3D Conference, online, 10–12 December 2020.
10. Della Santina, C.; Bianchi, M.; Grioli, G.; Angelini, F.; Catalano, M.; Garabini, M.; Bicchi, A. Controlling Soft Robots: Balancing Feedback and Feedforward Elements. *IEEE Robot. Autom. Mag.* **2017**, *24*, 75–83. [CrossRef]
11. Yip, M.C.; Camarillo, D.B. Model-Less Feedback Control of Continuum Manipulators in Constrained Environments. *IEEE Trans. Robot.* **2014**, *30*, 880–889. [CrossRef]
12. Kapadia, A.; Walker, I.D. Task-space control of extensible continuum manipulators. In Proceedings of the 2011 IEEE/RSJ International Conference on Intelligent Robots and Systems, San Francisco, CA, USA, 25–30 September 2011; pp. 1087–1092. [CrossRef]
13. Deutschmann, B.; Ott, C.; Monje, C.A.; Balaguer, C. Robust Motion Control of a Soft Robotic System Using Fractional Order Control. In *Advances in Service and Industrial Robotics*; Ferraresi, C., Quaglia, G., Eds.; Springer International Publishing: Cham, Switzerland, 2018; pp. 147–155. [CrossRef]
14. Chang, X.H.; Xiong, J.; Park, J.H. Fuzzy robust dynamic output feedback control of nonlinear systems with linear fractional parametric uncertainties. *Appl. Math. Comput.* **2016**, *291*, 213–225. [CrossRef]
15. Barambones, O.; Gonzalez de Durana, J.; De la Sen, M. Robust speed control for a variable speed wind turbine. *Int. J. Innov. Comput. Inf. Control IJICIC* **2010**, *8*, 7627–7640.
16. Iqbal, J.; Ullah, M.; Khan, S.G.; Khelifa, B.; Ćuković, S. Nonlinear control systems-A brief overview of historical and recent advances. *Nonlinear Eng.* **2017**, *6*, 301–312. [CrossRef]
17. Bode, H.W. *Network Analysis and Feedback Amplifier Design*; Bell Telephone Laboratory Series; Van Nostrand: New York, NY, USA, 1945.
18. Barbosa, R.S.; Machado, J.A.T.; Ferreira, I.M. Tuning of PID Controllers Based on Bode's Ideal Transfer Function. *Nonlinear Dyn.* **2004**, *38*, 305–321. [CrossRef]
19. Chen, Y.; Moore, K.L. Relay Feedback Tuning of Robust PID Controllers with Iso-damping Property. *IEEE Trans. Syst. Man Cybern. Part B (Cybern.)* **2005**, *35*, 23–31. [CrossRef]
20. Sabatier, J.; Agrawal, O.P.; Machado, J.A.T. (Eds.) *Advances in Fractional Calculus: Theoretical Developments and Applications in Physics and Engineering*; Springer: Dordrecht, The Netherlands, 2007.
21. Monje, C.A.; Chen, Y.; Vinagre, B.M.; Xue, D.; Feliu-Batlle, V. *Fractional-Order Systems and Controls: Fundamentals and Applications*; Springer Science & Business Media: Dordrecht, The Netherlands, 2010.
22. Oustaloup, A. *La Dérivation non Entière Théorie, Synthèse et Applications*; Hermès: Paris, France, 1995; p. 507.
23. Podlubny, I. Fractional-order systems and $PI^\lambda D^\mu$-controllers. *IEEE Trans. Autom. Control* **1999**, *44*, 208–214. [CrossRef]
24. Monje, C.A.; Vinagre, B.M.; Santamaría, G.E.; Tejado, I. Auto-tuning of fractional order $PI^\lambda D^\mu$ controllers using a PLC. In Proceedings of the 2009 IEEE Conference on Emerging Technologies Factory Automation, Mallorca, Spain, 22–25 September 2009; pp. 1–7. [CrossRef]
25. Petras, I. Fractional order feedback control of a DC motor. *J. Electr. Eng.* **2009**, *60*, 117–128.
26. Martín, F.; Monje, C.A.; Moreno, L.; Balaguer, C. DE-based tuning of $PI^\lambda D^\mu$ controllers. *ISA Trans.* **2015**, *59*, 398–407. [CrossRef]
27. Manti, M.; Pratesi, A.; Falotico, E.; Cianchetti, M.; Laschi, C. Soft assistive robot for personal care of elderly people. In Proceedings of the 2016 6th IEEE International Conference on Biomedical Robotics and Biomechatronics (BioRob), Singapore, 26–29 June 2016; pp. 833–838. [CrossRef]
28. Ljung, L. Experiments with Identification of Continuous Time Models. *IFAC Proc. Vol.* **2009**, *42*, 1175–1180. [CrossRef]
29. Muñoz, J.; Copaci, D.; Monje, C.A.; Blanco, D.; Balaguer, C. Iso-m based adaptive fractional order control with application to a soft robotic neck. *IEEE Access* **2020**, *1*. [CrossRef]
30. Muñoz, J.; Monje, C.A.; Nagua, L.F.; Balaguer, C. A graphical tuning method for fractional order controllers based on iso-slope phase curves. *ISA Trans.* **2020**. [CrossRef] [PubMed]
31. Martín Barrio, A.; Terrile, S.; Barrientos, A.; del Cerro, J. Robots Hiper-Redundantes: Clasificación, Estado del Arte y Problemática. *Rev. Iberoam. Autom. Inform. Industrial* **2018**, *15*, 351–362. [CrossRef]
32. Martin, A.; Barrientos, A.; del Cerro, J. The Natural-CCD Algorithm, a Novel Method to Solve the Inverse Kinematics of Hyper-redundant and Soft Robots. *Soft Robot.* **2018**, *5*, 242–257. [CrossRef] [PubMed]
33. Ansari, Y.; Manti, M.; Falotico, E.; Mollard, Y.; Cianchetti, M.; Laschi, C. Towards the development of a soft manipulator as an assistive robot for personal care of elderly people. *Int. J. Adv. Robot. Syst.* **2017**, *14*, 1729881416687132. [CrossRef]
34. Nise, N.S. Frequency response techniques. In *Control Systems Engineering*; Wiley: Hoboken, NJ, USA, 2019; Chapter 10, pp. 525–612.
35. Levy, E.C. Complex-Curve Fitting. *IRE Trans. Autom. Control* **1959**, *AC-4*, 37–43. [CrossRef]
36. Valerio, D.; da Costa, J.S. *An Introduction to Fractional Control*; Control, Robotics and Sensors; Institution of Engineering and Technology: London, UK, 2012. [CrossRef]

Article

Tuning Rules for Active Disturbance Rejection Controllers via Multiobjective Optimization—A Guide for Parameters Computation Based on Robustness

Blanca Viviana Martínez *, Javier Sanchis, Sergio García-Nieto and Miguel Martínez

Instituto Universitario de Automática e Informática Industrial, Universitat Politècnica de València, 46022 Valencia, Spain; jsanchis@isa.upv.es (J.S.); sgnieto@isa.upv.es (S.G.-N.); mmiranzo@isa.upv.es (M.M.)
* Correspondence: blamarca@doctor.upv.es

Abstract: A set of tuning rules for Linear Active Disturbance Rejection Controller (LADRC) with three different levels of compromise between disturbance rejection and robustness is presented. The tuning rules are the result of a Multiobjective Optimization Design (MOOD) procedure followed by curve fitting and are intended as a tool for designers who seek to implement LADRC by considering the load disturbance response of processes whose behavior is approximated by a general first-order system with delay. The validation of the proposed tuning rules is done through illustrative examples and the control of a nonlinear thermal process. Compared to classical PID (Proportional-Integral-Derivative) and other LADRC tuning methods, the derived functions offer an improvement in either disturbance rejection, robustness or both design objectives.

Keywords: active disturbance rejection control (ADRC); multiobjective optimization; time delay systems; tuning rules

1. Introduction

Active Disturbance Rejection Control (ADRC) [1] was proposed as an alternative for PID (Proportional-Integral-Derivative) control and has become a new control paradigm. It inherits from the PID controller its independence from the plant model and seeks to compensate its weaknesses through the concept of disturbance estimation and rejection.

The ADRC lumps together the non-modeled dynamics and non-manipulable external signals affecting the system in a single total perturbation. This signal is treated as an extended state to be estimated by an Extended State Observer (ESO) and its impact on the output is rejected by the control action. As a result, the ADRC loop induces the real plant to behave like a set of cascade integrators facilitating the control design.

The fact that the extended observer jointly treats external perturbations and modeling uncertainties highlights its attractiveness in the engineering field, since the knowledge of the process model is kept to minimum in order to design the control loop. What is more, in contrast with model-based approaches, the ADRC assumes a canonical form regardless of the process dynamics and unifies the unknown discrepancies between the canonical form and the real plant in the total perturbation [2]. The effectiveness of the ADRC has been tested in a variety of fields including power electronics, motion and process control. A summary of recent experimental studies in the aforementioned areas can be found in [3]. The emergence of innovative ADRC solutions, particularly in industrial control, is a motivation to consider this control approach for processes where a precise dynamic model is difficult to obtain and a simplified approximation could be used instead.

The implementation of the ADRC requires the order of the system and the nominal value of its critical gain; being the latter the parameter that usually relates the control input with the highest order derivative of the output. When the ESO and the control law are designed by evaluation of nonlinear functions, the algorithm is called NADRC (Nonlinear

ADRC). On the other hand, if a linear observer together with a linear control law are used, the control strategy is called LADRC (Linear ADRC).

The LADRC has gained popularity due to its simple structure and the reduced amount of parameters to be tuned in comparison with the NADRC. The bandwidth parameterization [4] formulates the observer and control law gains as functions of two main parameters—the observer bandwidth and the controller bandwidth. Usually, their selection is based on the closed loop desired behavior and is adjusted by trial and error turning the tuning problem into an empirical process.

The tuning of the LADRC is considered a research area of interest. It has been addressed taking as a starting point PID controllers operating in the control system [5,6] or strictly proper controllers with integrator [7], which state the desired disturbance rejection performance.

The inclusion of the nominal value of the process critical gain as the third tuning parameter (in addition to the two bandwidths) has been discussed in [8,9]. In [10], the nominal value of the critical gain is tuned through an online optimization process for a tank level control problem. The main disadvantage of this approach is the time required to perform the optimization search on the loop.

To avoid the computational cost related to the online tuning, some researches have determined a set of functions to obtain the three main parameters (nominal value of critical gain, observer bandwidth and controller bandwidth). In [11] a tuning method for LADRC suitable for the control of a type of high-order systems is presented. It is based on the interpretation of the maximum sensitivity (M_S) in the Nyquist diagram of the loop transfer function.

High-order plants can be used as approximations for some industrial processes. Nevertheless, the First Order Plus Dead Time (FOPDT) model is also a very common approximation which takes into account delays due to mass or energy transport, or limitations related to measuring and energy conversion devices [12]. The interest in the control of the FOPDT processes has inspired control strategies as the fractional order internal model controller (FO-IMC) from [13], where phase margin and gain crossover frequency specifications are employed to formulate a system of nonlinear equations which needs to be solved for the controller design.

On the other hand, tuning rules for the second order LADRC applied to FOPDT plants have been proposed in [14] through formulation of an optimization problem following the Aggregate Objective Function (AOF) approach. This is, two performance indices of interest as the settling time and the Integral of Squared Error (ISE) were merged in the Integral of Time Weighted Squared Error (ITSE) for minimization. In addition, a robustness measure was used as a fixed constraint.

The aforementioned work pointed out the importance of balancing the disturbance rejection performance with the closed loop robustness. However, including the robustness just as a constraint for the optimization problem could result in solutions offering an optimized performance (in terms of the index selected) but with a robustness measure that tends to be in the upper limit allowed. This may be enough for some designers, but for others, given the complexity of the process, robustness also becomes a design objective and a balance among all performance indices is required. As alternative, in the Generate-First Choose-Later (GFCL) multiobjective approach the objectives are optimized simultaneously providing a set of solutions, with different compromise, to be examined by the designer who makes the final decision.

Some contributions to the LADRC tuning have been made in the GFCL context. Nevertheless, they use the multiobjective approach to select some of the LADRC parameters to control a particular system or the optimization process needs to be performed for each design. For example, in [15] the Integral of Absolute Error (IAE) and the M_S are simultaneously minimized to select the LADRC bandwidths for the control of a power plant. In [16] a tuning scheme for the modified ADRC (MADRC) [17] for unstable time delay systems has been formulated as a multiobjective optimization problem regarding the setpoint following and disturbance rejection. This methodology is intended to be performed adapting the problem according to the system to be controlled. It means that the proper MADRC order should be selected and the optimization and decision making stages need to be carried out for each study case in order to obtain the control law and observer gains.

Motivated by the above, this paper explores the GFCL approach to provide a set of tuning rules for the second-order LADRC parameters computation applicable to the control of FOPDT systems. A Multiobjective Optimization Design (MOOD) procedure is used over a group of nominal plants to obtain a set of Pareto optimal solutions with a compromise between the step load disturbance response and robustness. Then, the LADRC parameters are fitted to functions of the normalized delay and finally, these functions are scaled to make them suitable for the control of a general first order system with delay. Even though the LADRC has a certain level of robustness because it addresses the differences between the actual system and the assumed plant in the total perturbation, its tuning considering the robustness as an objective design balances this feature with the closed loop performance and this is reflected in the derived tuning rules.

The tuning rules presented here have prominent advantages for the control engineer:

- They can be used to control systems approximated by a FOPDT model because only the static gain, apparent time constant and apparent delay are required as prior information. The FOPDT is also known as the three-parameter model and is widely accepted in the control of industrial processes.
- The LADRC main parameters, this is, the nominal value of control gain, the controller bandwidth, and the observer bandwidth are automatically computed through the substitution of the model parameters in the given formulae.
- The designer can select a robustness quality (low, medium or high) for the parameters computation which allows his/her involvement as a decision maker, but eliminates the time and complexity of performing an entire optimization process for the controller design. This is possible because the robustness was included as a design objective in the optimization process formulation, in contrast with other approaches from literature where robustness is imposed just as a constraint, and also, different Pareto optimal solutions were used for the rules derivation.
- The parameters computed through the proposed rules ensure closed loop stability as well as a reasonable compromise between disturbance rejection and loop robustness.
- The designer could use the rules to obtain intervals for each LADRC parameter and adjust the selection according to the preferred performance. An LADRC tuning Matlab app (available at Matlab central [18]) was created for this purpose. Within this tool, the user can also vary the robustness level to visualize the performance with the corresponding calculated parameters.

The paper is organized as follows—in Section 2 the time domain and frequency domain formulation of the second-order LADRC as well as the loop parameterization are presented. In Section 3, a concise description of the Multiobjective Optimization Design procedure is given and the pertinence of this approach in the tuning of LADRC is addressed. Section 4 describes in detail the tuning of LADRC by means of the MOOD procedure whose results were fitted to the rules presented in Section 5. A summary guide for the LADRC parameters computation based on the proposed rules and the interactive tuning tool are also provided in this Section. Section 6 presents the validation of the proposal by the simulation of two examples. Performance comparison with classical PID tuning methods and the LADRC tuning rules from [11,14] are also presented. In Section 7 a nonlinear thermal process is controlled by the LADRC designed according to the proposed tuning method and, finally, Section 8 draws the conclusions.

2. Linear Active Disturbance Rejection Control

This section introduces the Linear Active Disturbance rejection control (LADRC) algorithm for single-input single-output systems.

The LADRC loop is mainly comprised of three blocks as shown in Figure 1.

- Tracking differentiator: It is used to generate a transient profile r_1 for the reference $\tilde{r}$ and the corresponding derivatives $\dot{r}_1, \ddot{r}_1, \ldots, r_1^{(n)}$.
- Extended State Observer (ESO): It estimates the system states $z_1, z_2, \ldots, z_n$ and the additional state z_{n+1} representing the nonmodeled dynamics and perturbations.
- Controller: It provides a state feedback control law u_0 for the disturbance-free modified plant. Therefore, the control law $u = (u_0 - z_{n+1})/b_0$ is generated to act on the real plant and through which the disturbance information is rejected.

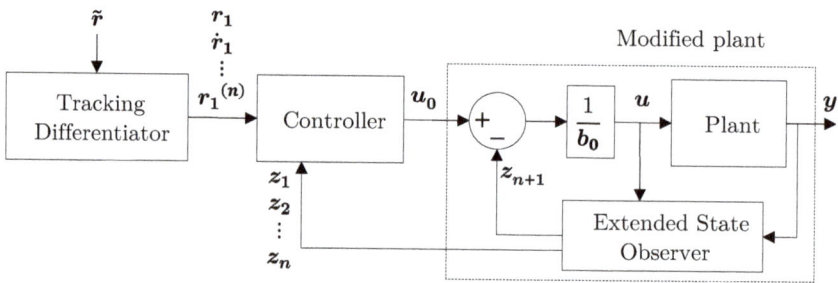

Figure 1. Active Disturbance Rejection Control (ADRC) loop.

For the LADRC implementation, the system order n and the nominal value of its critical gain b_0 are required. Many practical applications can be approximated through first or second order models. Moreover, if the plant is open loop stable, a low order LADRC can be implemented and closed loop stability can be achieved by proper selection of the LADRC parameters [5].

In this work, the second-order LADRC was selected as control algorithm for FOPDT systems. The LADRC theoretical formulation in time domain and frequency domain, as well as the closed loop parameterization used for the development of the tuning rules are explained next.

2.1. Time Domain Formulation

Consider the following input-output model of a second order system.

$$\ddot{y} = -a_1 \dot{y} - a_0 y + bu, \tag{1}$$

where y is the controlled output, u is the control action, a_0 and a_1 are constants determining the location of the system poles and b is known as critical gain.

The state space representation of (1) is given by (2), where w has been included to indicate the load disturbances acting on the system.

$$\begin{cases} \dot{x}_1 = x_2 \\ \dot{x}_2 = -a_0 x_1 - a_1 x_2 + bu + w \\ y = x_1. \end{cases} \tag{2}$$

In the case that a_0 and a_1 are unknown, the first two terms in the right side of the expression for $\dot{x}_2$ can be lumped in a function called total perturbation which also includes load disturbances and the difference between the real value of b and its known nominal value denoted by b_0. Thus,

$$f = -a_0 x_1 - a_1 x_2 + (b - b_0)u + w. \tag{3}$$

The model (4) is obtained by replacing (3) in (2).

$$\begin{cases} \dot{x}_1 = x_2 \\ \dot{x}_2 = f + b_0 u \\ y = x_1. \end{cases} \quad (4)$$

As the total perturbation is an unknown function, f is treated as an additional state that must be estimated and compensated by the control loop. The resulting extended state space model with $x_3 \triangleq f$ and $h = \dot{f}$ unknown is

$$\begin{cases} \dot{x}_1 = x_2 \\ \dot{x}_2 = x_3 + b_0 u \\ \dot{x}_3 = h \\ y = x_1. \end{cases} \quad (5)$$

The estimation of states in (5) is achieved through the Linear Extended State Observer (LESO) (6) whose inputs are the measured output y and the control action u. The z_i correspond to the estimated states and L_i are the observer gains. Note that, although the LESO has a similar structure to a traditional observer, it estimates not only the system states but also the information of the total perturbation contained in z_3. In contrast with the traditional observer, the LESO keeps the required amount of plant information to a minimum. The analysis of convergence and experimental validation of LESO are addressed in [19].

$$\begin{cases} \dot{z}_1 = z_2 + L_1(y - z_1) \\ \dot{z}_2 = z_3 + b_0 u + L_2(y - z_1) \\ \dot{z}_3 = L_3(y - z_1). \end{cases} \quad (6)$$

According to Figure 1, the control law acting on the real plant is

$$u = \frac{u_0 - z_3}{b_0}. \quad (7)$$

Therefore, the double integrator (8) is obtained by replacing (7) in (4) and assuming that $z_3 \approx f$.

$$\begin{cases} \dot{x}_1 = x_2 \\ \dot{x}_2 = u_0 \\ y = x_1 \end{cases} \quad (8)$$

Equation (8) represents a disturbance-free modified plant which is controlled by the feedback law

$$u_0 = k_1(\tilde{r} - z_1) - k_2 z_2, \quad (9)$$

where $\tilde{r}$ is the setpoint and k_1 and k_2 are gains selected taking into account the desired closed loop performance. Note that $\tilde{r}$ has been set as the reference in (9). This can be done in practice if the tracking differentiator is omitted or the setpoint derivatives are unbounded [20].

2.2. Frequency Domain Formulation

The block diagram from Figure 1 can be reformulated as the two degree-of-freedom configuration of Figure 2. The direct loop transfer function $G_C(s)$ and the feedback transfer function $G_F(s)$ are derived as follows.

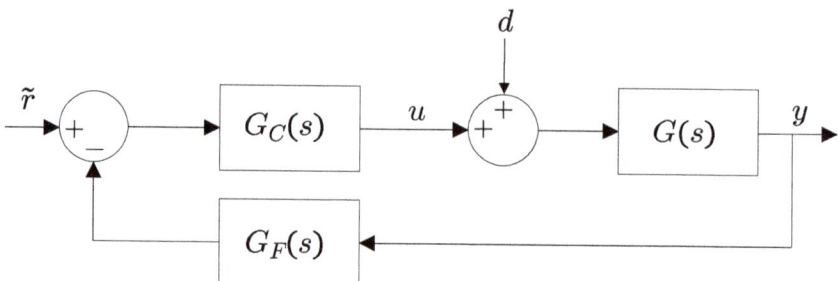

Figure 2. 2-degree-of-freedom (DOF) configuration of ADRC.

The linear extended state observer (6) in frequency domain is given by

$$\begin{cases} sZ_1 = Z_2 + L_1(Y - Z_1) \\ sZ_2 = Z_3 + b_0 U + L_2(Y - Z_1) \\ sZ_3 = L_3(Y - Z_1), \end{cases} \quad (10)$$

where s is the complex variable, Y is the Laplace transform of the output, U is the Laplace transform of the control action and Z_i are the Laplace transforms of the states.

The expressions (11)–(13) are obtained by solving the system of Equation (10).

$$Z_1 = \frac{b_0 s}{s^3 + L_1 s^2 + L_2 s + L_3} U + \frac{(L_1 s^2 + L_2 s + L_3)}{s^3 + L_1 s^2 + L_2 s + L_3} Y \quad (11)$$

$$Z_2 = \frac{b_0(s^2 + sL_1)}{s^3 + L_1 s^2 + L_2 s + L_3} U + \frac{(L_2 s^2 + L_3 s)}{s^3 + L_1 s^2 + L_2 s + L_3} Y \quad (12)$$

$$Z_3 = \frac{-L_3 b_0}{s^3 + L_1 s^2 + L_2 s + L_3} U + \frac{L_3 s^2}{s^3 + L_1 s^2 + L_2 s + L_3} Y. \quad (13)$$

The control action (14) is deduced by combining the frequency domain expressions of (7) and (9), with R being the Laplace transform of the reference.

$$U = \frac{1}{b_0}(k_1 R - k_1 Z_1 - k_2 Z_2 - Z_3). \quad (14)$$

Therefore, substituting (11)–(13) in (14) and reorganizing terms, U is rewritten as

$$U = \frac{k_1}{b_0}\left[\frac{s^3 + L_1 s^2 + L_2 s + L_3}{s^3 + (L_1 + k_2)s^2 + (k_2 L_1 + L_2 + k_1)s}\right]R \\ - \left[\frac{(k_1 L_1 + k_2 L_2 + L_3)s^2 + (k_1 L_2 + k_2 L_3)s + k_1 L_3}{b_0(s^3 + (L_1 + k_2)s^2 + (k_2 L_1 + L_2 + k_1)s)}\right]Y. \quad (15)$$

From Figure 2 and in the absence of load disturbance ($d = 0$)

$$U = G_C(s)R - G_C(s)G_F(s)Y. \quad (16)$$

Hence, the resulting direct loop transfer function (17) and the feedback transfer function (18) are obtained comparing the factors of R and Y in (15) with those in (16).

$$G_C(s) = \frac{k_1}{b_0}\left(\frac{s^3 + L_1 s^2 + L_2 s + L_3}{s^3 + (L_1 + k_2)s^2 + (k_2 L_1 + L_2 + k_1)s}\right) \quad (17)$$

$$G_F(s) = \frac{(k_1 L_1 + k_2 L_2 + L_3)s^2 + (k_2 L_3 + k_1 L_2)s + k_1 L_3}{k_1(s^3 + L_1 s^2 + L_2 s + L_3)}. \quad (18)$$

Finally, the transfer function from output to load disturbance is

$$G_D(s) = \frac{G(s)}{1 + G(s)G_C(s)G_F(s)} \tag{19}$$

and the transfer function from control action to output is

$$G_U(s) = -G_C(s)G_F(s). \tag{20}$$

Equation (19) describes the system response to a load disturbance and (20) represents the LADRC transfer function for disturbance rejection.

2.3. Control Loop Parameterization

The control loop parameterization seeks a set of parameters that allows the computation of the complete set of LADRC gains. In addition, if an LADRC is designed for the control of a nominal system (e.g., a nominal FOPDT system), the loop parameterization also allows the parameters scaling in order to make the controller suitable for other systems of the same nature.

Consider the following theorem related to the scaling and bandwidth parameterization of the LADRC loop.

Theorem 1. *[4] Assuming $G_a(s)$ is a stabilizing controller for plant $G_n(s)$ and the loop gain crossover frequency is ω_c, then the controller*

$$\bar{G}_a(s) = \frac{1}{k} G_a\left(\frac{s}{\omega_p}\right) \tag{21}$$

will stabilize the plant $\bar{G}_n(s) = kG_n(s/\omega_p)$ and the new loop gain $\bar{\mathcal{L}}(s) = \bar{G}_n(s)\bar{G}_a(s)$ will have a bandwidth of $\omega_c\omega_p$, and the same stability margins of $\mathcal{L}(s) = G_n(s)G_a(s)$.

In (21), k represents the gain scaling of plant $kG_n(s)$ respect to $G_n(s)$ and ω_p is the frequency scaling of plant $G_n(s/\omega_p)$ respect to $G_n(s)$.

Let $G_A(s)$ be the transfer function obtained by multiplying $G_C(s)$ and $G_F(s)$ in the right hand side of (20). This is,

$$G_A(s) = \frac{(k_1 L_1 + k_2 L_2 + L_3)s^2 + (k_2 L_3 + k_1 L_2)s + k_1 L_3}{b_0(s^3 + (L_1 + k_2)s^2 + (L_2 + k_2 L_1 + k_1)s)}. \tag{22}$$

Equation (22) is function of b_0, the observer gains L_i and the controller gains k_i. The bandwidth parameterization is used to reduce the calculation of the L_i to the selection of the parameter ω_o named observer bandwidth. Likewise, the k_i values are made dependent on the parameter ω_c known as controller bandwidth.

Consider the state space representation of the extended model (5)

$$\begin{bmatrix} \dot{x}_1 \\ \dot{x}_2 \\ \dot{x}_3 \end{bmatrix} = \underbrace{\begin{bmatrix} 0 & 1 & 0 \\ 0 & 0 & 1 \\ 0 & 0 & 0 \end{bmatrix}}_{A} \begin{bmatrix} x_1 \\ x_2 \\ x_3 \end{bmatrix} + \underbrace{\begin{bmatrix} 0 \\ b_0 \\ 0 \end{bmatrix}}_{B} u + \underbrace{\begin{bmatrix} 0 \\ 0 \\ 1 \end{bmatrix}}_{E} h \tag{23}$$

$$y = \underbrace{\begin{bmatrix} 1 & 0 & 0 \end{bmatrix}}_{C} \begin{bmatrix} x_1 \\ x_2 \\ x_3 \end{bmatrix},$$

whose matrix form is

$$\dot{x} = Ax + Bu + Eh \tag{24}$$
$$y = Cx,$$

with $x = [x_1\ x_2\ x_3]^\top$. Similarly, the matrix form of observer (6) is given by (25) with $z = [z_1\ z_2\ z_3]^\top$ and $L = [L_1\ L_2\ L_3]^\top$.

$$\dot{z} = Az + Bu + L(Cx - Cz). \tag{25}$$

Let $e = x - z$ be the estimation error. Its dynamic behavior is given by (26) and it is obtained after subtracting (25) from (24).

$$\dot{e} = (A - LC)e + Eh. \tag{26}$$

Assuming that h, even it is unknown, it is also differentiable and bounded, the observer gains can be calculated through pole placement. In [4], it is proposed that the three poles be located at position $-\omega_o$ in the left semi-plane such as

$$sI - (A - LC) = (s + \omega_o)^3. \tag{27}$$

Thus, the parameterization of the observer gains (28) is obtained as a function of ω_o by solving for both sides of (27) and comparing factors.

$$L_1 = 3\omega_o \quad L_2 = 3\omega_o^2 \quad L_3 = \omega_o^3. \tag{28}$$

On the other hand, the controller gains design takes into account the frequency representation of the modified plant (8) and the control action (9) to obtain the closed loop transfer function

$$G_Y(s) = \frac{k_1}{s^2 + k_2 s + k_1}. \tag{29}$$

According to the characteristic equation of (29), the closed loop poles depends on selection of the gains k_1 and k_2. Then, following the approach from [4], the poles are located at $-\omega_c$ as in (30) and the controller gains parameterization of (31) is derived.

$$s^2 + k_2 s + k_1 = (s + \omega_c)^2 \tag{30}$$
$$k_1 = \omega_c^2 \quad k_2 = 2\omega_c. \tag{31}$$

The bandwidth parameterization from (28) and (31) is used in (22) to obtain

$$G_A(s) = \frac{(3\omega_c^2 \omega_o + 6\omega_c \omega_o^2 + \omega_o^3)s^2 + (2\omega_c \omega_o^3 + 3\omega_c^2 \omega_o^2)s + \omega_c^2 \omega_o^3}{b_0[s^3 + (3\omega_o + 2\omega_c)s^2 + (3\omega_o^2 + 6\omega_c \omega_o + \omega_c^2)s]}. \tag{32}$$

Therefore, by proper selection of b_0, ω_c and ω_o, the second-order LADRC estimates and rejects the load disturbances acting on the loop.

Now, let the following FOPDT system be the plant to be controlled

$$G(s) = \frac{K}{Ts + 1} e^{-ls}, \tag{33}$$

where K is the static gain, T is the apparent time constant and l is the apparent delay or dead time [21].

If $G_n(s)$ is considered as a nominal FOPDT plant, then, following the scaling and bandwidth parameterization theorem [see (21)], the model (33) can be treated as a scaled version of (34) in which $k = K$, $\omega_p = 1/T$ and $\Theta = l/T$ as shown in (35).

$$G_n(s) = \frac{1}{s+1} e^{-\Theta s} \tag{34}$$

$$G(s) = K\left(\frac{1}{\frac{s}{1/T}+1}\right)e^{-\frac{L}{T}\frac{s}{1/T}}. \tag{35}$$

Hence, through some mathematical manipulation, the scaled controller $\bar{G}_A(s) = (1/k)G_A(s/\omega_p)$ leads to the definition of the new set of LADRC parameters

$$\bar{b}_0 = \frac{Kb_0}{T^2} \quad \bar{\omega}_c = \frac{\omega_c}{T} \quad \bar{\omega}_o = \frac{\omega_o}{T}. \tag{36}$$

In conclusion, if a stable second-order LADRC with parameters b_0, ω_c and ω_o is designed for the nominal system (34), then, the scaled LADRC with parameters $\bar{b}_0$, $\bar{\omega}_c$ and $\bar{\omega}_o$ is suitable for the control of the general FOPDT plant (33).

3. Multiobjective Optimization Design Procedure

In this section, the generalities of the multiobjective optimization approach used to address the LADRC tuning problem are presented. Particularly, the steps of a Multiobjective Optimization Design (MOOD) procedure are briefly explained and the pertinence of this approach for the tuning problem is explored by means of a numerical example.

When designing a controller, the tuning process or solution obtained is strongly dependent on the desired performance for the closed loop. The behavior of the output, control action, and any other signals of interest is usually measured through some performance indices or design objectives. If these indices are wanted to be minimized or maximized, then, an optimization statement can be formulated.

For each minimized or maximized index, a particular solution is obtained. Therefore, if different design objectives are optimized simultaneously, then, multiple solutions can be suitable for the tuning of the same controller, not implying that one is better than the other, but suggesting that a solution can be selected with a particular trade-off among the aforementioned conflicting objectives. In this case, if the designer is interested, for example, in the simultaneous minimization of two performance indices, a MOOD procedure could aid in the tuning problem.

A MOOD procedure comprises three fundamental steps [22].

1. Multiobjective Problem (MOP) definition: The design objectives of interest are stated as well as the decision variables and the possible constraints.
2. Optimization Process (OP): An algorithm is selected to search throughout the decision space for the approximations of the optimal solutions (Pareto Set) and their corresponding objective values (Pareto Front). This algorithm should fulfill some desirable characteristics in order to provide the designer with useful solutions.
3. Multicriteria Decision Making (MCDM): Specialized visualization techniques are employed to analyze the Pareto Front and Pareto Set approximations. The best solution is the one that meets the designer's preferences.

As an example, Figure 3 illustrates the concepts of Pareto dominance, Pareto Front and Pareto Set for the biobjective optimization problem $\min_\theta J(\theta) = [J_1(\theta), J_2(\theta)]$ with decision variables $\theta = [\theta_1, \theta_2]$. The decision vectors $\theta^1, ..., \theta^5$ dominates the vectors θ^6 and θ^7 because the objective vectors $J(\theta^1), ..., J(\theta^5)$ are not worse than $J(\theta^6), J(\theta^7)$ in both objectives and are better in at least one objective.

In order to explore the suitability of the multiobjective optimization approach for the LADRC tuning problem, the responses to an unitary step load disturbance ($\tilde{r} = 0, d = 1$) and to an unitary step setpoint ($\tilde{r} = 1, d = 0$) of the closed loop of Figure 2 with $G(s)$ as (37) were obtained for different combinations of the three LADRC tuning parameters in the search space: $b_0 \in [5, 35]$, $\omega_c \in [1, 25]$ rad/s, $\omega_o \in [1, 25]$ rad/s, and following a grid method with $\Delta b_0 = 1$ and $\Delta \omega_o = \Delta \omega_c = 0.2$ rad/s.

$$G_e(s) = \frac{1}{s+1}e^{-s}. \tag{37}$$

Initially, the LADRC stability region was analyzed. Figure 4 shows the pairs (ω_c, ω_o) for the critical gain nominal values $b_0 = 5, 15, 35$ that produce a stable output in system (37). From this figure, it is noted that as the nominal value of the critical gain increases, more pairs (ω_c, ω_o) appears in the stability region which represent more possible combinations for the LADRC tuning. In other words, there exists a stability bound that moves in the (ω_c, ω_o) increasing direction as a higher value of b_0 is selected.

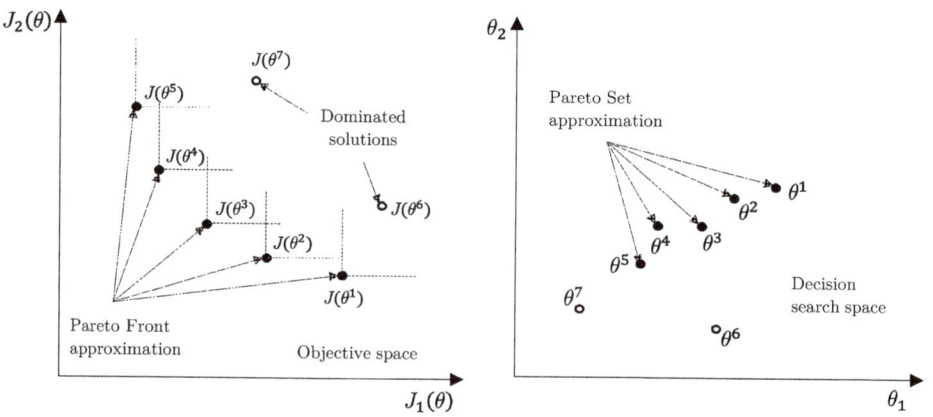

Figure 3. Pareto dominance, Pareto Front and Pareto set in a bidimensional case. There are no solution vectors dominating $\theta^1, \ldots, \theta^5$ so these solutions are the approximation of the Pareto Set and their corresponding objective vectors $J(\theta^1), \ldots, J(\theta^5)$ are the approximation of the Pareto Front.

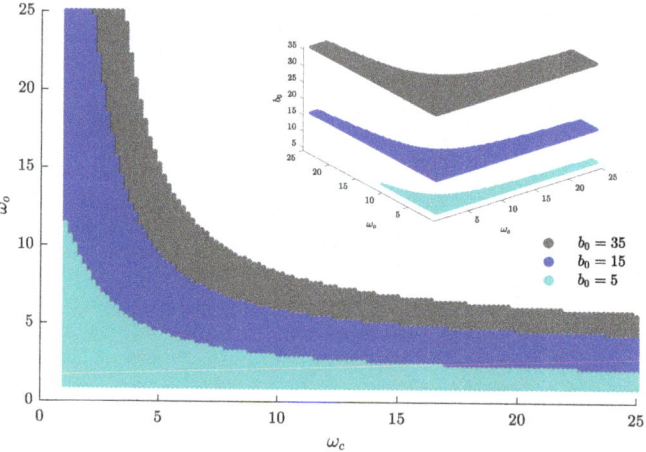

Figure 4. Closed loop stability regions for $G_e(s)$. Each point in the region represents a combination of parameters producing a stable output. For each value of $b_0 \in [5, 35]$ there exist pairs (ω_c, ω_o) that produce a stable output. The $b_0 = 5, 15, 35$ values are plotted as examples to illustrate the shape and behavior of the stability region.

Once the LADRC stability region was obtained, interest was put in the performance computed with those combinations of parameters. Particularly, the ITSE for load disturbance rejection, the robustness, and the Total Variation of control action (TV) were defined as design objectives as stated in Table 1.

Table 1. Design objectives for the performance evaluation of the Linear ADRC (LADRC).

Index/Design Objective	Definition
Integral of the time weighted squared error value	$\text{ITSE} = \int_{t=0}^{t_{98\%}} t \cdot (r(t) - y(t))^2 \, dt$
Total variation of the control action	$\text{TV} = \sum_{i=1}^{t_{98\%}} |u_{i+1} - u_i|$
Mixed robustness	$\varepsilon = \sup_{\omega} (|S(j\omega)| + |T(j\omega)|)$

Closed loop robustness is usually measured through the maximum peak of the sensitivity function M_S and the maximum peak of the complementary sensitivity function M_T such as $1.3 < M_S < 2$ and $M_T < 1.25$ [23]. In this work, a robustness measure denoted by ε is adopted which is defined in [24] as the structured singular value of matrix M from a $M - \Delta$ configuration with a diagonal block structure.

The ε index has been previously used in [14] to quantify the robust stability of the closed loop system with the LADRC and is computed as the maximum peak of the sum of the magnitudes of the frequency responses of the sensitivity function $S(j\omega)$ and the complementary sensitivity function $T(j\omega)$. The lower the value of ε, the more robust the closed loop system.

A first look at the minimum ITSE value inside the stability region shows that $\text{ITSE}_{min1} = 0.82$ for the solution $b_{01} = 17$, $\omega_{c1} = 1.8$ rad/s, $\omega_{o1} = 23.6$ rad/s. However, the associated robustness of $\varepsilon_1 = 5.93$ is regarded as poor. If the constraint $\varepsilon \leq 3$ is imposed on the robustness index, then a new solution $b_{02} = 24$, $\omega_{c2} = 2$ rad/s, $\omega_{o2} = 21$ rad/s is found with an $\text{ITSE}_{min2} = 1.13$ and a corresponding robustness of $\varepsilon_2 = 2.99$.

On the other hand, a search for the most robust controller results in the parameters $b_{03} = 15$, $\omega_{c3} = 19.8$ rad/s, $\omega_{o3} = 1$ rad/s which produce $\varepsilon_{min3} = 1.38$ but with a extremely high ITSE value of $\text{ITSE}_3 = 113.51$. Also, if the ITSE is constrained such as $\text{ITSE} \leq 2$, then the new solution is $b_{04} = 19$, $\omega_{c4} = 21$ rad/s, $\omega_{o4} = 2.8$ rad/s with a robustness $\varepsilon_{min4} = 2.02$ and a time performance index $\text{ITSE}_4 = 1.99$.

Table 2 comprises the solutions and performance comparison discussed above. Some additional indices as M_S, M_T, total variation of control action for disturbance rejection (TV_d), and total variation of control action for setpoint following (TV_s) are included as complementary information. Note that each of the LADRC set of parameters can be considered as optimal only respect to the corresponding minimized index. For example, the solution $(b_{02}, \omega_{c2}, \omega_{o2})$ is optimal respect to the ITSE, but the robustness obtained is the maximum allowed according to the constraint.

Table 2. Comparison of LADRC performance in control of $G_e(s)$.

Desired Performance	LADRC Parameters	M_S	M_T	ε	ITSE	TV_d	TV_s
min ITSE	$b_{01} = 17$ $\omega_{c1} = 1.8$ rad/s $\omega_{o1} = 23.6$ rad/s	3.45	2.48	5.93	0.82	3.14	2.50
min ITSE $\varepsilon \leq 3$	$b_{02} = 24$ $\omega_{c2} = 2$ rad/s $\omega_{o2} = 21$ rad/s	1.98	1.16	2.99	1.13	1.40	1.32
min ε	$b_{03} = 15$ $\omega_{c3} = 19.8$ rad/s $\omega_{o3} = 1$ rad/s	1.19	1.00	1.38	113.51	1.02	33.87
min ε ITSE ≤ 2	$b_{04} = 19$ $\omega_{c4} = 21$ rad/s $\omega_{o4} = 2.8$ rad/s	1.50	1.01	2.02	1.99	1.10	29.25

In addition to the solutions reported in Table 2, there are other sets of LADRC parameters within the stability region that offer a compromise between disturbance rejection, quantified by ITSE, and robustness. To search for these alternatives, the Pareto dominance definition was applied over the total of parameters combinations, restricting the robustness measure to the range $\varepsilon \in [2,3]$ which represents a maximum sensitivity in the range $M_S \in [1.3, 2]$ and a maximum complementary sensitivity in the interval $M_T \in [1, 1.4]$.

Figure 5a shows the Pareto Front approximation for the simultaneous minimization of ITSE for disturbance rejection and robustness. As expected, the ITSE can not be improved (decreased) without weakening the robustness. Likewise, a more robust closed loop system is possible as long as the ITSE value is allowed to increase. The solutions $(b_{02}, \omega_{c2}, \omega_{o2})$ and $(b_{04}, \omega_{c4}, \omega_{o4})$ from Table 2 would be located around the upper and bottom ends of the Pareto Front approximation, respectively.

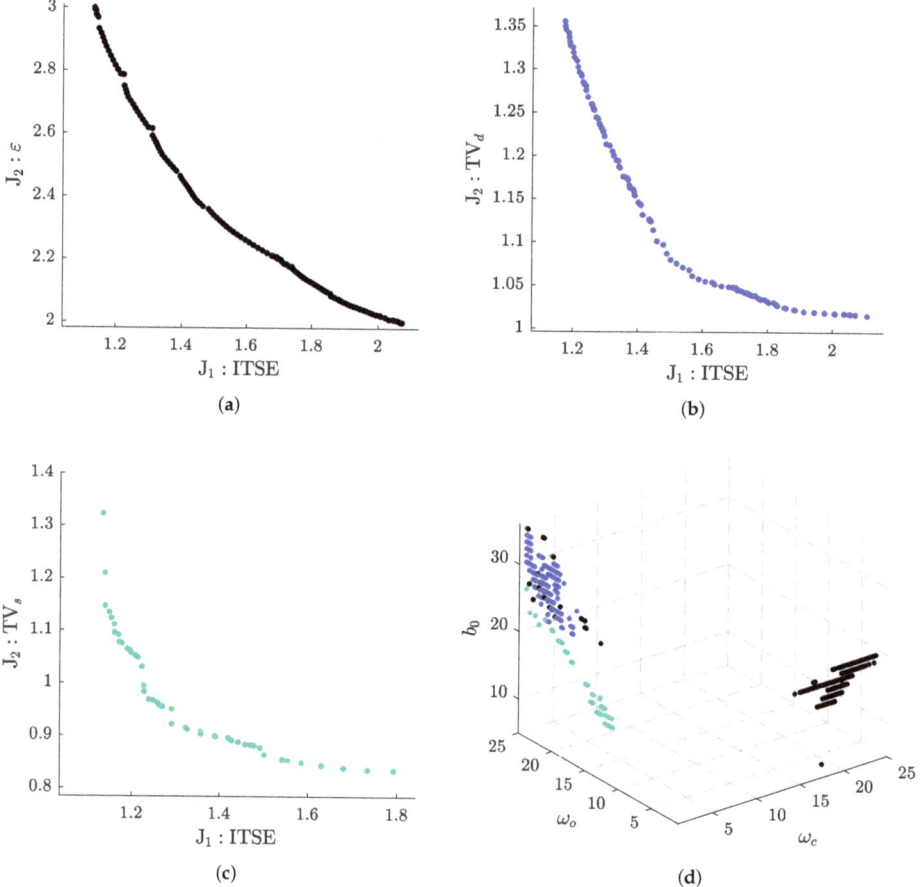

Figure 5. Pareto Fronts and Pareto Sets approximations for minimization of two design objectives J_1 and J_2. (**a**) the Pareto Front approximation for the simultaneous minimization of ITSE for disturbance rejection and robustness. (**b**) the Pareto Front approximation for minimization of ITSE and TV for disturbance rejection (TV_d). (**c**) the approximation of the Pareto Front when the ITSE for disturbance rejection is minimized simultaneously with the TV of the unitary setpoint (TV_s). (**d**) the Pareto Sets approximations for the three said cases.

From other point of view, Figure 5b is the Pareto Front approximation for minimization of ITSE and TV for disturbance rejection (TV_d) and Figure 5c is the approximation of the

Pareto Front when the ITSE for disturbance rejection is minimized simultaneously with the TV of the unitary setpoint (TV$_s$). This figures show that there is also a compromise between the ITSE performance and the control efforts.

Finally, the Pareto Sets approximations for the three said cases are presented in Figure 5d. Note that the optimal values for the nominal critical gain are higher than $b_0 = 1$, which would be the nominal value ($b_0 = K/T$) computed from the model (37), as is commonly suggested in literature. Moreover, in the solutions with a compromise between ITSE and robustness, the controller bandwidth can be selected to be greater than the observer bandwidth ($\omega_c > \omega_o$) or vice versa ($\omega_c < \omega_o$). Nevertheless, for a compromise between ITSE and the total variation of the control action, a selection of parameters in which $\omega_c < \omega_o$ seems more appropriated.

The case study addressed in this section gave some insight into the LADRC performance in the control of a FOPDT system. In summary, there exist a trade-off between the disturbance rejection performance of the LADRC and its robustness. The LADRC parameters that produce this compromise are Pareto optimal and can be searched through an optimization process were the objectives related to disturbance rejection and robustness are minimized simultaneously. Besides, the definition of constraints over the objective and search spaces could drive the optimization process to solutions that meet some desired additional performance. If the aforementioned optimization procedure is applied over a group of plants of the same kind, then the Pareto optimal alternatives could be used to derive tuning rules reflecting the desired trade-off.

4. LADRC Tuning by Multiobjective Optimization

For the tuning problem of the second-order LADRC related to the control of FOPDT systems, a MOOD procedure was applied to a group of nominal plants in the form of (34) which was obtained by varying the nominal delay from $\Theta = 0.5$ to $\Theta = 5$ with a change of $\Delta\Theta = 0.1$.

The FOPDT systems can be characterized based on the normalized dead time $\tau = l/(l + T)$ with $0 \leq \tau \leq 1$ [25]. Particularly, a system is lag-dominated if τ is small, balanced if τ is around 0.5 and delay-dominated if τ is large [26]. In terms of the nominal delay, τ can be written as

$$\tau = \frac{\Theta}{\Theta + 1}. \tag{38}$$

Thus, the MOOD procedure was applied to plants with τ ranging from 0.09 to 0.83, which includes lag-dominated, balance, and delay-dominated processes. The MOOD results were used to fit the optimal solutions for the LADRC parameters and the fitting curves were scaled to obtain the tuning rules as functions of the known FOPDT parameters. In this section, each step of the MOOD procedure and the data processing of solutions are explained in depth.

4.1. MultiObjective Problem Definition

The first stage of the MOOD procedure implies the definition of the decision space, the objective space, and the possible constraints. The decision variables are selected from the parametric controller; the objective space is related to the desired performance, and finally, constrains are the design limitations imposed on the overall concept.

The plant to be controlled corresponds to the FOPDT nominal model (34). Note that any controller designed for this plant can be scaled afterwards according to (36).

The following scaling for observer bandwidth were also adopted:

$$\omega_o = k_o \omega_c, \; k_o > 1, \tag{39}$$

which indicates that LADRC parameters meeting the relation $\omega_c < \omega_o$ are preferred. This additional scaling is commonly suggested in literature (e.g., in [4,11,12]).

The transfer function (40) is obtained by substituting (39) in (32).

$$G_A(s) = \frac{(3k_o\omega_c^3 + 6k_o^2\omega_c^3 + k_o^3\omega_c^3)s^2 + (2k_o^3\omega_c^4 + 3k_o^2\omega_c^4)s + k_o^3\omega_c^5}{b_0[s^3 + (3k_o\omega_c + 2\omega_c)s^2 + (3k_o^2\omega_c^2 + 6k_o\omega_c^2 + \omega_c^2)s]}. \quad (40)$$

Choosing a value of $k_o = 10$, the corresponding controller to tune is

$$G_A(s) = \frac{1630\omega_c^3 s^2 + 2300\omega_c^4 s + 1000\omega_c^5}{b_0(s^3 + 32s^2 + 361\omega_c^2 s)}, \quad (41)$$

with the decision variables:

$$\theta = [b_0, \omega_c]. \quad (42)$$

Two design objectives were selected: the ITSE for the response to a unitary step load disturbance and the mixed robustness index ε. Thus, the complete multiobjective problem is stated as

$$\min_\theta J(\theta) = [J_1(\theta), J_2(\theta)] \quad (43)$$

$$J_1(\theta) = \text{ITSE}(\theta) \quad (44)$$

$$J_2(\theta) = \varepsilon(\theta) \quad (45)$$

$$\theta = [b_0, \omega_c], \quad (46)$$

subject to

$$\begin{array}{rcl} & & \text{Stable in closed loop} \\ & & J_1(\theta) \leq \text{ITSE}_{\text{SIMC}} \\ 2 & \leq J_2(\theta) \leq & 3 \\ 1 & \leq b_0 \leq & 200 \\ 0.1 & \leq \omega_c \leq & 20 \end{array} \quad (47)$$

The constraints on design objectives were selected taking into account the performances offered over the group of nominal plants by classical PID tuning rules as IMC [27], SIMC [28], and AMIGO [29], and the LADRC tuning method from [14]. The upper limit of $J_1(\theta)$ was set as the ITSE value obtained with the SIMC approach such that the desired closed loop time constant was equal to the apparent delay l. The SIMC tuning produced the highest ITSE for each plant compared to the LADRC from [14] and the other PID controllers.

Similarly, the lower limit of $J_2(\theta)$ is the approximation of the robustness obtained with the AMIGO tuning rules, and its upper limit is approximately the robustness computed with the IMC method. The other controllers offer a robustness measure between these limits for all plants. What is more, the $\varepsilon(\theta)$ limits are related to the commonly adopted limits for maximum sensitivity and maximum complementary sensitivity.

The search space for decision variables was specified following the results from Section 3 where it was shown that to increase b_0 contributes to a bigger stability region in terms of the bandwidths and, as a consequence, lower performance indices can be computed.

4.2. Optimization Process

The evolutionary multiobjective algorithm $\varepsilon^\nearrow$–MOGA [30] was used to perform the optimization process. This algorithm uses the epsilon-dominance concept to obtain Pareto Front and Pareto Set approximations with limited memory resources and preserving the diversity of the Front by adjusting its limits dynamically [31]. The algorithm parameters were set to 200 individuals for main population, 8 individuals for auxiliary population, 1000 generations and 1000 divisions per dimension.

The Pareto Fronts and Pareto Sets approximations obtained for the complete group of nominal plants are presented in Figure 6. From Figure 6b, an interesting behavior is observed. The range of the decision variables for plants with $\tau \leq 0.5$ is wider than for plants with $\tau > 0.5$. For instance, a robustness measure between 2 and 3 can be obtained for the plant with $\tau = 0.09$ if the LADRC parameters are selected in the ranges $b_0 \in [86, 115]$, $\omega_c \in [11.6, 6.6]$, whereas the same variation in robustness for plant with $\tau = 0.833$ is achieved with $b_0 \in [6.2, 9.2]$, $\omega_c = [0.73, 0.71]$. Another important feature is the decreasing trend in the decision variables as the normalized delay increases. However, the rate of change in both parameters tends to be greater for plants with $\tau \leq 0.5$ than for plants with $\tau > 0.5$.

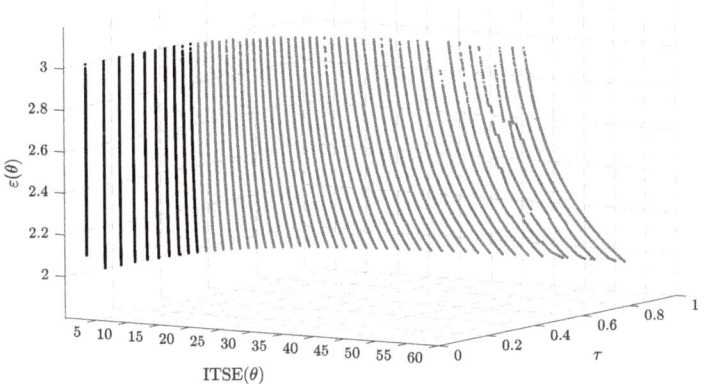

(a) Pareto Fronts approximations

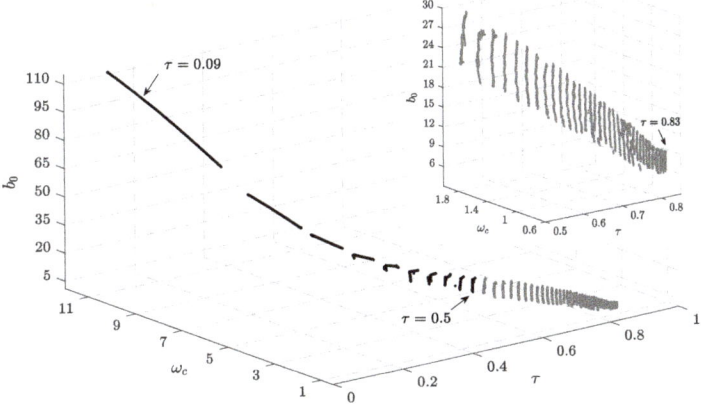

(b) Pareto Sets approximations

Figure 6. Results from the optimization process for the complete group of nominal plants. Pareto Sets approximations for plants with $\tau \leq 0.5$ (black) show that the LADRC parameters for this group have a wider range of variation and the rate of change in both parameters is greater compared to plants with $\tau > 0.5$ (gray). An inset showing the Pareto Sets approximations for plants with $\tau > 0.5$ is included for better visualization.

4.3. Multicriteria Decision Making

Once the Pareto Fronts and Pareto Sets approximations have been obtained, the last step in the MOOD procedure is the selection of the solution or candidate solutions preferred by the designer. Even if most of the preferences were taken into account in the optimization process, a final selection is needed. Depending on the number of design objectives, the visualization and graphical interpretation of the Pareto Front approximation is crucial. Some novel ideas to rank the potential solutions obtained by evolutionary algorithms in application to engineering problems are exposed in [32]. Likewise, an approach to the knee solution of the Pareto Front approximation for optimization problems with many objectives is addressed in [33].

According to the results from the optimization process, the following aspects were considered for the decision making stage.

- For data processing, two main groups were defined: *Group 1* containing data related to plants with a normalized delay $\tau \leq 0.5$ and *Group 2* with data belonging to plants with $\tau > 0.5$.
- From each Pareto Front approximation, three design alternatives distributed along the front were selected.
- For *Group 1*, the selection was made using the entire Pareto Front approximation.
- For *Group 2*, the selection was made limiting the upper end of the front such that the highest value for $\varepsilon(\theta)$ is 2.5. This criterion is based on the fact that the difficulty in controlling a process increases as its normalized delay increases [25]. Thus, for this group of plants, lower values of $\varepsilon(\theta)$ are preferred which correspond to more robust closed loop systems.
- Selected solutions are compared in the objective space with other alternatives related to PID and LADRC tuning rules.

Consider the first group of nominal plants (*Group 1*). In order to select the three desired design alternatives, let the Pareto Fronts to be divided in two regions according to bounds imposed on the mixed robustness measure. The upper region comprises solutions for which $2.5 \leq \varepsilon(\theta) \leq 3$ and the lower region includes those with $2 \leq \varepsilon(\theta) < 2.5$.

On each region, a point corresponding to the Nash solution was calculated by solving the problem [34]:

$$\max_{(J_1(\theta), J_2(\theta))} \left(J_1\left(\theta^2\right) - J_2(\theta)\right)\left(J_2\left(\theta^1\right) - J_1(\theta)\right), \tag{48}$$

where $J_1(\theta^2)$ is the optimal value (minimum) of the first design objective and $J_2(\theta^1)$ is the point that minimizes the second cost function. The Nash solution $(J_1(\theta), J_2(\theta))$ is considered a *fair* selection because it dominates the larger number of points in the rectangular area $(J_1(\theta^2) - J_2(\theta))(J_2(\theta^2) - J_1(\theta))$ [34].

The third solution for *Group 1* was selected as the midpoint of the Pareto Fronts. This is, the solution meeting the condition $\varepsilon(\theta) = 2.5$.

For the second group of plants (*Group 2*) the three selected solutions corresponds to the two ends of the front and the Nash solution.

Figure 7 illustrates the concepts explained and solutions selected taking as an example the Pareto Fronts approximations of the nominal plants with $\tau = 0.5$ (*Group 1*) and $\tau = 0.75$ (*Group 2*).

The complete set of Pareto Fronts approximations and selected solutions are presented in Figure 8. For comparison purposes, the performance obtained with the PID tuning methods IMC, SIMC, AMIGO, and the rules from [34] (SNS) are included for *group 1*. For *group 2*, the Pareto alternatives are compared with the SIMC and AMIGO approaches. Performance corresponding to the LADRC tuning rules from [14] (ADRC$_Z$) are also shown for both groups. Note that the fronts move to the right in the objective space as the normalized delay increases. From this figure, the following remarks are derived.

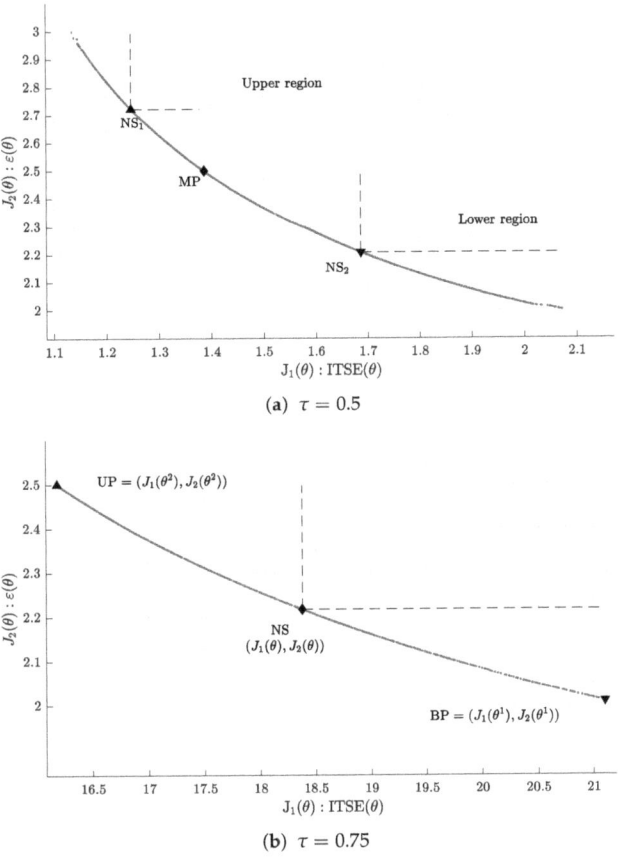

Figure 7. Location of the selected solutions into the Pareto Fronts approximations taking as example two nominal plants. (**a**) For plants in *Group 1*, selected solutions are the Nash solution from upper region NS$_1$, the midpoint MP, and the Nash solution from lower region NS$_2$. (**b**) For plants in *Group 2* the selected solutions are the upper end UP, the Nash solution NS, and the bottom end BP.

- The performance obtained with the PID controllers tuned by the IMC, SIMC and SNS rules are in the dominance area of the Pareto Fronts belonging to plants from *Group 1*. Particularly, the SIMC points are dominated by the optimal solutions in all cases.
- For plants from *Group 2*, the performance obtained with the AMIGO tuning method is outside the Pareto Fronts approximations due to the constraint imposed on $\varepsilon(\theta)$. However, the alternative solutions corresponding to the bottom end of the Fronts have better disturbance rejection with a reasonable level of robustness.
- The performance obtained with the ADRC$_Z$ tuning rules is in the dominance area of the approximated Pareto Fronts for the entire set of nominal plants. Even though the ADRC$_Z$ points are the results of fitting curves, they tend to move away from the Fronts as τ increases which highlights their suboptimal feature.

With the MOOD procedure developed for the tuning problem of the second-order LADRC applied to FOPDT nominal systems, a set of Pareto optimal solutions with a trade-off between disturbance rejection and robustness was obtained. The distribution of these solutions in the decision search space can lead to different fitting curves depending on the preferred level of compromise between objectives. This idea is the core of the fitting procedure presented in the next section.

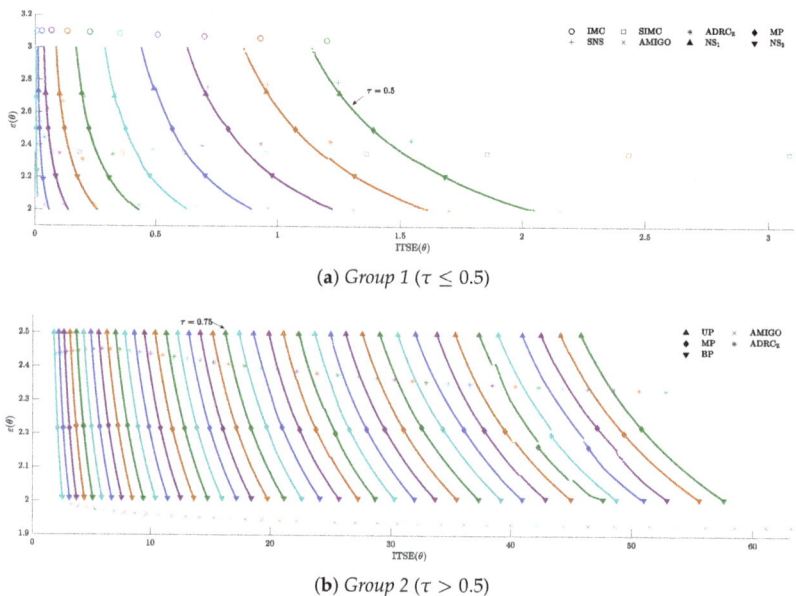

(a) *Group 1* ($\tau \leq 0.5$)

(b) *Group 2* ($\tau > 0.5$)

Figure 8. Pareto Fronts approximations and selected solutions for the complete set of nominal plants. Performance points obtained with the PID tuning methods IMC, SIMC, AMIGO, and SNS [34] as well as the LADRC tuning rules from [14] (ADRC$_Z$) are included for comparison. The SIMC points have been excluded from (**b**) for proper visualization because these alternatives are always dominated by the Pareto optimal solutions. Information related to the same plant has been plotted in the same color.

5. Tuning Rules for LADRC

The solutions obtained from the MOOD procedure correspond to the Pareto optimal LADRC parameters suitable to control FOPDT plants in the form of (34). These data were initially fitted to functions of the normalized delay τ. Afterwards, the resulting expressions were scaled to obtain the LADRC tuning rules applicable to the control of the general FOPDT system (33).

Data were fitted separately for the two previously defined groups of plants. This was mainly because of the behavior observed in the rate of change of the parameters with respect to the variation in the normalized delay (see Figure 6b). Additionally, in each group, the three optimal solutions selected were used to fit three curves related to different levels of robustness taking τ as independent variable. These levels of robustness were defined as follows.

- Low level (ε_{low}): The LADRC tuned by these approximation will offer a robustness around 2.7 for processes with $\tau \leq 0.5$ and around 2.5 for plants with $\tau > 0.5$. For *Group 1*, the tuning rule was approximated using the Nash solutions of the upper regions of the Pareto Fronts (NS$_1$). For *Group 2*, the curve was fitted using the upper ends of the fronts (UP).
- Medium level (ε_{med}): Processes with $\tau \leq 0.5$ and controlled by LADRC tuned according to this formulae will have a robustness of approximately 2.5. In the case of plants with $\tau > 0.5$, the robustness of the closed loop will be around 2.3. The midpoints of the Pareto Fronts (MP) were used to approximate the tuning function in the first group of systems and the Nash solutions (NS) were used for the second group.
- High level ($\varepsilon_{\text{high}}$): The highest robustness of the closed loop will approximately 2.2 for systems with $\tau \leq 0.5$ and 2.0 for plants meeting $\tau > 0.5$. In *Group 1* the approximation was done using the Nash solutions of the lower regions of the Pareto Fronts (NS$_2$) and in *Group 2*, the bottom ends of the fronts (BP) were used instead.

The nominal values for the critical gain were fitted to power functions in the case of systems with $\tau \leq 0.5$ and to polynomial functions for systems with $\tau > 0.5$ such as

$$b_0 = \begin{cases} k_b \left(\dfrac{\tau}{1-\tau}\right)^{n_b}, & \tau \leq 0.5 \\ a_b \left(\dfrac{\tau}{1-\tau}\right)^2 + b_b \left(\dfrac{\tau}{1-\tau}\right) + c_b, & \tau > 0.5, \end{cases} \quad (49)$$

where k_b, n_b, a_b, b_b and c_b are constants.

On the other hand, the controller bandwidth values were fitted for both groups to power functions of the form

$$\omega_c = k_\omega \left(\dfrac{\tau}{1-\tau}\right)^{n_\omega}, \quad (50)$$

with k_ω and n_ω as constants.

The resultant fitting functions are presented in Figures 9 and 10, and the corresponding parameters for expressions (49) and (50) are reported in Table 3.

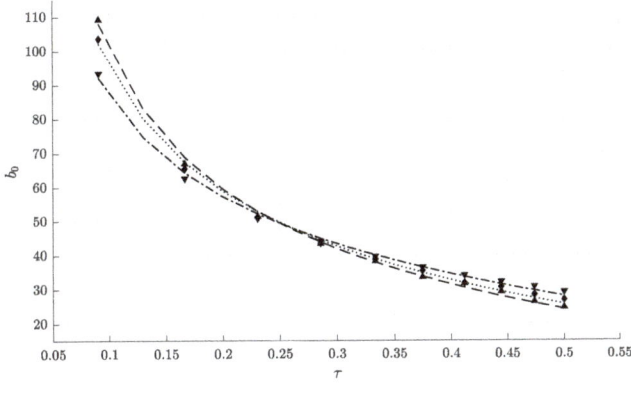

(a) Fitting for *Group 1* ($\tau \leq 0.5$)

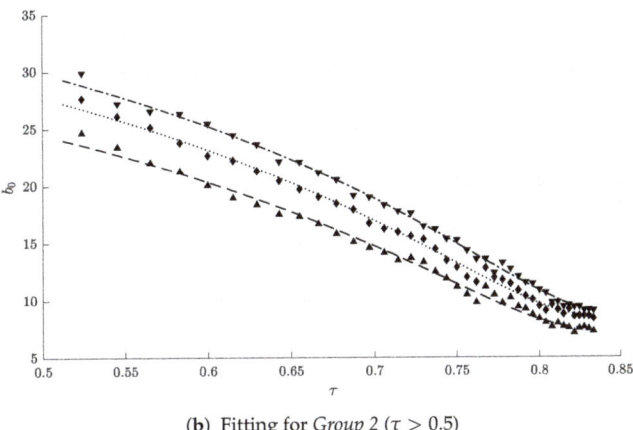

(b) Fitting for *Group 2* ($\tau > 0.5$)

Figure 9. Tuning for nominal values of the LADRC critical gain. Markers indicate the optimal solutions NS$_1$ (▲), MP (♦), NS$_2$ (▼). Lines are the fitting functions for robustness levels ε_{low} (— —), ε_{med} (· · ·), $\varepsilon_{\text{high}}$ (-··-).

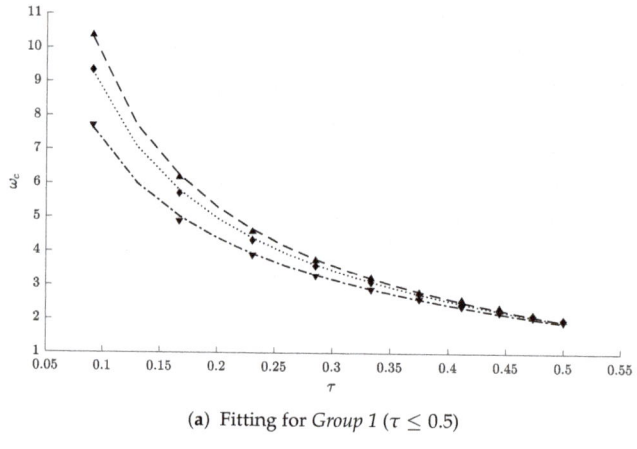

(a) Fitting for *Group 1* ($\tau \leq 0.5$)

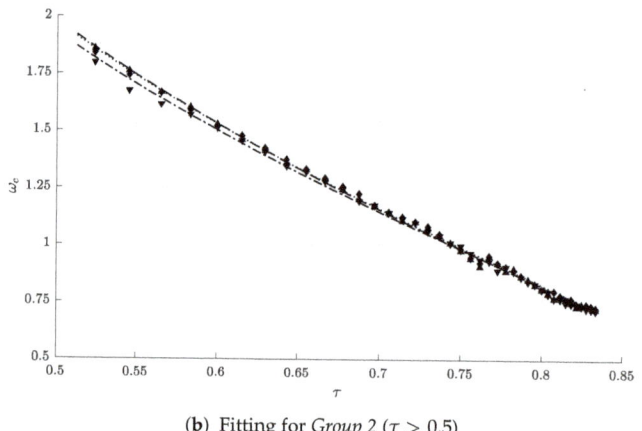

(b) Fitting for *Group 2* ($\tau > 0.5$)

Figure 10. Tuning for LADRC controller bandwidth. Markers indicate the optimal solutions UP (▲), NS (♦), and BP (▼). Lines are the fitting functions for robustness levels ε_{low} (− −), ε_{med} (· · ·), and $\varepsilon_{\text{high}}$ (-·-).

As last step in the data processing, (49) and (50) were substituted in the corresponding scaled parameters of (36) to obtain the general LADRC tuning rules

$$\bar{b}_0 = \begin{cases} \dfrac{K}{T^2}\left[k_b\left(\dfrac{\tau}{1-\tau}\right)^{n_b}\right], & \tau \leq 0.5 \\ \dfrac{K}{T^2}\left[a_b\left(\dfrac{\tau}{1-\tau}\right)^2 + b_b\left(\dfrac{\tau}{1-\tau}\right) + c_b\right], & \tau > 0.5 \end{cases} \quad (51)$$

$$\bar{\omega}_c = \dfrac{1}{T}\left[k_\omega\left(\dfrac{\tau}{1-\tau}\right)^{n_\omega}\right] \quad (52)$$

$$\bar{\omega}_o = \dfrac{10}{T}\left[k_\omega\left(\dfrac{\tau}{1-\tau}\right)^{n_\omega}\right]. \quad (53)$$

Equations (51)–(53) are now dependent on the three FOPDT plant parameters which can be easily obtained for many processes by identification techniques.

As a summary, in Table 3 a guide for the tuning of the LADRC for the control of FOPDT plants is presented. Each of the defined levels of robustness represents a compromise between this objective and the disturbance rejection performance. This way, the designer is provided with three closed loop stable candidate controllers that could be tested on the system for the final decision.

Table 3. LADRC tuning guide.

1. Approximate the process dynamics with the First Order Plus Dead Time (FOPDT) model

$$G(s) = \frac{K}{Ts+1}e^{-ls}.$$

2. Compute the normalized dead time. Note that the resulting normalized dead time meets the condition $0 \leq \tau \leq 1$

$$\tau = \frac{l}{T+l}.$$

3. Decide whether the process belongs to Group 1: $\tau \leq 0.5$ or Group 2: $\tau > 0.5$ according to the normalized dead time computed in step 2. This classification indicates the level of robustness (quantified by ε) of each of the three candidate controllers.

4. Use the tables given below to select the appropriate coefficients for the tuning rules according to preferences on the robustness quality.

Group 1: $\tau \leq 0.5$				Group 2: $\tau > 0.5$			
Robustness level	ε_{low}	ε_{med}	ε_{high}	Robustness level	ε_{low}	ε_{med}	ε_{high}
Robustness, ε	2.7	2.5	2.2	Robustness, ε	2.5	2.3	2.0
k_b	24.129	25.632	27.952	a_b	1.145	1.238	1.121
n_b	−0.651	−0.601	−0.518	b_b	−11.110	−12.192	−11.921
k_ω	1.946	1.938	1.903	c_b	34.443	38.682	40.601
n_ω	−0.724	−0.681	−0.604	k_ω	1.982	1.972	1.927
				n_ω	−0.635	−0.625	−0.612

5. Substitute the coefficients selected in step 4, the static gain, and the apparent time constant in the following rules to compute the LADRC parameters

$$\bar{b}_0 = \begin{cases} \frac{K}{T^2}\left[k_b\left(\frac{\tau}{1-\tau}\right)^{n_b}\right], & \tau \leq 0.5 \\ \frac{K}{T^2}\left[a_b\left(\frac{\tau}{1-\tau}\right)^2 + b_b\left(\frac{\tau}{1-\tau}\right) + c_b\right], & \tau > 0.5 \end{cases} \quad \bar{\omega}_c = \frac{1}{T}\left[k_\omega\left(\frac{\tau}{1-\tau}\right)^{n_\omega}\right] \quad \bar{\omega}_o = \frac{10}{T}\left[k_\omega\left(\frac{\tau}{1-\tau}\right)^{n_\omega}\right].$$

6. Implement the second-order LADRC using the time domain or the frequency domain formulation.

Furthermore, the designer could vary the values of the LADRC parameters in the intervals obtained based on the proposed rules to adjust the performance according to the preferences. To help in this task, the tuning tool of Figure 11 has been developed in Matlab App Designer and is available at Matlab Central [18]. It requires as inputs the FOPDT model and through interaction with robustness level and manual tuning sliders, the user can visualize the closed loop response and evaluate the second-order LADRC performance with the aid of some measures.

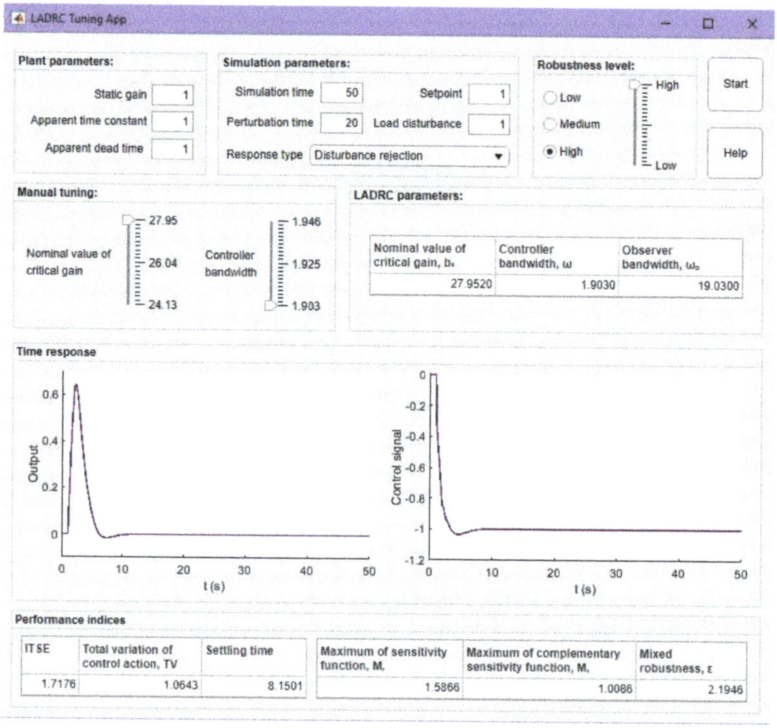

Figure 11. LADRC tuning tool. This Matlab App allows the automatic computation of the nominal value of the critical gain b_0, the controller bandwidth ω_c, and the observer bandwidth ω_o of the second-order LADRC for the control of a system approximated by a FOPDT model. Available at [18].

The tuning rules proposed in this section together with the developed tuning tool allow some degree of the designer involvement in the final selection of the LADRC parameters, but eliminates the time and complexity of performing the entire optimization process. The parameters computed by the proposed rules ensure closed loop stability as well as a reasonable compromise between disturbance rejection and loop robustness.

6. Validation of the LADRC Tuning Rules

In this section, two examples are presented to validate the proposed tuning rules. The load disturbance and setpoint responses are compared with the performance obtained from other controllers such as PID and LADRC tuned by different methods.

The performance indices in frequency domain M_S, M_T, ε and in time domain ITSE, TV, and settling time ($t_{98\%}$, in seconds) were calculated.

6.1. Example 1: A Lag-Dominated System

Consider the FOPDT lag-dominated system.

$$G_1(s) = \frac{1}{10s+1}e^{-2s}. \tag{54}$$

The Tuning Guide is used to illustrate the parameters computation. Following the steps from Table 3:

1. From (54), $K = 1$, $T = 10$, and $l = 2$.

2. The normalized dead time is

$$\tau = \frac{2}{10+2} = 0.17. \tag{55}$$

3. According to the normalized dead time from step 2, (54) belongs to Group 1 and thus, the three candidate controllers have robustness of approximately 2.7 (ε_{low}), 2.5 (ε_{med}), and 2.2 (ε_{high}).
4. For example, if a controller with a high robustness is preferred, the corresponding coefficients for the tuning rules are $k_b = 27.952$, $n_b = -0.518$ for computation of b_0; $k_\omega = 1.903$, $n_\omega = -0.604$ for computation of ω_c and ω_o.
5. The nominal value of critical gain, the controller bandwidth, and the observer bandwidth are computed by substituting the coefficients from step 4 and the FOPDT parameters in the tuning rules. This is,

$$\bar{b}_0 = \frac{1}{100}\left[27.952\left(\frac{0.17}{1-0.17}\right)^{-0.518}\right] = 0.643 \tag{56}$$

$$\bar{\omega}_c = \frac{1}{10}\left[1.093\left(\frac{0.17}{1-0.17}\right)^{-0.604}\right] = 0.503 \tag{57}$$

$$\bar{\omega}_o = 1.903\left(\frac{0.17}{1-0.17}\right)^{-0.604} = 5.031 \tag{58}$$

6. The parameters computed in step 5 can be used in the second-order LADRC for the control of plant (54).

Note that steps 4 and 5 from the above procedure must be repeated if a different robustness is desired. The LADRC parameters for the three levels of robustness (ε_{low}, ε_{med}, ε_{high}) are listed in Table 4. Parameters obtained with the tuning rules proposed in [14] (ADRC$_Z$) are also listed together with those corresponding to the PID controllers tuned by the IMC, SNS (from [34]), SIMC and AMIGO methods. Figures 12 and 13 show the time responses.

Table 4. Parameters for the control of $G_1(s)$.

LADRC	b_0	ω_c	ω_o	PID	K_p	T_i	T_d
ADRC$_Z$	0.349	1.950	0.960	IMC	4.320	10.800	0.751
ε_{low}	0.688	0.624	6.243	SNS	3.420	5.475	0.970
ε_{med}	0.674	0.580	5.795	SIMC	2.500	10.000	0
ε_{high}	0.643	0.503	5.031	AMIGO	2.450	5.867	0.943

The resulting values for the performance indices are reported in Table 5. It can be seen that each of the proposed controllers offers a robustness level similar to one of the PID alternatives with a lower ITSE for disturbance rejection. Also, the output backs to steady state faster than with the IMC and SIMC.

Compared with the ADRC$_Z$ tuning rules, the three proposed controllers have a lower ITSE value and return the output to steady state faster in the case of a load disturbance. Note that with the ε_{high} controller, a higher robustness level and better disturbance rejection performance can be achieved. Also, the total variation of the control action is lower for this alternative.

On the other hand, for setpoint following operation, a similar ITSE than ADRC$_Z$ is obtained with the ε_{low} controller. However, it is worth noting that control actions produced by the three alternatives are smoother, which is reflected in the total variations indices calculated. This is mainly because the initial values of the control signals (sometimes referred in literature as *proportional kick*) are significantly lower than those reached by the ADRC$_Z$ controller.

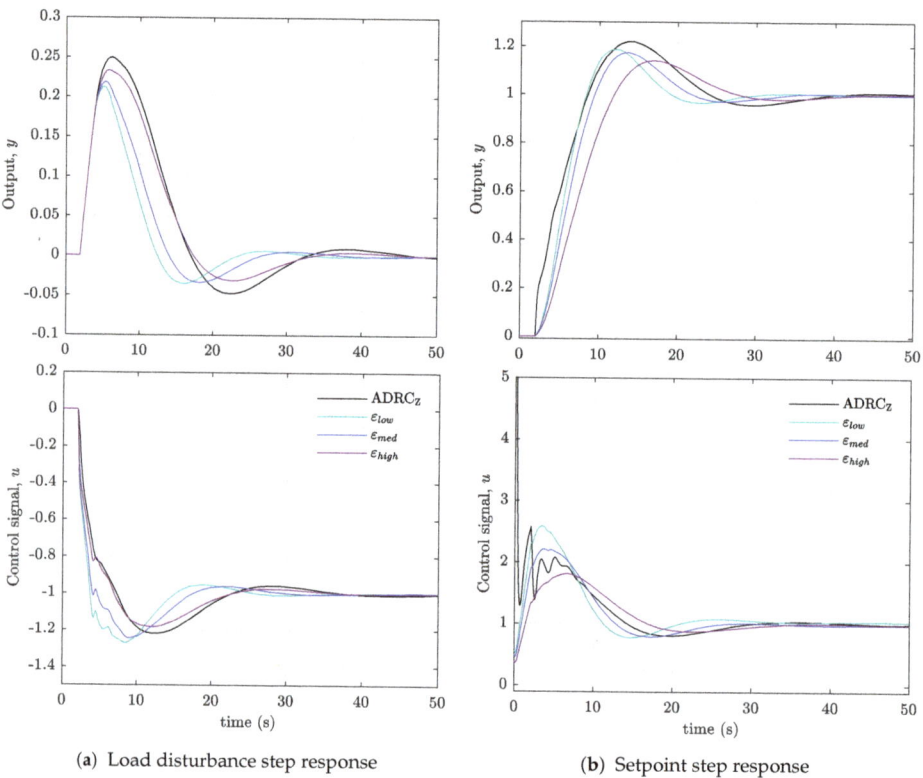

(a) Load disturbance step response (b) Setpoint step response

Figure 12. Closed loop time response of $G_1(s)$ with the second-order LADRC tuned with the proposed rules. Comparison with the performance of ADRC$_z$ controller.

Table 5. Performance comparison of proposed tuning rules with other tuning methods for control of $G_1(s)$. The ε_{high} controller is more robust and offers a lower ITSE for disturbance rejection than the ADRC$_Z$ controller.

	M_S	M_T	ε	Disturbance Rejection			Setpoint Following		
				ITSE	TV	$t_{98\%}$	ITSE	TV	$t_{98\%}$
IMC	2.032	1.097	3.103	2.485	1.358	44.1	2.788	60.985	8.7
SNS	1.767	1.181	2.545	1.738	1.331	26.7	4.023	48.914	18.5
SIMC	1.590	1.000	2.353	6.870	1.082	46.2	6.307	2.591	12.1
AMIGO	1.446	1.135	2.029	3.770	1.252	33.6	5.702	31.503	23.3
ADRC$_Z$	1.583	1.345	2.447	3.688	1.537	43.3	12.488	14.699	35.4
ε_{low}	1.842	1.489	2.771	1.227	1.743	29.9	12.672	4.298	26.4
ε_{med}	1.735	1.392	2.544	1.652	1.636	33.1	14.685	3.520	29.1
ε_{high}	1.598	1.258	2.236	2.982	1.438	31.4	19.404	2.562	25.9

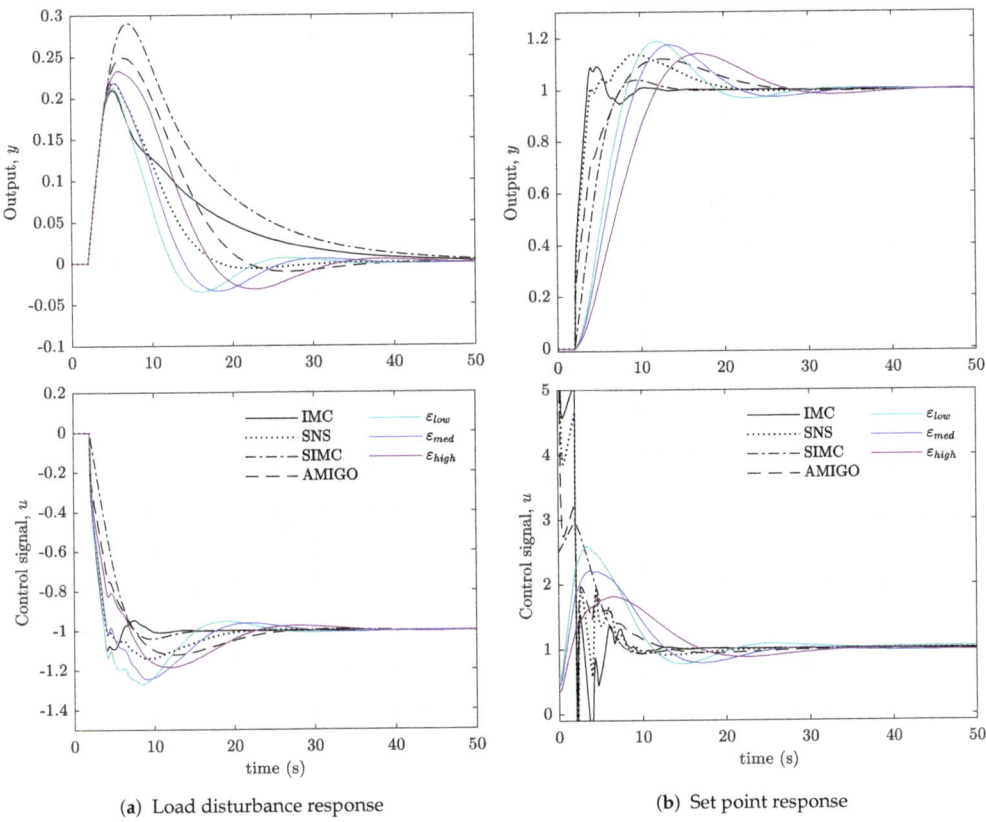

(a) Load disturbance response

(b) Set point response

Figure 13. Closed loop time response of $G_1(s)$ with the second-order LADRC tuned with the proposed rules. Comparison with the performance of PID controllers.

6.2. Example 2: A Delay-Dominated System

As second example, the following FOPDT delay-dominated system is analyzed:

$$G_2(s) = \frac{3}{0.25s+1}e^{-s}. \tag{59}$$

The normalized delay for this plant is $\tau = 0.80$. The PID tuning rules IMC, SIMC and AMIGO, and the LADRC tuning rules from [14] (ADRC$_z$) were used for comparison. In addition, the tuning rules for the second-order LADRC from [11] (ADRC$_H$) were also taken into account. The latter are proposed for the control of high order plants, but can be used for self-regulatory FOPDT systems with nominal delay (τ/T) above 0.46 by approximating the plant into the form $K/(Ts+1)^n$ (Note that K and T have a different meaning than in (33)).

Figures 14 and 15 show the closed loop time response of $G_2(s)$ with the LADRC and the PID controllers, respectively. The computed parameters are listed in Table 6 and performance indices are reported in Table 7.

Table 6. Parameters for the control of $G_2(s)$.

LADRC	b_0	ω_c	ω_o	PID	K_p	T_i	T_d
$ADRC_Z$	359.316	5.029	16.140	IMC	0.173	0.650	0.195
$ADRC_H$	345.819	2.521	33.007	SIMC	0.042	0.250	0
ε_{low}	399.747	3.288	32.876	AMIGO	0.104	0.585	0.227
ε_{med}	466.508	3.317	33.172				
ε_{high}	521.205	3.301	33.007				

According to the indices obtained for disturbance rejection, the proposed controllers can improve the performance in at least one of the design objectives when compared to the PIDs. For example, The ε_{med} controller is more robust and produces a lower ITSE than the PID tuned by the SIMC method. The same controller offers an improvement in robustness and disturbance rejection in comparison with $ADRC_Z$.

On the other hand, the ITSE calculated from the load disturbance response with the three proposed controllers are lower than the ITSE obtained with the $ADRC_H$ controller. The corresponding total variations of control signals are also lower and the system output stabilizes faster, even in the case of a setpoint change. It should be noted that LADRC parameters for $ADRC_H$ were obtained setting an required additional tuning parameter k as 3.25 after some trial and error tests to guarantee the stability.

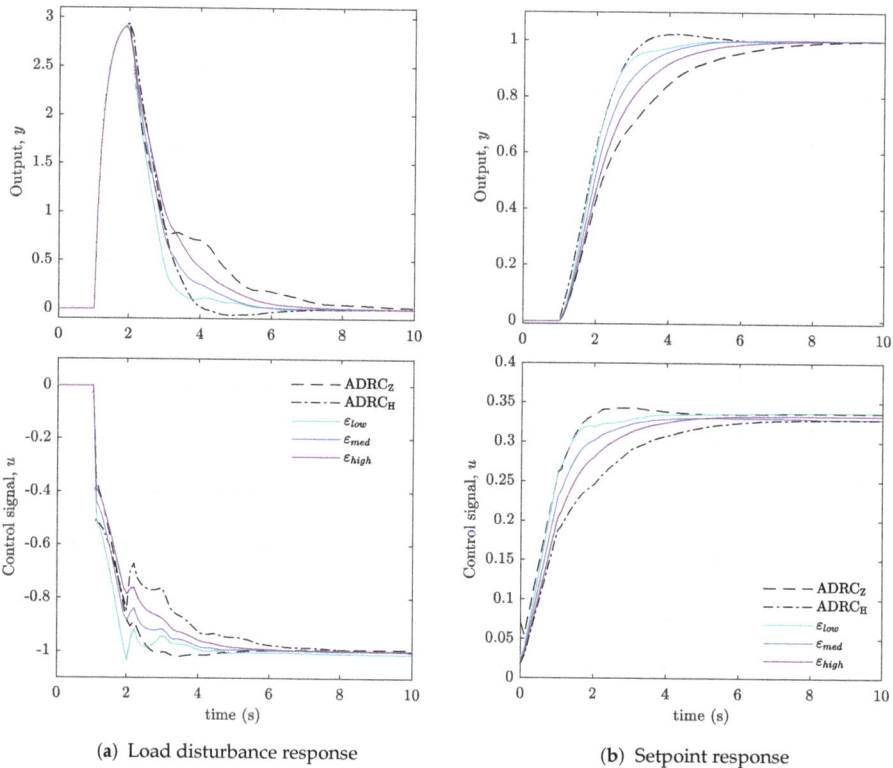

(a) Load disturbance response   (b) Setpoint response

Figure 14. Closed loop time response of $G_2(s)$ with the second-order LADRC tuned with the proposed rules. Comparison with the performance of $ADRC_z$ and $ADRC_H$ controllers.

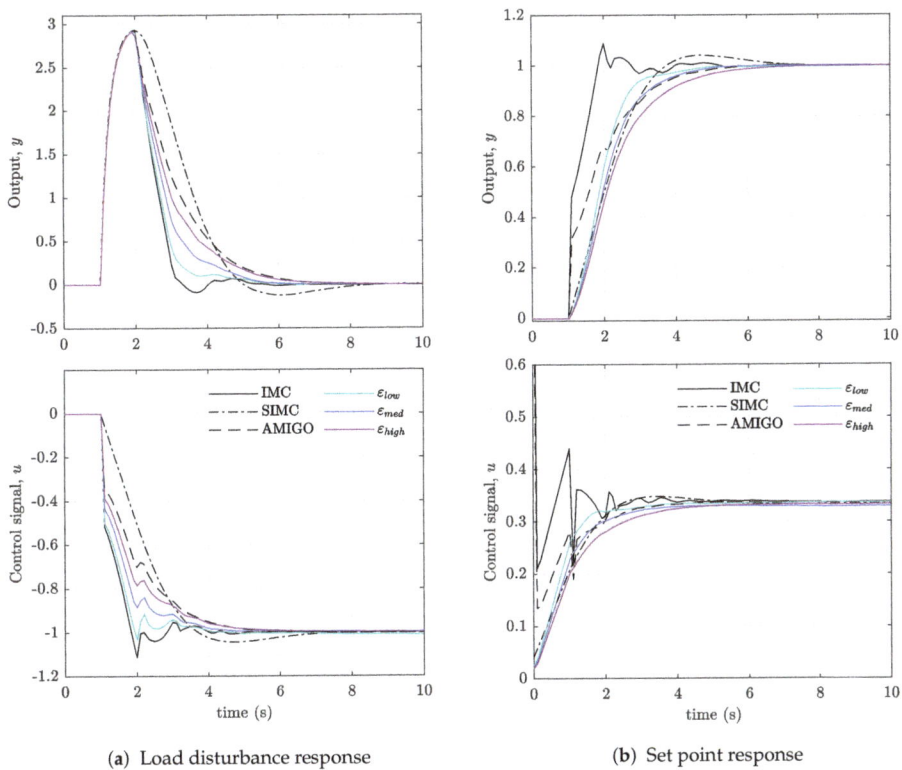

(a) Load disturbance response   (b) Set point response

Figure 15. Closed loop time response of $G_2(s)$ with the second-order LADRC tuned with the proposed rules. Comparison with the performance of PID controllers.

Table 7. Performance comparison of proposed tuning rules with other tuning methods for control of $G_2(s)$. For all controllers $M_T = 1$. The ε_{med} controller is more robust and offers a lower ITSE for disturbance rejection than the ADRC$_Z$ controller. The three proposed alternatives offers a better disturbance rejection performance than the ADRC$_H$ controller with similar or better robustness.

	M_S	ε	Disturbance Rejection			Setpoint Following		
			ITSE	TV	$t_{98\%}$	ITSE	TV	$t_{98\%}$
IMC	1.873	2.774	15.447	1.427	3.9	0.666	2.474	3.7
SIMC	1.590	2.353	30.885	1.082	7.3	1.559	0.319	6.1
AMIGO	1.401	1.933	22.487	1.041	6.0	1.087	1.415	4.7
ADRC$_Z$	1.622	2.357	18.875	1.069	4.0	1.329	0.310	4.3
ADRC$_H$	1.792	2.612	20.278	1.361	7.6	2.122	0.310	7.3
ε_{low}	1.798	2.615	15.817	1.321	4.9	1.381	0.312	4.1
ε_{med}	1.638	2.296	17.551	1.102	5.3	1.550	0.308	4.7
ε_{high}	1.526	2.073	19.930	1.038	6.1	1.775	0.312	5.5

7. Control of a Peltier Thermoelectric Module

The proposed tuning rules were used to design a second-order LADRC for the control of a thermoelectric module operating on the Peltier principle. It is assumed that the real behavior of the Peltier cell is modeled by the nonlinear differential equations presented in [35].

The thermal balance in the cold face is described by

$$Q_{cf} = 9.2\dot{T}_c$$
$$Q_{cf} = Q_{acf} - Q_{pcf} - Q_j + Q_{cond}$$
$$Q_{acf} = 11.75 - 0.5T_c \tag{60}$$
$$Q_{pcf} = 0.041 T_c I_p$$
$$Q_j = 0.41 I_p^2$$
$$I_p = \frac{1}{0.82}[V_{in} - 0.041(T_h - T_c)]$$
$$Q_{cond} = 0.2(T_h - T_c).$$

The thermal balance in the hot face is

$$Q_{hf} = 13\dot{T}_h$$
$$Q_{hf} = Q_{rhf} + Q_{phf} + Q_j - Q_{cond}$$
$$Q_{rhf} = 9.59(T_r - T_h) \tag{61}$$
$$Q_{phf} = 0.041 T_h I_p.$$

And finally, the radiator equilibrium corresponds to

$$Q_{rf} = 722.55\dot{T}_r$$
$$Q_{rf} = Q_{acc} - Q_{rhf} \tag{62}$$
$$Q_{acc} = 167.09 - 7.11 T_r.$$

The controlled output is the temperature on cold face $T_c \in [-12.0, 6.0]$ °C and the manipulated input is the applied voltage V_{in} in percentage of its range. A block diagram representing (60)–(62) is presented in Figure 16 and the corresponding description of variables is listed in Table 8.

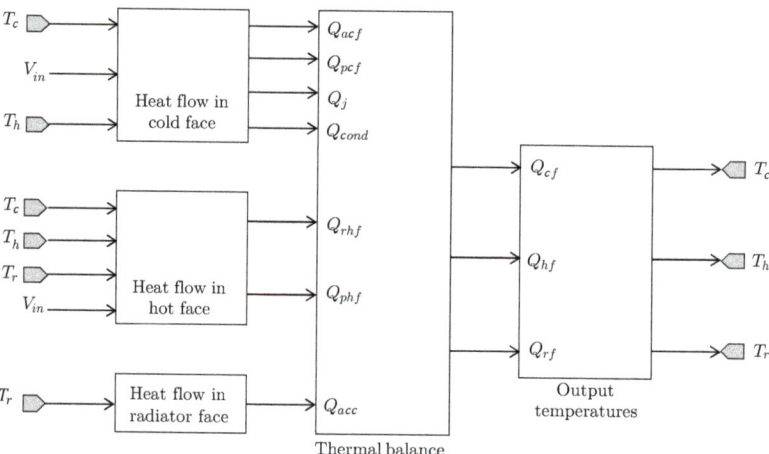

Figure 16. Block diagram of the thermoelectric module.

Table 8. Description of variables for the Peltier cell model.

Variable	Units	Description
T_c	°C	Temperature on the cold face
T_h	°C	Temperature on the hot face
T_r	°C	Temperature in the radiator
V_{in}	%	Voltage applied to the Peltier cell
I_p	A	Current flow in the Peltier cell
Q_{cf}	W	Net heat flow on the cold face
Q_{acf}	W	Heat flow transmitted by convection between the environment and the cold face
Q_{pcf}	W	Heat flow absorbed by the cold face due to the Peltier effect
Q_j	W	Heat flow generated by Peltier cell due to Joule effect
Q_{cond}	W	Heat flow transferred by conduction from the hot face to the cold face
Q_{hf}	W	Net heat flow on the hot face
Q_{rhf}	W	Heat flow transmitted by radiation between the hot face and radiator
Q_{phf}	W	Heat flow dissipated by the hot face due to Peltier effect
Q_{rf}	W	Net heat flow into the radiator
Q_{acc}	W	Heat flow transmitted by convection between the environment and the radiator

The Peltier cell behavior in the freeze zone (≈ -8.0 °C) can be approximated by the FOPDT nominal model [36]

$$G_p(s) = \frac{-0.315}{3.192s + 1} e^{-0.4s}. \tag{63}$$

The normalized delay for (63) is $\tau = 0.11$. By substituting this value in the corresponding tuning rules, the three second-order LADRC parameters sets (ε_{low}, ε_{med}, $\varepsilon_{\text{high}}$) from Table 9 are obtained. Two additional controllers are also included for comparison purposes: the LADRC tuned using the proposal from [14] (ADRC$_Z$) and a PID whose parameters were calculated by the SIMC method.

Table 9. Parameters for the control of a thermoelectric module.

LADRC	b_0	ω_c	ω_o	PID	K_p	T_i	T_d
ADRC$_Z$	−1.613	9.696	4.102	SIMC	12.667	3.192	0
ε_{low}	−2.885	2.744	27.439				
ε_{med}	−2.758	2.496	24.957				
$\varepsilon_{\text{high}}$	−2.532	2.090	20.905				

Consider that the cold face of the module is stable at −5.0 °C and a fault in power system reduces the input voltage 10% of its nominal value. The evolution of temperature T_c and the required voltage to reject the disturbance are shown in Figure 17a. The corresponding performance indices ITSE (°C$^2 \cdot$ s), TV (%) and $t_{98\%}$(s) are included in Table 10.

Table 10. Performance comparison of proposed tuning rules with reference controllers for the load disturbance response of the Peltier cell.

	M_S	M_T	ε	ITSE	TV	$t_{98\%}$
SIMC	1.590	1.000	2.353	2.373	10.810	12.0
ADRC$_Z$	1.545	1.455	2.607	0.773	14.001	7.0
ε_{low}	1.848	1.516	2.721	0.188	13.152	5.0
ε_{med}	1.749	1.425	2.511	0.302	13.152	5.8
$\varepsilon_{\text{high}}$	1.613	1.298	2.232	0.725	13.173	7.2

As expected, the ε_{low} controller produces the response with lower ITSE due to the relaxation in the robustness requirement. In addition, the total variation of control action and settling time are the lowest among the three proposals.

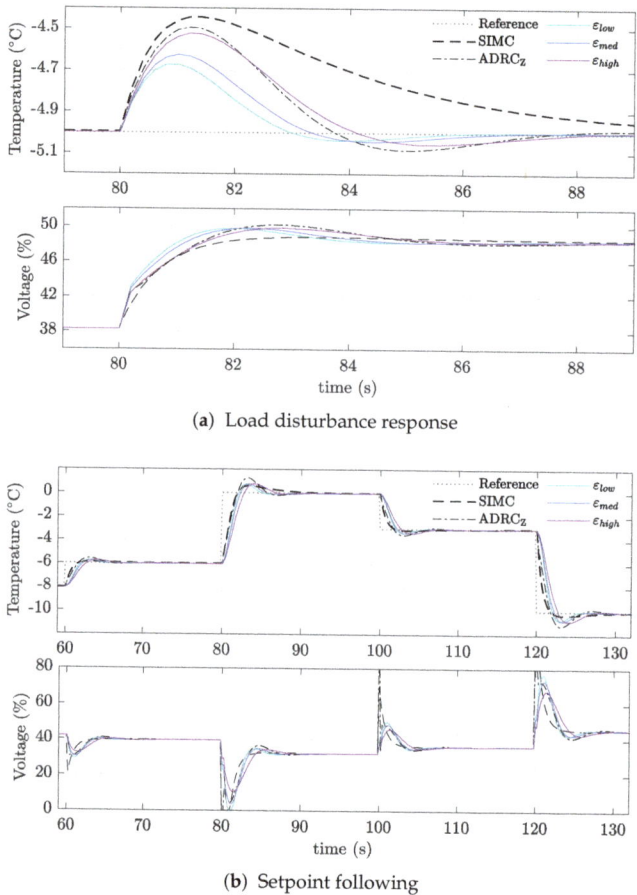

Figure 17. Closed loop time response of the Peltier thermoelectric module with the second-order LADRC tuned with the proposed rules. Comparison with the performance of ADRC$_Z$ and SIMC controllers.

On the other hand, the ε_{med} controller offers an improvement over the performance obtained with the ADRC$_Z$ tuning method. The robustness index is slightly lower which indicates a more robust closed loop system and the ITSE value reflects that the output stabilizes faster with less overshoot.

The most robust controller ε_{high} produces a time response similar to the ADRC$_Z$ but the ITSE and TV values are slightly lower. Note that this controller also has a better disturbance rejection and robustness level than the PID tuned by the SIMC method.

The thermoelectric module can be operated at different temperatures. Due to the nonlinearities, the transient temperature response shows different behavior depending on the magnitude and direction of the setpoint changes. An additional simulation was performed to test the LADRC alternatives under this scenario.

In Figure 17b the time response of the cold face temperature with different setpoints is presented. The corresponding indices are reported in Table 11.

The three controllers designed with the proposed tuning rules guarantee the setpoint following and the steady state is reached in less time than with the other controllers. However, the ITSE values are above those calculated for the PID and $ADRC_Z$. To clarify this behavior, the output overshoot (in % of the setpoint change) has been included in Table 11. As can be noticed, the SIMC method produces the lowest overshoot followed by the ε_{low}, ε_{med} and ε_{high} controllers. As expected, the overshoot in output increases for high changes in the magnitude of setpoint due to the nonlinear nature of the system.

Finally, in Figure 17b it is also shown that the three design alternatives can lead to a lower variation of the control action in contrast with the abrupt change produced by the other controllers when the setpoint changes. Note that this kind of peaks may be damaging for the system. The corresponding TV indices from Table 11 support this idea.

Table 11. Performance comparison of proposed tuning rules and reference controllers for the setpoint response of the Peltier cell.

Setpoint (°C)	SIMC	$ADRC_Z$	ε_{low}	ε_{med}	ε_{high}
Integral of the Time Weighted Squared Error					
−8 to −6	0.766	2.361	2.253	2.772	4.027
−6 to 0	14.331	25.004	20.384	24.379	34.836
0 to −3	1.429	4.369	4.727	5.722	8.342
−3 to −10	8.026	24.814	24.757	30.629	44.931
Total Variation of Control Action					
−8 to −6	39.184	24.751	24.053	21.442	17.562
−6 to 0	32.630	71.937	59.848	53.764	43.455
0 to −3	34.185	99.566	25.580	23.286	20.042
−3 to −10	54.513	65.692	57.137	52.347	45.522
Output Overshoot					
−8 to −6	2.849	22.343	9.651	9.052	9.425
−6 to 0	7.759	23.978	10.610	11.025	10.868
0 to −3	2.255	20.693	8.061	8.722	9.152
−3 to −10	2.161	20.077	8.565	8.100	8.579
Settling Time					
−8 to −6	7.4	8.2	4.8	5.2	6.4
−6 to 0	9.2	8.2	4.8	5.4	6.6
0 to −3	7.0	8.0	4.6	5.2	6.4
−3 to −10	7.0	8.0	4.8	5.2	6.4

8. Conclusions

In this paper, a set of tuning rules for the second-order LADRC which offer three different levels of compromise between disturbance rejection and robustness for the control of FOPDT systems were presented. A MOOD procedure was performed to address the tuning problem. It was focused on the simultaneous minimization of the integral of time weighted squared error and a robustness measure. The tuning rules were obtained by fitting a set of Pareto optimal solutions as functions of the normalized delay and the FOPDT model parameters. Hence, all the LADRC parameters: nominal value of critical gain, controller bandwidth, and observer bandwidth can be computed by selecting a desired quality of robustness (i.e., low, medium or high) and substituting the FOPDT parameters in the given rules.

An interactive tuning software was presented as complementary material. This tool is based on the proposed rules and allows the user to adjust the LADRC parameters by varying the robustness specification between the low and high levels. On the other hand, the designer can modify the LADRC parameters within predefined intervals to evaluate the overall performance of the closed loop.

The use and convenience of the tuning rules were exemplified with the control of lag-dominated and delay-dominated systems, as well as the control of the temperature in the cold face of a thermoelectric module. The examples showed that the proposed tuning method offers satisfactory performance for load disturbance rejection and setpoint following.

As part of the conceptual framework, an overall analysis on the conflicting objectives regarding the tuning of the LADRC was done. This allows to identify as future research the possibility of expand the objective space to include other performance criteria; for example, the total variation of the control signal. The parameterization adopted in this paper for the observer bandwidth oriented the optimization process to a particular area of the stability region and as a result, smooth manipulated signals were obtained. It would be of interest to analyze the trade-offs among other design objectives.

Author Contributions: Conceptualization, Methodology and Writing, B.V.M. and J.S.; Review and Supervision S.G.-N. and M.M.; and Software developed, B.V.M. All authors have read and agreed to the published version of the manuscript.

Funding: This work was supported in part by the Ministerio de Ciencia, Innovación y Universidades, Spain, under Grant RTI2018-096904-B-I00.

Conflicts of Interest: The authors declare no conflict of interest.

Abbreviations

In the following, the most important symbols and abbreviations used in this manuscript are listed.

n	System order
y	System output
u	Control law acting on the real plant
a_0, a_1	coefficients of the second-order model
b	Critical gain
$\tilde{r}$	System setpoint
d	Load disturbance
f	Total perturbation
b_0	Nominal value of critical gain
u_0	Estate feedback control law acting on the modified plant
x_i	i-th system real state
z_i	i-th estimated state
L_i	i-th observer gain
k_i	i-th control law gain
s	Complex variable
R	Laplace transform of the system setpoint
Y	Laplace transform of the system output
U	Laplace transform of the control law
Z_i	Laplace transform of the i-th estimated state
$G(s)$	Plant transfer function
$G_C(s)$	LADRC direct loop transfer function
$G_F(s)$	LADRC feedback transfer function
$G_A(s)$	Transfer function of controller
$G_D(s)$	Transfer function from output to load disturbance
$G_U(s)$	Transfer function to control action to output
$G_Y(s)$	Closed loop transfer function
k	Gain scaling of plant
ω_p	Frequency scaling of plant
ω_o	Observer bandwidth
ω_c	Controller bandwidth
$\bar{b}_0, \bar{\omega}_c, \bar{\omega}_o$	Scaled LADRC parameters
K	Static gain

T	Apparent time constant
l	Apparent delay or dead time
Θ	Nominal delay or dead time
τ	Normalized delay or dead time
$J_1(\theta), J_2(\theta)$	Design objectives
θ	Vector of decision variables
ITSE	Integral of Time Weighted Squared Error
TV	Total Variation of control action
$t_{98\%}$	Settling time
M_s	Maximum sensitivity
M_T	Complementary sensitivity
ε	Mixed robustness measure
K_p, T_i, T_d	PID controller parameters
$\varepsilon_{low}, \varepsilon_{med}, \varepsilon_{high}$	Low, medium, and high levels of robustness
k_b, n_b, a_b, b_b, c_b	Coefficients of the tuning rules for the nominal value of critical gain
k_ω, n_ω	Coefficients of the tuning rule for the controller and observer bandwidth

References

1. Han, J. From PID to Active Disturbance Rejection Control. *IEEE Trans. Ind. Electron.* **2009**, *56*, 900–906. [CrossRef]
2. Herbst, G. A Simulative Study on Active Disturbance Rejection Control (ADRC) as a Control Tool for Practitioners. *Electronics* **2013**, *2*, 246–279. [CrossRef]
3. Zheng, Q.; Gao, Z. Active disturbance rejection control: Some recent experimental and industrial case studies. *Control Theory Technol.* **2018**, *16*, 301–313. [CrossRef]
4. Gao, Z. Scaling and bandwidth-parameterization based controller tuning. In Proceedings of the 2003 American Control Conference, Denver, CO, USA, 4–6 June 2003; pp. 4989–4996. [CrossRef]
5. Zhao, C.; Li, D. Control design for the SISO system with the unknown order and the unknown relative degree. *ISA Trans.* **2014**, *53*, 858–872. [CrossRef]
6. Sun, L.; Zhang, Y.; Li, D.; Lee, K.Y. Tuning of Active Disturbance Rejection Control with application to power plant furnace regulation. *Control Eng. Pract.* **2019**, *92*, 104122. [CrossRef]
7. Zhou, R.; Tan, W. Analysis and Tuning of General Linear Active Disturbance Rejection Controllers. *IEEE Trans. Ind. Electron.* **2019**, *66*, 5497–5507. [CrossRef]
8. Tan, W.; Fu, C. Linear Active Disturbance-Rejection Control: Analysis and Tuning via IMC. *IEEE Trans. Ind. Electron.* **2016**, *63*, 2350–2359. [CrossRef]
9. Ahi, B.; Haeri, M. Linear Active Disturbance Rejection Control From the Practical Aspects. *IEEE/ASME Trans. Mechatron.* **2018**, *23*, 2909–2919. [CrossRef]
10. Madonski, R.; Gao, Z.; Lakomy, K. Towards a turnkey solution of industrial control under the active disturbance rejection paradigm. In Proceedings of the 2015 54th Annual Conference of the Society of Instrument and Control Engineers of Japan (SICE), Hangzhou, China, 28–30 July 2015. [CrossRef]
11. He, T.; Wu, Z.; Li, D.; Wang, J. A Tuning Method of Active Disturbance Rejection Control for a Class of High-Order Processes. *IEEE Trans. Ind. Electron.* **2020**, *67*, 3191–3201. [CrossRef]
12. Li, D.; Chen, X.; Zhang, J.; Jin, Q. On Parameter Stability Region of LADRC for Time-Delay Analysis with a Coupled Tank Application. *Processes* **2020**, *8*, 223. [CrossRef]
13. Muresan, C.I.; Birs, I.R.; Dulf, E.H. Event-Based Implementation of Fractional Order IMC Controllers for Simple FOPDT Processes. *Mathematics* **2020**, *8*, 1378. [CrossRef]
14. Zhang, B.; Tan, W.; Li, J. Tuning of linear active disturbance rejection controller with robustness specification. *ISA Trans.* **2019**, *85*, 237–246. [CrossRef] [PubMed]
15. Sun, L.; Hua, Q.; Shen, J.; Xue, Y.; Li, D.; Lee, K.Y. Multi-objective optimization for advanced superheater steam temperature control in a 300 MW power plant. *Appl. Energy* **2017**, *208*, 592–606. [CrossRef]
16. Srikanth, M.; Yadaiah, N. Optimal parameter tuning of Modified Active Disturbance Rejection Control for unstable time-delay systems using an AHP combined Multi-Objective Quasi-Oppositional Jaya Algorithm. *Appl. Soft Comput.* **2020**, *86*, 105881. [CrossRef]
17. Sun, L.; Li, D.; Gao, Z.; Yang, Z.; Zhao, S. Combined feedforward and model-assisted active disturbance rejection control for non-minimum phase system. *ISA Trans.* **2016**, *64*, 24–33. [CrossRef]
18. Martínez, B.V. LADRC Automatic Parameters Computation Based on Robustness. Version 1.0.0. 2021. Available online: https://es.mathworks.com/matlabcentral/fileexchange/86403 (accessed on 26 January 2021).
19. Zheng, Q.; Gao, L.Q.; Gao, Z. On Validation of Extended State Observer Through Analysis and Experimentation. *J. Dyn. Syst. Meas. Control* **2012**, *134*. [CrossRef]
20. Zhao, S.; Gao, Z. Modified active disturbance rejection control for time-delay systems. *ISA Trans.* **2014**, *53*, 882–888. [CrossRef]

21. Visioli, A. Identification and Model Reduction Techniques. In *Practical PID Control-Advances in Industrial Control*; Grimble, M.J., Johnson, M.A., Eds.; Springer: London, UK, 2006; pp. 165–207.
22. Reynoso-Meza, G.; Blasco-Ferragud, X.; Sanchis-Saez, J.; Herrero-Durá, J.M. Background on Multiobjective Optimization for Controller Tuning. In *Controller Tuning with Evolutionary Multiobjective Optimization. Intelligent Systems, Control and Automation: Science and Engineering*; Tzafestas, S.G., Ed.; Springer: Cham, Switzerland, 2017; Volume 85, pp. 23–58. [CrossRef]
23. Skogestad, S.; Postlethwaite, I. Classical Feedback Control. In *Multivariable Feedback Control-Analysis and Design*; John Wiley and Sons: Hoboken, NJ, USA, 2001; pp. 15–62.
24. Tan, W.; Liu, J.; Chen, T.; Marquez, H.J. Comparison of some well-known PID tuning formulas. *Comput. Chem. Eng.* **2006**, *30*, 1416–1423. [CrossRef]
25. Åström, K.J.; Hägglund, T. Process models. In *Pid Controllers: Theory, Design and Tuning*; Instrument Society of America: Pittsburgh, PA, USA, 1995; pp. 5–58.
26. Garpinger, O.; Hägglund, T.; Åström, K.J. Performance and robustness trade-offs in PID control. *J. Process Control* **2014**, *24*, 568–577. [CrossRef]
27. Lee, Y.; Park, S.; Lee, M.; Brosilow, C. PID controller tuning for desired closed-loop responses for SI/SO systems. *AIChE J.* **1998**, *44*, 106–115. [CrossRef]
28. Skogestad, S.; Grimholt, C. The SIMC Method for Smooth PID Controller Tuning. In *PID Control in the Third Millennium*; Vilanova, A.V.R., Ed.; Springer: Berlin/Heidelberg, Germany, 2012; pp. 147–175. [CrossRef]
29. Åström, K.; Hägglund, T. Revisiting the Ziegler–Nichols step response method for PID control. *J. Process Control* **2004**, *14*, 635–650. [CrossRef]
30. Reynoso-Meza, G.; Sanchis, J.; Blasco, X.; Herrero, J.M. Multiobjective evolutionary algorithms for multivariable PI controller design. *Expert Syst. Appl.* **2012**, *39*, 7895–7907. [CrossRef]
31. Herrero, J.M.; Martínez, M.; Sanchis, J.; Blasco, X. Well-Distributed Pareto Front by Using the $\wp^\lambda-MOGA$ Evolutionary Algorithm. In *Computational and Ambient Intelligence*; Sandoval, F., Prieto, A., Cabestany, J., Graña, M., Eds.; Springer: Berlin/Heidelberg, Germany, 2007; pp. 292–299. [CrossRef]
32. Méndez, M.; Frutos, M.; Miguel, F.; Aguasca-Colomo, R. TOPSIS Decision on Approximate Pareto Fronts by Using Evolutionary Algorithms: Application to an Engineering Design Problem. *Mathematics* **2020**, *8*, 2072. [CrossRef]
33. Cuate, O.; Schütze, O. Pareto Explorer for Finding the Knee for Many Objective Optimization Problems. *Mathematics* **2020**, *8*, 1651. [CrossRef]
34. Sánchez, H.S.; Visioli, A.; Vilanova, R. Optimal Nash tuning rules for robust PID controllers. *J. Frankl. Inst.* **2017**, *354*, 3945–3970. [CrossRef]
35. Huilcapi, V.; Herrero, J.M.; Blasco, X.; Martínez-Iranzo, M. Non-linear identification of a Peltier cell model using evolutionary multi-objective optimization. *IFAC-PapersOnLine* **2017**, *50*, 4448–4453. [CrossRef]
36. Reynoso-Meza, G.; Blasco-Ferragud, X.; Sanchis-Saez, J.; Herrero-Durá, J.M. Multiobjective Optimization Design Procedure for Controller Tuning of a Peltier Cell Process. In *Controller Tuning with Evolutionary Multiobjective Optimization. Intelligent Systems, Control and Automation: Science and Engineering*; Tzafestas, S.G., Ed.; Springer: Cham, Switzerland, 2017; Volume 85, pp. 187–199. [CrossRef]

Article

Study on the Intelligent Modeling of the Blade Aerodynamic Force in Compressors Based on Machine Learning

Mingming Zhang [1], Shurong Hao [1] and Anping Hou [2,*]

1 Faculty of Science, Beijing University of Technology, Beijing 100124, China; mmzhang@bjut.edu.cn (M.Z.); HaoSR@emails.bjut.edu.cn (S.H.)
2 School of Energy and Power, Beihang University, Beijing 100191, China
* Correspondence: houap@buaa.edu.cn; Tel.: +86-010-8231-6624

Abstract: In order to obtain the aerodynamic loads of the vibrating blades efficiently, the eXterme Gradient Boosting (XGBoost) algorithm in machine learning was adopted to establish a three-dimensional unsteady aerodynamic force reduction model. First, the database for the unsteady aerodynamic response during the blade vibration was acquired through the numerical simulation of flow field. Then the obtained data set was trained by the XGBoost algorithm to set up the intelligent model of unsteady aerodynamic force for the three-dimensional blade. Afterwards, the aerodynamic load could be gained at any spatial location during blade vibration. To evaluate and verify the reliability of the intelligent model for the blade aerodynamic load, the prediction results of the machine learning model were compared with the results of Computation Fluid Dynamics (CFD). The determination coefficient R^2 and the Root Mean Square Error (RMSE) were introduced as the model evaluation indicators. The results show that the prediction results based on the machine learning model are in good agreement with the CFD results, and the calculation efficiency is significantly improved. The results also indicate that the aerodynamic intelligent model based on the machine learning method is worthy of further study in evaluating the blade vibration stability.

Keywords: machine learning; eXterme Gradient Boosting; Computation Fluid Dynamics; blade vibration; unsteady aerodynamic model

Citation: Zhang, M.; Hao, S.; Hou, A. Study on the Intelligent Modeling of the Blade Aerodynamic Force in Compressors Based on Machine Learning. *Mathematics* 2021, 9, 476. https://doi.org/10.3390/math9050476

Academic Editor: Eva H. Dulf

Received: 28 December 2020
Accepted: 22 February 2021
Published: 25 February 2021

Publisher's Note: MDPI stays neutral with regard to jurisdictional claims in published maps and institutional affiliations.

Copyright: © 2021 by the authors. Licensee MDPI, Basel, Switzerland. This article is an open access article distributed under the terms and conditions of the Creative Commons Attribution (CC BY) license (https://creativecommons.org/licenses/by/4.0/).

1. Introduction

With the development of high load and high efficiency in compressors, the centrifugal load and aerodynamic load are endured by the blade due to the strong unsteady flow in the field. Also, the problem of blade vibration has become increasingly prominent. Therefore, the accurate prediction of the internal flow and blade aerodynamic force in the compressor is of great significance for evaluating the reliability of blade vibration in the design stage.

The traditional Computation Fluid Dynamics (CFD) technology can perform a high-fidelity simulation of the linear or non-linear blade vibration in the flow field [1–3]. However, it requires high computational expenses for the large-scale calculation. This is not suitable for the rapid evaluation of blade vibration reliability. To overcome the shortcomings of calculation costs [4], the reduced order models of unsteady flow field are proposed here based on the CFD model [5–12]. Proper Orthogonal Decomposition (POD) and Dynamic Mode Decomposition (DMD) are two typical modal decomposition methods, which are based on the flow field feature extraction technology. The complex unsteady flow field is represented with a set of characteristic modes of low-dimensional variables [5–9]. Another kind of reduced-order model based on the system identification technique has been used for the fluid problem [10–12]. Simple mathematical mapping was employed to describe the relationship between flow disturbances and aerodynamic characteristics.

In recent years, research on knowledge extraction and data visualization has promoted the exploration of artificial intelligence methods for crossing with fluid mechanics. Machine learning builds a powerful information processing framework with accurate algorithms and generalization capabilities. Efforts have been made for the application of machine learning in fluid mechanics [13–21]. The interaction of fluid mechanics and machine learning is summarized by Brunton [13], as well as the development trend of the interdisciplinary approach. It is believed that the application of machine learning can enhance the current fluid mechanics research. Deep Neural Networks (DNN) were stated to play a key role on modeling complex flow by Kutz [14]. A reduction model by DNN was designed based on the data of Direct Numerical Simulation (DNS) by Zhang [15]. The results show that DNN can predict the anisotropic Reynolds stress effectively. Chen [16] proposed the use of a deep Convolutional Neural Network (CNN) to extract flow information, and established a composite network to solve the problem of input with different variables. The hybrid deep neural network framework was used by Han [17] to directly capture the characteristics of unsteady flow in the field. The field predicted by DNN was in agreement with the result calculated by CFD solver. Hasegawa [18] constructed a reduced-order model combined with a CNN auto encoder and Long Short-Term Memory network (LSTM). The model proved to be able to predict the unsteady flow of bluff bodies. Also, the multi-core neural network was adopted by Kou [19] to achieve the correction from the low-order model to the high-fidelity results. A model was constructed with a combination of the Adaptive Simulated Annealing algorithm (ASA) and Recursive Radial Basis Function neural network (RRBF) for the cascade by Hu [20]. It was proven that the ASA-RRBF model has a higher accuracy than the single RRBF model.

The data-driven optimization of machine learning and the application of regression technology can map a high-dimensional flow field to a low-dimensional space, which can effectively solve the high-dimensional nonlinear problems. The ability of machine learning could simplify the treatment of the exploration and visualization of the high-dimensional database, which can greatly improve performance optimization and reduce the convergence cost [13]. The intelligent method provides a useful technology to extract relevant information, which promotes a rapid development of flow dynamics. The constructed reduced-order aerodynamic force model based on the machine learning can predict the unsteady aerodynamic force of the blade with a reasonable accuracy and a low computation cost [21].

Note that the current applications of artificial intelligence method in the fluids are mostly focused on the modeling of flow characteristics, while the modeling of blade aerodynamic force in vibration rarely involves the intelligence method. For this paper, the XGBoost algorithm was applied for the first time to the aerodynamic modeling of an actual compressor blade. A reduced-order intelligent model of the three-dimensional unsteady aerodynamic force of the blade was established for consideration of machine learning and CFD. By learning a small amount of CFD sample data, the trained low-dimensional XGBoost model could effectively capture the characteristics of the unsteady flow. The aerodynamic load of the compressor blade during the vibration process can be obtained by the intelligence model through the input and output mathematical mapping. Compared with the deep learning, the XGBoost model is suitable for data with a small number of variables. It has the advantages of model interpretability and invariance of input data. Also, it is convenient for parameter adjustment to achieve default predictions through automatic iteration. Under the premise of ensuring the accuracy, this reduced-order model presented can greatly reduce the calculation costs.

2. Description of the Machine Learning Algorithm

The eXtreme Gradient Boosting algorithm is an integrated machine learning algorithm based on a decision tree, which is in the foundation of gradient boosting framework. It is proposed to build an efficient and flexible algorithm by Chen [22] according to the second-order information [23]. This algorithm is a scalable machine learning system in the lifting method, which is integrated by multiple regression trees to form a strong classifier. The problem of overfitting in tree model can be effectively avoided [24]. After parallelization, it is more than one order of magnitude faster than similar algorithms under the same conditions [25]. The excellent performance in high-dimensional data analysis shows a strong ability in modeling the complex process [26]. Because of its high performance and low requirement, XGBoost has been widely used in disease prediction, credit debt default risk prediction, driving evaluation, route planning and so on [27–30].

The principle of its algorithm is to update iteratively the parameters of the previous classifier to reduce the gradient of the loss function and generate a new classifier [31]. By reducing the error of prediction through several regression trees, the regression tree group is guaranteed to have the maximum generalization ability. The regular term is added to the loss function of the model. Then the second-order Taylor expansion of the loss function is solved to determine the split node on the basis of the minimum loss function. The second-order derivative information and the addition of regularization method have improved the performance of generalization and calculation [32]. The structure of the XGBoost algorithm is indicated in Figure 1.

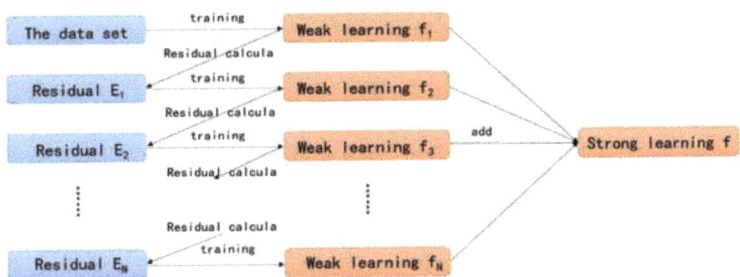

Figure 1. The Structure of the eXterme Gradient Boosting (XGBoost) Algorithm.

The given sample data set is:

$$\eta = \{(x_{i\_CFD}, y_{i\_CFD})\} \ (i = 1, 2, \cdots, n, x_{i\_CFD} \in R^m, y_{i\_CFD} \in R) \quad (1)$$

where x_{i_CFD} represents the i-th feature value of the sample data, y_{i_CFD} represents the experiment value of the i-th label of the sample data and x_{i-Pre} represents the predicted value of the i-th label of the model. Define the loss functions of y_{i_CFD} and y_{i_Pre}:

$$l(y_{i\_CFD}, y_{i\_Pre}) = (y_{i\_CFD} - y_{i\_Pre})^2 \quad (2)$$

Where y_{i_Pre} is the prediction in the integration model of the XGBoost system, which uses the sum of the predicted value of each tree (the total number of trees is K) for the sample. Assuming that the tree model to be trained in the k-th iteration is $f_k(x)$, the prediction function was defined as follows:

$$\begin{array}{l} y_{i\_Pre} = \sum\limits_{k=1}^{K} f_k(x_i), \ f_k \in \Gamma \\ \Gamma = f(x) = \omega_q(x) \ \ (q : R_m \to T, \omega \in R_T) \end{array} \quad (3)$$

As Γ is the space of Classification and Regression Trees (CART) numbers, q represents the score of the structure of each tree mapping each sample to the corresponding leaf node; $\omega_q(x)$ represents the set of scores for all leaf nodes of tree q. The optimized parameter in the XGBoost algorithm is defined as the function of $f(x)$. While a tree is added into the model each time, the loss of the objective function was expected to be decreased. The iteration functions could then be expressed as:

$$\begin{aligned}
y_{i\_Pre}{}^{(0)} &= 0 \\
y_{i\_Pre}{}^{(1)} &= f_1(x_i) = y_{i\_Pre}{}^{(0)} + f_1(x_i) \\
y_{i\_Pre}{}^{(2)} &= f_1(x_i) + f_2(x_i) = y_{i\_Pre}{}^{(1)} + f_2(x_i) \\
&\cdots \\
y_{i\_Pre}{}^{(t)} &= \sum_{k=1}^{t} f_k(x_i) = y_{i\_Pre}{}^{(t-1)} + f_t(x_i)
\end{aligned} \quad (4)$$

The objective function could then be expressed as:

$$\begin{aligned}
obj &= \sum_{i=1}^{n} l(y_{i\_CFD}, y_{i\_Pre}) + \sum_{i=1}^{K} \Omega(f_i) \\
\Omega(f) &= \gamma T + \tfrac{1}{2}\lambda\|\omega\|^2
\end{aligned} \quad (5)$$

where $\sum_{i=1}^{K} \Omega(f_i)$ indicates the regularization term of the loss function, which is the sum of the complexity of all K trees. The number of leaf nodes T is limited with a penalty term $\Omega(f_i)$ so as to prevent overfitting. ω represents the set of scores for all the leaf nodes of each tree, while γ and λ represent the coefficients. In order to solve the optimal objective function, the second-order Taylor expansion of the t-th tree $f_t(x_i)$ in Equation (4) is performed with bringing into the objective function. As the loss function $l(y_{i\_CFD}, y_{i\_Pre}{}^{(t-1)})$ is a constant, it can be ignored. And the leaf nodes of all trees can be regrouped. Then the node number and leaf weights are used to optimize the regularization term of the loss function. All samples x_i of leaf nodes are divided into a sample set, denoted as $I_j = \{i|q(x_i) = j\}$. The objective function could be rewritten as:

$$\begin{aligned}
obj &= \sum_{i=1}^{n}[l(y_{i\_CFD}, y_{i\_Pre}{}^{(t-1)}) + g_i f_t(x_i) + \tfrac{1}{2}h_i f_t^2(x_i)] + \Omega(f_t) + \text{constant} \\
&\approx \sum_{i=1}^{n}[g_i f_t(x_i) + \tfrac{1}{2}h_i f_t^2(x_i)] + \Omega(f_t) \\
&= \sum_{i=1}^{n}[g_i f_t(x_i) + \tfrac{1}{2}h_i f_t^2(x_i)] + \gamma T + \tfrac{1}{2}\lambda \sum_{j=1}^{T} \omega_j^2 \\
&= \sum_{i=1}^{n}[g_i \omega_q(x_i) + \tfrac{1}{2}h_{ii}\omega_q^2(x_i)] + \gamma T + \tfrac{1}{2}\lambda \sum_{j=1}^{T} \omega_j^2 \\
&= \sum_{j=1}^{T}[(\sum_{i\in I_j} g_i)\omega_j + \tfrac{1}{2}(\sum_{i\in I_j} h_i + \lambda)\omega_j^2] + \gamma T \\
&= \sum_{j=1}^{T}[G_j \omega_j + \tfrac{1}{2}(H_j + \lambda)\omega_j^2] + \gamma T
\end{aligned} \quad (6)$$

where $g_i = \frac{\partial l(y_{i\_CFD}, y_{i\_Pre}{}^{(t-1)})}{\partial y_{i\_Pre}{}^{(t-1)}}$, $h_i = \frac{\partial^2 l(y_{i\_CFD}, y_{i\_Pre}{}^{(t-1)})}{\partial (y_{i\_Pre}{}^{(t-1)})^2}$, $G_j = \sum_{i\in I_j} g_i$, $H_j = \sum_{i\in I_j} h_i$, $f(t) = \omega_q(x)$, $\omega \in R^T$ $q: R^d \to \{1, 2, \cdots, T\}$.

The smaller the value of the objective function is, the smaller the prediction error is, with a better generalization ability and robustness of the model. By using the highest value formula of the quadratic function, the weight ωj^* of each leaf node could be obtained. The optimal objective function can then be expressed as:

$$\omega j^* = -\frac{G_j}{H_i + \lambda}, \quad obj = -\frac{1}{2}\sum_{j=1}^{T} \frac{G_j^2}{H_i + \lambda} + \gamma T. \quad (7)$$

The modeling advantage of in XGBoost method can be concluded to the adjunction of regularization items displayed in the objective function. The regularization items are related to the number and the value of leaf nodes in the tree. In addition, the sparse value of the training data in the XGBoost algorithm should be noted. The default direction of the branch is specified for missing values, which greatly improves the efficiency of the algorithm [33]. As an advanced machine learning method developed in recent years, this method has a good performance in processing high-dimensional data with the reduction of the overfitting.

3. Methodology of Aerodynamic Intelligent Model

3.1. Data Collection for Machine Learning

The high accuracy data is the key to establishing an accurate unsteady aerodynamic model of blade in compressor. The training process of the XGBoost model in this paper was mainly driven by the database obtained from the CFD fluid-structure coupling computation. The research object was a 1.5 stage axial compressor, including struts, inlet guide vanes, first stage rotor and stator. Detailed introductions for the rig are presented in Zhang [34].

The unsteady flow field in the compressor was solved by using the numerical solution of 3-D Navier-Stokes equations adopted in software ANSYS Package. The spatial discretization of the flow governing equations was employed on an upwind scheme, and a second-order backward differencing was integrated for the time-accurate solution [35]. Boundary conditions imposed on the inlet consist of total pressure and total temperature. A specified average static pressure was implemented at the exit boundary. Smooth, adiabatic and no-slip wall boundary conditions were applied for the flow field solution [36]. While considering the fluid-structure interaction, the blade vibration was computed under the response to the flow. The detail simulation process of the compressor is described in reference [34]. The structural equations for mechanical blade were solved by the finite element method. Within each time step, the flow equations and the structural equations were solved simultaneously, exchanging information on the fluid-structure interface. This procedure was repeated until the flow and displacements were converged, before proceeding to the next time step. The numerical model of the 1.5-stage turbocompressor is shown in Figure 2.

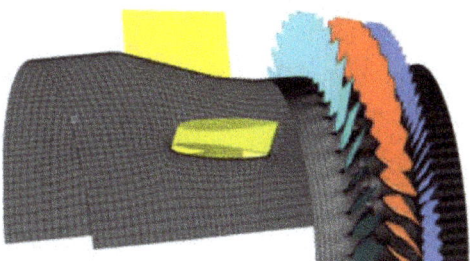

Figure 2. Numerical Model of the 1.5 Stage Turbocompressor.

After the convergence of the simulation, the results computed by the commercial CFD software were used for the current data learning, including the spatial unsteady flow data and aerodynamic force on blade surface in time domains. The snapshot data of the unsteady flow was captured at each time step, including pressure and aerodynamic force of the blades at modal coordinates. The data set for training/testing was composed with five variables, such as Cartesian coordinates, pressure and aerodynamic force. The three-dimensional coordinate (X, Y, Z) of the structure space was taken as an input, and the aerodynamic force was taken as an output to form the sample data $S = (X, Y, Z, Force)$. The flow snapshot data extracted from CFD was arranged in time series as a sequence $\{S_1, S_2, S_3, \cdots, S_N\}$, where $S_i = \{X_{ij}, Y_{ij}, Z_{ij}, Force_{ij}\}, i = 1, 2, \cdots, N, j = 1, 2, \cdots, n$. The distribution of the aerodynamic force on the blade surface is shown in Figure 3, which was extracted in the CFD fluid-structure coupling simulation at a single time. It can be seen that the distribution of aerodynamic force was not uniform on both the pressure side

(PS) and suction side (SS). Because of the unsteady flow in the field, the aerodynamic force that acted on the blade appears in a non-linear state, which resulted in the vibration blade indicating complex dynamic behaviors.

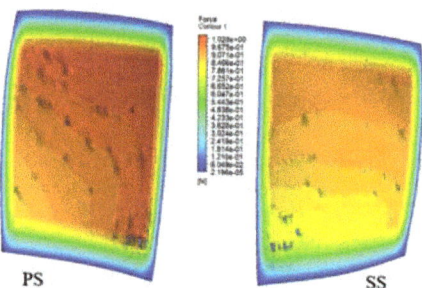

Figure 3. Distribution of the aerodynamic force on the blade.

3.2. Procedure of Aerodynamic Modeling Based on the XGBoost Algorithm

In this part, the methodology of aerodynamic modeling based on XGBoost algorithm is introduced in detail. The procedure can be concluded as follows.

Step 1: Data preprocessing.

After the data collection from CFD, the features of acquired data may have different magnitudes. When the gradient is updated, it may oscillate back and forth, and take a long time to reach the local optimal value or the global optimal value. In order to improve the training efficiency and avoid the numerical error caused by the size difference of the features, the data were handled in normalization. This ensured that the same dimension was achieved for different features, so that the descent of gradient could be a quick convergence. The normalized function form used in this article is shown as follows:

$$X = \frac{x_k - x_{\min}}{x_{\max} - x_{\min}}. \tag{8}$$

In machine learning algorithms, feature engineering is an important step in the process of modeling. The original data were transformed into the training data with feature engineering, providing the training model with a better robustness and generalization ability. This paper provides three characteristics of index, distance and average value of three-dimensional coordinates based on data information.

Step 2: Training set construction.

The training set was used to estimate the parameters in the intelligent model. As a result, the accuracy and efficiency of the model were determined by the selection of the training set. In order to optimize the effect of the model, the dichotomy process was adopted to partition the training set. That is to say for N samples, each segment was divided into the length of $[C/2]$, where $C = [N], [N/2], [N/2^2], \cdots, 2$. Then take a representative data set from each segment to form a training set. Taking into account of the accuracy and running time in calculation, two snapshots as $\{S_1, S_{[N/2]}\}$ were selected to form the training set to train the model in this paper.

Step 3: Training process.

The training set after data preprocessing was substituted into the initial XGBoost model established for training. The effect of prediction by the model was evaluated by the comparison to the $[N/2] + 1$ snapshot data, which were selected as the test set. The establishment of the model required the setting of hyper-parameters. The hyper-parameters used in this article were defined as: Max-depth (the maximum depth of the tree), Learning-rate (the learning rate), n-estimators (the number of sub-models) and objective (the given loss function). The hyper-parameters for the initialization model are given in Table 1.

Table 1. Parameters for XGBoost Model Initialization.

Max-Depth	Learning-Rate	n-Estimators	Objective
40	0.35	60	Reg:gamma

Step 4: Parameter Adjustment.

The adjustment of the hyper-parameters in the XGBoost model played a key role in affecting the training performance of the XGBoost algorithm. So, the GridSearchCV function was employed to adjust the parameters of the XGBoost model. The hyper-parameters after seeking are shown in Table 2.

Table 2. Parameters in XGBoost Model after Adjustment.

Max-Depth	Learning-Rate	n-Estimators	Objective
19	0.1	160	Reg:gamma

Step 5: Evaluation Criteria.

The indicators as the coefficient of determination R^2 and the root mean square error (RMSE) were introduced to evaluate the accuracy of the established XGBoost model. The fitness of the prediction to the observation can be represented by the coefficient of determination R^2, which was defined as the ratio of the regression sum of squares to the total sum of squares. This coefficient is often used to evaluate the merits and demerits of a regression model. If the coefficient of determination R^2 is calculated to be close to 1, it indicates that the regression model is effective. RMSE is the square root of the ratio, which is the square sum of the errors of prediction values to the number of observations. The optimal parameters of the model and the optimal prediction results were obtained through model training

$$R^2 = 1 - \frac{\sum_i (CFD_i - Pre_i)^2}{\sum_i (CFD_i - \overline{CFD_i})^2}, \quad (9)$$

$$RMSE = \sqrt{\frac{\sum_{i=1}^n (CFD_i - Pre_i)^2}{num}}, \quad (10)$$

where CFD_i represents the i-th label in the sample data which is captured from CFD simulation, $\overline{CFD_i}$ represents the average value of the label in the sample set, and Pre_i represents the predicted value of the i-th label in the XGBoost model. And prediction error is defined as: $error = Pre - CFD$, which is the difference between the prediction of the XGBoost model and the CFD result. The whole procedure of aerodynamic modeling based on the XGBoost algorithm is shown in Figure 4.

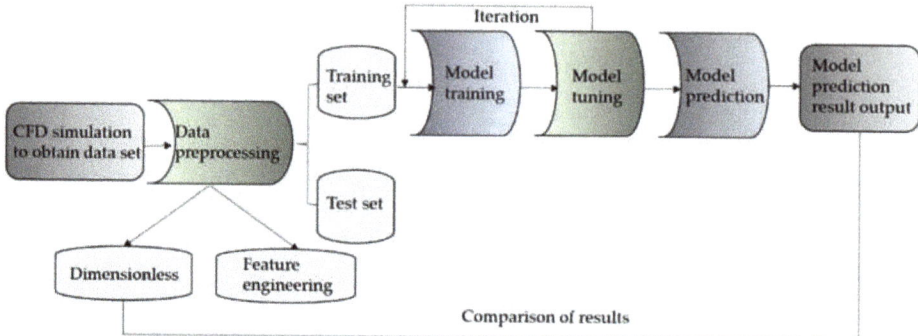

Figure 4. Procedure of Aerodynamic Modeling.

4. Modeling of Blade Aerodynamic Pressure Based on Machine Learning

In this section, the intelligent modeling is first performed for the three-dimensional unsteady pressure of the blade during the vibration process. Also, the effectiveness and accuracy of the model are evaluated based on the XGBoost algorithm. The three-dimension coordinate (X, Y, Z) of the state space in blade vibration was taken as the input, and the pressure data was taken as the output to form sample data $S = (X, Y, Z, Pressure)$. According to the procedure of aerodynamic modeling described above, the gradient descent method was used to find the optimal solution. After the dimensionless processing on pressure data, the training set was collected to train the XGBoost model. Also, the test set was brought into the trained optimal prediction model for comparison. Finally, the prediction on the blade aerodynamic pressure was obtained with the XGBoost model established.

The prediction results of the aerodynamic pressure of blade are shown in Figure 5 at a certain time. The predicted values of pressure in the XGBoost model were compared with the data in CFD simulation at 80% of the blade span. It can be seen that the curves predicted are in good accordance with each other, indicating the accuracy of the XGBoost model.

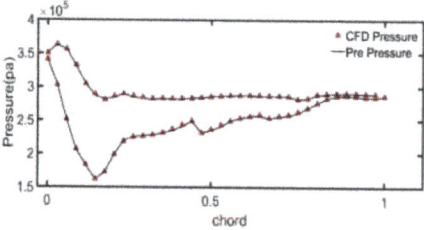

Figure 5. Comparison between XGBoost Model and Computation Fluid Dynamics (CFD).

We unfolded the three-dimensional compressor blade along the leading edge, and displayed the pressure surface (PS Side) and suction surface (SS Side) of the compressor blade on the same coordinate plane. The pressure contour predicted by the XGBoost model is exhibited in Figure 6, as well as the result simulated by CFD. The two contours look almost the same, but there are still errors located under 40% of span, which are revealed in Figure 7. In addition, under the program running with 0.3 s, the coefficient of determination R^2 was computed to be 0.99947. The RMSE was obtained as 1012.4 by the model, which is approximately a 0.3% error rate to the average pressure of the blade. Compared with CFD simulation data, the three-dimensional aerodynamic pressure model of the blade based on the XGBoost intelligent method reflects a good accuracy and efficiency. The current study demonstrates that it is sufficient to predict the blade aerodynamic force by capturing the characteristic of flow based on the machine learning method.

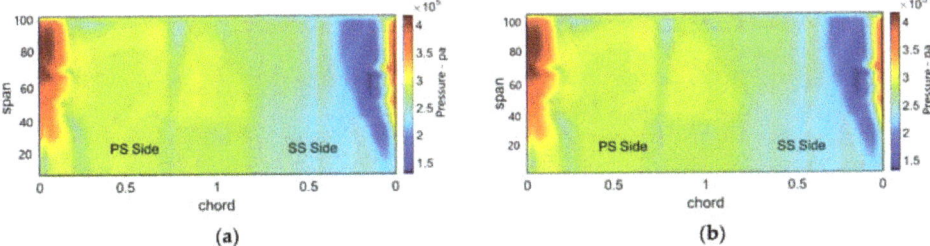

Figure 6. Pressure Contours of the Blade: (a) CFD; (b) XGBoost Model.

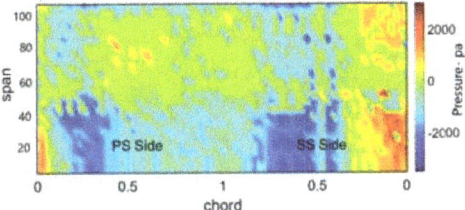

Figure 7. Error Contour for Pressure.

5. Modeling of Blade Aerodynamic Force Based on Machine Learning

Under the verification for the effectiveness of the intelligent modeling method, the XGBoost algorithm was then used to model the unsteady aerodynamic force for the three-dimensional blade in this section. The aerodynamic force on the blade surface was obtained based on the integral of the pressure over the mesh grid area in CFD. Because of the micro size of grid at the blade edge, the value of force at the blade edge was much smaller. In order to restore the distribution of force on the blade, the process of dimensionless was performed for the aerodynamic force Df on the blade surface

$$Df = \frac{F}{\overline{S} \times P},\qquad(11)$$

where F is the aerodynamic force data on the blade surface in CFD, $\overline{S}$ is the average area of the blade surface mesh and P equals the standard atmospheric pressure. Next, the distribution of aerodynamic force on the blade surface was predicted by the XGBoost model at any position during the vibration process.

The predicted values of aerodynamic force in the XGBoost model were chosen here to compare with the data in CFD simulation at 3%, 80% and 90% of the blade span, respectively, as shown in Figure 8. It can be seen that the aerodynamic force of the blade increases sharply from the leading edge, and decreases at the trailing edge. It was found that the values of aerodynamic force appear to have significant differences along the variation of the blade span. The aerodynamic forced distributed along the direction of the blade spanwise presents a nonlinear characteristic.

This appearance can also be observed at the distribution of force at the three-dimension surface of the blade. The 3D plots are adopted here to show the distribution of aerodynamic force at a blade modal location. As indicated from Figure 9, it can be seen that the distribution of the aerodynamic force predicted by the XGBoost model is accordant with the CFD data on the pressure surface of the blade. The load of blade is mainly concentrated in the middle part of the blade, corresponding to the region of high aerodynamic force. Because of the non-linear feature, the unsteady force is not easy to express. According to the errors displayed in Figure 10, the nodes of aerodynamic force modeling by XGBoost method show good agreement with the CFD data. Although the existence of error was discovered at certain points, the effectiveness of the aerodynamic force model is still verified through the comparison.

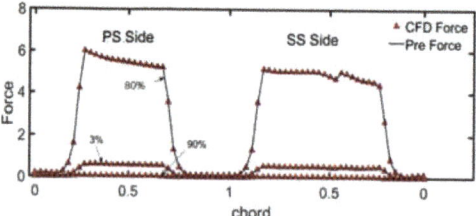

Figure 8. Unsteady aerodynamic value of the blade section.

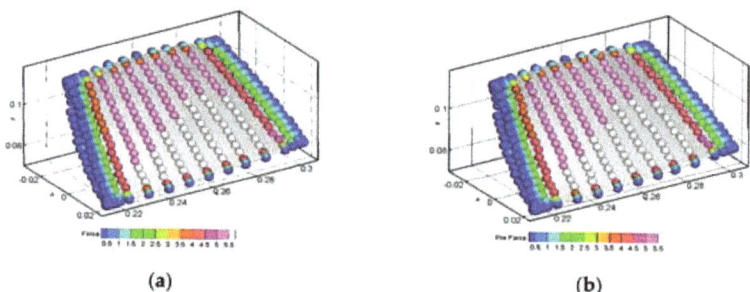

Figure 9. Aerodynamic Force at the Blade Pressure Side: (**a**) CFD; (**b**) XGBoost Model.

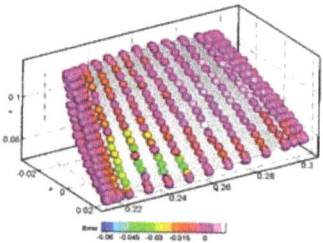

Figure 10. Aerodynamic Force Error on Blade.

The dimensionless aerodynamic force contour predicted by the XGBoost model is expressed in Figure 11, along with the contour simulated by CFD. The two aerodynamic clouds coincide exactly with each other. Also, the errors inevitably appear in the comparison of XGBoost model with CFD, as indicated in Figure 12. But it can be seen that the errors emerge mostly in the region with a large gradient. The values of error oscillate around 0 with the maximum value as 0.06, which is relatively small in contrast to the dimensionless aerodynamic force of the blade. At the running of program with 0.23 s, the coefficient of determination R^2 of the XGBoost model was computed to be 0.99998, which is very close to 1. Also, the RMSE was obtained as 0.005846 by the model, which is approximately a 0.1% error to the average dimensionless aerodynamic force of the blade. With the comparison to the CFD simulation data, this shows a good accuracy and reliability of predicting the aerodynamic force by the three-dimension aerodynamic force model of the blade based on the XGBoost intelligent method.

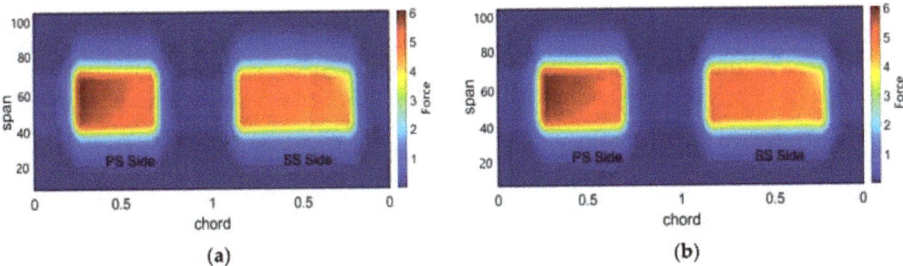

Figure 11. Aerodynamic Force on the Blade: (**a**) CFD; (**b**) XGBoost Model.

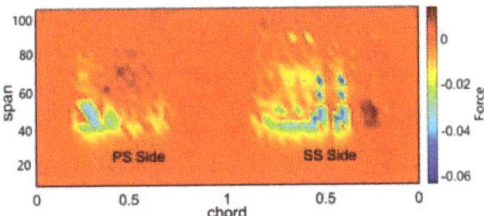

Figure 12. Error Contour for Aerodynamic Force.

To check the generalization ability and robustness of the XGBoost model, snapshot data sets with the blade vibration at different times were used as the testing set. The trained XGBoost model was also used to predict the aerodynamic force of each snapshot data in the testing set. The results of prediction are revealed in Figure 13, as represented by the coefficient of determination R^2 and RMSE for different testing data. For the trained aerodynamic force XGBoost model, the prediction accuracy also shows a slight discrepancy compared to different positions of blade vibration. The maximum coefficient of determination R^2 of the prediction model is 0.99999, with the minimum value as 0.99987. The maximum RMSE value is 0.01852, with the minimum value to be 0.00519. The coefficients of determination R^2 are all above 0.9998, and the RMSE values are all less than 0.0186. This means that the XGBoost model reflects a good generalization ability with high robustness. From all analysis above, it can be concluded that the three-dimension aerodynamic model based on the XGBoost algorithm can accurately predict the aerodynamic force of the blade on the basis of any spatial position in the blade vibration process.

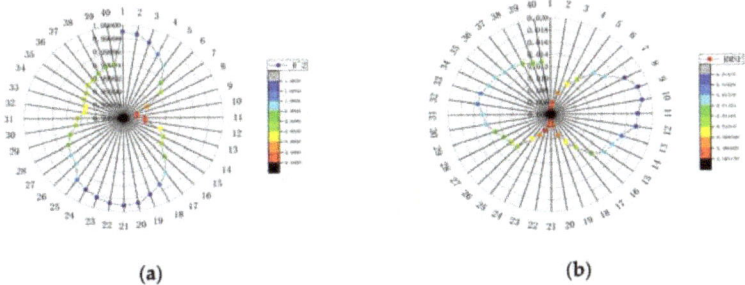

Figure 13. Graphs of Error at different testing data: (**a**) CFD; (**b**) XGBoost Model.

6. Discussion

With the assistance of CFD technology, the unsteady flow field simulation of the compressor is considered as a full-order solution to the system. Although the data obtained is regarded as being accurate, it is not convenient for the rapid qualitative analysis of the system with a high time cost and low efficiency [4]. With the simple control equation of the reduced-order model, the data can significantly reduce computation expenses and improve calculation efficiency [10]. The mathematical mapping between the input and output can be set up by solving the complex Navier-Stokes equations once for training. Then the fluid-structure coupled solution in the CFD solver can be replaced by the intelligent model.

Recently there has been research conducted on the aerodynamic reduction modeling of wings by artificial intelligent methods. But there are considerable differences between blades and wings. Compared with isolated wings, there is an obvious unsteady aerodynamic interference effect in the blade row [36]. Therefore, the aeroelastic analysis of blades is different from the traditional vibration analysis of wings in outflow. Since the internal flow is a very complex full three-dimensional unsteady viscous flow field, the aerodynamic interference between the blades is very prominent. It is impossible to use theory to predict the unsteady aerodynamic force of vibrating blades with so many parameters [37]. In nonlinear dynamic analysis, it is assumed that the blade is flat with no thickness, reducing the real three dimensions to two dimensions. However, an efficient and accurate aerodynamic model of the three-dimension blade is the basis for the analysis of the nonlinear dynamic system. The vibration modeling of the actual three-dimensional blades was rarely used in the previous research.

In this paper, the XGBoost algorithm of machine learning was used to establish a reduced-order model of the unsteady aerodynamic force for a vibrating blade. By learning from the high-fidelity sample data, the aerodynamic distribution of a three-dimensional blade could be quickly predicted accurately at any spatial position of the blade during the vibration process. This provides a basis for the further nonlinear dynamic analysis of the blade. But how to incorporate this into the nonlinear dynamics equations with an appropriate format remains a question. It is also worth conducting further integration with fluid mechanics to evaluate the blade vibration stability.

7. Conclusions

The internal field of the compressor is essentially a three-dimensional unsteady flow. The flow around the blade is very complex. In order to achieve an unsteady aerodynamic load on the blade, a reduced-order intelligent model of the three-dimensional blade in compressor was established in this paper based on a machine learning algorithm for the first time. The main conclusions are as follows:

(1) With the combination of the intelligent algorithm in machine learning and CFD technology, the modeling for the aerodynamic force can be performed for a three-dimensional blade of compressor in vibration. Also, the procedure for aerodynamic modeling based on the XGBoost algorithm was established, which is described as data collection, data preprocessing, training set construction, model training and parameter adjustment.

(2) The high-fidelity data for model training can be set up by solving the complex Navier-Stokes equations once for the flow field. Then the information of the unsteady flow can be effectively captured based on the XGBoost model training for the mathematical mapping between the input and output. The rapid identification was achieved for the three-dimensional aerodynamic force on the blade, which improves the efficiency of calculation.

(3) Based on the data of blade vibration in CFD simulation, an intelligent model based on the XGBoost algorithm was established for the prediction of the three-dimensional unsteady aerodynamic pressure and force. With the comparison to the CFD data, it showed a good accuracy and reliability on the prediction in the XGBoost intelligent method. The distribution of an unsteady aerodynamic load on the blade can be accurately predicted on the basis of any spatial position in the blade vibration process. It provides a new perspective for the analysis of blade nonlinear dynamics. The aerodynamic intelligent model based on the machine learning is worthy of further integration with fluid mechanics for evaluating the blade vibration stability.

Author Contributions: Conceptualization, M.Z. and S.H.; methodology, M.Z. and S.H.; software, S.H.; validation, M.Z. and S.H.; formal analysis, S.H.; writing—original draft preparation, S.H.; writing—review and editing, M.Z.; supervision, A.H.; project administration, A.H.; funding acquisition, A.H. All authors have read and agreed to the published version of the manuscript.

Funding: This research was funded by the National Science and Technology Major Project, grant number 2017-II-0009-0023.

Acknowledgments: The authors would like to acknowledge the funding support from BIC-ESAT in Peking University.

Conflicts of Interest: The authors declare no conflict of interest.

Abbreviations

The following abbreviations are used in this manuscript:

XGBoost	eXterme Gradient Boosting
CFD	Computation Fluid Dynamics
R^2	the Coefficient of Determination
RMSE	the Root Mean Square Error
POD	Proper Orthogonal Decomposition
DMD	Dynamic Mode Decomposition
DNN	Deep Neural Networks
DNS	Direct Numerical Simulation
CNN	Convolutional Neural Network
LSTM	Long Short-Term Memory network
ASA	Adaptive Simulated Annealing
RRBF	Recursive Radial Basis Function network
CART	Classification and Regression Trees
PS	Pressure Side
SS	Suction Side

References

1. Chao, D.; Wang, P. Numerical Calculation of Three-Dimensional Flow Field in Blade Row of Turbomachine. *Therm. Power Eng.* **1994**, *9*, 230–233.
2. Zhang, Z.; Liu, X. Numerical Simulation of Unsteady Flow in Three-Dimensional Oscillating Cascades. In Proceedings of the 2009 Chinese Society of Engineering Thermophysics, Aerothermodynamics and Fluid Machinery Academic Conference, Dalian, China, 18–20 October 2009.
3. Hu, Y.; Zhou, X. Numerical Analysis and Application of Unsteady Aerodynamic Forces in Vibrating Cascades. *J. Appl. Mech.* **2004**, *21*, 49–52.
4. Chen, G.; Li, Y. Research Progress and Prospect of reduced-order Model of unsteady flow Field and its Application. *Prog. Mech.* **2011**, *41*, 686–701.
5. Kou, J.; Zhang, W. Modal Analysis of Transonic Chattering based on POD and DMD. *Chin. J. Aeronaut.* **2016**, *37*, 2679–2689.
6. Qiu, R.; Huang, R.; Wang, Y.; Huang, C. Dynamic Mode Mecomposition and Reconstruction of Transient Cavitating Flows around a Clark-Y Hydrofoil. *Theor. Appl. Mech. Lett.* **2020**, *10*, 327–332. [CrossRef]
7. Liu, M.; Tan, L. Dynamic Mode Decomposition of Cavitating Flow around ALE 15 Hydrofoil. *Renew. Energy* **2019**, *139*, 214–227. [CrossRef]
8. Li, C.Y.; Tse, T.K.; Hu, G. Dynamic Mode Decomposition on Pressure Flow Field Analysis: Flow Field Reconstruction, Accuracy, and Practical Significance. *J. Wind Eng. Ind. Aerodyn.* **2020**, *205*, 104278. [CrossRef]
9. Hu, J.; Wang, Y.; Liu, H. Comparison of Modal Decomposition Methods for Unsteady Flow Separation in Compressor Cascades. *J. Northwest. Polytech. Univ.* **2020**, *38*, 121–129. [CrossRef]
10. Cowan, T.; Arena, A.; Gupta, K. Accelerating Computational Fluid Dynamics Based Aeroelastic Predictions Using System Identification. *J. Aircr.* **2001**, *38*, 81–87. [CrossRef]
11. Su, D.; Zhang, W.; Zhang, C.; Ye, Z. Unsteady Aerodynamic Modeling Method of Turbine Based on System Identification Technology. *Aeronaut. J.* **2012**, *33*, 242–248.
12. Zhang, J.; Li, L.; Yuan, M. Aerodynamic Order Reduction Model Optimization of Airfoils Parameterized by Radial Basis Function. *Appl. Math. Mech. J.* **2019**, *40*, 250–258.
13. Brunton, S.L.; Noack, B.R.; Koumoutsakos, P. Machine Learning for Fluid Mechanics. *Annu. Rev. Fluid Mech.* **2020**, *52*, 477–508. [CrossRef]
14. Kutz, J.N. Deep Learning in Fluid Dynamics. *J. Fluid Mech.* **2017**, *814*, 1–4. [CrossRef]

15. Zhang, Z.; Song, X.; Ye, S. Application of Deep Learning Method to Reynolds Stress Models of Channel Flow Based on Reduced-Order Modeling of DNS Data. *J. Hydrodyn.* **2019**, *31*, 58–65. [CrossRef]
16. Wang, Y.; Li, D.; Chen, G. A Reduced-Order Model of Aerodynamic Deep Learning Based on Flow Field Characteristics. In Proceedings of the 4th National Conference on Unsteady Aerodynamics, Hefei, China, 10–13 May 2018.
17. Han, R.; Wang, Y.; Zhang, Y.; Chen, G. A Novel Spatial-Temporal Prediction Method for Unsteady Wake Flows Based on Hybrid Deep Neural Network. *Phys. Fluids* **2019**, *31*, 127101.
18. Hasegawa, K.; Fukami, K.; Murata, T.; Fukagata, K. Machine-Learning-Based Reduced-Order Modeling for Unsteady Flows around Bluff Bodies of Various Shapes. *Theor. Comput. Fluid Dyn.* **2020**, 1–17.
19. Kou, J.; Zhang, W. Multi-Fidelity Modeling Framework for Nonlinear Unsteady Aerodynamics of Airfoils. *Appl. Math. Model.* **2019**, *76*, 832–855. [CrossRef]
20. Hu, J.; Liu, H.; Wang, Y. Reduced Order Model for Unsteady Aerodynamic Performance of Compressor Cascade Based on Recursive RBF. *Chin. J. Aeronaut.* **2020**, *8*, 22. [CrossRef]
21. Li, K.; Kou, J.; Zhang, W. Deep Neural Network for Unsteady Aerodynamic and Aeroelastic Modeling across Multiple Mach Numbers. *Nonlinear Dyn.* **2019**, *96*, 2157–2177. [CrossRef]
22. Chen, T.; Guestrin, C. XGBoost: A Scalable Tree Boosting System. In Proceedings of the 22nd ACM SIGKDD International Conference, San Francisco, CA, USA, 13–17 August 2016.
23. Liu, Y.; Qiao, M. Prediction of Heart Disease Based on Clustering and XGboost Algorithm. *Comput. Syst. Appl.* **2019**, *28*, 228–232.
24. Ye, Q.; Rao, H.; Ji, M. Commercial Sales Forecast Based on Xgboost. *J. Nanchang Univ. Sci. Ed.* **2017**, *41*, 275–281.
25. Huang, Q.; Zheng, Y.; Deng, Y. Research on Traffic Forecast of Holiday Road Network based on XGBoost. *Highway* **2018**, *63*, 234–238.
26. Ma, X.; Sha, J.; Wang, D.; Yu, Y.; Yang, Q.; Niu, X. Study on a Prediction of P2P Network Loan Default Based on the Machine Learning Light GBM and XGboost Algorithms According to Different High Dimensional Data Cleaning. *Electron. Commer. Res. Appl.* **2018**, *31*, 24–39. [CrossRef]
27. Zhou, R.; Peng, H.; Li, X.; Yan, Y. Prediction Model of Credit Default Based on XGBoost Algorithm. *Bond* **2019**, *10*, 61–68.
28. Wang, X.; Wang, L.; Wang, S.; Chen, J.; Wu, C. An XGBoost-Enhanced Fast Constructive Algorithm for Food Delivery Route Planning Problem. *Comput. Ind. Eng.* **2020**, *152*, 107029. [CrossRef]
29. Shi, X.; Wong, Y.; Li, M.; Palanisamy, C.; Chai, C. A Feature Learning Approach Based on XGBoost for Driving Assessment and Risk Prediction. *Accid. Anal. Prev.* **2019**, *129*, 170–179. [CrossRef]
30. Budholiya, K.; Shrivastava, S.; Sharma, V. An Optimized XGBoost Based Diagnostic System for Effective Prediction of Heart Disease. *J. King Saud Univ. Comput. Inf. Sci.* **2020**, *10*, 13. [CrossRef]
31. Shen, C. XGBoost Principle and its Application. *Comput. Prod. Circ.* **2019**, *3*, 086.
32. Zhang, D.; Gong, Y. Predictive Diagnosis of Acute Liver Failure by XGBoost Compared with Neural Network and Stochastic Forest Coupling Factor Analysis. *Pract. Underst. Math.* **2020**, *13*, 141–152.
33. Dong, W.; Huang, Y.; Lehane, B.; Ma, G. XGBoost Algorithm-Based Prediction of Concrete Electrical Resistivity for Structural Health Monitoring. *Autom. Constr.* **2020**, *114*, 103155. [CrossRef]
34. Zhang, M.; Hou, A. Investigation on the Flow Field Entropy Structure of Non-Synchronous Blade Vibration in an Axial Turbocompressor. *Entropy* **2020**, *22*, 1372. [CrossRef]
35. Zhang, M.; Hou, A.; Zhou, S.; Yang, X. Analysis on Flutter Characteristics of Transonic Compressor Blade Row by a Fluid-Structure Coupled Method. In Proceedings of the ASME Turbo Expo 2012, Copenhagen, Denmark, 11–15 June 2012; American Society of Mechanical Engineers: New York, NY, USA, 2012.
36. Zhang, M.; Hou, A.; Li, J. Analysis of Blade Vibration Response Induced by Rotating Stall in Axial Compressor. *J. Aerosp. Power* **2012**, *27*, 2269–2277.
37. Du, C. Study on Unsteady Aerodynamic Force of Vibratory Blade in Turbine. *J. Appl. Mech.* **1997**, *14*, 25–29.

Article

Flow towards a Stagnation Region of a Vertical Plate in a Hybrid Nanofluid: Assisting and Opposing Flows

Iskandar Waini [1,2], **Anuar Ishak** [2,*] **and Ioan Pop** [3]

1. Fakulti Teknologi Kejuruteraan Mekanikal dan Pembuatan, Universiti Teknikal Malaysia Melaka, Hang Tuah Jaya, Durian Tunggal, Melaka 76100, Malaysia; iskandarwaini@utem.edu.my
2. Department of Mathematical Sciences, Faculty of Science and Technology, Universiti Kebangsaan Malaysia, Bangi, Selangor 43600, Malaysia
3. Department of Mathematics, Babeş-Bolyai University, 400084 Cluj-Napoca, Romania; ipop@math.ubbcluj.ro
* Correspondence: anuar_mi@ukm.edu.my

Abstract: This study investigates a hybrid nanofluid flow towards a stagnation region of a vertical plate with radiation effects. The hybrid nanofluid consists of copper (Cu) and alumina (Al_2O_3) nanoparticles which are added into water to form $Cu-Al_2O_3$/water nanofluid. The stagnation point flow describes the fluid motion in the stagnation region of a solid surface. In this study, both buoyancy assisting and opposing flows are considered. The similarity equations are obtained using a similarity transformation and numerical results are obtained via the boundary value problem solver (bvp4c) in MATLAB software. Findings discovered that dual solutions exist for both opposing and assisting flows. The heat transfer rate is intensified with the thermal radiation (49.63%) and the hybrid nanoparticles (32.37%).

Keywords: hybrid nanofluid; dual solutions; mixed convection; stagnation point; radiation; stability analysis

Citation: Waini, I.; Ishak, A.; Pop, I. Flow towards a Stagnation Region of a Vertical Plate in a Hybrid Nanofluid: Assisting and Opposing Flows. *Mathematics* **2021**, *9*, 448. https://doi.org/10.3390/math9040448

Academic Editor: Eva H. Dulf

Received: 4 February 2021
Accepted: 20 February 2021
Published: 23 February 2021

Publisher's Note: MDPI stays neutral with regard to jurisdictional claims in published maps and institutional affiliations.

Copyright: © 2021 by the authors. Licensee MDPI, Basel, Switzerland. This article is an open access article distributed under the terms and conditions of the Creative Commons Attribution (CC BY) license (https://creativecommons.org/licenses/by/4.0/).

1. Introduction

The phenomenon of the flow on a stagnation region commonly occurs in aerodynamic industries and engineering applications. To name a few, such applications are polymer extrusion, drawing of plastic sheets, and wire drawing. Hiemenz [1] was the first researcher to consider the boundary layer flow toward a stagnation point on a rigid surface. Besides this, the axisymmetric flow was considered by Homann [2], whereas the oblique stagnation-point flow was studied by Chiam [3]. Further, Merkin [4] studied a similar problem by considering the mixed convection flow. He discovered that the solution is not unique for the opposing flow case. However, Ishak et al. [5] exposed that the dual solutions exist for both opposing and assisting flows, and these behaviours were also reported by several researchers [6–9].

In 1995, Choi and Eastman [10] presented a new type of heat transfer fluid called nanofluid, which is a mixture of single type nanoparticles and the base fluid, to enhance the thermal conductivity. Some works on such fluids can be found in [11–16]. Recently, some studies have shown that advanced nanofluids composed of other types of nanoparticles mixed with regular nanofluids could improve their thermal properties, and this mixture is termed "hybrid nanofluid". The earlier experimental works on the hybrid nanofluid have been done by Turcu et al. [17], Jana et al. [18], and Suresh et al. [19]. Besides, the numerical studies on the hybrid nanofluid flow were studied by Devi and Devi [20]. They observed that the heat transfer rate of the hybrid nanofluid is higher than that of the regular nanofluid. Moreover, the non-uniqueness of the solutions in the hybrid nanofluid flow was examined by Waini et al. [21–27] Other physical aspects were considered by several authors [28–35]. Furthermore, the review papers can be found in [36–41].

Different from the above-mentioned studies, this paper considers the assisting and opposing buoyant flows of a hybrid nanofluid containing Al_2O_3-Cu hybrid nanoparticles when the effect of thermal radiation is taken into consideration. The governing equations along with the boundary conditions are transformed into a system of ordinary differential equations using a similarity transformation. The system of equations is then solved numerically using the boundary value problem solver (bvp4c) in MATLAB software. Most importantly, in this study, two solutions are discovered for both opposing and assisting flows. Then, further analysis is performed to study the temporal stability of these solutions as time evolves.

2. Mathematical Formulation

Consider the flow configuration as shown in Figure 1. The free stream velocity is $U(x) = ax$ and the surface temperature is $T_w(x) = T_\infty + bx$, where a and b are constants. Meanwhile, the ambient temperature T_∞ is assumed to be constant. Accordingly, the hybrid nanofluid equations are as follows ([5,14]):

$$\frac{\partial u}{\partial x} + \frac{\partial v}{\partial y} = 0 \tag{1}$$

$$u\frac{\partial u}{\partial x} + v\frac{\partial u}{\partial y} = U\frac{dU}{dx} + \frac{\mu_{hnf}}{\rho_{hnf}}\frac{\partial^2 u}{\partial y^2} + \frac{(\rho\beta)_{hnf}}{\rho_{hnf}}(T - T_\infty)g \tag{2}$$

$$u\frac{\partial T}{\partial x} + v\frac{\partial T}{\partial y} = \frac{k_{hnf}}{(\rho C_p)_{hnf}}\frac{\partial^2 T}{\partial y^2} - \frac{1}{(\rho C_p)_{hnf}}\frac{\partial q_r}{\partial y} \tag{3}$$

subject to

$$\begin{array}{c} v = 0, \quad u = 0, \quad T = T_w(x) = T_\infty + bx \quad \text{at} \quad y = 0 \\ u \to U(x) = ax, \quad T \to T_\infty \quad \text{as} \quad y \to \infty \end{array} \tag{4}$$

where u and v represent the velocity components along the x- and y- axes. Besides, g and q_r are the acceleration caused by the gravity and the radiative heat flux, respectively. Meanwhile, the temperature of the hybrid nanofluid is given by T.

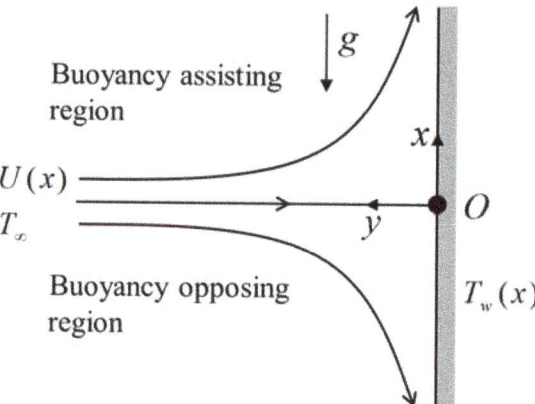

Figure 1. The flow configuration.

The expression of the radiative heat flux is ([42,43]):

$$q_r = -\frac{4\sigma^*}{3k^*}\frac{\partial T^4}{\partial y} \tag{5}$$

where σ^* and k^* denote the Stefan-Boltzmann constant and the mean absorption coefficient, respectively. Following Rosseland [42], after employing a Taylor series, one gets $T^4 \cong 4T_\infty^3 T - 3T_\infty^4$. Then, the Equation (3) turns to [43]:

$$u\frac{\partial T}{\partial x} + v\frac{\partial T}{\partial y} = \frac{1}{(\rho C_p)_{hnf}}\left[k_{hnf} + \frac{16\sigma^* T_\infty^3}{3k^*}\right]\frac{\partial^2 T}{\partial y^2} \quad (6)$$

Further, the thermophysical properties can be referred to in Tables 1 and 2. Data from these tables are adapted from Oztop and Abu-Nada [13], Devi and Devi [20], and Waini et al. [21]. Note that φ_1 (Al$_2$O$_3$) and φ_2 (Cu) are the nanoparticles volume fractions, and the subscripts $n1$ and $n2$ are corresponded to their solid components, while the subscripts f, nf, and hnf signify the base fluid, nanofluid, and hybrid nanofluid, respectively. To get a similarity solution, we employ the following similarity transformation ([5,14]):

$$\psi = \sqrt{a\nu_f}xf(\eta), \quad \theta(\eta) = \frac{T - T_\infty}{T_w - T_\infty}, \quad \eta = y\sqrt{\frac{a}{\nu_f}} \quad (7)$$

where ψ is the stream function defined as $u = \partial\psi/\partial y$ and $v = -\partial\psi/\partial x$, then one gets

$$u = axf'(\eta), \quad v = -\sqrt{a\nu_f}f(\eta) \quad (8)$$

Table 1. Thermophysical properties of nanoparticles and water.

Properties	Base Fluid	Nanoparticles	
	Water	Al$_2$O$_3$	Cu
ρ (kg/m^3)	997.1	3970	8933
$\beta \times 10^{-5}$ (1/K)	21	0.85	1.67
C_p (J/kgK)	4179	765	385
k (W/mK)	0.613	40	400
Prandtl number, Pr	6.2		

Table 2. Thermophysical properties of nanofluid and hybrid nanofluid.

Properties	Nanofluid	Hybrid Nanofluid
Dynamic viscosity	$\mu_{nf} = \frac{\mu_f}{(1-\varphi_1)^{2.5}}$	$\mu_{hnf} = \frac{\mu_f}{(1-\varphi_1)^{2.5}(1-\varphi_2)^{2.5}}$
Density	$\rho_{nf} = (1-\varphi_1)\rho_f + \varphi_1\rho_{n1}$	$\rho_{hnf} = (1-\varphi_2)[(1-\varphi_1)\rho_f + \varphi_1\rho_{n1}] + \varphi_2\rho_{n2}$
Thermal expansion	$(\rho\beta)_{nf} = (1-\varphi_1)(\rho\beta)_f + \varphi_1(\rho\beta)_{n1}$	$(\rho\beta)_{hnf} = (1-\varphi_2)[(1-\varphi_1)(\rho\beta)_f + \varphi_1(\rho\beta)_{n1}] + \varphi_2(\rho\beta)_{n2}$
Heat capacity	$(\rho C_p)_{nf} = (1-\varphi_1)(\rho C_p)_f + \varphi_1(\rho C_p)_{n1}$	$(\rho C_p)_{hnf} = (1-\varphi_2)[(1-\varphi_1)(\rho C_p)_f + \varphi_1(\rho C_p)_{n1}] + \varphi_2(\rho C_p)_{n2}$
Thermal conductivity	$\frac{k_{nf}}{k_f} = \frac{k_{n1}+2k_f-2\varphi_1(k_f-k_{n1})}{k_{n1}+2k_f+\varphi_1(k_f-k_{n1})}$	$\frac{k_{hnf}}{k_{nf}} = \frac{k_{n2}+2k_{nf}-2\varphi_2(k_{nf}-k_{n2})}{k_{n2}+2k_{nf}+\varphi_2(k_{nf}-k_{n2})}$ where $\frac{k_{nf}}{k_f} = \frac{k_{n1}+2k_f-2\varphi_1(k_f-k_{n1})}{k_{n1}+2k_f+\varphi_1(k_f-k_{n1})}$

Furthermore, the continuity equation, i.e., Equation (1), is identically satisfied. Now, Equations (2) and (6) respectively reduce to:

$$\frac{\mu_{hnf}/\mu_f}{\rho_{hnf}/\rho_f}f''' + ff'' - f'^2 + 1 + \frac{(\rho\beta)_{hnf}/(\rho\beta)_f}{\rho_{hnf}/\rho_f}\lambda\theta = 0 \quad (9)$$

$$\frac{1}{\Pr}\frac{1}{(\rho C_p)_{hnf}/(\rho C_p)_f}\left(\frac{k_{hnf}}{k_f}+\frac{4}{3}R\right)\theta''+f\theta'-f'\theta=0 \qquad (10)$$

subject to the boundary conditions:

$$f(0)=0, \quad f'(0)=0, \quad \theta(0)=1,$$
$$f'(\infty)=1, \quad \theta(\infty)=0 \qquad (11)$$

where $(')$ represents the differentiation with respect to η, Pr is the Prandtl number, R and λ signify the radiation and the mixed convection parameters, given by:

$$\Pr=\frac{(\mu C_p)_f}{k_f}, \quad R=\frac{4\sigma^* T_\infty^3}{k^* k_f}, \quad \lambda=\frac{g\beta_f b}{a^2}=\frac{Gr_x}{Re_x^2} \qquad (12)$$

Further, $Gr_x=g\beta_f(T_w-T_\infty)x^3/\nu_f^2$ corresponds to the local Grashof number and $Re_x=ax^2/\nu_f$ stands for the local Reynold's number. Note that $\lambda<0$ signifies the opposing and $\lambda>0$ signifies the assisting flows, while the forced convection flow (no buoyancy effects) is given by $\lambda=0$.

The skin friction coefficient C_f and the local Nusselt number Nu_x are defined as [43]:

$$C_f=\frac{\mu_{hnf}}{\rho_f U^2}\left(\frac{\partial u}{\partial y}\right)_{y=0}, \quad Nu_x=\frac{x}{k_f(T_w-T_\infty)}\left(-k_{hnf}\left(\frac{\partial T}{\partial y}\right)_{y=0}+(q_r)_{y=0}\right) \qquad (13)$$

By employing Equation (7), one gets:

$$Re_x^{1/2}C_f=\frac{\mu_{hnf}}{\mu_f}f''(0), \quad Re_x^{-1/2}Nu_x=-\left(\frac{k_{hnf}}{k_f}+\frac{4}{3}R\right)\theta'(0) \qquad (14)$$

3. Stability Analysis

The temporal stability of the dual solutions as time evolves is studied. This analysis was first introduced by Merkin [44] and then followed by Weidman et al. [45] Firstly, consider the new variables as follows:

$$\psi=\sqrt{a\nu_f}xf(\eta,\tau), \quad \theta(\eta,\tau)=\frac{T-T_\infty}{T_w-T_\infty}, \quad \eta=y\sqrt{\frac{a}{\nu_f}}, \tau=at \qquad (15)$$

Now, the unsteady form of Equations (2) and (3) are considered, while Equation (1) remains unchanged. On using (15), one obtains:

$$\frac{\mu_{hnf}/\mu_f}{\rho_{hnf}/\rho_f}\frac{\partial^3 f}{\partial \eta^3}+f\frac{\partial^2 f}{\partial \eta^2}-\left(\frac{\partial f}{\partial \eta}\right)^2+1+\frac{(\rho\beta)_{hnf}/(\rho\beta)_f}{\rho_{hnf}/\rho_f}\lambda\theta-\frac{\partial^2 f}{\partial \eta \partial \tau}=0 \qquad (16)$$

$$\frac{1}{\Pr}\frac{1}{(\rho C_p)_{hnf}/(\rho C_p)_f}\left(\frac{k_{hnf}}{k_f}+\frac{4}{3}R\right)\frac{\partial^2\theta}{\partial\eta^2}+f\frac{\partial\theta}{\partial\eta}-\frac{\partial f}{\partial\eta}\theta-\frac{\partial\theta}{\partial\tau}=0 \qquad (17)$$

subject to:

$$f(0,\tau)=0, \quad \frac{\partial f}{\partial \eta}(0,\tau)=0, \quad \theta(0,\tau)=1,$$
$$\frac{\partial f}{\partial \eta}(\infty,\tau)=1, \quad \theta(\infty,\tau)=0 \qquad (18)$$

Then, consider the following perturbation functions [45]:

$$f(\eta,\tau)=f_0(\eta)+e^{-\gamma\tau}F(\eta), \quad \theta(\eta,\tau)=\theta_0(\eta)+e^{-\gamma\tau}G(\eta) \qquad (19)$$

Here, Equation (19) is introduced to apply a small disturbance on the steady solution $f=f_0(\eta)$ and $\theta=\theta_0(\eta)$ of Equations (9)–(11). The functions $F(\eta)$ and $G(\eta)$ in

Equation (19) are relatively small compared to $f_0(\eta)$ and $\theta_0(\eta)$. The sign (positive or negative) of the eigenvalue γ determines the stability of the solutions. By employing (19), Equations (16)–(18) become:

$$\frac{\mu_{hnf}/\mu_f}{\rho_{hnf}/\rho_f} F''' + f_0 F'' + f_0'' F - 2 f_0' F' + \frac{(\rho\beta)_{hnf}/(\rho\beta)_f}{\rho_{hnf}/\rho_f} \lambda G + \gamma F' = 0 \qquad (20)$$

$$\frac{1}{\Pr} \frac{1}{(\rho C_p)_{hnf}/(\rho C_p)_f} \left(\frac{k_{hnf}}{k_f} + \frac{4}{3} R \right) G'' + f_0 G' + \theta_0' F - f_0' G - \theta_0 F' + \gamma G = 0 \qquad (21)$$

subject to:
$$F(0) = 0, \quad F'(0) = 0, \quad G(0) = 0,$$
$$F'(\infty) = 0, \quad G(\infty) = 0 \qquad (22)$$

Without loss of generality we set $F''(0) = 1$ [46] to get the eigenvalues γ in Equations (20) and (21). The stability of the solutions as time evolves is determined by examining the values of the smallest eigenvalue that was obtained. As time passes, there is an initial decay of disturbance if γ is positive (see Equation (19)), and thus the solution is stable and physically reliable in the long run. On the other hand, if γ is negative, there is an initial growth of disturbance, hence the solution is unstable.

4. Results and Discussion

Equations (9)–(11) were solved numerically by utilising the boundary value problem solver (bvp4c) in MATLAB software, which employs the 3-stage Lobatto IIIa formula [47]. This is a collocation formula and provides a continuous solution with fourth-order accuracy. The effectiveness of this solver ultimately counts on our ability to provide the algorithm with an initial guess for the solution. Moreover, the suitable value of the boundary layer thickness must be chosen depending on the values of the parameters applied. To solve this boundary value problem, it is necessary to first reduce the equations to a system of first-order ordinary differential equations. The effects of the physical parameters such as Al_2O_3 (φ_1) and Cu (φ_2) nanoparticles volume fractions, the Prandtl number Pr, the radiation parameter R, and the mixed convection parameter λ on the flow behaviour are examined.

The values of the skin friction coefficient $f''(0)$ and the local Nusselt number $-\theta'(0)$ for several values of Pr when $R = 0$, $\lambda = 1$, and $\varphi_1 = \varphi_2 = 0$ (regular fluid) are compared with published results of Ishak et al. [5], as presented in Table 3. It should be mentioned that Ishak et al. [5] solved their problem by the Keller-box method. Meanwhile, the boundary value problem solver (bvp4c) is employed in this study. It is found that the results are in excellent agreement. This gives confidence to the validity and accuracy of the numerical results for other values of parameters. Besides, the values of $f''(0)$ show a decreasing behaviour, while the values of $-\theta'(0)$ increase for larger Pr. Additionally, Table 4 describes the values of $Re_x^{1/2} C_f$ and $Re_x^{-1/2} Nu_x$ for Cu/water nanofluid when $\varphi_1 = R = 0$ and $Pr = 6.2$ with different values of λ and φ_2. Here, we note that the values of both $Re_x^{1/2} C_f$ and $Re_x^{-1/2} Nu_x$ increase with the increasing of λ and φ_2. Besides, dual solutions are found for opposing ($\lambda = -1$) and assisting ($\lambda = 1$) flows, whereas the unique solution is obtained for $\lambda = 0$ (force convection flow). Furthermore, the values of $Re_x^{1/2} C_f$ for $\lambda = 0$ provided in the same table are compared with those of Bachok et al. [14], and the results are in excellent agreement, which thus gives confidence to the results for other values of λ.

Table 3. Values of $f''(0)$ and $-\theta'(0)$ for different values of Pr when $\varphi_1 = \varphi_2 = 0$ (regular fluid), $R = 0$, and $\lambda = 1$.

Pr	$f''(0)$		$-\theta'(0)$	
	Ishak et al. [5]	Present Results	Ishak et al. [5]	Present Results
0.7	1.7063 [1.2387]	1.70632 [1.23873]	0.7641 [1.0226]	0.76406 [1.02263]
1	1.6754 [1.1332]	1.67544 [1.13319]	0.8708 [1.1691]	0.87078 [1.16913]
6.2		1.52677 [0.61317]		1.65242 [2.13399]
7	1.5179 [0.5824]	1.51791 [0.58240]	1.7224 [2.2192]	1.72238 [2.21919]
10	1.4928 [0.4958]	1.49284 [0.49578]	1.9446 [2.4940]	1.94462 [2.49403]
20	1.4485 [0.3436]	1.44848 [0.34364]	2.4576 [3.1646]	2.45759 [3.16461]

Results in "[ ]" are the lower branch (second) solutions.

Table 4. Values of $Re_x^{1/2} C_f$ and $Re_x^{-1/2} Nu_x$ for Cu/water nanofluid when $\varphi_1 = R = 0$ and Pr = 6.2 under various values of λ and φ_2.

λ	φ_2	$Re_x^{1/2} C_f$		$Re_x^{-1/2} Nu_x$
		Bachok et al. [14]	Present Results	Present Results
−1	0.1		1.5811 [−0.1602]	1.8967 [−2.3965]
	0.2		2.3161 [0.1908]	2.2872 [−3.8078]
0	0.1	1.8843	1.8843	1.9692
	0.2	2.6226	2.6227	2.3494
1	0.1		2.1725 [0.8884]	2.0336 [3.7324]
	0.2		2.9183 [1.2445]	2.4064 [5.7802]

Results in "[ ]" are the lower branch (second) solutions.

Moreover, Table 5 shows the effect of λ, R and φ_2 on $Re_x^{1/2} C_f$ and $Re_x^{-1/2} Nu_x$ when Pr = 6.2 for nanofluid (Cu/water) and hybrid nanofluid (Cu-Al$_2$O$_3$/water). For the first solutions, we found that the values of $Re_x^{1/2} C_f$ are accelerated with the increasing of λ and φ_2; however, they are decelerated with R. Besides, the values of $Re_x^{-1/2} Nu_x$ enhance with increasing values of these parameters. The local Nusselt number $Re_x^{-1/2} Nu_x$ enhance up to 32.37% for Cu-Al$_2$O$_3$/water ($\varphi_1 = 0.1$, $\varphi_2 = 0.04$) compared to the regular fluid ($\varphi_1 = \varphi_2 = 0$) when $\lambda = -1$, $R = 0$, and Pr = 6.2. Meanwhile, the values of $Re_x^{-1/2} Nu_x$ are prominent for larger radiation ($R = 1$) with 49.63% enhancement compared to the non-radiant case ($R = 0$) when $\lambda = -1$, $\varphi_1 = 0.1$, $\varphi_2 = 0.04$, and Pr = 6.2. Moreover, the rise in λ from −1 to 1 contributes to the increment in the values of $Re_x^{-1/2} Nu_x$ up to 8.66% when $R = 1$, $\varphi_1 = 0.1$, $\varphi_2 = 0.04$, and Pr = 6.2.

Table 5. Values of $Re_x^{1/2} C_f$ and $Re_x^{-1/2} Nu_x$ when Pr = 6.2 for different physical parameters.

λ	R	φ_2	Cu/Water ($\varphi_1 = 0$)		Cu-Al$_2$O$_3$/Water ($\varphi_1 = 0.1$)	
			$Re_x^{1/2} C_f$	$Re_x^{-1/2} Nu_x$	$Re_x^{1/2} C_f$	$Re_x^{-1/2} Nu_x$
−1	0	0	0.9131 [−0.3719]	1.4779 [−1.1835]	1.2896 [−0.3019]	1.7766 [−1.8984]
		0.02	1.0475 [−0.3432]	1.5672 [−1.4219]	1.4271 [−0.2546]	1.8673 [−2.1625]
		0.04	1.1801 [−0.3071]	1.6528 [−1.6605]	1.5656 [−0.2208]	1.9563 [−2.4313]
	1	0	0.8342 [−0.3440]	2.5034 [−1.5160]	1.2300 [−0.2235]	2.7541 [−2.3870]
		0.02	0.9755 [−0.2936]	2.6044 [−1.8351]	1.3717 [−0.1513]	2.8430 [−2.6909]
		0.04	1.1136 [−0.2334]	2.6950 [−2.1359]	1.5137 [−0.0716]	2.9272 [−2.9891]
1	0	0	1.5268 [0.6132]	1.6524 [2.1340]	1.8958 [0.7439]	1.9328 [3.0137]
		0.02	1.6524 [0.6617]	1.7309 [2.4197]	2.0304 [0.8031]	2.0173 [3.3718]
		0.04	1.7789 [0.7136]	1.8079 [2.7224]	2.1675 [0.8656]	2.1010 [3.7485]
	1	0	1.5928 [0.8445]	2.8771 [3.7973]	1.9481 [0.9837]	3.0441 [4.7696]
		0.02	1.7140 [0.9041]	2.9422 [4.1738]	2.0797 [1.0510]	3.1132 [5.2068]
		0.04	1.8367 [0.9661]	3.0043 [4.5633]	2.2140 [1.1210]	3.1807 [5.6584]

Results in "[ ]" are the lower branch (second) solutions.

The variations of the skin friction coefficient $Re_x^{1/2}C_f$ and the local Nusselt number $Re_x^{-1/2}Nu_x$ against λ for several values of φ_2 and R are illustrated in Figures 2–5. The dual solutions of Equations (9)–(11) are possible for both assisting ($\lambda > 0$) and opposing ($\lambda < 0$) flows. The flow is accelerated for $\lambda > 0$ because there is a favourable pressure gradient induced by the buoyancy forces, which results in larger heat transfer and skin friction coefficients rather than the case of $\lambda = 0$ (non-buoyant case). We note that the separation of the boundary layer occurs when $\lambda < 0$. The dual solutions happen for $\lambda > \lambda_c$ and no solution for $\lambda < \lambda_c$. The curve terminates at $\lambda = \lambda_c$ (critical value) and this point is known as the bifurcation point of the solutions. Separately, Figures 2 and 3 display the variations of $Re_x^{1/2}C_f$ and $Re_x^{-1/2}Nu_x$ against λ for different values of φ_2 when $\text{Pr} = 6.2$ and $\varphi_1 = 0.1$ in the absence of R. It is observed that the values of $Re_x^{1/2}C_f$ and $Re_x^{-1/2}Nu_x$ enhance with the rising of φ_2. Moreover, it is noticed that the boundary layer separation is delayed with the added hybrid nanoparticles. The critical values are $\lambda_c = -4.6983, -5.1215,$ and -5.5404 for $\varphi_2 = 0, 0.02$ and 0.04, respectively. Apart from that, the variations of $Re_x^{1/2}C_f$ and $Re_x^{-1/2}Nu_x$ with λ for different values of R when $\text{Pr} = 6.2$, $\varphi_1 = 0.1$, and $\varphi_2 = 0.04$ are illustrated in Figures 4 and 5. In the presence of R, we found that the skin friction coefficient $Re_x^{1/2}C_f$ decreases for $\lambda < 0$ but increases for $\lambda > 0$, whereas the local Nusselt number $Re_x^{-1/2}Nu_x$ enhances for both cases. Besides, we notice that the domain of λ for the existence of the dual solutions decreases for larger values of R where the critical values of λ slightly increase. Note that the critical values λ_c for $R = 0, 1,$ and 2 are $\lambda_c = -5.5404, -4.7843,$ and -4.4030, respectively. It is observed in Figures 3 and 5, the second solutions of $Re_x^{-1/2}Nu_x$ are boundless as $\lambda \to 0^-$ and as $\lambda \to 0^+$.

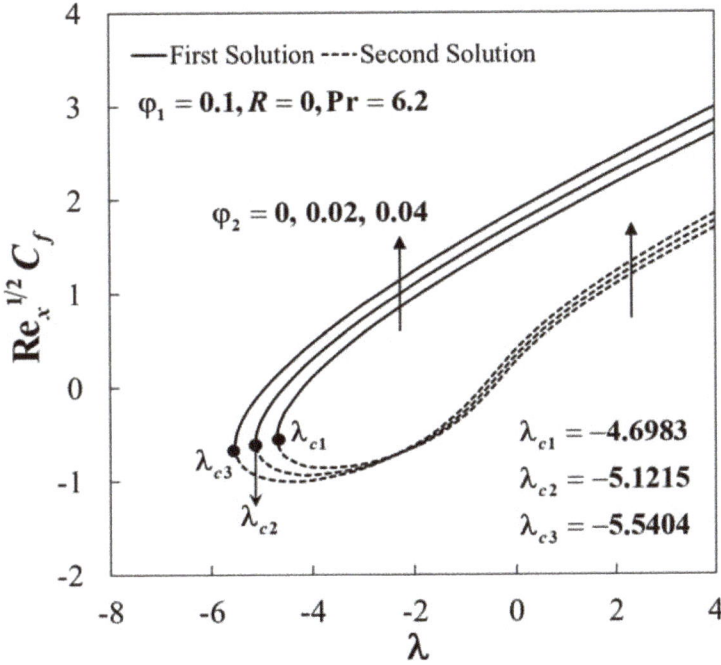

Figure 2. The variations of the skin friction coefficient $Re_x^{1/2}C_f$ against the mixed convection parameter λ for different values of the Cu nanoparticle volume fractions φ_2.

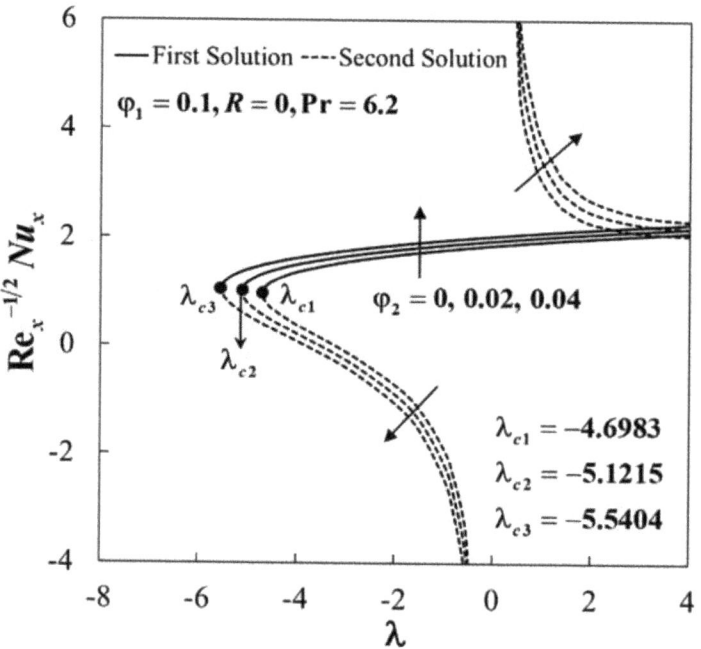

Figure 3. The variations of the local Nusselt number $Re_x^{-1/2} Nu_x$ against the mixed convection parameter λ for different values of the Cu nanoparticles volume fractions φ_2.

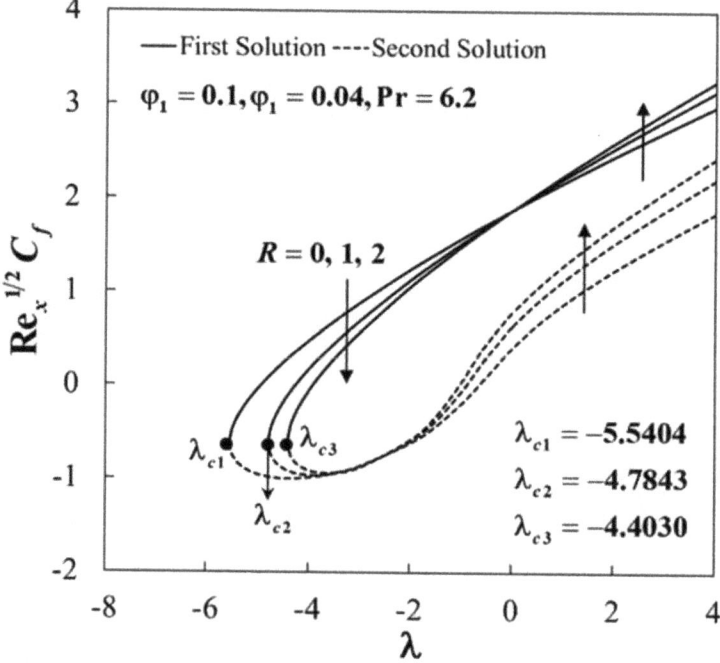

Figure 4. The variations of the skin friction coefficient $Re_x^{1/2} C_f$ against the mixed convection parameter λ for different values of the radiation parameter R.

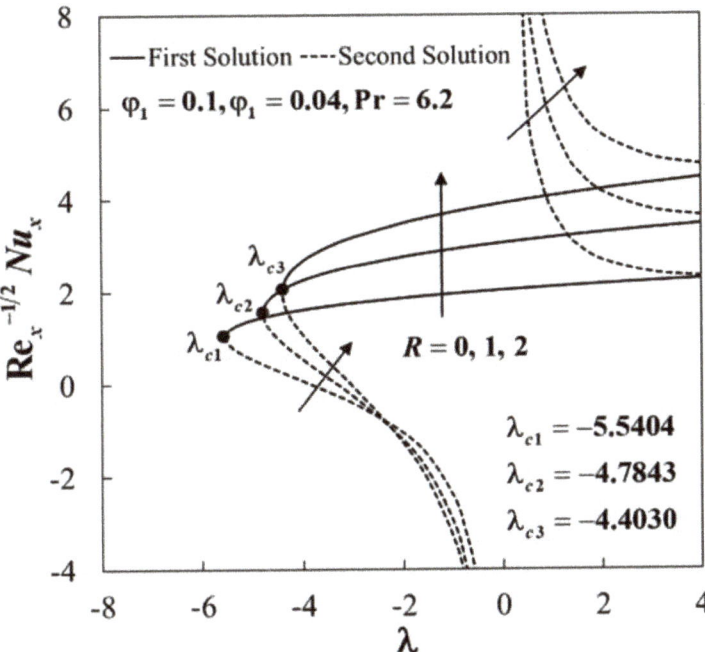

Figure 5. The variations of the local Nusselt number $Re_x^{-1/2}Nu_x$ against the mixed convection parameter λ for different values of the radiation parameter R.

The impact of φ_2 and R on the velocity $f'(\eta)$ and the temperature $\theta(\eta)$ profiles for the case of the opposing ($\lambda = -1$) and assisting ($\lambda = 1$) flows are presented in Figures 6–13. There exist dual solutions for $f'(\eta)$ and $\theta(\eta)$ which satisfy the infinity boundary conditions (11) asymptotically. The rising of φ_2 leads to an upsurge in the values of $f'(\eta)$ and $\theta(\eta)$ on the first solutions for both cases when $Pr = 6.2$, $\varphi_1 = 0.1$, and $R = 0$ as shown in Figures 6–9. Meanwhile, the velocity $f'(\eta)$ decreases when $\lambda = -1$ but increases when $\lambda = 1$ on the first solutions for larger values of R when $Pr = 6.2$, $\varphi_1 = 0.1$, and $\varphi_2 = 0.04$. The effect of R is to increase the temperature $\theta(\eta)$ inside the boundary layer for both cases as displayed in Figures 10–13. The radiation is dominant over conduction for larger values of R, causing a rise in the fluid temperature. It is also noticed that the solutions of the lower branch for the velocity have negative values ($f'(\eta) < 0$), which implies that the reverse flow occurs away from the wall, and these behaviors are displayed in Figures 6, 8, 10 and 12. The behaviors of $\theta(\eta)$ with different values of φ_2 and R for both cases are given in Figures 7, 9, 11 and 13. The overshoot of the temperature $\theta(\eta)$ near the wall is observed when $\lambda = -1$, and $\theta(\eta) < 0$ when $\lambda = 1$ for the second solution.

The variations of γ against λ when $Pr = 6.2$, $\varphi_1 = 0.1$, $\varphi_2 = 0.04$ and $R = 1$ are described in Figure 14. For positive values of γ, it is noted that $e^{-\gamma\tau} \to 0$ as time evolves ($\tau \to \infty$). In the meantime, for the negative value of γ, $e^{-\gamma\tau} \to \infty$. These behaviors show that the first solution is stable and physically reliable, while the second solution is unstable in the long run.

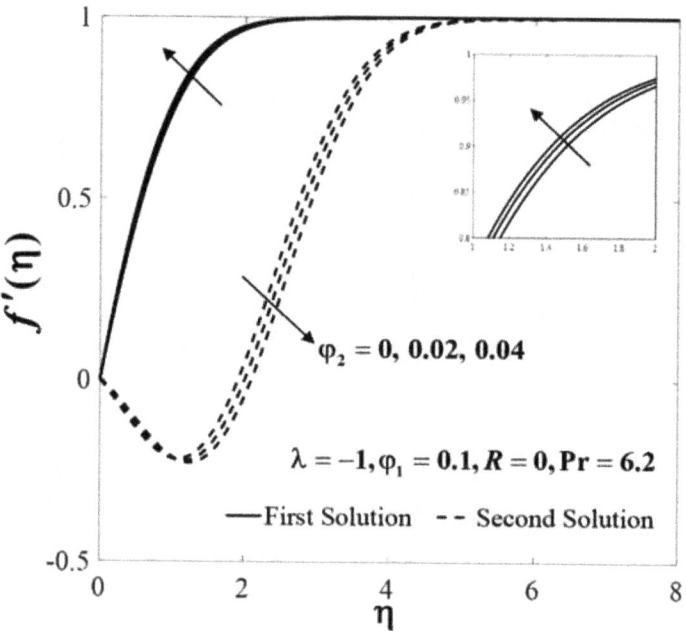

Figure 6. The velocity profiles $f'(\eta)$ for different values of the Cu nanoparticles volume fractions φ_2 when $\lambda = -1$ (opposing flow).

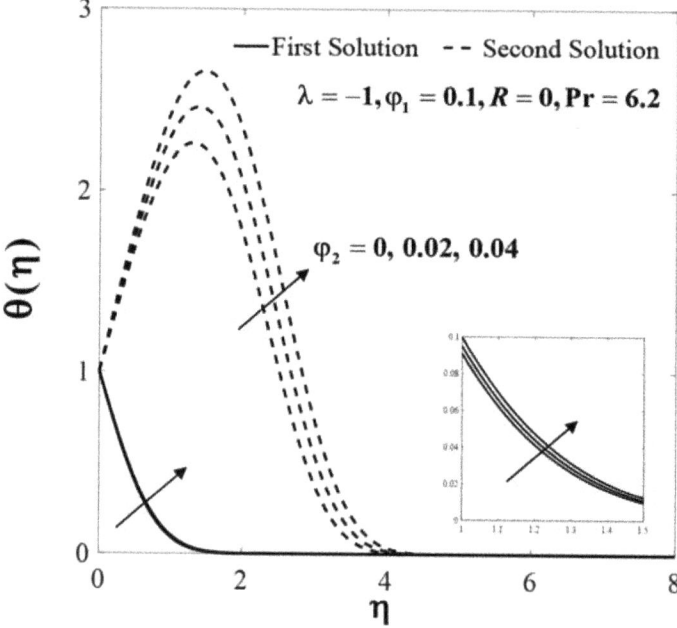

Figure 7. The temperature profiles $\theta(\eta)$ for different values of the Cu nanoparticles volume fractions φ_2 when $\lambda = -1$ (opposing flow).

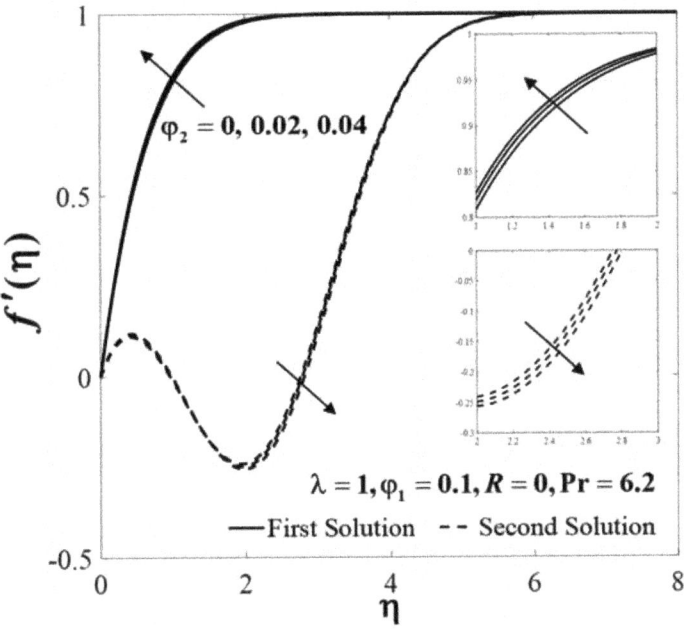

Figure 8. The velocity profiles $f'(\eta)$ for different values of the Cu nanoparticles volume fractions φ_2 when $\lambda = 1$ (assisting flow).

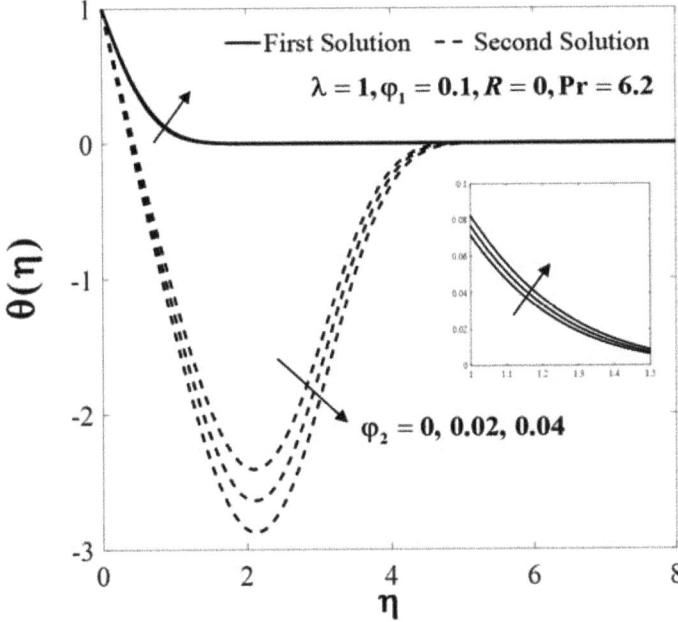

Figure 9. The temperature profiles $\theta(\eta)$ for different values of the Cu nanoparticle volume fractions φ_2 when $\lambda = 1$ (assisting flow).

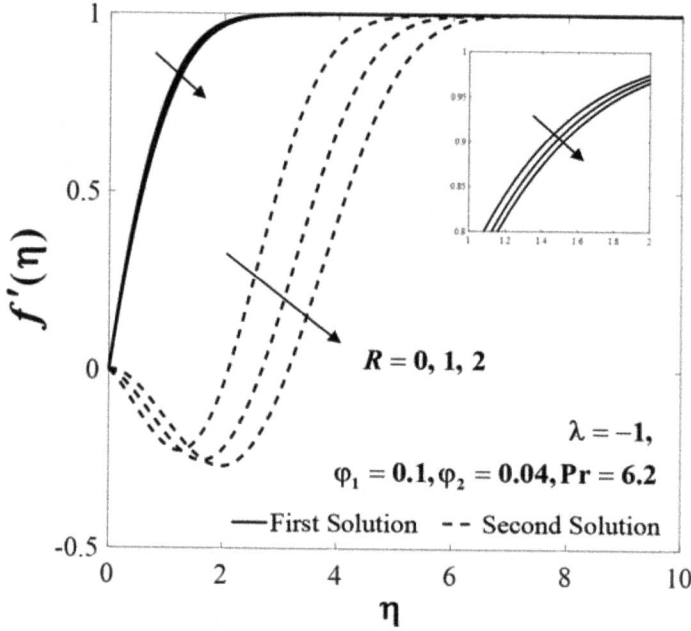

Figure 10. The velocity profiles $f'(\eta)$ for different values of the radiation parameter R when $\lambda = -1$ (opposing flow).

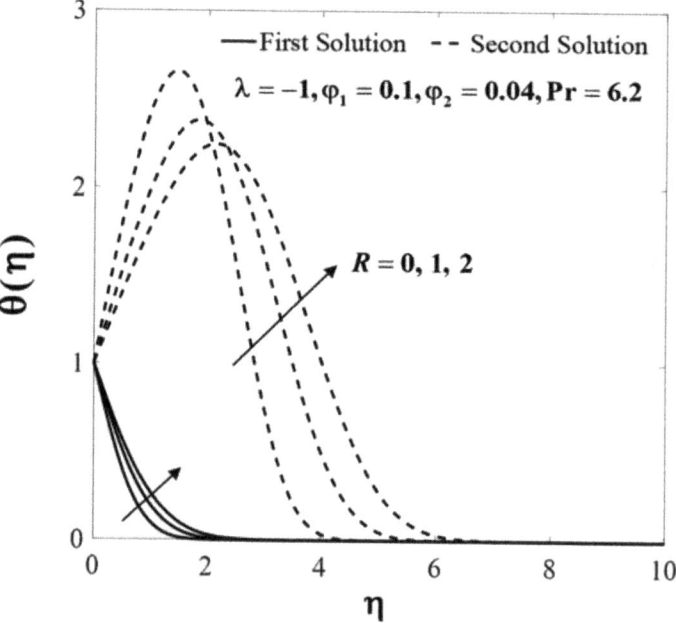

Figure 11. The temperature profiles $\theta(\eta)$ for different values of the radiation parameter R when $\lambda = -1$ (opposing flow).

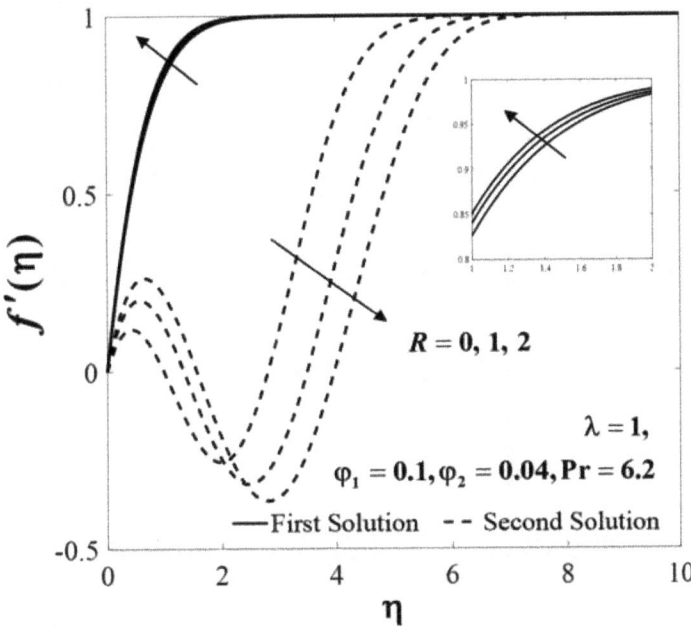

Figure 12. The velocity profiles $f'(\eta)$ for different values of the radiation parameter R when $\lambda = 1$ (assisting flow).

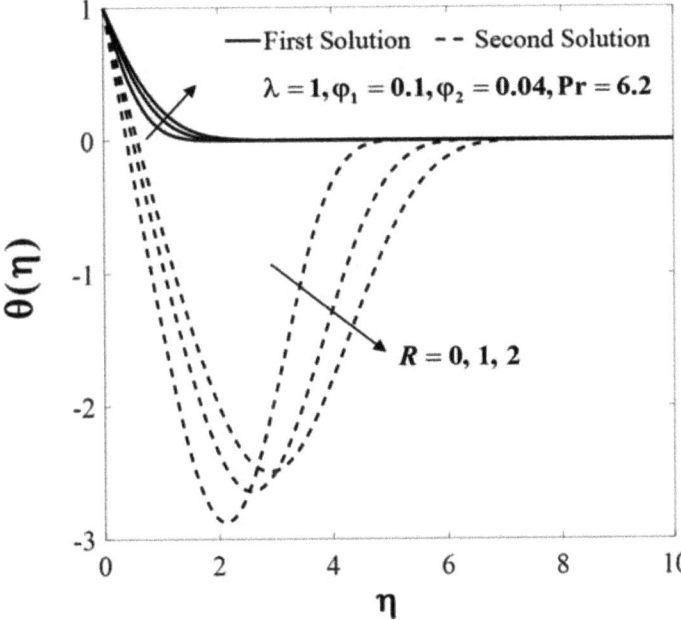

Figure 13. The temperature profiles $\theta(\eta)$ for different values of the radiation parameter R when $\lambda = 1$ (assisting flow).

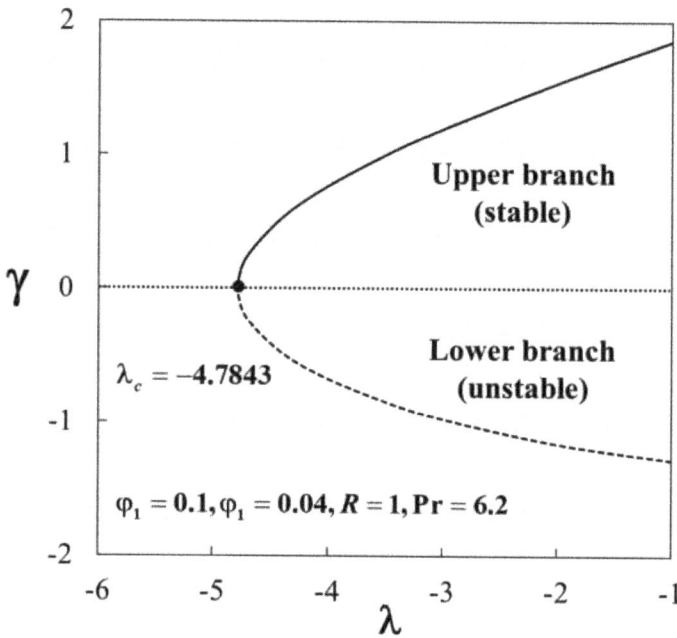

Figure 14. Variations of the smallest eigenvalues γ against the mixed convection parameter λ.

5. Conclusions

The stagnation point flow towards a vertical plate in a hybrid nanofluid with thermal radiation was examined in the present paper. Findings revealed that dual solutions appeared for both assisting ($\lambda > 0$) and opposing ($\lambda < 0$) flows. The dual solutions were found for $\lambda > \lambda_c$ and no solution for $\lambda < \lambda_c$, while the solutions bifurcated at $\lambda = \lambda_c$. In addition, the consequence of the copper nanoparticle volume fractions φ_2 is to enhance the skin friction coefficient $Re_x^{1/2} C_f$ and the local Nusselt number $Re_x^{-1/2} Nu_x$ for both cases. However, the values of $Re_x^{1/2} C_f$ decreased for $\lambda < 0$, but increased for $\lambda > 0$, whereas the values of $Re_x^{-1/2} Nu_x$ were intensified for both cases in the presence of the radiation parameter R. From these findings, the increments of the local Nusselt number $Re_x^{-1/2} Nu_x$ are observed in the range of 8.66% to 49.63% for the pertinent physical parameters considered. Besides, we noticed that the domain of the mixed convection parameter λ where the dual solutions are in existence decreased for larger R. Further, the first solution of the velocity $f'(\eta)$ and the temperature $\theta(\eta)$ profiles enlarged with the increase of the copper nanoparticles volume fractions φ_2. Moreover, the effect of the radiation parameter R is to increase the temperature $\theta(\eta)$ inside the boundary layer for both cases. Lastly, it was discovered that between the two solutions, the solution with lower boundary layer thickness is stable and thus physically reliable in the long run.

Author Contributions: Conceptualization, I.P.; funding acquisition, A.I.; methodology, I.W.; Project administration, A.I.; supervision, A.I. and I.P.; validation, I.P.; writing—original draft, I.W.; writing—review and editing, A.I., I.P. All authors have read and agreed to the published version of the manuscript.

Funding: This research was funded by Universiti Kebangsaan Malaysia (Project Code: DIP-2020-001).

Acknowledgments: The authors would like to thank the anonymous reviewers for their constructive comments and suggestions. The financial supports received from the Universiti Kebangsaan Malaysia (Project Code: DIP-2020-001) and the Universiti Teknikal Malaysia Melaka are gratefully acknowledged.

Conflicts of Interest: The authors declare no conflict of interest.

References

1. Hiemenz, K. Die Grenzschicht an einem in den gleichförmigen Flüssigkeitsstrom eingetauchten geraden Kreiszylinder. *Dinglers Polytech. J.* **1911**, *326*, 321–410.
2. Homann, F. Der Einfluß großer Zähigkeit bei der Strömung um den Zylinder und um die Kugel. *Z. Angew. Math. Mech.* **1936**, *16*, 153–164. [CrossRef]
3. Chiam, T.C. Stagnation-point flow towards a stretching plate. *J. Phys. Soc. Jpn.* **1994**, *63*, 2443–2444. [CrossRef]
4. Merkin, J.H. Mixed convection boundary layer flow on a vertical surface in a saturated porous medium. *J. Eng. Math.* **1980**, *14*, 301–313. [CrossRef]
5. Ishak, A.; Nazar, R.; Arifin, N.M.; Pop, I. Dual solutions in mixed convection flow near a stagnation point on a vertical porous plate. *Int. J. Therm. Sci.* **2008**, *47*, 417–422. [CrossRef]
6. Subhashini, S.V.; Samuel, N.; Pop, I. Effects of buoyancy assisting and opposing flows on mixed convection boundary layer flow over a permeable vertical surface. *Int. Commun. Heat Mass Transf.* **2011**, *38*, 499–503. [CrossRef]
7. Roşca, A.V.; Roşca, N.C.; Pop, I. Note on dual solutions for the mixed convection boundary layer flow close to the lower stagnation point of a horizontal circular cylinder: Case of constant surface heat flux. *Sains Malays* **2014**, *43*, 1239–1247.
8. Khashi'ie, N.S.; Arifin, N.M.; Rashidi, M.M.; Hafidzuddin, E.H.; Wahi, N. Magnetohydrodynamics (MHD) stagnation point flow past a shrinking/stretching surface with double stratification effect in a porous medium. *J. Therm. Anal. Calorim.* **2019**, *8*, 1–14. [CrossRef]
9. Ali, F.M.; Naganthran, K.; Nazar, R.; Pop, I. MHD mixed convection boundary layer stagnation-point flow on a vertical surface with induced magnetic field. *Int. J. Numer. Methods Heat Fluid Flow* **2020**, *30*, 4697–4710. [CrossRef]
10. Choi, S.U.S.; Eastman, J.A. *Enhancing Thermal Conductivity of Fluids with Nanoparticles*; Argonne National Lab: Lemont, IL, USA, 1995; Volume 66, pp. 99–105.
11. Khanafer, K.; Vafai, K.; Lightstone, M. Buoyancy-driven heat transfer enhancement in a two-dimensional enclosure utilizing nanofluids. *Int. J. Heat Mass Transf.* **2003**, *46*, 3639–3653. [CrossRef]
12. Tiwari, R.K.; Das, M.K. Heat transfer augmentation in a two-sided lid-driven differentially heated square cavity utilizing nanofluids. *Int. J. Heat Mass Transf.* **2007**, *50*, 2002–2018. [CrossRef]
13. Oztop, H.F.; Abu-Nada, E. Numerical study of natural convection in partially heated rectangular enclosures filled with nanofluids. *Int. J. Heat Fluid Flow* **2008**, *29*, 1326–1336. [CrossRef]
14. Bachok, N.; Ishak, A.; Pop, I. Stagnation-point flow over a stretching/shrinking sheet in a nanofluid. *Nanoscale Res. Lett.* **2011**, *6*, 623. [CrossRef]
15. Yacob, N.A.; Ishak, A.; Pop, I.; Vajravelu, K. Boundary layer flow past a stretching/shrinking surface beneath an external uniform shear flow with a convective surface boundary condition in a nanofluid. *Nanoscale Res. Lett.* **2011**, *6*, 314. [CrossRef] [PubMed]
16. Waini, I.; Ishak, A.; Pop, I. Dufour and Soret effects on Al_2O_3-water nanofluid flow over a moving thin needle: Tiwari and Das model. *Int. J. Numer. Methods Heat Fluid Flow* **2020**, in press. [CrossRef]
17. Turcu, R.; Darabont, A.; Nan, A.; Aldea, N.; Macovei, D.; Bica, D.; Vekas, L.; Pana, O.; Soran, M.L.; Koos, A.A.; et al. New polypyrrole-multiwall carbon nanotubes hybrid materials. *J. Optoelectron. Adv. Mater.* **2006**, *8*, 643–647.
18. Jana, S.; Salehi-Khojin, A.; Zhong, W.H. Enhancement of fluid thermal conductivity by the addition of single and hybrid nano-additives. *Thermochim. Acta* **2007**, *462*, 45–55. [CrossRef]
19. Suresh, S.; Venkitaraj, K.P.; Selvakumar, P.; Chandrasekar, M. Synthesis of Al_2O_3-Cu/water hybrid nanofluids using two step method and its thermo physical properties. *Colloids Surfaces A Physicochem. Eng. Asp.* **2011**, *388*, 41–48. [CrossRef]
20. Devi, S.P.A.; Devi, S.S.U. Numerical investigation of hydromagnetic hybrid Cu-Al_2O_3/water nanofluid flow over a permeable stretching sheet with suction. *Int. J. Nonlinear Sci. Numer. Simul.* **2016**, *17*, 249–257. [CrossRef]
21. Waini, I.; Ishak, A.; Pop, I. Mixed convection flow over an exponentially stretching/shrinking vertical surface in a hybrid nanofluid. *Alex. Eng. J.* **2020**, *59*, 1881–1891. [CrossRef]
22. Waini, I.; Ishak, A.; Pop, I. Hybrid nanofluid flow past a permeable moving thin needle. *Mathematics* **2020**, *8*, 612. [CrossRef]
23. Waini, I.; Ishak, A.; Pop, I. Squeezed hybrid nanofluid flow over a permeable sensor surface. *Mathematics* **2020**, *8*, 898. [CrossRef]
24. Waini, I.; Ishak, A.; Pop, I. Hiemenz flow over a shrinking sheet in a hybrid nanofluid. *Results Phys.* **2020**, *19*, 103351. [CrossRef]
25. Waini, I.; Ishak, A.; Pop, I. Hybrid nanofluid flow on a shrinking cylinder with prescribed surface heat flux. *Int. J. Numer. Methods Heat Fluid Flow* **2020**, in press. [CrossRef]
26. Waini, I.; Ishak, A.; Pop, I. Melting heat transfer of a hybrid nanofluid flow towards a stagnation point region with second-order slip. *Proc. Inst. Mech. Eng. Part E J. Process Mech. Eng.* **2020**, in press. [CrossRef]
27. Waini, I.; Ishak, A.; Pop, I. Unsteady hybrid nanofluid flow on a stagnation point of a permeable rigid surface. *ZAMM Z. Angew. Math. Mech.* **2020**, in press. [CrossRef]
28. Aly, E.H.; Pop, I. MHD flow and heat transfer over a permeable stretching/shrinking sheet in a hybrid nanofluid with a convective boundary condition. *Int. J. Numer. Methods Heat Fluid Flow* **2019**, *29*, 3012–3038. [CrossRef]
29. Khan, U.; Zaib, A.; Khan, I.; Baleanu, D.; Nisar, K.S. Enhanced heat transfer in moderately ionized liquid due to hybrid MoS_2/SiO_2 nanofluids exposed by nonlinear radiation: Stability analysis. *Crystals* **2020**, *10*, 142. [CrossRef]

30. Khan, U.; Zaib, A.; Khan, I.; Baleanu, D.; Sherif, E.S.M. Comparative investigation on MHD nonlinear radiative flow through a moving thin needle comprising two hybridized AA7075 and AA7072 alloys nanomaterials through binary chemical reaction with activation energy. *J. Mater. Res. Technol.* **2020**, *9*, 3817–3828. [CrossRef]
31. Khashi'ie, N.S.; Arifin, N.M.; Hafidzuddin, E.H.; Wahi, N. Thermally stratified flow of Cu-Al_2O_3/water hybrid nanofluid past a permeable stretching/shrinking circular cylinder. *J. Adv. Res. Fluid Mech. Therm. Sci.* **2019**, *63*, 154–163.
32. Waini, I.; Ishak, A.; Pop, I. Hybrid nanofluid flow towards a stagnation point on a stretching/shrinking cylinder. *Sci. Rep.* **2020**, *10*, 9296. [CrossRef] [PubMed]
33. Khashi'ie, N.S.; Arifin, N.M.; Wahi, N.; Pop, I.; Nazar, R.; Hafidzuddin, E.H. Thermal marangoni flow past a permeable stretching/shrinking sheet in a hybrid Cu-Al2O3/water nanofluid. *Sains Malays* **2020**, *49*, 211–222. [CrossRef]
34. Zainal, N.A.; Nazar, R.; Naganthran, K.; Pop, I. Unsteady three-dimensional MHD nonaxisymmetric Homann stagnation point flow of a hybrid nanofluid with stability analysis. *Mathematics* **2020**, *8*, 784. [CrossRef]
35. Zainal, N.A.; Nazar, R.; Naganthran, K.; Pop, I. Impact of anisotropic slip on the stagnation-point flow past a stretching/shrinking surface of the Al_2O_3-Cu/H_2O hybrid nanofluid. *Appl. Math. Mech.* **2020**, *41*, 1401–1416. [CrossRef]
36. Sarkar, J.; Ghosh, P.; Adil, A. A review on hybrid nanofluids: Recent research, development and applications. *Renew. Sustain. Energy Rev.* **2015**, *43*, 164–177. [CrossRef]
37. Sidik, N.A.C.; Adamu, I.M.; Jamil, M.M.; Kefayati, G.H.R.; Mamat, R.; Najafi, G. Recent progress on hybrid nanofluids in heat transfer applications: A comprehensive review. *Int. Commun. Heat Mass Transf.* **2016**, *78*, 68–79. [CrossRef]
38. Babu, J.A.R.; Kumar, K.K.; Rao, S.S. State-of-art review on hybrid nanofluids. *Renew. Sustain. Energy Rev.* **2017**, *77*, 551–565. [CrossRef]
39. Sajid, M.U.; Ali, H.M. Thermal conductivity of hybrid nanofluids: A critical review. *Int. J. Heat Mass Transf.* **2018**, *126*, 211–234. [CrossRef]
40. Huminic, G.; Huminic, A. Entropy generation of nanofluid and hybrid nanofluid flow in thermal systems: A review. *J. Mol. Liq.* **2020**, *302*, 112533. [CrossRef]
41. Yang, L.; Ji, W.; Mao, M.; Huang, J. An updated review on the properties, fabrication and application of hybrid-nanofluids along with their environmental effects. *J. Clean. Prod.* **2020**, *257*, 120408. [CrossRef]
42. Rosseland, S. *Astrophysik und Atom-Theoretische Grundlagen*; Springer: Berlin/Heidelberg, Germany, 1931.
43. Cortell, R. Heat and fluid flow due to non-linearly stretching surfaces. *Appl. Math. Comput.* **2011**, *217*, 7564–7572. [CrossRef]
44. Merkin, J.H. On dual solutions occurring in mixed convection in a porous medium. *J. Eng. Math.* **1986**, *20*, 171–179. [CrossRef]
45. Weidman, P.D.; Kubitschek, D.G.; Davis, A.M.J. The effect of transpiration on self-similar boundary layer flow over moving surfaces. *Int. J. Eng. Sci.* **2006**, *44*, 730–737. [CrossRef]
46. Harris, S.D.; Ingham, D.B.; Pop, I. Mixed convection boundary-layer flow near the stagnation point on a vertical surface in a porous medium: Brinkman model with slip. *Transp. Porous Media* **2009**, *77*, 267–285. [CrossRef]
47. Shampine, L.F.; Gladwell, I.; Thompson, S. *Solving ODEs with MATLAB*; Cambridge University Press: Cambridge, UK, 2003.

Article

Improving the Gridshells' Regularity by Using Evolutionary Techniques

Marjan Goodarzi [1], Ali Mohades [1],* and Majid Forghani-elahabad [2]

[1] Laboratory of Algorithms and Computational Geometry, Department of Mathematics and Computer Science, Amirkabir University of Technology (Tehran Polytechnic), Tehran 1591639675, Iran; marjangoodarzi@aut.ac.ir
[2] Center of Mathematics, Computing, and Cognition—Federal University of ABC, Santo André, SP 09210-580, Brazil; m.forghani@ufabc.edu.br
* Correspondence: mohades@aut.ac.ir; Tel./Fax: +98(21)-6454-5657

Abstract: Designing and optimizing gridshell structures have been very attractive problems in the last decades. In this work, two indexes are introduced as "length ratio" and "shape ratio" to measure the regularity of a gridshell and are compared to the existing indexes in the literature. Two evolutionary techniques, genetic algorithm (GA) and particle swarm optimization (PSO) method, are utilized to improve the gridshells' regularity by using the indexes. An approach is presented to generate the initial gridshells for a given surface in MATLAB. The two methods are implemented in MATLAB and compared on three benchmarks with different Gaussian curvatures. For each grid, both triangular and quadrangular meshes are generated. Experimental results show that the regularity of some gridshell is improved more than 50%, the regularity of quadrangular gridshells can be improved more than the regularity of triangular gridshells on the same surfaces, and there may be some relationship between Gaussian curvature of a surface and the improvement percentage of generated gridshells on it. Moreover, it is seen that PSO technique outperforms GA technique slightly in almost all the considered test problems. Finally, the Dolan–Moré performance profile is produced to compare the two methods according to running times.

Keywords: gridshell structures; shape ratio; length ratio; regularity; particle swarm optimization; genetic algorithm

1. Introduction

Gridshells which are also called lattice shells or reticulated shells are generally defined as structures with the shape and rigidity of a double curvature shell consisting of a grid not a continuous surface [1]. Although gridshells come to several forms, they are usually designed with triangular, quadrilateral, or hexagonal faces (or grid cells) [1–10]. Forming and optimizing gridshell structures have been very attractive problems in the past decades. Several approaches, such as inversion method [1], dynamic relaxation [2,4,11], force density method [3,12], and so forth [10,13], have been studied so far in the literature to address the problem of forming a grid shell structure. Moreover, various techniques from gradient-based to evolutionary methods have been employed for optimization of gridshells taking into account various aspects of a gridshell such as economic, structural, or aesthetic [11–18]. The focus of this work is on the optimization problem, and it is assumed that the initial forms of the desired gridshells are given.

Bouhaya et al. [1] coupled genetic algorithms with a geometric technique, which is called compass method, to present a novel approach for generating elastic gridshells on an imposed shape with boundary conditions. The authors used three benchmarks with different Gaussian curvature to illustrate the proposed technique. These benchmarks are also employed in the present work for generating the numerical results as they contain a variety of conditions with different Gaussian curvatures. Richardson et al. [11] presented a two-phase design technique. Using a multi-objective genetic algorithm, Winslow et al. [17]

established a design tool for synthesis of optimal gridshell structures taking into account two or more load cases such as wind load. Feng et al. [5] considered three categories of indexes including mechanical, geometry, and economic criteria for optimization of free-form cable-braced gridshells. Focusing on triangular gridshells and optimization over a free-form surface, Wang et al. [16] presented a framework to generate gridshells.

We note that usually the researchers have taken into account the structural aspects of gridshells in optimization phase and less attention has been heeded to improvement of the gridshells' regularity while the later can affect directly on the economy and aesthetic indexes. In fact, improving the regularity of a gridshell may lead to decreasing the number of different elements' types as well as enhancing the aesthetic aspect of the desired grids. Hence, in this work, improvement of the regularity of gridshells is considered as the main aim. To this end, two indexes are introduced to measure the regularity of a gridshell. The indexes are called "length ratio" and "shape ratio" and defined as the standard deviation of all the elements' lengths and all the inner angles in the gridshells' faces, respectively. There are a few studies in the literature which have worked on improving the regularity of gridshells. In fact, to the best of our knowledge, this is the first time that these two indexes are introduced for measuring the free-form gridshells' regularity. Considering the geodesic domes, Nooshin and his coworkers in [19] have proposed some measures for making the regularity of such structures quantifiable. Here, we compare our introduced indexes with the proposed ones in [19] on some benchmark for illustrating the practical efficiency of the introduced indexes in this work.

It is noted that although gradient-based techniques guarantee a superior convergence rate for the cases with a few number of design variables, in the cases with many design variables, generally, evolutionary methods work more appropriately. This is why we employ evolutionary techniques in this work. Among the evolutionary techniques, genetic algorithms (GAs) have been used the most in optimization of gridshells [1,3,5,8,10,11,13–18]. Another well-known evolutionary method is particle swarm optimization (PSO) to which less attention has been paid for improving the gridshell structures so far. However, it is a very powerful technique and has been applied to many other optimization problems [20–27]. Thus, these two techniques are considered in the present work.

We first present an approach, Algorithm 1, for generating initial triangular and quadrangular gridshells in MATLAB. Then, it is explained how the nodal positions in a given grid can be represented as birds (or particles) in PSO technique and as chromosomes in GA, and how these techniques can be used to improve the regularity of gridshells by bettering the nodal positions. We introduce two indexes to make the gridshells' regularity quantifiable and use them in the improvement process. Moreover, providing the numerical results, it is illustrated that our introduced indexes are practically more efficient than the existing indexes in the literature. In addition, the performances of GA and PSO techniques are compared through experimental results generated on eighteen test problems by which several interesting observations are obtained. Finally, we produce the Dolan and Moré performance profile [28] to have a more intuitive comparison between GA and PSO technique based on running times.

The rest of this work is organized as follows. Section 2 presents an algorithm to generate the initial gridshells for a given surface in MATLAB, explains briefly the GA and PSO techniques stage by stage, describes how these techniques can be used for improving the gridshells' regularity, and moreover provides the mathematical model for the problem. In Section 3, first our introduced indexes are compared to two existing ones in the literature on some benchmark, and then the performances of GA and PSO techniques are compared on three benchmarks. Finally, Section 4 provides the concluding remarks and some directions for the future works.

2. Main Block

Here, an approach, Algorithm 1, is presented to generate initial gridshells. Then, we describe the genetic algorithm (GA) and particle swarm optimization (PSO) technique briefly for making it more convenient to read this work without having previous knowledge of these methods. Next, two indexes are introduced for making the gridshells' regularity quantifiable, and finally the mathematical model of the problem is provided.

2.1. Generating An Initial Gridshell

There are several different approaches in the literature to generate an initial form of gridshell. Bouhaya et al. [1] used the compass method which is a geometric approach and allows creating a network of parallelograms on any surface. The authors then coupled the compass method with genetic algorithms to propose an optimization approach. In optimization of the cable-braced grid shells, Feng et al. [5] created the initial forms by translating the generatrix and directrix which allows the mesh to be parallelogram. The authors then proposed an optimization technique based on the usage of generatrix and directrix. Wang et al. [16] formed the surface model by using NURBS (non-uniform rational B-splines), then generated a set of uniformly distributed random points on the surface, and finally connected the points by using Delaunay-based triangularization to generate an initial triangular gridshell.

It is noted that usually the proposed optimization methods in the literature are based on some generating techniques for the initial form. However, the focus of this work is on the regularity improvement of a given gridshell without taking into account how the initial form is obtained, and hence the presented techniques here can be used for improving the regularity of any given initial form of a gridshell which is given as two sets. A set with the (Cartesian) coordinates of nodes (or vertices), which is denoted by V here, and another set which shows the faces in the grid and is denoted by F here. In fact, F is a matrix, each row of which states which vertices form the corresponding face (or grid cell). Having the set V of the initial gridshell, the proposed approach here can be employed to improve the regularity of the gridshell by bettering the positions of the nodes without making any change in the matrix F. Moreover, one can add some restrictions such as fixing the position of some nodes to the improvement process.

In this work, for convenience and not being involved in the complication of generating the initial gridshells which is not of our focus, it is assumed that the given gridshell is based on a surface $F(x, y, z) = 0$. In this way, to keep the nodes (vertices) on the desired surface, the improvement process is made on the first two coordinates of the vertices, i.e., x and y, and the third coordinate is always obtained by using the surface equation. Any way, we observe that the detailed approach here can be employed for any given initial gridshell.

Another aspect to notice is that although gridshells with the triangular to hexagonal grid cells have been studied in the literature, triangular grids are the most widely used in reality as they can describe any free-form shape [16]. In addition, as the triangular gridshells have less economic advantages in construction than the equivalent structures built of quadrilateral grids, many researchers have studied quadrangular gridshells [5]. As a result, one can say that triangular and quadrangular gridshells are the most important cases, and hence the focus of this work is on these two groups of gridshells.

Given a surface $F(x, y, z) = 0$, to build an initial gridshell, usually some uniformly distributed random points are generated on the surface and then those points are connected to construct the triangular or quadrangular grid cells. However, as our main object here is to improve the regularity of the initial gridshells, we consider the equidistant points on the surface at the beginning which is the most regular initial case, and then we see that it can be still improved by using our introduced indexes. After generating the equidistant points on the surface, we use a very nice function in MATLAB, i.e., *surf2patch*(), to transform the coordinates (x, y, z) to two matrices V and F. However, it is possible to have duplicate vertices in V as any vertex in a grid cell may belong to some other cells as well. Hence, there may also be some duplicativeness in matrix F which should be removed.

This way, we present the following algorithm which generates both triangular and quadrangular initial gridshells (matrices F, faces, and V, vertices) for any given surface $F(x, y, z) = 0$.

Some explanations are given in Appendix A Section on the used MATLAB functions and commands in Algorithm 1. We note that Algorithm 1 works with any generated set of points in Step 1 including the uniformly distributed random points or the equidistant points on the surface. However, in our usage of this algorithm, we generate the equidistant points on x−axis and y−axis (and then the third coordinates of the points are determined by using the given surface F(x, y, z) = 0) rather than the randomly generated points. This is only to have the most regular initial gridshell.

Algorithm 1. Generating initial triangular or quadrangular gridshells (matrices F and V) for a given surface F(x, y, z) = 0 with some specified domain (in MATLAB)

Step 1. Generating some points on the surface in the domain. This way, we obtain three matrices X, Y, and Z including the coordinates of the points. Set *ind* (0 for triangular or 1 for quadrangular case).
Step 2. Determine the matrices F, faces, and V, vertices, as follows.
  if *ind*
  [F, V] = surf2patch(X, Y, Z)
  else,
  [F, V] = surf2patch(X, Y, Z, 'triangles').

Step 3. Remove the redundant vertices in matrices F and V as follows.
  [V, ∼, I] = unique (V, 'rows');
  F = I(F);

2.2. Genetic Algorithm

Here, we explain briefly the genetic algorithm (GA) to eliminate the need for previous knowledge of this technique for the readers. The genetic algorithm (GA) is an iterative search method which relies on bio-inspired operators such as selection, crossover, and mutation. This technique was developed by Holland [29] and contains six main stages explained below. It is noted that our main aim is not proposing an improved GA or tuning the best parameters of GAs on the desired problems but rather is to show how this technique can be employed to improve the gridshells' regularity by using our introduced indexes in MATLAB. By the way, we tested some of the very common values for the parameters in GA taken from the literature, and then considered the best ones in our primary generated numerical results.

(1) Generating an initial population: Based on Darwinism, in this technique it is always assumed that there is an initial population of individuals which can change partially in each generation (iteration in the algorithm) according to the fitness (or cost) function. In fact, in each iteration the weak individuals are normally removed and instead some new stronger offspring are added to the population according to the crossover and mutation processes. We note that the number of individuals in each generation are the same as the number of individuals in the initial population. To generate an initial population, usually some random solutions are generated within a certain reasonable domain. Moreover, each solution, which is a member of the initial population, should be represented as a string (vector or matrix), which is called chromosome in this technique.

It is noted that although both matrices F and V are required to draw (or drive) the desired gridshells, the matrix F does not change during the improvement process, and we only need to improve the nodal positions according to the desired cost function. Hence, only the matrix V is improved. Moreover, with the first two coordinates of each vertex, i.e., x and y, the third coordinate can be obtained by using the given surface, that is F (x, y, z) = 0. Therefore, in this work, the first two columns of the initially generated matrix V are considered as a basic solution and denoted by V_{new}. Then, a population of N_{pop} individuals is randomly generated between $V_{new} - t$ and $V_{new} + t$ as the initial population, where t is a tolerance. As the initially generated matrix F, which shows the faces in the gridshell, does not change in the improvement process, the same matrix is used to evaluate every individual in the population.

(2) Evaluating each solution: An important part in every improvement process is determining the fitness function (in the case of maximizing) or cost function (in the case of minimizing), and then adopting it to the process. After generating the initial population, every individual is evaluated. The newly obtained solutions in crossover and mutation processes are also evaluated.

Then, usually in the merging stage, all the solutions are sorted and the weak ones are removed. Here, the cost function is considered as one of the presented indexes or their combination. More details on the cost function in our work is given in Section 2.5.

(3) Parent selection: In genetic algorithm, two current solutions (called parents) are selected in order to create two new solutions (called offspring or children) in crossover stage. There are various methods to select parents among which the random selection, tournament selection, and roulette wheel selection are the most used [30–32]. Here, we use the roulette wheel selection method. For improving this method, we first generate a vector of probability, i.e., P = ($p_1, \cdots, p_{Npop}$), based on the Boltzmann selection technique [33,34] and using a selection pressure β as follows.

$$\begin{aligned}&(i)\ P = exp(-\beta \times C/C_{max})\\&(ii)\ P = P/\sum_{j=1}^{N_{pop}} p_j\end{aligned} \quad (1)$$

where C is the vector of the solutions' costs and C_{max} is the cost of the worst solution in the current population. According to experimental results, to improve the convergence of GA technique, for each problem, we set the selection pressure β so that the summation of the first half of components in probability vector P stays between 0.7 and 0.8. This way, we obtain some probabilities whose summation is 1. Calculating the accumulated vector and generating a random number from zero to 1, the first component in the accumulated vector which is equal to or greater than this random number gives the desired parent. The crossover percentage in GA is usually considered between 0.5 and 1 [31]. Here, it is set to pc = 0.8. It is noted that in our primary numerical experiments on desired problems here, changing the crossover percentage from 0.7 to 0.9 led to a negligible change in the final results, and so the average of pc = 0.8 is considered. Therefore, as the number of parents should be even, in each iteration N_c = 2 × [pc × N_{pop}/2] parents are selected and the same number of offspring (children) are generated, where [•] is the nearest integer number to •.

(4) Crossover: This is an important operator in GA which mimics mating in biological populations. It propagates the good features from the current population to the next one leading to better fitness (or cost) value on average. There are several strategies to do the crossover including single point, double-point, uniform, and arithmetic crossover. Here, as the points can move continuously on the surface, we use the arithmetic crossover. In this way, we first generate a vector α of random numbers from the continuous uniform distribution, and then considering X1 and X2, the new children Y1 and Y2 are generated as follows.

$$Y_1 = \alpha \times X_1 + (1-\alpha) \times X_2 \quad (2)$$

$$Y_2 = \alpha \times X_2 + (1-\alpha) \times X_1. \quad (3)$$

(5) Mutation: This operator allows for global search of the solution's space, promoting the diversity in population characteristics. It also prohibits getting trapped in local minima. The mutation percentage in GA is usually chosen between 0.001 and 0.5 [31], and hence according to our primary generated numerical results pm = 0.3 is considered in this work. This way, in each generation (from the second generation), Nm = [pm × N_{pop}] of mutants are generated. Moreover, to generate a mutant, a mutation rate less than 0.1 is usually considered in GA [31,32]. Comparing the numerical results for different values 0.01, 0.02, $\cdots$, 0.08, we consider μ = 0.02 as mutation rate which determines the number of components (or genes) which are changed in the selected solution (or chromosome), and we do the mutation by using the standard normal distribution to change the values of the selected components in each solution. In fact, $\lceil \mu \times N_v \rceil$ component are changed in the selected

solution for mutation, where N_θ is the number of vertices in the gridshells and $\lceil \bullet \rceil$ is the smallest integer number not less than $\bullet$.

(6) Merging: After selecting parents, generating new children, and mutating some solutions, we merge all the current and newly obtained solutions leading to $N_{pop} + N_c + N_m$ solutions. Then, as the number of solutions in each generation should be the same, all the solutions are sorted and arranged ascendingly according to the costs, and the first N_{pop} ones are selected as the next population.

We note that in GA the stages (3)–(6), explained above, are repeated until some considered stop criteria is satisfied.

Stop criteria. In fact, in all the iterative processes such as GA and PSO, the algorithm needs some stopping criteria. Some common stopping criteria used in the literature are: (i) stop by exceeding the given maximum number of iterations, (ii) stop when the improvement of solution in a given number of iterations is less than a given limit, (iii) stop when a satisfactory solution is determined, and (iv) stop when the cost function slope is almost zero. Here, for both GA and PSO, we consider a maximum number of M = 2000 iteration as the stop criterion.

2.3. Particle Swarm Optimization

Here, we explain briefly the particle swarm optimization (PSO) for the reader not being required to have any previous knowledge of this technique as well. This technique is also an iterative search method, inspired from social behavior, which was initially proposed by Kennedy and Eberhart [30]. There are four main stages in PSO explained as follows.

(1) Generating an initial population: Like GA, it is assumed that there is some initial population of individuals, called particles, in PSO. Usually, all the particles in PSO move from the current positions to some new positions based on the swarm intelligence in each iteration. Here, we consider the same initial population for PSO as the GA method. It is noted that each row in the matrix V_{new}, defined in Section 2.3, shows the position of a particle in the initial population and the matrix is updated whenever the position of the particles are changed. We note that the initially generated particles move toward the positions of the so-called Pbest and Gbest, explained below.

Thus, after generating the initial population, we need to evaluate them to determine Pbest and Gbest in the population.

(2) Velocity Updating: In this technique, the movement of each particle in every iteration is determined by its velocity. Let x_i^k and v_i^k respectively denote the position and velocity of particle i in the kth iteration in the search space. The velocity of particle i for the next iteration is calculated as follows.

$$v_i^{k+1} = wv_i^k + c_1 r_1 \left(Pbest_i^k - x_i^k \right) + c_2 r_2 \left(Gbest^k - x_i^k \right) \qquad (4)$$

where w is the inertia factor which controls the flying dynamics, c_1 and c_2 are the acceleration factors for the experiences of Pbest and Gbest, respectively, r_1 and r_2 are random variables in the interval [0,1] which provide the ability of stochastic searching for PSO. The accelerating factors c_1 and c_2 compromise the trade-off between exploitation and exploration. It is noted that $Pbest_i^k$ is the best experienced position for particle i until the kth iteration, and $Gbest^k$ is the best experienced position among all the particles so far. We also note that the velocities for the particles in the initial population are set initially to be zero.

There are three parameters in Equation (4) which are w, c_1, and c_2. Many studies have been done so far to determine the best parameters for PSO technique. On the inertia weight, i.e., w, the studies show that a fixed w will not get to good results, and hence several techniques have been proposed in the literature by which w is lessened along with iteration times [26]. Some researchers suggested the interval [0.9,1.2] and some others the interval [0,1] for w [25,26,30]. Similarly, several researchers have studied the acceleration factors c_1 and c_2, and suggested different values for these factors. As our main aim is not tuning the best parameters for the PSO method in this area, we simply compared the four more common strategies taken from literature [26] numerically to select the best one. The strategies are (1) $w = 1$, $wdamp = 0.999$, $c_1 = 2$ and $c_2 = 2$, (2) $w = 1$, $wdamp = 0.999$, $c_1 = 2.8$ and $c_2 = 1.3$, (3) $w = 1$, $wdamp = 0.999$, $c_1 = 1.49445$ and $c_2 = 1.49445$ and (4) Letting $\varphi_1 = 2.05$, $\varphi_2 = 2.05$, $\varphi = \varphi_1 + \varphi_2$, and $\xi = 2/(\varphi - 2 + \sqrt{\varphi^2 - 4\varphi})$, then we have $w = \xi$, $c_1 = \xi \times \varphi_1$, and $c_2 = \xi \times \varphi_2$. We note that in the first three strategies, the inertia weight is updated in each iteration by $w = w.wdamp$, and this is why $wdamp$ is called the damping factor. It is also noted that the values of c_1 and c_2 in the strategies 3 and 4 are the same. We found the first strategy as the best among these four strategies for our desired problem, and hence we consider its parameters in this work.

We note that after updating velocities and before updating the particles' positions, it is important to check if the velocities are within a pre-specified range. In fact, to avoid violent random walking and control the global exploration of the particles, some lower and upper speed limits for each particle are determined and when the velocity of a particle exceeds one of the limits, it is replaced with the related limit. These limits do not impact on the particle position, and only lessen the step size of velocity, and hence the limits control the particles' moves and the aspects of exploration and exploitation [25,30]. Moreover, greater (smaller) speed limits lead to global (local) exploration [25,30]. The process of controlling velocity is called velocity clamping.

Although the movement of particles are controlled by velocity limits, sometimes even by using the lower limit of velocity, the new position of a particle is obtained out of the search area or feasible space. In fact, when the current position of a particle is close enough to the borders of the search area, according to the direction of velocity, even with small velocity, the new position of the particle will be out of the search area. This shows that in the next iterations, according to the inertia, this will happen again that the new position of the particle stays out of search space. Hence, to avoid such an event, the direction of velocity is changed to the opposite direction, that is its sign will be changed. This process is called "velocity mirror effect".

(3) Position Updating: Unlike the genetic technique in which usually not all the members of the population are replaced with some new children in each iteration, in the PSO method, all the particles in the population move and change in each iteration. To do so, after calculating the velocity of the particle i, its position is updated as follows.

$$x_i^{k+1} = x_i^k + v_i^{k+1} \qquad (5)$$

Apart from all the modifications and limits on the velocities, the new positions of some particles may be out of the search area. Hence, the updated positions should be checked for being within the allowed domain. If the position of a particle exceeds the lower or upper bounds, the position of the particle is replaced with the associated bound.

(4) Memory updating: In this technique, in each iteration, the position of every particle may change, and it is one of the differences between GA and PSO. Therefore, after updating the positions of the particles, we need to evaluate all the particles in order to check if it is required to update the P best and Gbest variables as they play an essential role in movement of the particles. However, it is not required to sort the particles after the evaluation process, and we only need to update the memory of P best and Gbest, if it is required. In this work, the same cost function is considered for both GA and PSO techniques.

The above-mentioned stages (2)–(4) in PSO are repeated until some considered stop criteria is satisfied. As stated in the previous section, we consider a maximum number of $M = 2000$ iteration as stop criteria for both GA and PSO in this work.

2.4. The Regularity Indexes

Here, we explain in more detail our introduced indexes, which are length ratio and shape ratio, as well as two similar indexes introduced in [19]. As the indexes introduced by Nooshin and his coworkers have been also called length and shape ratios, to make the four indexes distinguishable, we denote our length and shape ratios by OLR and OSR, respectively, and the introduced length and shape ratios by Nooshin and his coworkers by NLR and NSR, respectively. We recall that V is a matrix containing the position of all the vertices in Cartesian coordinates with no redundant vertices. In fact, V is an $N_v \times 3$ matrix, where N_v is the number of vertices in the gridshell. Each row in V provides the Cartesian coordinates of a vertex in the grid, and the vertices are numbered in the order of appearance in V.

In the order of appearance in V. For instance, the vertex whose coordinates are given in the third row of V is numbered 3. The matrix F gives the vertices of each face. In fact, in a triangular (quadrangular) gridshell, F is an $N_f \times 3$ ($N_f \times 4$) matrix, where N_f is the number of faces in the grid. Each row in F shows which vertices are in the corresponding face. To have a better understanding, a simple grid (pyramid) is given in Figure 1. The matrices F and V for this figure are as follows.

$$V = \begin{bmatrix} 0 & 0.65 & 0 \\ 0.35 & 0.45 & 0.7 \\ 0.35 & 0 & 0 \\ 0.75 & 0.65 & 0 \end{bmatrix} \quad F = \begin{bmatrix} 1 & 2 & 3 \\ 4 & 2 & 1 \\ 3 & 2 & 4 \\ 3 & 4 & 1 \end{bmatrix}$$

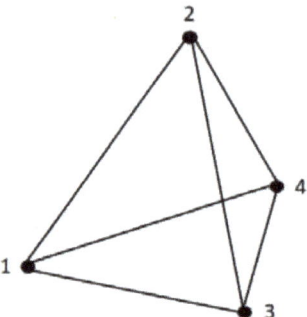

Figure 1. A simple grid (pyramid).

As it is seen, there are four vertices and also four faces in Figure 1. The ith row in V gives the Cartesian coordinates of the ith vertex. For example, the coordinates of vertex (1) are (0, 0.65, 0).

The ith row in F gives the vertices of the ith face in the grid. For example, the first face in the grid is the triangle consisting of vertices (1), (2) and (3).

Length ratios: Our introduced length ratio (OLR) is defined as the standard deviation of lengths of all the elements in the grid. Hence, after calculation of all the lengths, OLR can be obtained by computing the standard deviation of all the lengths. The introduced length ratio by Nooshin and his coworkers [19] (NLR) is (the shortest element's length/the longest element's length) for each face in the gridshell, and then for the gridshell is the mean value of all the calculated values for the faces.

Hence, for both OLR and NLR, one first needs to calculate the length of all the elements in the grid. To do so, we use the rows in F to find the beginning and end vertices of each element, and then use matrix V to find the coordinates of the desired vertices to compute the distance between them. For example, in Figure 1, according to the first row in F, we calculate the distance between the vertices (1) and (2), (2) and (3), and (3) and (1) by using their given coordinates in matrix V.

This way, we obtain the row [0.8078 0.8322 0.7382] as the lengths of elements in the first face in this figure. The matrix of lengths, which is denoted by L, for this figure is given below in which each row gives the lengths of the elements in the corresponding face in matrix F.

$$L = \begin{bmatrix} 0.8078 & 0.8322 & 0.7382 \\ 0.8307 & 0.8078 & 0.7500 \\ 0.8322 & 0.8307 & 0.7632 \\ 0.7632 & 0.7500 & 0.7382 \end{bmatrix}$$

Now, OLR can be simply computed as the standard deviation of all the lengths calculated in matrix L above which is equal to 0.0398. For computing NLR, we first calculate (the shortest element's length/the longest element's length) for each face. This way, the vector [0.8870, 0.9029, 0.9171, 0.9672] is obtained in which for example the first component corresponds to the first face in the grid, that is the first row in matrix L above. Then, NLR is equal to the mean of all the calculated values in this vector, which is 0.9186 in this example. As a result, we have OLR = 0.0398 and NLR = 0.9186 in this example.

Shape ratios: Our introduced shape ratio (OSR) is defined as the standard deviation of all the angles between the elements in all the faces. Similar to the length ratio, the NSR, which is the introduced shape ratio in [19], is (the smallest internal angle/the largest internal angle) for each face in the gridshell, and then for the gridshell is the mean value of all the calculated values for the faces. This time, one needs to compute the inner angles in all the faces of the gridshell. To do so, as we have the Cartesian coordinates of the vertices from V and the faces from F, in each iteration, we consider a face, which is triangle or quadrangle, and compute its angles by using the formula $\theta = \arccos\left(\frac{\vec{a} \times \vec{b}}{\|a\| \times \|b\|}\right)$, where $\vec{a}$ and $\vec{b}$ are two vectors and θ is the angle between them. This way, a matrix of size of F is obtained as a matrix of angles, denoted by A here. For example, the angles for the first face in Figure 1 are [1.1336 0.9335 1.0746] in radian and the matrix of angles (in radian) in this figure is as follows.

$$A = \begin{bmatrix} 1.1336 & 1.9335 & 1.0746 \\ 1.0684 & 1.9506 & 1.1227 \\ 1.0922 & 1.9537 & 1.0957 \\ 1.0456 & 1.0191 & 1.0769 \end{bmatrix}$$

Now, in this example, OSR can be simply calculated as the standard deviation of all the angles calculated in matrix A above which is equal to 0.0684. To compute NSR, one first needs to calculate (the smallest internal angle/the largest internal angle) for each face. This way, the vector [0.8235, 0.8467, 0.8704, 0.9463] is obtained. Then, NSR is equal to the mean value of all the calculated values in this vector, which equals 0.8717 in this example. Briefly, we have OSR = 0.0684 and NSR0 = 0.8717 in this example.

We note that in both introduced ratios in [19], the smallest item is divided by the largest item for each face, and the average of all the calculated values is considered as the corresponding ratio. However, as the standard deviation is a very popular measure and has several advantages, we considered it instead of simply dividing the smallest value by the largest value. In fact, standard deviation measures the deviation from the mean and is based on all the items (not only the smallest and largest). Moreover, as the square is a nice function in which the numbers smaller than one become smaller and the numbers larger than one become larger, and hence we can ignore the small deviations and consider the larger ones more clearly. To show the practical efficiency of our introduced indexes, we compare the indexes on some benchmarks in Section 3.

2.5. Mathematical Model of the Problem

Now that the problem has been completely described, the general mathematical model of the problem is provided in this section as follows.

$$\min f = \alpha \left(\frac{\sum_{i=1}^{N_e}(l_i - \bar{l})^2}{N_e - 1} \right) + \beta \left(\frac{\sum_{i=1}^{N_a}(\theta_i - \bar{\theta})^2}{N_a - 1} \right)$$

s.t. $V_{new} \in \Omega$, and F is given and fixed,

where α and β are constants, Ne the number of elements, li the length of the ith element, $\bar{l}$ the mean value of all the elements' lengths, the number of all the angles, θi the ith angle, and θ is the mean value of all the angles in the given gridshell. Moreover, Ω is a feasible region in the xy plane, V new an $N_V \times 2$ matrix which contains the first two columns of V, which contains the Cartesian coordinates of the vertices, and F is a matrix which gives the faces in the desired gridshell.

Moreover, Ω is a feasible region in the xy plane, V new an $N_V \times 2$ matrix which contains the first two columns of V, which contains the Cartesian coordinates of the vertices, and F is a matrix which gives the faces in the desired gridshell.

One can set the constants α and β according to a specific aim in the improvement process. For example, setting α = 1, β = 0 the gridshell's regularity is improved according to the length ratio while setting α = 0, β = 1 the gridshell's regularity is improved taking into account the shape ratio. Additionally, setting 0 < α, β < 1, we have a multi-objective case. We note that having V $_{new}$ and F, one can first calculate the third coordinates of each vertex by using the formula of the surface, and then the elements' lengths and the inner angles of all the faces. Therefore, the mathematical model of the problem is well-defined and the function f can be minimized by moving V $_{new}$ in Ω.

Now, having described all the processes, we are ready to provide the experimental results.

3. Experimental Results

Here, several numerical results are provided in two sections. In the first one, the practical efficiency of our introduced ratios and the presented ones in [19] are compared, and in the second one the performances of GA and PSO methods in improving the regularity of gridshells are compared.

An important stage of generating numerical results is to choose some benchmarks. The following criteria have been considered to choose some known benchmarks from the literature. (1) As the indexes in [19] are introduced for geodesic domes, to have a fairer comparison, we need some gridshell benchmark which is somehow similar to domes, and hence we consider Hemisphere surface taken from the literature [1] as one of the benchmarks in this work. (2) As the effect of Gaussian curvature of a gridshell on the other structural aspects of the gridshell and vice versa have been widely studied in the literature [1,35], for observing if the change in the Gaussian curvature has some effect on the regularity improvement process, we need to choose gridshells with different Gaussian curvatures (positive, negative, and both). (3) As it is not the focus of this work to be involved in the complication of generating the initial gridshells, we consider the gridshells which are associated with some surface equations.

This way, along with Hemisphere surface with positive Gaussian curvature, we consider two other gridshells associated with surfaces of sinusoidal, which is a surface with Gaussian positive and negative curvature, and Hyperbolic paraboloid, which is a surface with Gaussian negative curvature. All the three chosen gridshells are taken from the literature [1] and depicted in Figure 2.

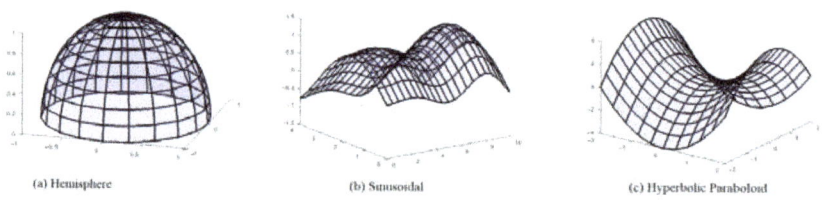

(a) Hemisphere (b) Sinusoidal (c) Hyperbolic Paraboloid

Figure 2. Three benchmark gridshells (quadrangular) with different Gaussian curvature taken from [1].

To determine the coordinate matrices of the hemisphere gridshell, which is depicted in Figure 2a, we use the built-in sphere () command which generates the $x-$, $y-$, and $z-$coordinates of a unit sphere consisting of 20-by-20 faces. The first 10 faces situate under or on the $xy-$plane, and thus we only consider the faces numbered from 11 to 20, which are above the $xy-$plane, as the hemisphere.

To determine the coordinate matrices of the sinusoidal gridshell, which is depicted in Figure 2b, the equation $z = 0.05x\sin(x) + \sin(y)$ is used for $0 \leq x \leq 10$ and $0 \leq y \leq 4$. In fact, on the $x-$axis the equidistant points with distance of 0.5 and on $y-$axis the equidistant points with distance of 0.4 are considered.

To determine the coordinate matrices of hyperbolic paraboloid gridshell, which is depicted in Figure 2c, the equation $z = x2 - y2$ is used for $-2 \leq x \leq 2$ and $-2 \leq y \leq 2$. In fact, on the $x-$axis the equidistant points with distance of 0.4 and on $y-$axis the equidistant points with distance of 0.25 are considered. In the last two cases, i.e., sinusoidal and hyperbolic paraboloid, the $x-$ and $y-$ coordinates are generated by using the *meshgrid()* command, and then the $z-$ coordinates matrix is obtained by using the surfaces' formulas.

As our focus is on improving the regularity of gridshells, to have very regular initial grids the equidistant points are considered rather than randomly generated points. We note that the gridshells depicted in Figure 2 are the quadrangular ones. As the triangular gridshells are the same as the quadrangular ones but with the triangle faces, they are not given here to avoid the prolongation of the paper. In fact, Algorithm 1 generates the triangular gridshells by adding the diagonal of the faces in quadrangular gridshells.

To generate the numerical results, both GA and PSO algorithms and Algorithm 1, our presented approach to generate the initial gridshells, are implemented in MATLAB programming environment and the program is executed on a PC with Intel(R) Core (TM) i5-2400S Duo CPU 2.50 GHz, with GB of RAM. We note that when the primary shape of the structural appearance is ascertained, it is not allowed to make major changes and one can make only minor changes in the shape improvement process [5]. Therefore, to generate the initial populations in both GA and PSO algorithms, the feasible solution domain is restricted to a small interval around the initial equidistant generated gridshells. To this end, considering V_{new} as an $N_v \times 2$ matrix which contains the first two columns of V, the matrices $V_{max} = V_{new} + 0.02 \times \mathbf{1}_{N_v \times 2}$ and $V_{min} = V_{new} - 0.02 \times \mathbf{1}_{N_v \times 2}$ are considered as the upper and lower bounds of the feasible domain, respectively, where $\mathbf{1}_{N_v \times 2}$ is an $N_v \times 2$ matrix with all the entities being 1. Therefore, we recall that (i) the initially generated equidistant gridshells are very regular and (ii) the feasible solution domain is restricted.

As some initial conditions, we suppose that all the vertices which situate on the edges of the gridshell should be fixed which is makeable by considering the same maximum and minimum ranges for those vertices in matrices V_{max} and V_{min}. The upper and lower bounds on the velocities of particles in the PSO method are respectively assigned as $Vel_{max} = 0.1 \times (V_{max} - V_{min})$ and $Vel_{min} = -Vel_{max}$. The population size in GA and PSO algorithms is usually specified between 20–60 [25,26,31,32], among which we set the average value of $N_{pop} = 40$ as the population size in both methods. Note that our primary experimental results did not show notable changes in the results varying the population size from 40 to 60; however, we obtained better results in both methods by increasing N_{pop} from 20 to 40. Moreover, the stop criteria are set to a maximum number of M = 2000 iterations for both algorithms. It is recalled that the details and the selected values of parameters for both algorithms have been discussed in Sections 2.3 and 2.4, and hence we do not restate them here. The other details on the implementation of algorithms and the generated numerical results are given in the next sections.

3.1. Comparison of the Regularity Indexes

Here, our two introduced indexes, OLR and OSR, are compared to two corresponding introduced indexes by Nooshin et al. [19], i.e., NLR and NSR. As Nooshin and his coworkers have introduced the indexes for geodesic domes and not the free-form gridshells, we compare the indexes only on the hemisphere's gridshell, depicted in Figure 2a which is the closer one to a dome. As the main aim in this section is to compare the regularity indexes, we only use the PSO method. To have a more complete comparison, both triangular and quadrangular gridshells associated with the hemisphere are generated. This way, we improve the regularity of the considered gridshells four times separately, each time applying one of the indexes as the cost function in the PSO algorithm.

The final results rounded to three decimal places on comparing the length and shape ratios are respectively stated in Tables 1 and 2. It is seen that using our proposed indexes as the cost function in the PSO algorithm, the Nooshin et al.'s indexes are also lessened (improved). For example, in Table 1, for a triangular case, the initial values of OLR and NLR are respectively 0.087 and 0.611. By applying OLR as the cost function, after 2000 iterations, the OLR and NLR are respectively equal to 0.083 and 0.615 which shows improvement in both ratios. However, by applying NLR as the cost function, after 2000 iterations, the OLR and NLR are respectively equal to 0.096 and 0.633 which shows worsening in OLR and improvement in NLR. Some similar observations can be made for quadrangular cases in this table and both cases in Table 2. We note that according to the definition, the best value of NLR or SLR is 1, and so they are improving as they are approaching 1. Therefore, as Tables 1 and 2 show, the regularity of the gridshells is practically worsened by using the introduced indexes by Nooshin et al. [19] which shows clearly the practical efficiency of our introduced indexes.

Table 1. The final results on comparing the length ratios.

Applied Indexes	Triangular Grid			Quadrangular Grid		
	Initial Value	Final Value		Initial Value	Final Value	
		OLR	NLR		OLR	NLR
↓ OLR	0.087	0.083	0.615	0.064	0.061	0.754
NLR	0.611	0.096	0.633	0.753	0.066	0.767

Table 2. The final results on comparing the shape ratios.

Applied Indexes	Triangular Grid			Quadrangular Grid		
	Initial Value	Final Value		Initial Value	Final Value	
		OSR	NSR		OSR	NSR
↓ OSR	0.403	0.397	0.430	0.085	0.085	0.905
NSR	0.426	0.461	0.473	0.914	0.090	0.904

3.2. Improving the Regularity

Here, the performances of GA and PSO techniques are compared together in improving the regularity of the three gridshells with different Gaussian curvature taken from the literature [1]. Both triangular and quadrangular gridshells associated with each surface are generated. The regularity of each generated gridshell is improved by applying our introduced indexes and using both GA and PSO methods. Three cost functions have been considered, (1) length ratio, (2) shape ratio, and (3) a multi-objective case by combining both length and shape ratios with the same weight. Therefore, having three surfaces, two generated gridshells on each surface, and considering three cost functions to improve the regularity makes eighteen test problems on which the performances of GA and PSO algorithms are compared. To have a better reading, all the diagrams of these two methods on each surface are grouped and depicted in a figure. This way, Figures 3–5 provide the diagrams of GA and PSO techniques on Hemisphere, Sinusoidal, and Hyperbolic Paraboloid, respectively. Note that in these figures, TH, TS, THP, QH, QS, and QHP stand for triangular hemisphere, sinusoidal, hyperbolic parabolic, and quadrangular hemisphere, sinusoidal, and hyperbolic parabolic, respectively.

For convenience, a data box has been added into each diagram in which the following information is given. (1) The applied cost function and the considered gridshell, (2) the initial cost of each method which is the value of the cost function in the first iteration, (3) the final cost of each method which is the value of the cost function in the last iteration, and (4) the improvement percentage for each algorithm which is calculated by using Equation (6). Of note is that the cost values are rounded to three decimal places and the improvement percentages are rounded to one decimal place. Hence, this is why sometimes the diagrams are not matched at the end point, that is iteration 2000, while the given final costs are equal.

$$Improvement\ percentage = \frac{Initial\ cost\ -\ Final\ cost}{Initial\ cost} \quad (6)$$

In each of Figures 3–5, there are two rows of diagrams, three diagrams in each row. The first (second) row contains the diagrams associated with triangular (quadrangular) gridshells. The applied cost functions on the diagrams in each row are always in order of length ratio, shape ratio, and multi objective case which is length ratio + shape ratio. Next, several observations concluded from the GA and PSO diagrams are stated and discussed.

(1) In almost all the cases, it is seen that the initial cost for triangular gridshells of the same surfaces are higher than the cost for the corresponding quadrangular gridshells. This is due to the higher number of elements in the triangular gridshells. To have a better understanding, the number of vertices, denoted by N_v, and faces, denoted by N_f, in the gridshells are given in Table 3. It is noted that the number of vertices for both triangular and quadrangular cases are equal because Algorithm 1 uses the same procedure for generating the initial gridshells, and the number of faces in the triangular cases are double of the one in the corresponding quadrangular one because the algorithm generates the triangular gridshells by adding the diagonal of the faces in quadrangular gridshells.

Table 3. The number vertices and faces in all the generated gridshells.

Gridshell Type	Triangular		Quadrangular	
	N_v	N_f	N_v	N_f
Hemisphere	201	400	201	200
Sinusoidal surface	231	400	231	200
Hyperbolic parabolic	187	320	187	160

(2) Although the number of vertices and faces are smaller for the hyperbolic parabolic surface (see Table 3), the costs of either the initial or the final grids for this surface are greater than the ones for the other surfaces. It seems that as the Gaussian curvature is negative in this surface, considering the equidistant points do not lead to a gridshell as regular as the surfaces with positive (or positive and negative) Gaussian curvature. However, this observation needs further investigation to be validated which will be studied in our future works.

(3) It is seen that the costs of either the initial or the final gridshells in almost all the cases for the sinusoidal surface are less than the costs for the corresponding cases with the other surfaces even when the number of vertices and faces in this surface are the most ones. It seems that to generate the most regular gridshells, the surfaces with positive and negative Gaussian curvature are better than the surfaces with merely positive Gaussian curvature or the surfaces with merely negative Gaussian curvature. This observation also needs further investigations which is of our interest for the future works.

(4) Although the initial gridshells are greatly regular because of consideration of equidistant points and the feasible solution domain is restricted, the regularity of gridshells in some cases has been improved more than 50% which is significant (see the second diagram in the second row in Figure 3).

(5) In all the cases, the (initial or final) costs increase from the length ratio to the shape ratio and from the shape ratio to the multi-objective case. It shows that arriving at similar angles in all the faces of gridshells is more difficult than designing the gridshells with similar lengths in all the faces, and also that improving in both directions simultaneously is even more difficult.

(6) It is seen that the improvement percentages of both algorithms are decreased from Hemisphere to Sinusoidal and then to Hyperbolic Paraboloid surfaces in almost all the cases. It may also have a relationship with the Gaussian curvature of the surfaces as it changes from positive in the first surface to positive and negative in the second one, and finally to negative in the third surface. Hence, it seems that the possibility of improving the regularity on the gridshells with positive Gaussian curvature is higher than the other cases. This observation also needs more investigation.

(7) The improvement percentages on all the three surfaces increase from the triangular gridshells to the quadrangular ones which shows a higher possibility of regularity improvement on the quadrangular gridshells.

(8) It is seen that the behavior of both algorithms are somehow similar on all the three gridshells. Hence, it seems that changing on the Gaussian curvature or changing from triangular to quadrangular gridshells do not affect the behavior of GA or PSO methods considerably. This observation also needs more investigation.

(9) Finally, it is seen that in almost all the test problems, PSO outperforms GA slightly. We note that the focus of this work is not to tune the best parameters for GA or PSO algorithms, and hence it is possible that one gets better results on GA by changing the values of its parameters or selecting some other strategies in the stages of this algorithm, and surely the same can be happened to PSO technique. What we see in the Figures 3–5 states that the performance of the PSO method with described parameters is slightly better than the performance of GA technique with detailed parameters here in almost all the cases.

Finally, to have a comparison of running times of GA and PSO, measured in CPU seconds, on the same Np = 18 test problems provided in this section, the running times of both methods on the test problems are considered to produce the performance profile of Dolan and Moré [28]. In this performance profile, for two algorithms, the ratio of the running times of the methods versus the minimum time of the two methods is considered.

Indeed, considering $t_{i,1}$ and $t_{i,2}$ as the running times of GA and PSO techniques, in CPU seconds, respectively, for $i = 1, 2, \cdots, 18$, the performance ratios in this performance profile are $r_{i,j} = \frac{t_{i,j}}{min\{t_{i,j}:j=1,2\}}$, for $j = 1, 2$ [28]. The performance of each technique is calculated as $Pr_j(T) = \frac{1}{N_p} size\{i | r_{i,j} \leq T\}, j = 1, 2$, where size is the number of test problems. This way, $Pr_j(t)$ is the probability for method j ($j = 1, 2$ corresponds to GA and PSO, respectively) that a performance ratio $r_{i,j}$ is within the factor T. Figure 6 is the resulting CPU time performance profiles for the two methods. In this figure, the horizontal axis T gives the outcome of dividing the running time of the PSO method into the one of the GA method. This axis states that in the best case, PSO technique solves some problems (almost 10% of the test problems) around 1.45 times faster than GA method. We note that the vertical axis, that is Per(T), at time T gives the percentage of problems solved by PSO, T times faster than the GA method. Using this profile, one algorithm is preferred to another when its performance diagram lies above the other [28]. Hence, Figure 6 shows clearly that PSO is preferred to GA based on the running times. However, the differences in running times are not so significant.

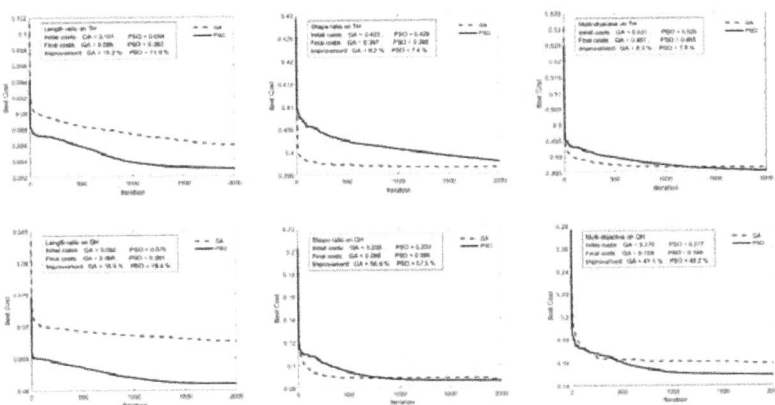

Figure 3. The particle swarm optimization (PSO) and genetic algorithm (GA) diagrams on Hemisphere gridshell considering different cost functions.

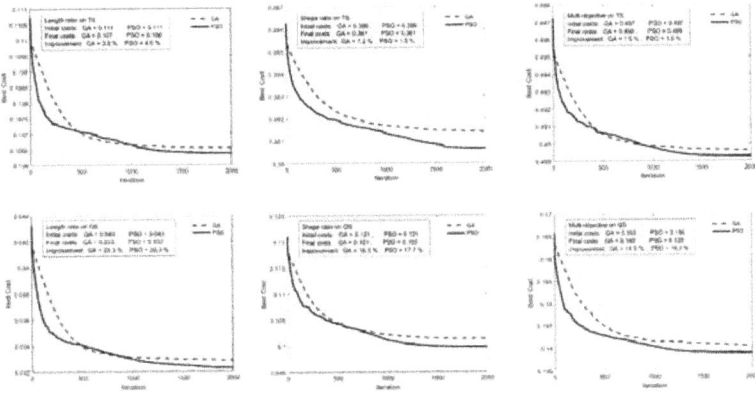

Figure 4. The PSO and GA diagrams on Sinusoidal gridshell considering different cost functions.

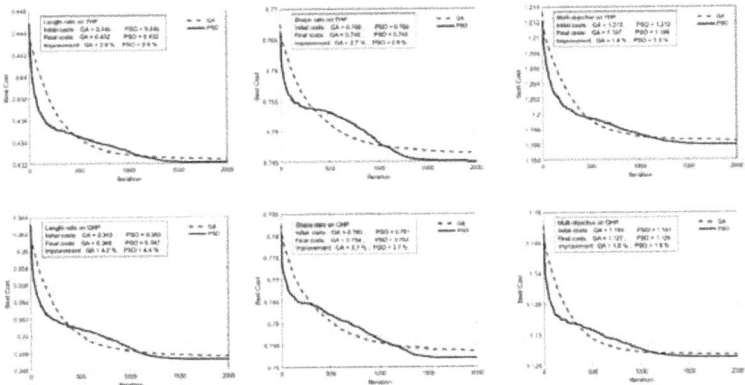

Figure 5. The PSO and GA diagrams on Hyperbolic Paraboloid gridshell considering different cost functions.

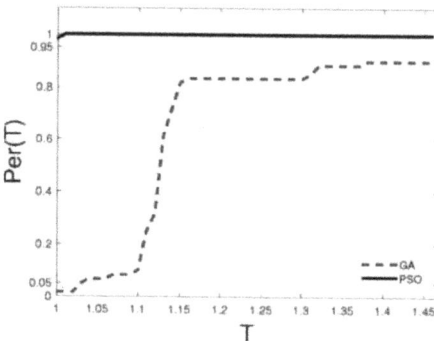

Figure 6. Dolan–Moré diagram related to comparing PSO and GAs in improving of regularity of gridshells.

4. Conclusions

Here, we presented two indexes as length ratio and shape ratio which were defined as the standard deviation of the lengths of all the elements and the standard deviation of the inner angles between all the elements in a gridshell, respectively. The practical efficiency of our introduced indexes was shown in comparison with some available indexes in the literature. We also showed how the genetic algorithm (GA) and particle swarm optimization (PSO) technique can be utilized for improving the gridshells' regularity based on the introduced ratios. To this end, an algorithm was presented to generate initial gridshells on a given surface. Three surfaces with different Gaussian curvatures were selected from the literature to provide the experimental results on the proposed approaches for regularity improvement of gridshells. On each surface, triangular and quadrangular gridshells were generated and the regularity of each gridshell was improved by using each ratio separately and also by using a combination of both ratios with the same weight as the cost function in both techniques. This way, PSO and GA methods were compared on eighteen test problems.

Through the experimental results, we saw that (1) the initial cost for triangular gridshells on the same surfaces are usually higher than the cost for the corresponding quadrangular gridshells, (2) even the regularity of the very regular initially generated gridshells by using the equidistant points can be improved up to 56% in some cases, (3) the initial and final costs increase from the length ratio to the shape ratio and from the shape ratio to the multi-objective case in which both length and shape ratios are combined with the same weight, (4) the percentage improvement on all the three surfaces increase from the triangular gridshells to the quadrangular ones showing a higher possibility of regularity improvement on the quadrangular gridshells, and that (5) PSO method slightly outperforms GA technique on almost all the test problems.

Moreover, some interesting relationships between the regularity improvement and Gaussian curvature of the selected surfaces were observed including (1) considering the equidistant points on the surfaces with negative Gaussian do not lead to a gridshell as regular as the surfaces with positive (or positive and negative) Gaussian curvature; (2) to generate the most regular gridshells, the surfaces with positive and negative Gaussian curvature are better than the surfaces with merely positive Gaussian curvature or the surfaces with merely negative Gaussian curvature; (3) the possibility of improving the regularity on the gridshells with positive Gaussian curvature is higher than the other cases; and (4) the behavior of GA and PSO techniques do not change considerably from triangular to quadrangular gridshells or from the positive Gaussian curvature to the negative one. However, these observations need more investigation which will be made in our future works. Another idea for a future work is to consider our introduced regularity indexes and some structural aspects of gridshells such as strain energy simultaneously for proposing some multi-objective optimization method to design gridshells. It would also be interesting to see how the regularity indexes and other structural aspects affect each other. Finally, to have an intuitive comparison between GA and PSO based on running times, the Dolan and Moré performance profile was produced taking into account the running times. This profile showed that PSO is also preferred to GA in accordance with running times.

Author Contributions: Conceptualization, M.G. and M.F.-e.; Data curation, M.G. and M.F.-e.; Formal analysis, M.F.-e.; Funding acquisition, M.G.; Supervision, A.M.; Writing—original draft, M.F.-e.; Writing—review and editing, M.F.-e. All authors have read and agreed to the published version of the manuscript.

Funding: This research received no external funding.

Acknowledgments: The authors are grateful to the three anonymous referees for their constructive comments and recommendations, which have significantly improved the presentation of this paper. Also, the first two authors thank Algocg group for supporting this work.

Conflicts of Interest: The authors declare no conflict of interest.

Appendix A

In Algorithm 1, the MATLAB function of surf2patch () is used for transforming the Cartesian coordinates (x, y, z) to two matrices V and F. However, as there is a possibility of generating duplicate vertices in V, the command unique () is used. This command in addition to remove the duplicate rows in V gives the positions of the removed rows which can be used to update the matrix F (please see Step 3 in Algorithm 1). We note that having the matrices V and F obtained by Algorithm 1, the desired gridshell can be drawn by using the command patch () in MATLAB as follows.

$$\text{Patch} = \text{patch (`faces', F, `vertices', V)} \tag{A1}$$

References

1. Bouhaya, L.; Baverel, O.; Caron, J.F. Optimization of gridshell bar orientation using a simplified genetic approach. *Struct. Multidiscip. Optim.* **2014**, *50*, 839–848. [CrossRef]
2. Adriaenssens, S.M.L.; Barnes, M.R. Tensegrity spline beam and grid shell structures. *Eng. Struct.* **2001**, *23*, 29–36. [CrossRef]
3. Basso, P.; Grosso, A.E.D.; Pugnale, A.; Sassone, M. Computational morphogenesis in architecture: Cost optimization of free-form grid shells. *J. Int. Assoc. Shell Spat. Struct.* **2009**, *50*, 143–150.
4. Day, A.S. An introduction to dynamic relaxation. *Engineer* **1965**, *29*, 218–221.
5. Feng, R.; Zhang, L.; Ge, J.-M. Multi-objective morphology optimization of free-form cable-braced grid shells. *Int. J. Steel Struct.* **2015**, *15*, 681–691. [CrossRef]
6. Feng, R.; Ge, J.-M. Shape optimization of free-form cable-braced grid shells based on the translational surfaces technique. *Int. J. Steel Struct.* **2013**, *13*, 435–444. [CrossRef]
7. Khorasani, A.M.; Goodarzi, M.; Forghani-elahabad, M. Particle Swarm Optimization Method in Optimization of Grid Shell Structures. In Proceedings of the CNMAC 2019—XXXIX Congresso Nacional de Matemática Aplicada e Computacional, Uberlândia, MG, Brazil, 16–20 September 2020; Series of the Brazilian Society of Computational and Applied Mathematics. [CrossRef]
8. Marino, E.; Salvatori, L.; Orlando, M.; Borri, C. Two shape parametrizations for structural optimization of triangular shells. *Comput. Struct.* **2016**, *166*, 1–10. [CrossRef]
9. Mueller, K.M.; Liu, M.; Burns, S.A. Fully stressed design of Frame structures and multiple load paths. *J. Struct. Eng.* **2002**, *128*, 806–814. [CrossRef]
10. Seifi, H.; Javan, A.R.; Xu, S.; Zhao, Y.; Xie, Y.M. Design optimization and additive manufacturing of nodes in gridshell structures. *Eng. Struct.* **2018**, *160*, 161–170. [CrossRef]
11. Richardson, J.N.; Adriaenssens, S.; Coelho, R.F.; Bouillard, P. Coupled form-finding and grid optimization approach for single layer grid shells. *Eng. Struct.* **2013**, *52*, 230–239. [CrossRef]
12. Schek, H.-J. The Force Density Method for Form Finding and Computation of General Networks. *Comput. Methods Appl. Mech. Eng.* **1974**, *3*, 115–134. [CrossRef]
13. Pugnale, A.; Sassone, M. Morphogenesis and structural optimization of shell structures with the aid of a genetic algorithm. *J. Int. Assoc. Shell Spat. Struct.* **2007**, *48*, 161–166.

14. Rombouts, J.; Lombaert, G.; De Laet, L.; Schevenels, M. A novel shape optimization approach for strained gridshells: Design and construction of a simply supported gridshell. *Eng. Struct.* **2019**, *192*, 166–180. [CrossRef]
15. Vincenti, A.; Ahmadian, M.R.; Vannucci, P. Bianca: A genetic algorithm to solve hard combinatorial optimization problems in engineering. *J. Glob. Optim.* **2010**, *48*, 399–421. [CrossRef]
16. Wang, Q.-S.; Ye, J.; Wu, H.; Gao, B.-Q.; Shepherd, P. A triangular grid generation and optimization framework for the design of free-form gridshells. *Comput. Des.* **2019**, *113*, 96–113. [CrossRef]
17. Winslow, P.; Pellegrino, S.; Sharma, S.B. Multi-objective optimization of free-form grid structures. *Struct. Multidiscip. Optim.* **2010**, *40*, 257–269. [CrossRef]
18. Czerniachowska, K.; Hernes, M. A genetic algorithm for the shelf-space allocation problem with vertical position effects. *Mathematics* **2020**, *8*, 1881. [CrossRef]
19. Nooshin, H.; Mohammadi, N.; Parke, G. Regularity of Geodesic Domes. In Proceedings of the 35th Annual Symposium of IABSE/52nd Annual Symposium of IASS/6th International Conference on Space Structures, London, UK, 20–23 September 2011.
20. Elbeltagi, E.; Hegazy, T.; Grierson, D. Comparison among five evolutionary-based optimization algorithms. *Adv. Eng. Inform.* **2005**, *19*, 43–53. [CrossRef]
21. Xu, G.; Yu, G. On convergence analysis of particle swarm optimization algorithm. *J. Comput. Appl. Math.* **2018**, *333*, 65–73. [CrossRef]
22. Engelbrecht, A.P.; Grobler, J.; Langeveld, J. Set based particle swarm optimization for the feature selection problem. *Eng. Appl. Artif. Intell.* **2019**, *85*, 324–336. [CrossRef]
23. Chen, Y.; Li, L.; Xiao, J.; Yang, Y.; Liang, J.; Li, T. Particle swarm optimizer with crossover operation. *Eng. Appl. Artif. Intell.* **2018**, *70*, 159–169. [CrossRef]
24. Xu, G.; Yang, Y.-Q.; Liu, B.-B.; Xu, Y.-H.; Wu, A.-J. An efficient hybrid multi-objective particle swarm optimization with a multi-objective dichotomy line search. *J. Comput. Appl. Math.* **2015**, *280*, 310–326. [CrossRef]
25. Zhang, L.; Yu, H.; Hu, S. Optimal choice of parameters for particle swarm optimization. *J. Zhejiang Univ. Sci.* **2005**, *6*, 528–534.
26. Wang, D.; Tan, D.; Liu, L. Particle swarm optimization algorithm: An overview. *Soft Comput.* **2018**, *22*, 387–408. [CrossRef]
27. Kim, T.-H.; Cho, M.; Shin, S. Constrained mixed-variable design optimization based on particle swarm optimizer with a diversity classifier for cyclically neighboring subpopulations. *Mathematics* **2020**, *8*, 2016. [CrossRef]
28. Dolan, E.D.; Moré, J.J. Benchmarking optimization software with performance profiles. *Math. Program.* **2002**, *91*, 201–213. [CrossRef]
29. Holland, J.H. *Adaptation in Natural and Artificial Systems*; University of Michigan Press: Ann Arbor, MI, USA, 1975.
30. Kennedy, J.; Eberhart, R.C. Particle swarm optimization. In Proceedings of the IEEE International Conference on Neural Networks IV, Piscataway, NY, USA, 4–6 October 1995; IEEE: Piscataway, NJ, USA, 1995; pp. 1942–1948.
31. Angelova, M.; Pencheva, T. Tuning genetic algorithm parameters to improve convergence time. *Int. J. Chem. Eng.* **2011**, *2011*, 1–7. [CrossRef]
32. Koza, J.R. *Genetic Programming: On the Programming of Computers by Means of Natural Selection*; The MIT Press: Cambridge, UK, 1992.
33. Lee, C.-Y. Entropy-boltzmann selection in the genetic algorithms. *IEEE Trans. Syst. Man Cybern. Part B (Cybern.)* **2003**, *33*, 138–149.
34. Maza, M.D.L.; Tidor, B. An analysis of selection procedures with particular attention paid to proportional and boltzmann selection. In Proceedings of the 5th International Conference on Genetic Algorithms, Urbana-Champaign, IL, USA, 17–21 July 1993; pp. 124–131.
35. Douthe, C.; Mesnil, R.; Orts, H.; Baverel, O. Isoradial meshes: Covering elastic gridshells with planar facets. *Autom. Constr.* **2017**, *83*, 222–236. [CrossRef]

Article

Mathematical Calculation of Stray Losses in Transformer Tanks with a Stainless Steel Insert

Serguei Maximov [1], Manuel A. Corona-Sánchez [1], Juan C. Olivares-Galvan [2,*], Enrique Melgoza-Vazquez [1], Rafael Escarela-Perez [2] and Victor M. Jimenez-Mondragon [2]

[1] Tecnológico Nacional de Mexico, Instituto Tecnológico de Morelia, PGIIE, Av. Tecnológico No. 1500, Lomas de Santiaguito, Morelia 58120, Mich., Mexico; sgmaximov@yahoo.com.mx (S.M.); manuel.cs@morelia.tecnm.mx (M.A.C.-S.); emv@ieee.org (E.M.-V.)

[2] Departamento de Energía, Universidad Autónoma Metropolitana Azcapotzalco, Ciudad de México 02200, Mexico; r.escarela@ieee.org (R.E.-P.); vmjm@azc.uam.mx (V.M.J.-M.)

* Correspondence: jolivares@azc.uam.mx; Tel.: +52-135-1519-3760

Abstract: At present it is claimed that all electrical energy systems operate with high values of efficiency and reliability. In electric power systems (EPS), electrical power and distribution transformers are responsible for transferring the electrical energy from power stations up to the load centers. Consequently, it is mandatory to design transformers that possess the highest efficiency and reliability possible. Considerable power losses and hotspots may exist in the bushing region of a transformer, where conductors pass through the tank. Most transformer tanks are made of low-carbon steel, for economical reasons, causing the induction of high eddy currents in the bushing regions. Using a non-magnetic insert in the transformer tank can reduce the eddy currents in the region and as a consequence avoid overheating. In this work, analytical formulations were developed to calculate the magnetic field distribution and the stray losses in the transformer region where bushings are mounted, considering a stainless steel insert (SSI) in the transformer tank. Previously, this problem had only been tackled with numerical models. Several cases were analyzed considering different non-magnetic insert sizes. Additionally, a numerical study using a two dimensional (2D) finite element (FE) axisymmetric model was carried out in order to validate the analytical results. The solved cases show a great concordance between models, obtaining relative errors between the solutions of less than two percent.

Keywords: power transformer; stray losses; analytical methods; finite element method

1. Introduction

Nowadays, electric power systems (EPS) are constantly changing. The use of new technologies such as smart-grids, micro-grids and renewable energy systems demand high flexibility, performance and efficiency in the EPS. The transformer is a fundamental component in these systems, which is present in different stages of the EPS, such as at the generation, transmission and distribution stages [1,2]. Losses in transformers appear in their different components, such as: windings, insulation, core and tank. These losses depend on the operating conditions of the transformer (nominal values, DC bias, presence of harmonics, etc.) and the electrical and magnetic properties of the materials. The study of transformer losses is an active area of investigation because there is a compromise between design and cost. Power losses in transformers can be separated into two types: no-load and load losses; the no-load losses originate in the transformer core and the load losses are composed by the ohmic losses in the windings and the stray losses. As the rated power increases in transformers, the stray losses increase significantly. The stray losses in structural components, such as the tank, decrement the efficiency of the transformer considerably [3]. Losses in structural components in power transformers are due to stray fluxes; when a time varying flux impinges on a conductive element it induces a current in it, generating Joule losses. Besides power losses, hotspots in structural components caused by this phenomenon may appear. For these reasons, there are diverse techniques to reduce losses without unduly increasing the final cost. A crucial structural part of a transformer is the tank region where the bushings are mounted. The field concentration and overheating in this zone can cause damage to the device [4].

Analytical and numerical models can be employed for the design of the transformer taking into account the well-known advantages of each one: analytic models provide accuracy, less calculation time and simplicity once the model has been developed, whereas numeric models allow one to solve complex geometries and deal with nonlinear materials. Several research works, analyzing the magnetic field distribution and the eddy current power losses in the bushing transformer region, using either numerical or analytical models, have been carried out previously [1,4–15].

An analytical model is proposed in [4,5] to determine the power losses based on Poynting's theorem. To obtain the solution some semi-empirical coefficients are required. Analytical expressions were developed in [1] to calculate the magnetic field and the stray losses in the transformer tank near the bushing. The configuration is modeled with a finite disk and a conductor in the center. It is considered that the axial component of the induced current density has only a small contribution to the power losses, and therefore it is disregarded. This automatically implies that the solution obtained will be an approximation. A study to reduce stray losses in a pad mounted transformer wall using an insert plate is presented in [6,7]. These studies were carried out using two dimensional (2D) and three dimensional (3D) finite element (FE) analysis. The results were validated with experimental tests in different combinations. Additionally, numerical results were compared with empirical formulas. In these works, it was verified that the use of non-magnetic inserts in the transformer tank reduce eddy current losses. A transient analysis was carried out in [8] to compute the power losses in the low-carbon steel tank of a current transformer, taking into account different insert configurations and materials. The analyses were done using a 3D FE model. It was also concluded that using non-magnetic inserts reduce power losses in the tank. In [9,11] the temperature distribution on transformer covers is considered. The stray losses in the tank were analytically calculated using Turowski's formula [4]. Maximov et al. [10] presented a study of eddy current losses in the tank of a transformer. Numerical results and analytical formulas were obtained for the losses as a function of the current. However, in the solution to this problem only two materials were considered: the tank and the air. In [12,13], the determination of eddy current losses and temperature distribution in the zone of the transformers bushing are presented. The study was carried out with the finite difference method. An analytical solution for the bushing regions, using the same model geometry of [10], is proposed in [14]. However in this research the presence of harmonics in the current is considered. Oliveira et al. [15] developed a time domain model to determine the eddy currents in the transformer tank walls considering different types of excitations. The results obtained with the proposed model were validated with a 3D FE solution. The tank wall was considered to be made of a single magnetic material.

As can be appreciated from the literature reviewed above, in previous research where a stainless steel insert (SSI) is added to reduce the power losses in the tank, a numerical model is employed in the analysis, whereas analytical models are developed only for the cases where the tank in the bushing region is made from a single material. In this paper a mathematical model is proposed to determine the magnetic field distribution in the tank wall bushing regions, considering a tank wall composed of two different materials, a stainless steel section representing the insert and a low-carbon steel section modeling the rest of the tank. An analytical calculation is also developed to determine the stray losses in the bushing region.

2. Model

The bushing transformer region is considered through an idealized model. Consider an infinite conductor that passes the transformer tank wall at a right angle across a circular hole. The tank is considered to have an annular shape with thickness h and inner and outer radii c and b, respectively. A SSI is also considered in the tank, with a radial length $c - a$. Figure 1 shows the geometry of the model. The analysis domain is divided into four regions: Ω_1 is the region existing between the conductor and the tank wall, namely, the circular hole; Ω_1 is defined by $r_0 \leq r \leq a$, $-h/2 \leq z \leq h/2$, where r_0 is the radius of the conductor. Region Ω_2 represents the carbon steel tank wall, that is, $c \leq r \leq b$, $-h/2 \leq z \leq h/2$. Region Ω_3 is the SSI, $a \leq r \leq c$, $-h/2 \leq z \leq h/2$. Finally, region Ω_4 is the medium at both sides of the tank wall, considered as air, $r \geq r_0$, $|z| > h/2$.

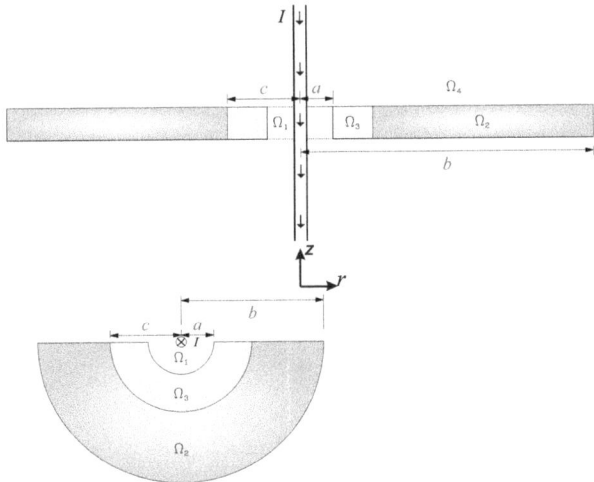

Figure 1. Geometry and parameters of the model.

A conductor passing through the hole transports alternating electric current of the form:

$$I(t) = I e^{j\omega t},$$

where I is the current amplitude. Maxwell's equations in the frequency domain, in each region Ω_k, have the following form:

$$\begin{aligned} \nabla \times \mathbf{E}_k &= -j\omega \mu_0 \mu_k \mathbf{H}_k, & \nabla \cdot \mathbf{H}_k &= 0, \\ \nabla \times \mathbf{H}_k &= \sigma_k \mathbf{E}_k, & \nabla \cdot \mathbf{E}_k &= 0, \end{aligned} \qquad (1)$$

where $k = 1$ corresponds to air and the hole (region Ω_1), $k = 2$ corresponds to the carbon steel (region Ω_2) and $k = 3$ is associated with the stainless insert (region Ω_3). Additionally, $\sigma_1 = 0$ and $\mu_1 = \mu_3 = 1$. Because of the axial symmetry of the system and the symmetry with respect to the plane $z = 0$, the solution to the system of Equations (1) can be sought as follows:

$$\mathbf{H}_k = H_{k\varphi}(r,z)\mathbf{e}_\varphi, \qquad \mathbf{E}_k = E_{kr}(r,z)\mathbf{e}_r + E_{k\varphi}(r,z)\mathbf{e}_\varphi, \qquad (2)$$

where $H_{k\varphi}(r,z)$ is an even function with respect to the variable z; i.e.,

$$H_{k\varphi}(r,-z) = H_{k\varphi}(r,z). \qquad (3)$$

Due to Ampère's circuital law and axial symmetry, the magnetic field in air and the hole is:

$$H_{1\varphi}(r) = \frac{I}{2\pi r}. \qquad (4)$$

Boundary conditions between regions Ω_1 and $\Omega_{2,3}$ after taking into account Equation (4) assume the following form (see [10,14]):

$$H_{3\varphi}\Big|_{r=a} = \frac{I}{2\pi a}, \qquad H_{2\varphi}\Big|_{r=b} = \frac{I}{2\pi b},$$

$$\frac{1}{r}\frac{\partial(rH_{3\varphi})}{\partial r}\Bigg|_{z=h/2} = 0, \qquad \frac{1}{r}\frac{\partial(rH_{2\varphi})}{\partial r}\Bigg|_{z=h/2} = 0, \qquad (5)$$

$$H_{3\varphi}\Big|_{z=h/2} = \frac{I}{2\pi r}, \qquad H_{2\varphi}\Big|_{z=h/2} = \frac{I}{2\pi r}.$$

Additionally, on the boundary that separates regions Ω_2 and Ω_3 we have:

$$H_{3\varphi}\Big|_{r=c} = H_{2\varphi}\Big|_{r=c}, \qquad (6)$$

$$\frac{1}{\sigma_3}\frac{1}{r}\frac{\partial(rH_{3\varphi})}{\partial r}\Bigg|_{r=c} = \frac{1}{\sigma_2}\frac{1}{r}\frac{\partial(rH_{2\varphi})}{\partial r}\Bigg|_{r=c}, \qquad (7)$$

3. Analytical Solution

Maxwell's equations in regions Ω_2 and Ω_3 reduce to:

$$\frac{1}{r}\frac{\partial}{\partial r}\left(r\frac{\partial H_{k\varphi}}{\partial r}\right) + \frac{\partial^2 H_{k\varphi}}{\partial z^2} - \frac{H_{k\varphi}}{r^2} - j\omega\sigma_k\mu_0\mu_k H_{k\varphi} = 0. \qquad (8)$$

Equation (8) has been previously solved [10,14]. The solution of this equation has the following form [14]:

$$H_{k\varphi}(r,z) = \left(\frac{A_k}{r} + B_k r\right)\cosh(\beta_k z)$$
$$+ \sum_{n=0}^{\infty}\Big\{C_{k,n}I_1(\lambda_{k,n}r) + D_{k,n}K_1(\lambda_{k,n}r)\Big\}\cos(\varkappa_n z), \qquad (9)$$

where $I_1(\lambda_{k,n}r)$ and $K_1(\lambda_{k,n}r)$ are the modified Bessel functions of the first order,

$$\beta_k^2 = j\omega\sigma_k\mu_0\mu_k, \qquad \varkappa_n = \frac{(2n+1)\pi}{h}$$
$$\lambda_{k,n}^2 = \varkappa_n^2 + \beta_k^2 \qquad k = 2,3, \qquad n \in \mathbb{Z}^+.$$

Constants A_k, B_k, $C_{k,n}$ and $D_{k,n}$ are to be obtained from boundary conditions (5)–(7). Substitution of solution (9) into the last boundary conditions of (5) yields:

$$B_k = 0, \qquad A_k = \frac{I}{2\pi\cosh(\beta_k h/2)}.$$

As a result,

$$H_{k\varphi}(r,z) = \frac{I}{2\pi r}\frac{\cosh(\beta_k z)}{\cosh(\beta_k h/2)}$$
$$+ \sum_{n=0}^{\infty}\Big\{C_{k,n}I_1(\lambda_{k,n}r) + D_{k,n}K_1(\lambda_{k,n}r)\Big\}\cos(\varkappa_n z), \qquad (10)$$

The convergence of the generalized Fourier-series (10) is provided by the general theory of linear partial differential equations, with Hermitian differential operators and boundary conditions (5)–(7). However, the convergence can be also proved explicitly, which has been done, for instance, in the Appendix section of [10].

At the same time, since function $\cosh(\beta_k z)/\cosh(\beta_k h/2)$ can be expanded in a Fourier series as follows [10]:

$$\frac{\cosh(\beta_k z)}{\cosh(\beta_k h/2)} = \sum_{n=0}^{\infty} \frac{4(-1)^n \varkappa_n}{\lambda_{k,n}^2 h} \cos(\varkappa_n z),$$

then, Equation (10) takes the form:

$$H_{k\varphi}(r,z) = \sum_{n=0}^{\infty} \left\{ C_{k,n} I_1(\lambda_{k,n} r) + D_{k,n} K_1(\lambda_{k,n} r) \right. \\ \left. + \frac{2I}{\pi r} \frac{(-1)^n \varkappa_n}{\lambda_{k,n}^2 h} \right\} \cos(\varkappa_n z), \qquad (11)$$

Let us substitute general solution (11) for $H_{3\varphi}(r,z)$ into boundary condition $H_{3\varphi}\big|_{r=a} = I/2\pi a$. We obtain within the interval $-h/2 \leq z \leq h/2$:

$$\sum_{n=0}^{\infty} \left\{ C_{3,n} I_1(\lambda_{3,n} a) + D_{3,n} K_1(\lambda_{3,n} a) + \frac{2I}{\pi a} \frac{(-1)^n \varkappa_n}{\lambda_{3,n}^2 h} \right\} \cos(\varkappa_n z) = \frac{I}{2\pi a}. \qquad (12)$$

On the other hand, within the same interval, we can write (see [10]):

$$\sum_{n=0}^{\infty} \frac{4(-1)^n}{\varkappa_n h} \cos(\varkappa_n z) = 1.$$

After substituting this result into (12) we come to the following equation:

$$\sum_{n=0}^{\infty} \left\{ C_{3,n} I_1(\lambda_{3,n} a) + D_{3,n} K_1(\lambda_{3,n} a) \right. \\ \left. + \frac{2I}{\pi a} \frac{(-1)^n \varkappa_n}{\lambda_{3,n}^2 h} \right\} \cos(\varkappa_n z) = \sum_{n=0}^{\infty} \frac{2I}{\pi a} \frac{(-1)^n}{\varkappa_n h} \cos(\varkappa_n z), \qquad (13)$$

which, in turn, leads to the following:

$$C_{3,n} I_1(\lambda_{3,n} a) + D_{3,n} K_1(\lambda_{3,n} a) + \frac{2I}{\pi a} \frac{(-1)^n \varkappa_n}{\lambda_{3,n}^2 h} = \frac{2I}{\pi a} \frac{(-1)^n}{\varkappa_n h}.$$

This equation, after some simple algebraic operations, becomes:

$$C_{3,n} I_1(\lambda_{3,n} a) + D_{3,n} K_1(\lambda_{3,n} a) = \frac{2I}{\pi a} \frac{(-1)^n \beta_3^2}{\lambda_{3,n}^2 \varkappa_n h}. \qquad (14)$$

A similar result can be obtained from boundary condition $H_{2\varphi}\big|_{r=b} = I/2\pi b$:

$$C_{2,n} I_1(\lambda_{2,n} b) + D_{2,n} K_1(\lambda_{2,n} b) = \frac{2I}{\pi b} \frac{(-1)^n \beta_2^2}{\lambda_{2,n}^2 \varkappa_n h}. \qquad (15)$$

Boundary conditions (6) and (7) result in the following equations:

$$C_{2,n} I_1(\lambda_{2,n} c) + D_{2,n} K_1(\lambda_{2,n} c) + \frac{2I}{\pi c} \frac{(-1)^n \varkappa_n}{\lambda_{2,n}^2 h}$$
$$= C_{3,n} I_1(\lambda_{3,n} c) + D_{3,n} K_1(\lambda_{3,n} c) + \frac{2I}{\pi c} \frac{(-1)^n \varkappa_n}{\lambda_{3,n}^2 h} \quad (16)$$

$$\frac{\lambda_{2,n}}{\sigma_2} \left\{ C_{2,n} I_0(\lambda_{2,n} c) - D_{2,n} K_0(\lambda_{2,n} c) \right\}$$
$$= \frac{\lambda_{3,n}}{\sigma_3} \left\{ C_{3,n} I_0(\lambda_{3,n} c) - D_{3,n} K_0(\lambda_{3,n} c) \right\} \quad (17)$$

The system of linear Equations (14)–(17), with respect to the constants $C_{k,n}$ and $D_{k,n}$, $k = \overline{1,n}$, is easy to solve. However, there is no necessity to solve this system of equations exactly, since the magnetic field rapidly decays for an increasing outer radius b. This principle can be formally taken into account in the system of Equations (14)–(17) by considering radius b to be sufficiently high (formally, $b \to \infty$). The validity of this assumption for the calculation of transformer tanks losses has been shown previously in [10,14]. By applying the limit to Equation (15) when $b \to \infty$ and taking into account the asymptotic behavior of the modified Bessel functions $I_1(x)$ and $K_1(x)$, namely,

$$I_1(x) \propto \frac{e^x}{\sqrt{2\pi x}} \left(1 + \mathcal{O}(x^{-1})\right),$$

$$K_1(x) \propto \sqrt{\frac{\pi}{2x}} e^{-x} \left(1 + \mathcal{O}(x^{-1})\right),$$

it follows that $C_{2,n} = 0$. Through substitution of the result into Equations (14)–(17) and solving this system of equations with respect to $C_{3,n}$, $D_{3,n}$ and $D_{2,n}$, we obtain:

$$C_{3,n} = \frac{2I}{\pi \Delta} \frac{(-1)^n}{\lambda_{3,n}^2 h} \left\{ \frac{\beta_3^2}{\varkappa_n a} \left[\frac{\lambda_{2,n}}{\sigma_2} K_0(\lambda_{2,n} c) K_1(\lambda_{3,n} c) - \frac{\lambda_{3,n}}{\sigma_3} K_1(\lambda_{2,n} c) K_0(\lambda_{3,n} c) \right] \right.$$
$$\left. - \frac{\varkappa_n (\beta_3^2 - \beta_2^2)}{\lambda_{2,n} \sigma_2 c} K_0(\lambda_{2,n} c) K_1(\lambda_{3,n} a) \right\}, \quad (18)$$

$$D_{3,n} = -\frac{2I}{\pi \Delta} \frac{(-1)^n}{\lambda_{3,n}^2 h} \left\{ \frac{\beta_3^2}{\varkappa_n a} \left[\frac{\lambda_{2,n}}{\sigma_2} K_0(\lambda_{2,n} c) I_1(\lambda_{3,n} c) + \frac{\lambda_{3,n}}{\sigma_3} K_1(\lambda_{2,n} c) I_0(\lambda_{3,n} c) \right] \right.$$
$$\left. - \frac{\varkappa_n (\beta_3^2 - \beta_2^2)}{\lambda_{2,n} \sigma_2 c} K_0(\lambda_{2,n} c) I_1(\lambda_{3,n} a) \right\} \quad (19)$$

and

$$D_{2,n} = \frac{2I}{\pi \Delta c} \frac{(-1)^n}{\lambda_{3,n} \sigma_3 h} \left\{ \varkappa_n \frac{\beta_3^2 - \beta_2^2}{\lambda_{2,n}^2} \left[K_1(\lambda_{3,n} a) I_0(\lambda_{3,n} c) + I_1(\lambda_{3,n} a) K_0(\lambda_{3,n} c) \right] - \frac{\beta_3^2}{\varkappa_n \lambda_{3,n} a} \right\}, \quad (20)$$

where

$$\Delta = \left[\frac{\lambda_{2,n}}{\sigma_2} K_0(\lambda_{2,n} c) K_1(\lambda_{3,n} c) - \frac{\lambda_{3,n}}{\sigma_3} K_1(\lambda_{2,n} c) K_0(\lambda_{3,n} c) \right] I_1(\lambda_{3,n} a)$$
$$- \left[\frac{\lambda_{2,n}}{\sigma_2} K_0(\lambda_{2,n} c) I_1(\lambda_{3,n} c) + \frac{\lambda_{3,n}}{\sigma_3} K_1(\lambda_{2,n} c) I_0(\lambda_{3,n} c) \right] K_1(\lambda_{3,n} a). \quad (21)$$

Then, constant $C_{2,n}$ can be approximately calculated from Equation (15) as follows:

$$C_{2,n} = \frac{1}{I_1(\lambda_{2,n}b)} \left\{ \frac{2I}{\pi b} \frac{(-1)^n \beta_2^2}{\lambda_{2,n}^2 \varkappa_n h} - D_{2,n} K_1(\lambda_{2,n}b) \right\}. \tag{22}$$

Solution (10) is an infinite sum that can be truncated at a term with number $N-1$ and by introducing the Lanczos sigma factor (see [16,17]):

$$\varsigma_n = \frac{\sin(\pi n/N)}{\pi n/N} \tag{23}$$

to suppress Gibbs' oscillations. Then, solution (11) takes the following form:

$$H_{k\varphi}(r,z) = \frac{I}{2\pi r} \frac{\cosh(\beta_k z)}{\cosh(\beta_k h/2)}$$
$$+ \sum_{n=0}^{N-1} \varsigma_n \left\{ C_{k,n} I_1(\lambda_{k,n} r) + D_{k,n} K_1(\lambda_{k,n} r) \right\} \cos(\varkappa_n z), \tag{24}$$

4. Electric Field and Eddy Current Losses

Eddy current losses in the transformer tank wall have an ohmic nature. The averaged power loss density over a period is as follows [10,14]:

$$P(r,z) = \frac{1}{T} \int_0^T \sigma(r,z) |\mathbf{E}(r,z,t)|^2 dt = \frac{|\mathbf{j}(r,z)|^2}{2\sigma(r,z)}, \tag{25}$$

where $T = 2\pi/\omega$ is the period. The tank wall conductivity $\sigma(r,z)$ is a function of the coordinates due to the insert in the tank wall, and $\mathbf{j}(r,z)$ is the current density in the frequency domain. Then, the total losses in the tank wall are [10]:

$$P_{tot} = \int_0^{2\pi} d\varphi \int_a^b r dr \int_{-h/2}^{h/2} dz\, P(r,z) = \pi \int_a^b r dr \int_{-h/2}^{h/2} dz \frac{|\mathbf{j}(r,z)|^2}{\sigma(r,z)}$$
$$= \frac{\pi}{\sigma_3} \int_a^c r dr \int_{-h/2}^{h/2} dz \left(|j_{3,r}(r,z)|^2 + |j_{3,z}(r,z)|^2 \right)$$
$$+ \frac{\pi}{\sigma_2} \int_c^b r dr \int_{-h/2}^{h/2} dz \left(|j_{2,r}(r,z)|^2 + |j_{2,z}(r,z)|^2 \right). \tag{26}$$

The current density can be obtained from Maxwell's Equations (1) as follows:

$$\mathbf{j}_k = \nabla \times \mathbf{H}_k = j_{k,r}(r,z)\mathbf{e}_r + j_{k,z}(r,z)\mathbf{e}_z, \tag{27}$$

where

$$j_{k,r}(r,z) = -\frac{I\beta_k}{2\pi r} \frac{\sinh(\beta_k z)}{\cosh(\beta_k h/2)}$$
$$+ \sum_{n=0}^{N-1} \varsigma_n \varkappa_n \left\{ C_{k,n} I_1(\lambda_{k,n} r) + D_{k,n} K_1(\lambda_{k,n} r) \right\} \sin(\varkappa_n z) \tag{28}$$

and

$$j_{k,z}(r,z) = \sum_{n=0}^{N-1} \varsigma_n \lambda_{k,n} \left\{ C_{k,n} I_0(\lambda_{k,n} r) - D_{k,n} K_0(\lambda_{k,n} r) \right\} \cos(\varkappa_n z) \tag{29}$$

5. Study Cases and Discussion

Several study cases are carried out in this section with different SSI sizes in the tank wall. In order to compare the results obtained with our analytical formulas, the solution in each case is also computed with a 2D FE model [18]. The low-carbon steel has the following properties: a relative permeability $\mu_r = 100$, a conductivity $\sigma = 7.0 \times 10^6$ S/m and a relative permittivity $\epsilon_r = 1$. The SSI has a relative permeability $\mu_r = 1.0$, a conductivity $\sigma = 1.1 \times 10^6$ S/m and a relative permittivity $\epsilon_r = 1$. The model dimensions in all the cases are $a = 8.5$ cm, $b = 34$ cm and $h = 12.7$ mm. The current I carried by the conductor is 5000 A at frequency $f = 60$ Hz. Various radial distances of the insert $(c - a)$ are considered, which are obtained by varying the percentage of the tank wall volume that corresponds to the SSI. This means that if the insert volume is 0 %, the tank wall is made only of low-carbon steel and therefore $c = a$. On other hand, if the insert volume is 100%, the tank wall would be made exclusively of the stainless steel and its radial distance would be given by $c - a = b - a = 255$ mm.

The numerical solution is obtained with a time-harmonic 2D eddy current axisymmetric FE model. A special 2D formulation is applied considering that field configuration is such that the magnetic field has only one component normal to the plane and the current density has its components in the plane, which matches with our problem. This formulation is incorporated into our FE code FLD and has been compared with analytical and 3D FE solutions, obtaining great accuracy in all cases [18]. FLD is a set of computer programs and routines, developed by the authors, for the analysis of electromagnetic problems using the FE method, which is programmed in Fortran 95 [19]. Using a 2D FE model permits a large number of simulations in a much shorter time and without the computational cost of a 3D model. Moreover, the geometry of the proposed model is represented faithfully by a 2D axisymmetric model.

The analysis domain considered in the FE axisymmetric model is composed by regions Ω_3 and Ω_4. A Dirichlet boundary condition, obtained by applying Ampère's circuital law (4), is assigned to nodes located at the periphery of the model. Figure 2 shows the FE meshes used to solve two different cases. Figure 2a is the mesh used for the case where the tank wall of the transformer lacks an SSI. Figure 2b shows a case where an insert exists, modeled by the left blue region. The mesh used for each case results from an automatic mesh adaptation procedure; regions with a rapid variation of the field, after the iterative procedure, will contain a higher density of elements. In all cases second order elements were employed. Details of the implementation of the automatic mesh adaptation are reported in [18]. In both cases, most of the elements are in the periphery of the region Ω_2, consisting of low-carbon steel, due to the skin effect in this material.

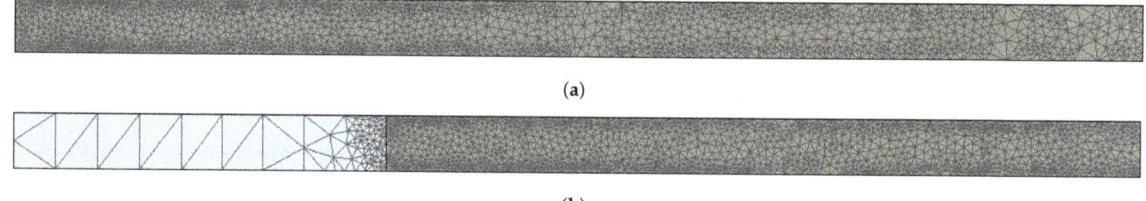

Figure 2. Mesh of the 2D FE axisymmetric model. (a) Tank wall without insert, case $c = a$. (b) Tank wall considering an insert, case $c - a = 85.0$ mm.

In Figure 3 the magnetic field distribution of $H_\varphi(r,z)$, obtained with (4) and (24), is presented for different radii of the SSI. Observe how the magnetic field penetrates the tank wall according with the SSI size. Since the low-carbon steel possesses greater permeability and conductivity than the stainless steel, it has a smaller depth of penetration (δ). Hence, the magnetic field decays rapidly in region Ω_2, whereas the magnetic field penetrates easily to region Ω_3.

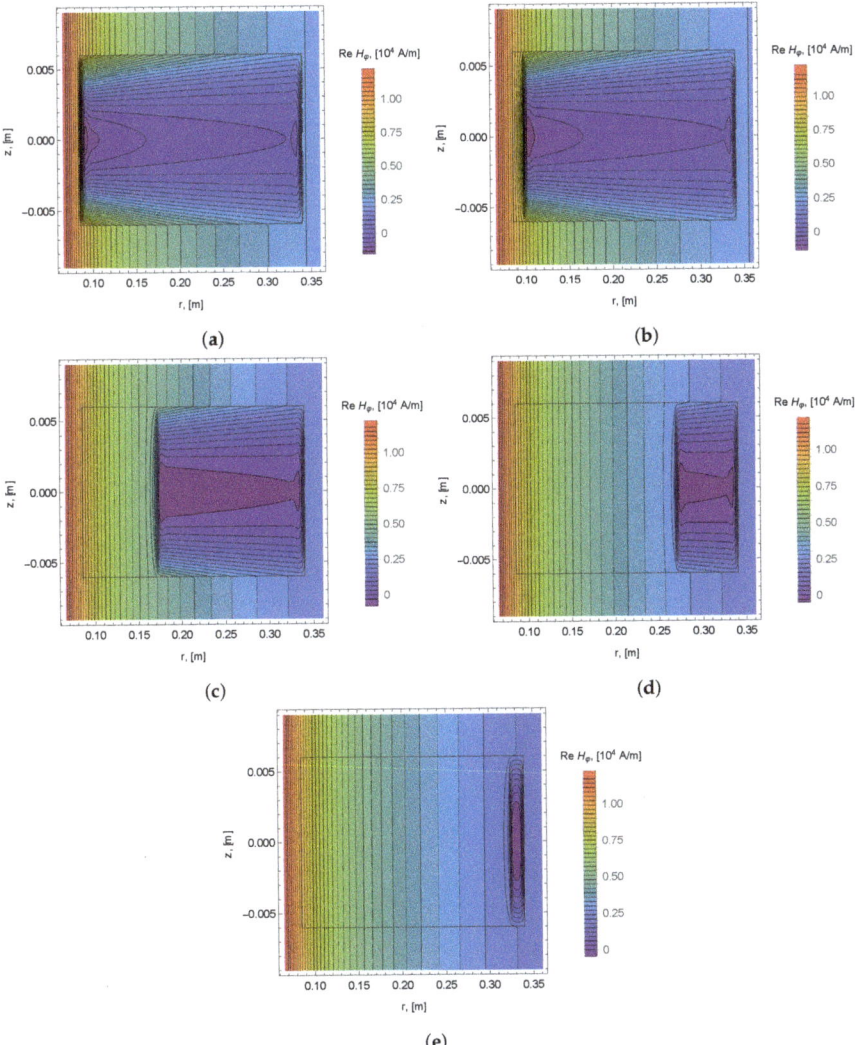

Figure 3. Magnetic field distribution inside the tank wall. (**a**) Case $c = a$. (**b**) Case $c − a = 11.914$ mm. (**c**) Case $c − a = 85$ mm. (**d**) Case $c − a = 183.89$ mm. (**e**) Case $c − a = 238.67$ mm.

Figure 4 shows two cases of the magnetic field penetration in the tank wall and the eddy current density obtained with 2D FE simulation. Figure 4a is the case where there is no insert. In this solution the skin effect phenomena in the tank wall can also be seen, producing the greater magnitude of H_φ in the inner radius, the nearest region to the conductor. Figure 4b presents the case in which the radial distance of the stainless insert is $c − a = 85$ mm. In this case, the closed path of the eddy currents in the tank can be seen. The magnitude of the eddy current density is greater in the low-carbon steel.

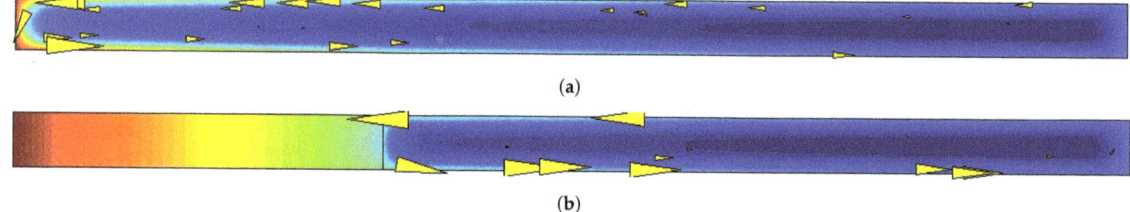

Figure 4. Magnitude of H_φ (color) and eddy current density (arrows) in the tank wall. (**a**) Case $c = a$. (**b**) Case $c - a = 85.0$ mm.

Figure 5 shows the magnetic field H_φ evaluated at the center of the tank wall ($z = 0$) for different insert radii. For all cases, the analytical solution, developed in this work and the numerical one obtained with the 2D FE model are compared. It can be seen that for all the cases, the values of H_φ calculated analytically match very closely the values obtained numerically, demonstrating the validity of the analytical formula (24). In these graphs, the behavior of the magnetic field inside the tank wall can be noticed more clearly: when H_φ penetrates in the low-carbon steel, it decreases rapidly, having a greater variation at the edges of the tank. The magnetic field outside of the tank wall decreases according to the distance to the conductor, as established in (4). Table 1 shows the relative error between the analytical solution and numerical solutions calculated for these cases. The maximum relative error obtained is 1.71%, which demonstrates the validity of the solutions.

Table 1. Relative error between solutions.

Case $c - a$ (mm)	Relative Error (%)
0	0.52
11.914	1.37
85	0.83
183.79	1.71
238.67	0.45

The relative error is calculated with

$$\text{relative error} = \frac{\max\limits_{i=1,n} |f_i - g_i|}{\max\limits_{i=1,n} |g_i|} \, 100\%$$

where f_i and g_i are the numerical and analytical solutions respectively, evaluated at point i, while n is the total number of points considered.

The eddy current losses P_e in the bushing region were calculated for several cases using the analytic expressions (26), (28) and (29), and were compared with the losses estimated using 2D FE simulations. Table 2 presents the eddy current power losses obtained for thirteen different configurations of the tank wall, varying the volume occupied by the stainless insert. In each case, the corresponding radial distance of the insert is shown. It can be observed that the losses calculated in all cases differ by less than 3%, which confirms the correctness of the analytical expressions presented previously. Figure 6 presents also the power losses in the tank wall, calculated with the two approaches, in a graphical way. This graph shows the reduction of the total power losses in the tank wall according to the increment of the radial distance $c - a$, as expected. Although these results could point to the use of transformer tanks made exclusively with stainless steel, at least in the bushing region, this material is more expensive than the low-carbon steel, meaning a greater investment to manufacture the transformer.

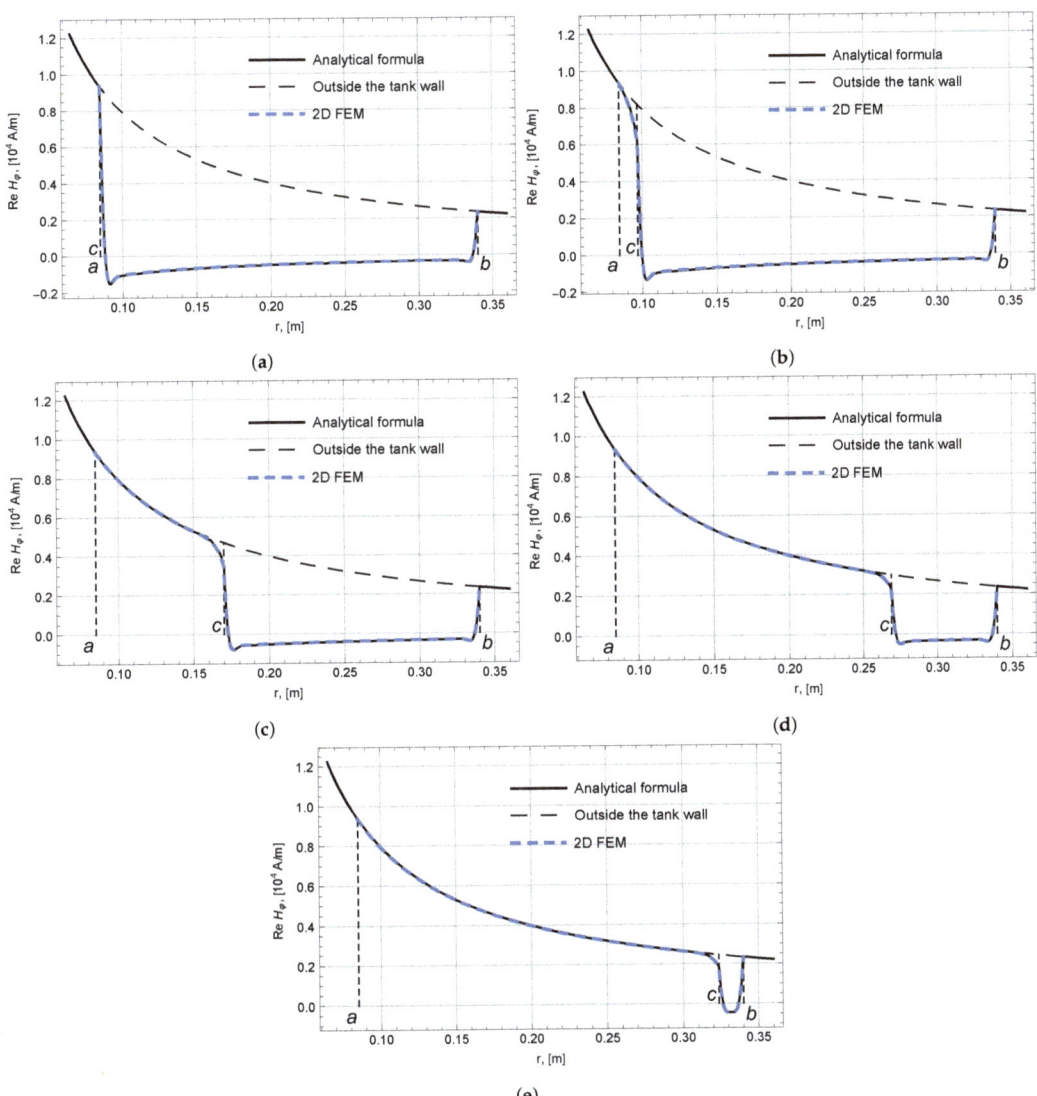

Figure 5. Magnetic field H_φ evaluated at $z = 0$ and $z = h/2$. (**a**) Case $c = a$. (**b**) Case $c - a = 11.914$ mm. (**c**) Case $c - a = 85$ mm. (**d**) Case $c - a = 183.89$ mm. (**e**) Case $c - a = 238.67$ mm.

Table 2. Eddy current losses in the tank wall.

Case	Insert Volume (%)	Radial Distance Insert $c - a$ (mm)	P_e (W) Analytical	P_e (W) Numerical	Relative Error (%)
1	0	0	334.345	333.313	0.308
2	1	6.152	317.872	316.813	0.333
3	2	11.914	303.079	302.003	0.355
4	5	27.444	267.309	266.387	0.344
5	10	49.396	224.551	223.852	0.311
6	20	85.000	168.447	168.018	0.254
7	30	114.34	130.565	130.330	0.18
8	40	139.88	101.941	101.836	0.103
9	50	162.81	78.918	78.965	0.059
10	60	183.89	59.669	59.799	0.217
11	70	203.24	43.135	43.380	0.568
12	80	221.47	28.630	28.975	1.205
13	90	238.67	15.724	16.142	2.658

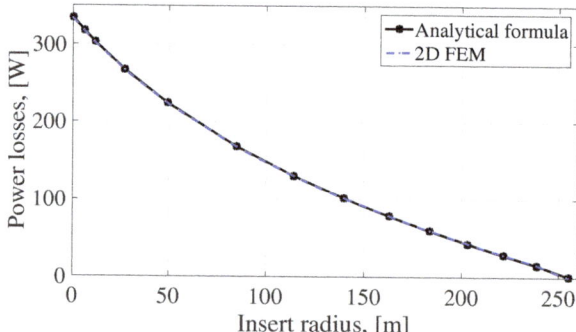

Figure 6. Power losses for different SSI dimensions.

6. Limits and Applicability of the Analytical Solution

Solution (24) was obtained under the assumption of axial symmetry. The edge effect on the external border was neglected by formally taking the limit: $b \to \infty$. This approach (axial symmetry) is expected to be applicable to other cases such as when the conductor crosses the covering plate not at the plate centre, but closer to the border of the transformer cover. This is due to the skin effect, making the magnetic field decay exponentially from the hole border (see Figure 7). However, in a layer near the external border, the magnetic field increases exponentially up to values of the magnetic field outside the plate (Figure 7). Therefore, if the conductor is situated too close to the border, the edge effect may become considerable.

Nevertheless, there is a case where this effect could become considerable. Just near the external border, the magnetic field increases exponentially up to values of the magnetic field outside the plate (see Figure 7). If the conductor is far away from the external border, the magnetic field quantity near the plate border is small enough so that the external border effect can be neglected. In the case of the conductor crossing the plate nearby the plate border, the edge effect becomes considerable. However, this effect is presented only in a small region shown in Figure 8, so that its contribution to the complete value of eddy current losses is small. Therefore, the only restriction of our analytical solution is $\delta_1 \ll d$, which is normally accomplished in actual transformers.

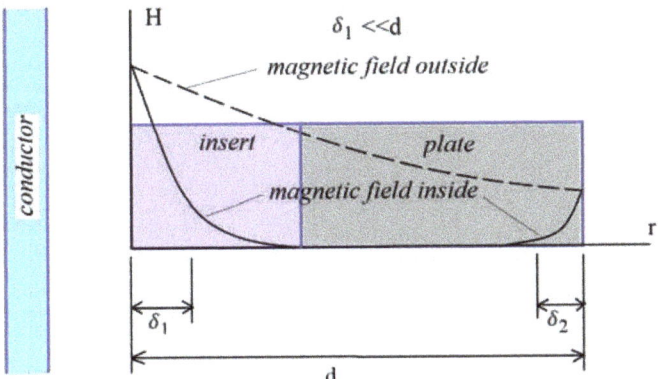

Figure 7. Magnetic field behaviour in the insert and the neighbouring plate material.

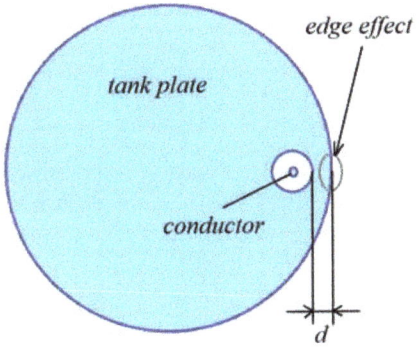

Figure 8. Limiting case of a conductor close to the plate border.

7. Conclusions

A new analytical model to determine the magnetic field distribution around the bushing region in the transformer tank, and considering the existence of a SSI, has been developed. The results obtained with the proposed model were validated with detailed 2D FE simulations. All the cases considered show great concordance between analytical and numerical solutions. A relative error was calculated in order to compare quantitatively the analytical model with the numerical one. This way, it was shown that the solutions differed by less than 2% in all the simulated cases. A formula to calculate the eddy current losses was also developed. The power losses were calculated for a total of thirteen different cases varying the radial distance of the insert, and again the analytical results were very close to the numerical ones. The results show that stray losses in the tank are reduced with increases in the SSI dimensions. The relative errors between stray losses, calculated with the analytical and the FE models, for all cases were less than 3%.

These equations can serve as a basis to develop a thermal analysis in the bushing regions or an economical analysis of the insert cost against the savings due to the power loss reduction. Therefore, the new analytical model can be a useful tool for transformer designers who are interested in obtaining the optimal size of non-magnetic inserts in the tank wall, according to the rated values and dimensions of the transformer.

Author Contributions: Conceptualization, J.C.O.-G., S.M. and M.A.C.-S.; methodology, S.M., M.A.C.-S. and J.C.O.-G.; software, M.A.C.-S., E.M.-V., R.E.-P. and V.M.J.-M.; validation, S.M. and M.A.C.-S.; formal analysis, S.M. and M.A.C.-S.; investigation, S.M. and M.A.C.-S.; resources, J.C.O.-G., R.E.-P. and V.M.J.-M.; data curation, S.M. and M.A.C.-S.; writing—original draft preparation, S.M. and M.A.C.-S.; writing—review and editing, J.C.O.-G., E.M.-V. and R.E.-P.; visualization, S.M., J.C.O.-G., E.M.-V., R.E.-P. and V.M.J.-M.; supervision, S.M., J.C.O.-G., E.M.-V. and R.E.-P.; project administration, J.C.O.-G. and V.M.-J.M.; funding acquisition, J.C.O.-G., R.E.-P. and V.M.J.-M. All authors have read and agreed to the published version of the manuscript.

Funding: The authors are grateful for the financial support provided by the following CONACYT projects: CBS-2015/256519 and CB-2015/257598.

Conflicts of Interest: The authors declare no conflict of interest.

References

1. M. Del Vecchio, R.; Poulin, B.; Feghali, P.T.; Shah, D.M.; Ahuja, R. *Transformer Design Principles: With Applications to Core-Form Power Transformers*; CRC Press: Boca Raton, FL, USA, 2017.
2. Huerta-Rosales, J.R.; Granados-Lieberman, D.; Amezquita-Sanchez, J.P.; Camarena-Martinez, D.; Valtierra-Rodriguez, M. Vibration Signal Processing-Based Detection of Short-Circuited Turns in Transformers: A Nonlinear Mode Decomposition Approach. *Mathematics* **2020**, *8*, 575. [CrossRef]
3. Karsai, K.; Kerényi, D.; Kiss, L. *Large Power Transformers*; Studies in Electrical and Electronic Engineering 25; Elsevier: Amsterdam, The Netherlands, 1987.
4. Turowski, J.; Pelikant, A. Eddy Current Losses and Hot-spot Evaluation in Cover Plates of Power Transformers. *EE Proc. Electr. Power Appl.* **1997**, *144*, 435–440. [CrossRef]
5. Turowski, J. Losses in Cover Plates of single and Three Phase Power Transformers. *Rozpr. Elektrotech.* **1959**, *1*, 87–119.
6. Olivares, J.C.; Escarela-Perez, R.; Kulkarni, S.; Leon, F.D.; Melgoza-Vasquez, E.; Hernández-Anaya, O. Improved Insert Geometry for Reducing Tank-wall Losses in Pad-mounted Transformers. *IEEE Trans. Power Deliv.* **2004**, *19*, 1120–1126. [CrossRef]
7. Olivares, J.C.; Escarela-Perez, R.; Kulkarni, S.; Leon, F.D.; Venegas-Vega, M. 2D Finite-element Determination of Tank Wall Losses in Pad-mounted Transformers. *Electr. Power Res.* **2004**, *71*, 179–185. [CrossRef]
8. Kumbhar, G.B.; Mahajan, S.M.; Collett, W.L. Reduction of Loss and Local Overheating in the Tank of a Current Transformer. *IEEE Trans. Power Deliv.* **2010**, *25*, 2519–2525. [CrossRef]
9. Lopez-Fernandez, X.M.; Penabad-Duran, P.; Turowski, J. Three-Dimensional Methodology for the Overheating Hazard Assessment on Transformer Covers. *IEEE Trans. Ind. Appl.* **2012**, *48*, 1549–1555. [CrossRef]
10. Maximov, S.; Olivares-Galvan, J.C.; Escarela-Perez, R.; Magdaleno-Adame, S.; Campero-Littlewood, E. New Analytical Formulae for Electromagnetic Field and Eddy Current Losses in Bushing Regions of Transformers. *IEEE Trans. Magn.* **2015**, *51*, 6300710. [CrossRef]
11. Penabad-Durán, P.; Barba, P.D.; Lopez-Fernandez, X.; Turowski, J. Electromagnetic and Thermal Parameter Identification Method for Best Prediction of Temperature Distribution on Transformer Tank Covers. *COMPEL Int. J. Comput. Math. Electr. Electron. Eng.* **2015**, *34*, 485–495. [CrossRef]
12. Zahedi, M.Z.; Iskender, I. Evaluation of Eddy Current Losses in the Cover Plates of Distribution Transformers. *Int. J. Tech. Phys. Probl. Eng.* **2018**, *10*, 27–33.
13. Zia Zahedi, M.; Iskender, I. Nonlinear Adaptive Magneto-Thermal Analysis at Bushing Regions of a Transformers Cover Using Finite Difference Method. *J. Therm. Sci. Eng. Appl.* **2018**, *11*. [CrossRef]
14. Khan, S.; Maximov, S.; Escarela-Perez, R.; Olivares-Galvan, J.C.; Melgoza-Vazquez, E.; Lopez-Garcia, I. Computation of Stray Losses in Transformer Bushing Regions Considering Harmonics in the Load Current. *Appl. Sci.* **2020**, *10*, 3527. [CrossRef]
15. De Oliveira, L.F.; Sadowski, N.; Cabral, S.H.L. Alternative Model for Computing Transformer Tank Induced Losses in the Time Domain. *IET Electr. Power Appl.* **2020**. [CrossRef]
16. Lanczos, C. Applied Analysis. *Phys. Today 10* **1957**, *6*, 44. [CrossRef]
17. Wilcox, D.J.; Conlon, M.; Hurley, W.G. Calculation of Self and Mutual Impedance for Coils on Ferromagnetic Cores. *IEE Proc.* **1988**, *135*, 470–476. [CrossRef]
18. Corona-Sánchez, M.A.; Melgoza-Vázquez, E.; Maximov, S.; Escarela-Perez, R. An Improved Time-Harmonic 2-D Eddy Current Finite-Element H Formulation. *IEEE Trans. Magn.* **2017**, *53*, 1–4. [CrossRef]
19. Melgoza, E.; Escarela-Pérez, R.; Corona-Sánchez, M.A. Applications of Computational Electromagnetism in Electric Power Engineering. In Proceedings of the 2016 IEEE PES Transmission Distribution Conference and Exposition-Latin America (PES T& D-LA), Morelia, Mexico, 20–24 September 2016; pp. 1–6.

Article

Determination of Aircraft Cruise Altitude with Minimum Fuel Consumption and Time-to-Climb: An Approach with Terminal Residual Analysis

Taehak Kang [1] and Jaiyoung Ryu [1,2,*]

1. Department of Mechanical Engineering, Chung-Ang University, Seoul 06974, Korea; taehak94@cau.ac.kr
2. Department of Intelligent Energy and Industry, Chung-Ang University, Seoul 06974, Korea
* Correspondence: jairyu@cau.ac.kr; Tel.: +82-2-820-5279

Abstract: A pandemic situation of COVID-19 has made a cost-minimization strategy one of the utmost priorities for commercial airliners. A relevant scheme may involve the minimization of both the fuel- and time-related costs, and the climb trajectories of both objectives were optimized to determine the optimum aircraft cruise altitude. The Hermite-Simpson method among the direct collocation methods was employed to discretize the problem domain. Novel approaches of terminal residual analysis (TRA), and a modified version, m-σ TRA, were proposed to determine the goals. The multi-objective cruise altitude (MOCA) was different by 2.5%, compared to the one statistically calculated from the commercial airliner data. The present methods, TRA and m-σ TRA were powerful tools in finding a solution to this complex problem. The value σ also worked as a transition criterion between a single- and multi-objective climb path to the cruise altitude. The exemplary MOCA was determined to be 10.91 and 11.97 km at σ = 1.1 and 2.0, respectively. The cost index (CI) varied during a flight, a more realistic approach than the one with constant CI. With validated results in this study, TRA and m-σ TRA may also be effective solutions to determine the multi-objective solutions in other complex fields.

Keywords: multi-objective optimization; cruise altitude; fuel consumption; time to climb; Hermite-Simpson method; trajectory optimization; terminal residual analysis (TRA); m-σ terminal residual analysis (m-σ TRA)

Citation: Kang, T.; Ryu, J. Determination of Aircraft Cruise Altitude with Minimum Fuel Consumption and Time-to-Climb: An Approach with Terminal Residual Analysis. *Mathematics* **2021**, 9, 147. https://doi.org/10.3390/math9020147

Received: 15 November 2020
Accepted: 7 January 2021
Published: 11 January 2021

Publisher's Note: MDPI stays neutral with regard to jurisdictional claims in published maps and institutional affiliations.

Copyright: © 2021 by the authors. Licensee MDPI, Basel, Switzerland. This article is an open access article distributed under the terms and conditions of the Creative Commons Attribution (CC BY) license (https://creativecommons.org/licenses/by/4.0/).

1. Introduction

The outbreak of COVID-19 made the year 2020 the worst year in the history of the airline industry, with a net loss of 84.3 billion dollars [1]. The official reports from the Bureau of Transportation Statistics (BTS) of U.S. already confirmed a net loss of 5.2 billion and 11 billion dollars in the first [2] and second [3] quarters, respectively, of the year 2020. Revenue passenger kilometer (RPK) fell by 55% and cargo & mail ton-kilometer (CTK) by 16.8%, and the year 2021 is also expected to be unfavorable [1]. Although Ref. [1] notes that the overall prediction of the performance of the airline industry is favorable in 2021, the end of this pandemic is still unpredictable. Amidst a harsh pandemic environment, tight cost management became even more crucial, and the authors propose a rational, cost-minimizing approach. Since fuel expenditure is almost a quarter (23.7%) of the total operating cost (TOC) [1], and the time of arrival is also as essential as fuel costs, the focus of our study is on the minimization of both the fuel consumption (FC) and the time of travel (TT).

The optimization of FC and TT indeed is one of the major priorities in the literature as multi-objective [4–10] and single-objective [11–15] studies using various methods of Gauss Pseudo-spectral method [16] (sometimes accompanied by Chebyshev direct method [4]), energy-state [11,12], genetic algorithm [5], particle swarm [10], direct collocation methods [17,18], and statistical approach [9]. The optimization of each phase in a flight profile, consisting of the climb, cruise, and descent, had distinct merits for improvements. First, the cruise phase contributes the most to the TOC, even more for a long-range flight. Aircraft under the descent phase uses all of their potential and kinetic energy for deceleration, but the optimization is primarily for safety [19,20]. During the climb phase, optimized ascension protocols such as continuous climb operations (CCOs) yield a considerable reduction in fuel consumption [21,22]. The authors chose the climb phase for multi-objective optimization because of its potential for further optimization [9,12].

Overall, the main objective of this study is to minimize both FC and TT in conjunction with the climb phase. In this case, TT became the time-to-climb (TTC) to describe the time spent to ascend to the desired optimum cruise altitude. From the perspective of aerodynamics, cruising at higher altitudes reduces the aerodynamic drag and increases cruise speed at a given thrust, but there is extra fuel consumption for ascension [23]. Cruising at a lower altitude implies more aerodynamic drag and lower aircraft speed, which resulted in increased TT. An optimally determined cruise altitude would save both fuel and time, potentially achievable by optimizing the aircraft performance during the ascending phase. This study does not adapt the conventional approach of total cost minimization involving constant cost index (CI) throughout a flight path [3,5,7] but shows that CI changes over the course of the climb path. Moreover, wind shear conditions were not considered, and only conventional International Standard Atmosphere (ISA) conditions were used. This ISA model is equivalent to the 1976 US standard atmosphere model [24], and hence the atmosphere model used in the present study was named "1976 US atmospheric model". This study presents a distinctive method, effectively determining the optimum cruise altitude of a generic supersonic aircraft using residual analysis arguments, incorporating two minimizing objectives, FC and TTC. The rationale for the use of supersonic jet data was given in the next paragraph and Section 4.

The contents start with a brief description of the indirect method and explain the difficulties involved in finding the analytic solution of complex problems in Section 2.1. Section 2.2 describes the direct methods, which had gained popularity with the increase in computing power in recent decades. The study justifies the use of direct methods over indirect methods. There are several direct methods such as the Gauss-Lobatto [25] and the Gauss Pseudo-spectral method [16], but the Hermite-Simpson method was chosen. The procedure with the chosen method for the discretization of the problem domain in Section 2.2 as well. The calculated cruise altitude result was compared to the statistically derived result from subsonic, commercial airplanes in Ref. [9]. The present study intended to calculate the cruise altitude of an aircraft by optimizing the climb trajectory in three-dimensional thrust-Mach-altitude space. Such data in the literature using subsonic commercial aircraft [9,22] was available only in stage-wise aerodynamic data having single column data at each flight stage, and hence, the authors used supersonic jet data [12,26] with boundary conditions set for subsonic operations. The details about the models were described in Section 2.3, and auxiliary models and novel methods used in this paper were noted in Sections 2.4–2.6. The optimum climb trajectories were obtained for each targeted objective, minimum FC and TTC. Then, the results (Section 3) were presented followed by discussions (Section 4) and conclusion (Section 5).

2. Methods

The following Sections 2.1 and 2.2 describe the general formulation of the optimal control problem (OCP).

2.1. Optimal Control in Continuous-Time Domain

In general, the formulation of continuous-time optimal control problem with no path constraints on the states or the control variables and fixed initial and final time t_0 and t_E can be defined as follows:

$$J = \Phi(x(t_E)) + \int_{t_0}^{t_E} L(x(t), u(t), t)\, dt \tag{1}$$

where the control vector trajectory $u : [t_0, t_E] \subset \mathbb{R} \mapsto \mathbb{R}^{n_u}$ is usually minimized in the performance index $J : [t_0, t_E] \times \mathbb{R}^{n_x} \times \mathbb{R}^{n_u} \times \mathbb{R} \mapsto \mathbb{R}$. Equation (1) is subject to:

$$\dot{x}(t) = f(x(t), u(t), t), x(t_0) = x_0 \tag{2}$$

where $[t_0, t_E]$ is the time interval of the problem domain, $x : [t_0, t_E] \mapsto \mathbb{R}^{n_x}$ is the state vector, $\Phi : \mathbb{R}^{n_x} \times \mathbb{R} \mapsto \mathbb{R}$ is a terminal cost function, $L : \mathbb{R}^{n_x} \times \mathbb{R}^{n_u} \times \mathbb{R} \mapsto \mathbb{R}$ is an intermediate cost function, and $f : \mathbb{R}^{n_x} \times \mathbb{R}^{n_u} \times \mathbb{R} \mapsto \mathbb{R}^{n_x}$ is a vector field. This formulation is expressed in the Bolza form where Equation (2) describes the dynamics of the system with the corresponding initial conditions. Here, $\Phi(x(t_E))$ is the Mayer term and $L(x(t), u(t), t)$ is the Lagrange term.

When constraints are involved, a time dependent Lagrange multiplier vector function $\lambda : [t_0, t_E] \mapsto \mathbb{R}^{n_x}$, also known as co-state, is introduced to define an augmented performance index $\bar{J}$, which is defined as:

$$\bar{J} = \Phi(x(t_E)) + \int_{t_0}^{t_E} \left(L(x(t), u(t), t) + \lambda^T(t) [f(x(t), u(t), t) - \dot{x}] \right) dt. \tag{3}$$

The Hamiltonian function $\mathcal{H}$ then is defined as:

$$\mathcal{H}(x(t), u(t), \lambda(t), t) = L(x(t), u(t), t) + \lambda^T(t) f(x(t), u(t), t), \tag{4}$$

which can re-write Equation (3) as:

$$\bar{J} = \Phi(x(t_E)) + \int_{t_0}^{t_E} \left(\mathcal{H}(x(t), u(t), \lambda(t), t) - \dot{\lambda}^T(t) \dot{x} \right) dt. \tag{5}$$

When time t_0 and t_E are fixed, an infinitesimal variation in $u(t)$, $x(t)$, and $\bar{J}$ can be considered which are denoted as $\delta u(t)$, $\delta x(t)$, and $\delta \bar{J}$. Such can be formulated as:

$$\delta \bar{J} = \left[\left(\frac{\partial \Phi}{\partial x} - \lambda^T \right) \delta x(t) \right]_{t=t_E} + \left[\lambda^T \delta x(t) \right]_{t=t_0} \\ + \int_{t_0}^{t_E} \left(\left(\frac{\partial \mathcal{H}}{\partial x} + \dot{\lambda}^T \right) \delta x(t) + \left(\frac{\partial \mathcal{H}}{\partial u} \right) \delta u(t) \right) dt. \tag{6}$$

The Lagrange multipliers $\lambda : [t_0, t_E] \mapsto \mathbb{R}^{n_x}$ can arbitrarily be chosen to make $\delta x(t)$ and $\delta x(t_E)$ coefficient equal to zero. Hence the multipliers chosen are:

$$\dot{\lambda}^T(t) = -\frac{\partial \mathcal{H}}{\partial x}, \tag{7}$$

$$\lambda^T(t_E) = \frac{\partial \Phi}{\partial x} \bigg|_{t=t_E}. \tag{8}$$

The chosen multipliers change the expression for $\bar{J}$. When the initial state is fixed at $\delta x(t_0) = 0$:

$$\delta \bar{J} = \int_{t_0}^{t_E} \left[\left(\frac{\partial \mathcal{H}}{\partial u} \right) \delta u \right] dt, \tag{9}$$

where the stationarity condition for the minimum at $\delta \bar{J} = 0$ becomes:

$$\frac{\partial \mathcal{H}^T}{\partial u} = 0. \tag{10}$$

Equations (2), (7), (8) and (10) are the necessary conditions in the first order for a minimum of J. Equations (7) and (8) are known as the adjoint equation, describing the co-states, and the transversality conditions, describing the initial states. These are necessary optimality conditions defining a two-point boundary value problem, useful for finding analytic solutions to certain types of optimal control problems. They also are used to find solutions in general cases using numerical algorithms. Further details are described in Refs. [27–29].

Definition 1. *(Terminal constraints).* *The above formulation can be given a set of terminal constraints defined as:*

$$\psi(x(t_E), t_E) = 0, \tag{11}$$

where $\psi : \mathbb{R}^{n_x} \times \mathbb{R} \mapsto \mathbb{R}^{n_\psi}$ *is in a vector form, variational analysis shows that Equations (2), (7) and (10) become necessary conditions for a minimum of J with the following terminal condition:*

$$\left(\frac{\partial \psi^T}{\partial x} + \frac{\partial \psi^T}{\partial x} \zeta - \lambda \right)^T \bigg|_{t_E} \delta x(t_E) + \frac{\partial \psi}{\partial t} + \frac{\partial \psi^T}{\partial t} \zeta + \mathcal{H} \bigg|_{t_E} \delta t_E = 0, \tag{12}$$

where $\zeta \in \mathbb{R}^{n_{psi}}$ *is Lagrange multiplier for the terminal constraint,* δt_E *is the infinitesimal variation in* t_E, *and* $\delta x(t_E)$ *is the infinitesimal variation in* $x(t_E)$.

Definition 2. *(Pontryagin's maximum principle).* *The adaptation of inequality constraints coupled to the input variables in optimal control problems is common under realistic conditions [30]. The input variable u then is restricted within the admissible compact region* Ω *, which is defined as:*

$$u(t) \in \Omega. \tag{13}$$

In this case, Equations (2), (7) and (8) become necessary conditions, and stationarity condition of Equation (10) is replaced with:

$$\mathcal{H}(x^*(t), u^*(t), \lambda^*(t), t) = \max_{u(t) \in U} \mathcal{H}(x^*(t), u^*(t), \lambda^*(t), t), \tag{14}$$

for all admissible $u(t)$. The superscript * denotes the optimal variables. The above, Pontryagin's maximum principle, Hamiltonian $\mathcal{H}$ must be maximized over all admissible $u(t)$ for optimal values of the state and co-state variables.

Definition 3. *(Path constraints).* *The problem domain in practical applications usually restricts the state and control trajectories where a set of constraints have to be satisfied within a time interval* $[t_0, t_E]$. *The constraint in the trajectory is given as:*

$$E(x(t), u(t), t) \leq 0, \tag{15}$$

where $E : \mathbb{R}^{n_x} \times \mathbb{R}^{n_u} \times [t_0, t_E] \mapsto \mathbb{R}^{n_p}$. *Additionally, equality constraints can be imposed at some intermediate point in time* t_e, $t_0 \leq t_e \leq t_E$. *These interior point constraints can be expressed as:*

$$D(x(t_e), t_e) = 0, \tag{16}$$

where $D : \mathbb{R}^{n_x} \times \mathbb{R} \mapsto \mathbb{R}^{n_q}$. *Further details are available in Ref.* [31].

Definition 4. *(Singular arcs).* *In singular arc problems, the matrix* $\partial^2 \mathcal{H} / \partial u^2$ *becomes singular with external arcs satisfying Equation (10). In this case, the optimality of the singular arc should be validated [31,32]. Usually, Hamiltonian function* $\mathcal{H}$ *becomes linear in at least one of the control variables for certain practical cases. This causes the control not to be determined with the state and co-state by Equation (10), but by the condition where the time derivative of* $\partial \mathcal{H} / \partial u$ *is zero along the singular arc. In such cases, supplementary conditions called the generalized Legendre-Clebsch conditions must be identified:*

$$(-1)^k \frac{\partial}{\partial u} \left[\frac{d^{2k}}{dt^{2k}} \frac{\partial \mathcal{H}}{\partial u} \right] \geq 0, k = 0, 1, 2, \ldots. \tag{17}$$

The computational difficulty involved in singular arc problems is the non-unique control variables. The inequality constraint domains, especially, have complications of:

1. An unknown number of constrained sub-arcs at the initial stage of formulation.
2. Unidentified locations of the junction points where the transition from the constrained to the unconstrained (and vice versa) occur.
3. Possible discontinuity of both the control variables u and the adjoint variable λ at the junction points which is related to the dissatisfaction of boundary conditions.

2.2. Direct Transcription Method

The complexity involved in solving the indirect methods questions its robustness. The direct methods, on the other hand, have advantages over indirect methods that good initial and any co-state guesses are not compulsory, making them robust by having a broad range for convergence even without optimally derived conditions and undetermined switching structure. Linear interpolation discretizes a continuous solution to a set of equations with state and control variables to solve the differential equations. As a result, this transforms an OCP into nonlinear programming (NLP) problem where the exact solution of OCP, having an infinite number of state and control variable combinations, is approximated to a finite number.

The typical methods are direct shooting, direct multiple shooting, and direct collocation. A direct collocation method is robust in problems with small perturbations, which is suitable for aircraft trajectory optimization. The present study uses the Hermite-Simpson methods with cubic polynomials to define the state trajectories and a piece-wise linear function for the control [18] (Figure 1). This method places the collocation points at the centers of the intervals and imposes constraints on the dynamic equations during the discretization process.

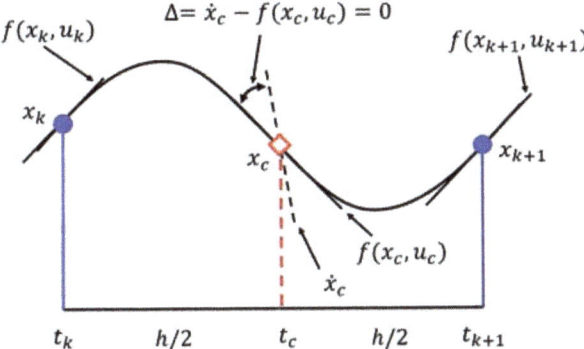

Figure 1. The main concept of Hermite collocation methods (reproduced from Figure 1, Ref. [33]). This method captures the local derivatives of the state variables to minimize the error between state derivatives from dynamics and polynomial differentiation.

First, time t_E is discretized into N intervals as:

$$t_0 = 0 \leq t_1 \leq t_2 \ldots \leq t_k \leq t_{k+1} \leq \ldots \leq t_N = t_E. \tag{18}$$

The states between t_k and t_{k+1} can be expressed in the form of cubic polynomial as:

$$x(t) = a_{k,0} + a_{k,1}t + a_{k,2}t^2 + a_{k,3}t^3, \tag{19}$$

which gives a derivative form of:

$$\dot{x}(t) = a_{k,1} + 2a_{k,2}t + 3a_{k,3}t^2, \tag{20}$$

where $a_{k,0}$, $a_{k,1}$, $a_{k,2}$, and $a_{k,3}$ are coefficients of the polynomial approximation in k^{th} interval. Since the collocation point is the midpoint of the interval:

$$t_{k,c} = \frac{t_k + t_{k+1}}{2}. \tag{21}$$

The values of the state and its derivatives are independent to the shifting of the interval from $[t_k, t_{k+1}]$ to $[0, h]$ with $h = t_{k+1} - t\_k$, time interval should be shifted. Let $x(0) = x_k$, $x(h) = x\_(k+1)$, $\dot{x}(0) = \dot{x}_k$, and $\dot{x}(z) = \dot{x}_{k+1}$, and combining Equations (19) and (20) gives:

$$\begin{bmatrix} x(0) \\ \dot{x}(0) \\ x(h) \\ \dot{x}(h) \end{bmatrix} = \begin{bmatrix} 1 & 0 & 0 & 0 \\ 0 & 1 & 0 & 0 \\ 1 & h & h^2 & h^3 \\ 0 & 1 & 2z & 3h^2 \end{bmatrix} \begin{bmatrix} a_{k,0} \\ a_{k,1} \\ a_{k,2} \\ a_{k,3} \end{bmatrix}. \tag{22}$$

In the inverse form gives:

$$\begin{bmatrix} a_{k,0} \\ a_{k,1} \\ a_{k,2} \\ a_{k,3} \end{bmatrix} = \begin{bmatrix} 1 & 0 & 0 & 0 \\ 0 & 1 & 0 & 0 \\ -\frac{3}{h^2} & -\frac{2}{h} & \frac{3}{h^2} & -\frac{1}{h} \\ \frac{2}{h^3} & \frac{1}{h^2} & -\frac{2}{h^3} & \frac{1}{h^2} \end{bmatrix} \begin{bmatrix} x(0) \\ \dot{x}(0) \\ x(h) \\ \dot{x}(h) \end{bmatrix}. \tag{23}$$

The substitution of coefficients $[a_{k,0}, a_{k,1}, a_{k,2}, a_{k,3}]$ into Equations (19) and (20) allows for the computation of the collocation points as:

$$x_c(t) = x\left(\frac{h}{2}\right) = \frac{1}{2}(x_k(t) + x_{k+1}(t)) + \frac{h}{8}[f(x_k(t), u_k(t)) - f(x_{k+1}(t), u_{k+1}(t))]. \tag{24}$$

In the time-derivative form, the points in the center of the interval is given by:

$$\dot{x}_c(t) = \dot{x}\left(\frac{h}{2}\right) = -\frac{3}{2h}(x_k - x_{k+1}) - \frac{1}{4}[f(x_k, u_k)]. \tag{25}$$

The above equation depends on the states and control at the intervals. Appropriate values of both state and controls should be chosen for the collocation points to represent the correct physics of the system. The control at the collocation point is given by:

$$u_c = \frac{u_k + u_{k+1}}{2}. \tag{26}$$

A defect Δ_k is then described as follows:

$$\begin{aligned} \Delta_k &= \dot{x}_c - f(x_c, u_c) \\ &= -\frac{3}{2h}(x_k - x_{k+1}) - \frac{1}{4}[f(x_k, u_k) + f(x_{k+1}, u_{k+1})] - f(x_c, u_c) \\ &= -\frac{3}{2h}\left[(x_k - x_{k+1}) + \frac{h}{6}[f(x_k, u_k) + 4f(x_c, u_c) + f(x_{k+1}, u_{k+1})]\right]. \end{aligned} \tag{27}$$

State constraints then are redefined as:

$$\Delta_k = \left[(x_k - x_{k+1}) + \frac{h}{6}[f(x_k, u_k) + 4f(x_c, u_c) + f(x_{k+1}, u_{k+1})]\right]. \tag{28}$$

The last term in the above expression is implicit Hermite integration of the system dynamics which is used to solve nonlinear functions. Here, the NLP solver would select $[x_k, u_k, x_{k+1}, u_{k+1}]$ to minimize Δ to zero for convergence of the solution.

The cost function now can be defined using numerical integration schemes. When trapezoid method is chosen, the cost function is written as:

$$J(u) = \Phi(x(t_E)) + \int_{t=0}^{t_E} L(t, x(t), u(t)) \, dt. \tag{29}$$

Using trapezoid integration along the interval:

$$J_{NLP} = \Phi(x_N) + \frac{1}{2}\sum_{k=1}^{N-1}(L(t_{k+1}, x_{k+1}, u_{k+1}) + L(t_k, x_k, u_k))(t_{k+1} - t_k). \tag{30}$$

For a special case with linear quadratic regulator and evenly discretized time points at the intervals h, the resultant equation is expressed as:

$$J_{NLP} = \Phi(x_N) + \frac{1}{2}\sum_{k=1}^{N-1}\left(x_{k+1}^T Q_1 x_{k+1} + u_{k+1}^T Q_2 u_{k+1} + x_k^T Q_1 x_k + u_k^T Q_2 u_k\right)h. \tag{31}$$

In general, the problem formulation using NLP in the OCP can be rewritten as:

$$\min_{x_k, u_k}\left(\Phi(x_N) + \frac{1}{2}\sum_{k=1}^{N-1}\left(x_{k+1}^T Q_1 x_{k+1} + u_{k+1}^T Q_2 u_{k+1} + x_k^T Q_1 x_k + u_k^T Q_2 u_k\right)h\right). \tag{32}$$

where Q_1 and Q_2 are linear quadratic regulators, and Equation (32) is subjected to:

$$\Delta_k = [(x_k - x_{k+1}) + \frac{h}{6}[f(x_k, u_k) + 4f(x_c, u_c) + f(x_{k+1}, u_{k+1})] = 0, \tag{33}$$

with constraints on states and controls defined as:

$$\begin{aligned} u_{min} &\leq u_k \leq u_{max}, \\ x_{min} &\leq x_k \leq u_{max}, \\ C_{eq}(x_k, u_k) &= 0, \\ C(x_k, u_k) &= 0, \end{aligned} \tag{34}$$

where C_{eq} and C represent equality and inequality constraints, respectively.

The procedure described in this section was used to solve the problem in this study which is already implemented in the MATLAB-compatible toolbox ICLOCS2 [34], where nonlinear problems were solved with interior point NLP solver IPOPT [35].

2.3. Aircraft Model

The multi-objective in this study is the achievement of both minimum FC and TTC to reach the optimum cruise altitude. To achieve these objectives, the climb path to the desired altitude should be optimized. First, the equation of motion (EOM) (Figure 2) in point-mass approximation adopted from Ref. [36] can be written as follows:

$$\dot{V} = \frac{(T(M,z)\cos\alpha - F_D)}{m} - \frac{\varepsilon}{r_t^2}\sin\gamma, \tag{35}$$

$$\dot{\gamma} = \frac{T(M,z)\sin\alpha + F_L}{mV} + \left[\frac{V}{r_t} - \frac{\varepsilon}{vr_t^2}\cos\gamma\right], \tag{36}$$

$$\dot{z} = V\sin\gamma, \tag{37}$$

$$\dot{W} = -\frac{T(M,z)}{I_{sp}}, \tag{38}$$

where V is airspeed, T is thrust, M is Mach number, F_D is drag, F_L is lift, W is weight, a product of mass m and gravitational acceleration g_0, ε is the gravitational constant, r_t is the sum of the radius of the Earth R_e and the altitude z, γ is the flight path angle, and α is angle of attack. They are subject to initial conditions of $V(t_i) = V_i$, $\gamma(t_i) = \gamma_i$, $z(t_i) = z_i$, and $m(t_i) = m_i$, where t_i is the initial time. The aerodynamic properties of the aircraft are approximated with functions of α:

$$F_L = qSC_{L_\alpha}\alpha, \tag{39}$$

$$F_D = qS\left(C_{D_0} + \eta C_{L_\alpha}\alpha^2\right), \tag{40}$$

where $q = \rho V^3/2$ denotes the dynamic pressure, S is the aerodynamic reference area, η is the efficiency factor ($0 \leq \eta \leq 1$). Both C_{L_α} and C_{D_0} are the slope coefficients of lift and zero-lift drag which are dependent on the Mach number, M, in most cases.

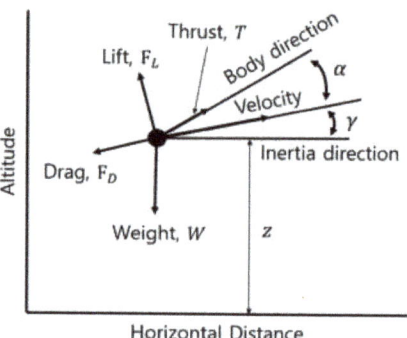

Figure 2. Longitudinal dynamics of an aircraft represented in point-mass approximation.

The objectives in this study are minimum FC and TTC. Hence, the cost functions for each objective are given as Equations (41) and (42), respectively:

$$\min_{x(t), u(t), t_E} -m(t_E), \tag{41}$$

$$\min_{x(t), u(t), t_E} t_E, \tag{42}$$

The altitudes as boundary conditions were given in 1 km interval and hence the $z(t_E)$ was given a range between 1000 m to 21,000 m with an interval of 1 km. The boundary conditions and the constants were given as:

$z(0) = 0$ [m], $\quad V(0) = 129$ [m/s], $\quad \gamma(0) = 0$ [deg], $\quad m(0) = 19{,}050$ [kg],
$V(t_E) = 295$ [m/s], $\quad \gamma(t_E) = 0$ [deg], $\quad S = 49.24$ [m^2],
$\varepsilon = 3.99 \times 10^{14}$ [m$^3/\text{s}^2$], $\quad I_{sp} = 1600$ [s], $\quad g_0 = 9.81$ [m/s^2], $\quad R_e = 6{,}378{,}145$ [m],

with the bounds on the variables, given as:

$0 \leq z \leq 21{,}000$ [m], $\quad 5 \leq V \leq 1200$ [m/s], $\quad 40 \leq \gamma \leq 40$ [deg],
$100 \leq m \leq 20{,}000$ [kg] $\quad -20 \leq \alpha \leq 20$ [deg], $\quad 0 \leq t_E \leq 600$ [s],

where the accuracy criteria for the numerical solution was given as Table 1.

Table 1. Accuracy criteria for the numerical solution.

	Accuracy Criteria			
	z [m]	V [m/s]	γ [deg]	m [kg]
ζ_{tol}	1	0.5	1	1
$\chi_{g tol}$	1	0.5	1	1
ζ_{tol}	Maximum absolute local error			
$\chi_{g tol}$	Maximum local constraint violation			

The thrust and aerodynamic data should be defined either with the experimental data or highly refined simulation data, preferably using computational fluid dynamics with direct numerical simulation (DNS) [37] sometimes coupled with linear interaction analysis (LIA) [38], or large eddy simulation (LES) [39] for an accurate description of an object moving in a compressible fluid like air. Simulation data describing the macroscopic motion of the aircraft [26,40] also are valuable resources. In the present study, typical supersonic aircraft aerodynamic profiles employed from Refs. [12,26] were used and reproduced in Tables 2 and 3, where adjustments were made by Ref. [41] with linear interpolation to fill in the "missing" data. According to the authors in Ref. [26], the aircraft is a generic-type supersonic interceptor.

Table 2. Thrust as a function of altitude z and Mach number M from Ref. [12] adjusted by Ref. [41] for Aircraft 1.

Mach No., M	Thrust, T (10^3 lb, 4.5×10^3 N)								
	Altitude, z (10^3 ft, 0.3×10^3 m)								
	0	5	10	15	20	25	30	40	50	70
0	24.2	24	20.3	17.3	14.5	12.2	10.2	5.7	3.4	0.1
0.2	28	24.6	21.1	18.1	15.2	12.8	10.7	6.5	3.9	0.2
0.4	28.3	25.2	21.9	18.7	15.9	13.4	11.2	7.3	4.4	0.4
0.6	30.8	27.2	23.8	20.5	17.3	14.7	12.3	8.1	4.9	0.8
0.8	34.5	30.3	26.6	23.2	19.8	16.8	14.1	9.4	5.6	1.1
1.0	37.9	34.3	30.4	26.8	23.3	19.8	16.8	11.2	6.8	1.4
1.2	36.1	38	34.9	31.3	27.3	23.6	20.1	13.4	8.3	1.7
1.4	36.1	36.6	38.5	36.1	31.6	28.1	24.2	16.2	10	2.2
1.6	36.1	35.2	42.1	38.7	35.7	32	28.1	19.3	11.9	2.9
1.8	36.1	33.8	45.7	41.3	39.8	34.6	31.1	21.7	13.3	3.1

Table 3. Lift and drag coefficients as a function of angle of attack α and Mach number M from Ref. [12] for Aircraft 1.

Aerodynamic Parameters	Mach No.,								
	0	0.4	0.8	0.9	1.0	1.2	1.4	1.6	1.8
C_{L_α}	3.44	3.44	3.44	3.58	4.44	3.44	3.01	2.86	2.44
C_{D_0}	0.013	0.013	0.013	0.014	0.031	0.041	0.039	0.036	0.035
η	0.54	0.54	0.54	0.75	0.79	0.78	0.89	0.93	0.93

A second aircraft model (Tables 4 and 5) was then employed using the same approach to compare the possible difference between two aircraft models. The same cost functions and altitude range as Aircraft 1 were given with slightly different boundary and constraint conditions as shown below:

$z(0) = 0$ [m], $V(0) = 129$ [m/s], $\gamma(0) = 0$ [deg], $m(0) = 16{,}329.3$ [kg],
$V(t_E) = 295$ [m/s], $\gamma(t_E) = 0$ [deg], $S = 46.45$ [m^2],
$\varepsilon = 3.99 \times 10^{14}$ [m^3/s^2], $I_{sp} = 2800$ [s], $g_0 = 9.81$ [m/s^2], $R_e = 6{,}378{,}145$ [m],

with the bounds on the variables, given as:

$0 \leq z \leq 21{,}000$ [m], $5 \leq V \leq 1200$ [m/s], $40 \leq \gamma \leq 50$ [deg],
$100 \leq m \leq 20{,}000$ [kg], $-20 \leq \alpha \leq 20$ [deg], $0 \leq t_E \leq 600$ [s],

Refs. [12,26] mentions that both models were based on typical supersonic interceptors. To the best of the authors' knowledge, supersonic interceptors have two types: one is heavy, long-range, and the other is lightweight, short-range, and the authors speculated that Aircraft 1 belonged to the former, and Aircraft 2 for the latter for having different initial mass (19,050 compared to 16,329.3 kg) and maximum thrust (205.65 compared to 134.55 kN). The authors note that the cruise altitude cannot be the same for all aircraft and may vary depending on the aircraft model. The results of the aircraft model 2 and the effects of such differences were presented and discussed in Section 4.5.

Table 4. Thrust as a function of altitude z and Mach number M from Ref. [12] for Aircraft 2.

Mach No., M	Thrust, T (10^3 lb, 4.5×10^3 N)										
	Altitude, z (10^3 ft, 0.3×10^3 m)										
	0	5	15	25	35	45	55	65	75	85	95	105
0	23.3	20.6	15.4	9.9	5.8	2.9	1.3	0.7	0.3	0.1	0.1	0.1
0.4	22.8	19.8	14.4	9.9	6.2	3.4	1.7	1.0	0.5	0.3	0.1	0.1
0.8	24.5	22.0	16.5	12.0	7.9	4.9	2.8	1.6	0.9	0.5	0.3	0.2
1.2	29.4	27.3	21.0	15.8	11.4	7.2	3.8	2.7	1.6	0.9	0.6	0.4
1.6	29.7	29.0	27.5	21.8	14.7	10.5	6.5	3.8	2.3	1.4	0.8	0.5
2.0	29.9	29.4	28.4	26.6	21.2	14.0	8.7	5.1	3.3	1.9	1.0	0.5
2.4	29.9	29.2	28.4	27.1	25.6	17.2	10.7	6.5	4.1	2.3	1.2	0.5
2.8	29.8	29.1	28.2	26.8	25.6	20.0	12.2	7.6	4.7	2.8	1.4	0.5
3.2	29.7	28.9	27.5	26.1	24.9	20.3	13.0	8.0	4.9	2.8	1.4	0.5

Table 5. Lift and drag coefficients as a function of angle of attack α and Mach number M from Ref. [11] for Aircraft 2.

Aerodynamic Parameters	Mach No.,							
	0	0.4	0.8	0.9	1.0	1.2	1.4	1.6	1.8
C_{L_α}	3.44	3.44	3.44	3.58	4.44	3.44	3.01	2.86	2.44
C_{D_0}	0.013	0.013	0.013	0.014	0.031	0.041	0.039	0.036	0.035
η	0.54	0.54	0.54	0.75	0.79	0.78	0.89	0.93	0.93

2.4. Atmospheric Model

For a realistic approach to the problem, a 1976 US atmospheric model [24] was used. This semi-empirical model effectively describes the atmospheric conditions around an aircraft at all altitudes and made the mathematical approach in this study more realistic. The formula for the model is given as:

$$T_M = T_b + L_b(z - z_b), \tag{43}$$

$$P = \begin{cases} P_b e^{\left[\frac{g_0 M_A (z - z_b)}{R T_b}\right]}, & L_b = 0; \\ P_b \left[\frac{T_b}{T_b + L_b(z - z_b)}\right]^{\left[\frac{g_0 M_A}{R L_b}\right]}, & L_b \neq 0; \end{cases} \tag{44}$$

$$\rho = \frac{P}{R_{sp} T}, \tag{45}$$

where T_M is the molecular temperature, T_b is the base temperature at each atmospheric level, L_b is the base temperature lapse rate, z is altitude, z_b is the base altitude at each atmospheric level, P is atmospheric pressure, P_b is the base pressure at each atmospheric level, g_0 is gravitational acceleration, R is universal gas constant, M_A is molar mass of Earth's air, ρ is the density of air, and R_{sp} is the specific gas constant of air. The atmospheric level, and the necessary boundary and parameter values are given in Table 6.

2.5. B-Spline Curve

The optimum trajectory solutions of each target objectives, minimum FC and TTC, were generated with an altitude interval of 1000 m, and B-spline was selected for the plotting of the curve connecting the property values at the end of the climb. B-spline is the general form of the Bézier spline which builds parametric curves around the polynomial expressions [42]. When a knot vector is $B = \beta_0, \beta_1, \ldots, \beta_k$, and control points $w_0, \ldots, w_n$ are defined where B is a non-decreasing sequence with $\beta_i \in [0,1]$, the basis functions are defined as below:

$$N_{i,0}(\beta) = \begin{cases} 1, \text{if } \beta_i \leq \beta \leq \beta_{i+1} \text{ and } \beta_i < \beta_{i+1}; \\ 0, \text{otherwise}; \end{cases} \tag{46}$$

where $j = 1, 2, \ldots, w$. The B-spline curve then is defined as:

$$N_{i,j}(\beta) = \frac{\beta - \beta_i}{\beta_{i+j} - \beta_i} N_{i,j-1}(\beta) + \frac{\beta_{i+j+1} - t}{\beta_{i+j+1} - \beta_{i+1}} N_{i+1,j-1}(\beta), \tag{47}$$

$$C(r) = \sum_{i=0}^{n} w_i N_{i,w}(\beta). \tag{48}$$

B-spline curves were used for having several advantages for the presentation of our results:

1. A curve can be generated in a complex solution even with a relatively small number of control points.
2. Depiction of a smooth, continuous curve joining the point data set would provide additional information to the physical transition between the solution points.
3. The data points from the fitted curve can be used for other aircraft path tracking problems for having a smooth transition along the curve.

Table 6. 1976 US atmosphere model: boundary and parameter values from Ref. [24].

Atmospheric Level	Altitude Range (km)	b *	z_b †	P_b (Pa) ‡	T_b (K) §	L_b (K/km) ‖
Troposphere	0–11	0	0	101325	288.15	−6.5
Tropopause (Stratosphere I)	20–32	1	11	22,632.06	216.65	0.0
Stratosphere II	20–32	2	20	5474.89	216.65	+1.0
Stratosphere III	32–47	3	32	868.02	228.65	+2.8
Stratopause (Mesosphere I)	47–51	4	47	110.91	270.65	0
Mesosphere II	51–71	5	51	66.94	270.65	−2.8
Mesosphere III	71–84.9	6	71	3.96	214.65	−2.0
—	—	7	84.852	0.37	186.87	—

Note: *, Interval (Layer) Number; †, Base Geopotential Altitude Above Mean Sea Level (MSL); ‡, Base Static Pressure; §, Base Temperature; ‖, Base Temperature Lapse Rate per Kilometer of Geopotential Altitude.

2.6. Residual-Based Approach

The Pareto-optimal nature in multi-objective optimization of cruise altitude makes the simultaneous achievement of both minimum FC and TTC impossible [43]. A Pareto-optimal set is a solution set where a trade-off between the target objectives should occur. Hence, the solution satisfying minimum FC and TTC should either be obtained or determined through other measures. Numerous approaches to solve such issues include the weighted sum method [44], the ε-constraint method [45], the particle swarm method [10], and the genetic algorithm hybrid [5]. The first two methods [44,45] have the advantage of having an easily acquirable set of optimal solutions with a superposition of single-objective solution sets, incorporating single-objective designs to obtain the desired range of optimal solutions.

The present study adopts the arguments in residual analysis to determine the optimal solution of the multi-objective problem. A novel method, terminal residual analysis (TRA), is proposed in this study for the selection of optimum cruise altitude achieving both the minimum FC and TTC objectives. The residual ω was the difference between the reference solution value y and the target solution value $\hat{y}$:

$$\omega(s) = y(z) - \hat{y}(z), \tag{49}$$

with the use of B-spline, a set of reference solution data K was defined as:

$$K = \{\xi_1, \xi_2, \ldots, \xi_{n-1}, \xi_n\}, \tag{50}$$

$$G = \{\chi_1, \chi_2, \ldots, \chi_{n-1}, \chi_n\}, \tag{51}$$

where n is the number of points in the B-spline curve of the solution curve assuming equivalent intervals in the reference curve and the subjected target solution curve. Here, an acceptable error level (AEL) σ is introduced to find the maximum, potentially the terminal value among the pool of residuals ω in Equation (49). The selection criterion for this approach is given as:

$$\max(\omega(z)) < \frac{\sigma}{100} y_n(z), \qquad (52)$$

where y_n was the reference solution data at t_E. Further, a modified approach also was introduced with a marginally acceptable error level (m-σ) TRA to compensate the low accuracy at the low σ value which will be discussed in Section 3.2. In the modified version, the determined altitude value was extended with linear extrapolation from the last turning point $d^2\nu/d\sigma^2 = 0$ towards the value of $\sigma = 0$. Here, ν is the multi-objective cruise altitude (MOCA) which details were explained at the end of this section.

In this study, the reference data model was set to be the minimum FC solution data, where the minimum TTC solution data was considered as a deviation from the minimum FC solution data. This not only is because of the cost of fuel, taking up about 15 to 20% of total operating cost [1,46–49] but also as low CI is recommended for optimal flight operation [50]. The equation for CI in this study is as follows:

$$CI = \frac{Cost\ of\ Time\ [\$/s]}{Cost\ of\ Fuel\ [\$/kg]}. \qquad (53)$$

CI is a dimensionless coefficient describing the relationship between the cost of time (CoT) and the cost of fuel (CoF). The unobtainable extreme values of CI are CI = 0, representing the minimum FC for the best range, and CI = max, representing the minimum TTC with maximum speed. CoT and CoF were assumed to be equal in the corresponding units and hence CI was calculated to represent the total cost ratio between FC and TTC during the climb.

The residuals between the minimum FC and TTC were subjected to σ, originating from the minimum FC curve. Such ensured the achievement of both the minimum FC and TTC conditions at a specific cruise altitude within σ, which was initially selected to be 1%. With the assumption that the determined cruise altitude falls within an error range of 1%. The cruise altitude was named a multi-objective cruise altitude (MOCA), ν, and the error range, 1%, was selected based on the statistical results from Ref. [9] that noted about 46.8% of the sample data already had less than 1% optimization potential in fuel consumption. The σ value of 1, therefore, was equal to additional fuel usage of 1%, of the fuel used for minimum FC climb to reach the desired altitude.

3. Results

The results in the following sections were mostly linked to the result of Aircraft 1. The results and discussion about Aircraft 2 were presented and discussed in Section 4.5.

3.1. Minimum FC and TTC

As mentioned in Section 2.6, the minimum FC and TTC solution paths were the bounds of Pareto fronts where trade-off optimal solutions lie within. The minimum FC results had almost equivalent values with the ones of minimum TTC that later diverged to take individual trajectories (Figure 3). One could even intuitively find that a diverging point of two solution trajectories occurred at around 10 to 12 km.

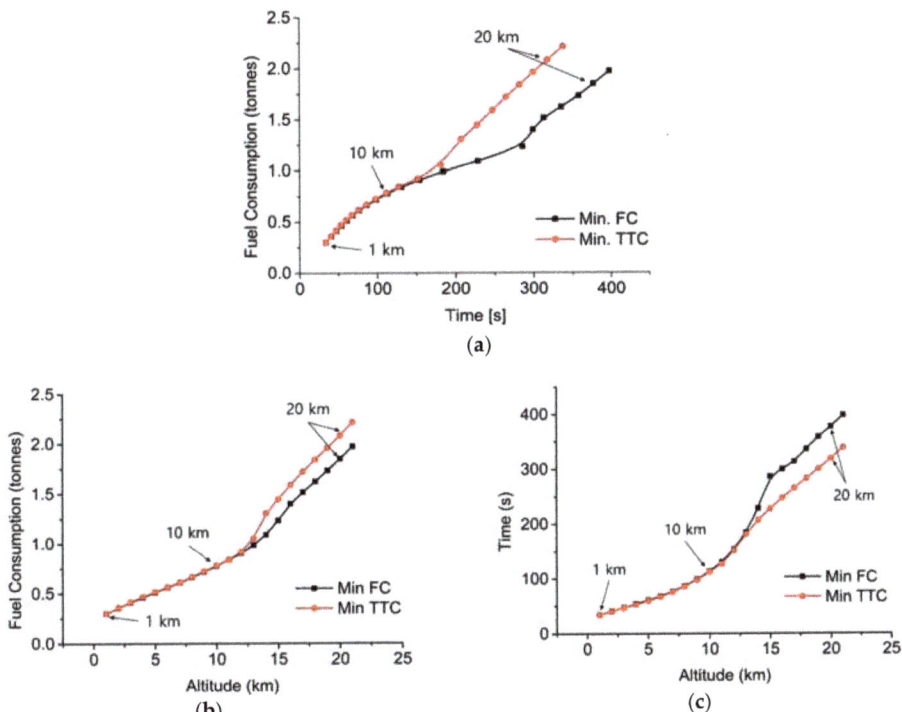

Figure 3. The minimum FC and TTC solutions at t_E in an altitude interval of 1 km. The plots of (**a**), TTC against FC; (**b**) altitude against FC; (**c**) altitude against TTC are illustrated.

3.2. Terminal Residual Analysis (TRA) and Modified TRA

The residual between B-splined minimum FC and TTC solution data at t_E was plotted with the criterion of $\sigma = 1$ from the minimum FC solution in Figure 4, as proposed in Section 2.6. The altitude for Aircraft 1 was 10.43 km at $\max(\omega(z))$, at which the residual value started to deviate dramatically from the σ range.

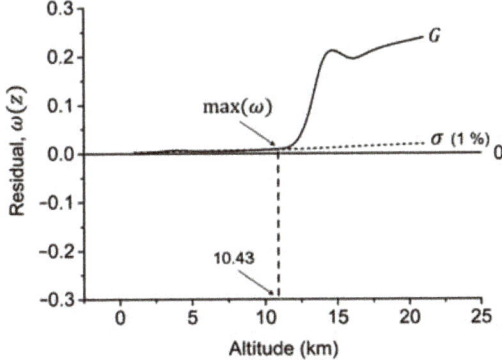

Figure 4. The plot of residuals between the minimum TTC and FC solution data which graphically illustrated the concept of the selection criterion $\max(\omega(z))$ described in the Section 2.6. G, residual between minimum FC and TTC solution; σ, acceptable error level (AEL); $\max(\omega(z))$, terminal residual.

It was clear that the optimum cruise altitude would vary depending on the value of σ, and the effect of the variation of σ was plotted in Figure 5a. Since v fell dramatically after a certain level of σ, an extrapolated line was drawn from a point at which the first local maximum $(d^2v/d\sigma^2 = 0)$ counting from $\sigma = 0$ was located at $\sigma = 1.4$. The linearly extrapolated segment was named a marginally acceptable error level (m-σ). The MOCA value with original TRA approaching towards $\sigma = 0$ fell dramatically after $\sigma = 1.4$ but the one with m-σ TRA gradually fell until $\sigma = 0$, adapting the trends in the previous solution segments. The modified approach added a feasible tendency for the low-σ region.

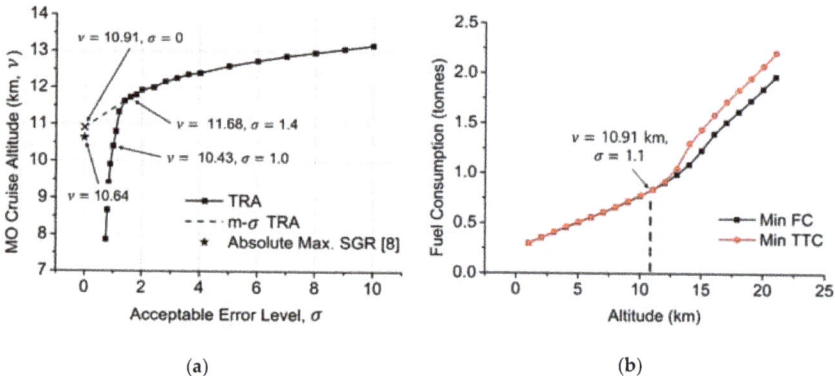

Figure 5. The effect of different σ on the MOCA (**a**) and the corresponding MOCA for the minimum TTC objective (**b**) of Aircraft 1. (**a**) The AEL and the m-σ TRA line segment were plotted against the MOCA, and along the TRA line, respectively. The optimum cruise altitude to achieve absolute specific ground range (SGR) from Ref. [9] was illustrated as a star (black). The m-σ TRA value and the absolute maximum SGR value from the reference have an almost identical fuel-saving optimum cruise altitude. (**b**) The altitude 10.91 was the predicted minimum FC cruise altitude with m-σ TRA. The corresponding altitude along the minimum TTC trajectory is achieved with $\sigma = 1.1$. AEL, acceptable error level, σ; MOCA, multi-objective cruise altitude; TRA, terminal residual analysis; SGR, specific ground range.

The m-σ TRA at $\sigma = 0$ resulted in the minimum fuel, optimum cruise altitude of 10.91 km. This value was close to the optimum cruise altitude of 10.64 km from the Ref. [9] for the absolute maximum specific ground range (SGR). The difference in the determined MOCA of the present study to the Ref. [9] was 2.5%. When the optimum cruise altitude MOCA of 10.91 km was considered, the corresponding σ value for Aircraft 1 was close to 1.1 in Figure 5b. It meant that a transition in the flight mode from minimum FC to minimum TTC was possible at about 11 km, with just additional fuel usage of 1.1%. The details of this argument were further discussed in Section 4.3.

3.3. Optimized Climb Trajectory to Given Altitudes

The climb trajectories of minimum FC and TTC were analyzed by plotting the individual trajectory curves by imposing equally spaced altitude values (Figure 6). The properties of altitude z, fuel consumption (FC), Mach number M, cost index (CI), flight path angle γ, and angle of attack α were plotted. All results similar trends of diverging solutions curves from one another at a certain near-intersecting point and the dramatic differences between the trajectories after the bifurcation. One exception was M in Figure 6g, where the final solution set was identical in both objectives. It may be because of fixed boundary conditions for the final speed of the aircraft ($V = 295$ [m/s]) and the same target altitude. The equation for M is $M = V/d$, where d is the speed of sound, which was constant at the given elevations, hence making the calculated final M in both objectives equivalent.

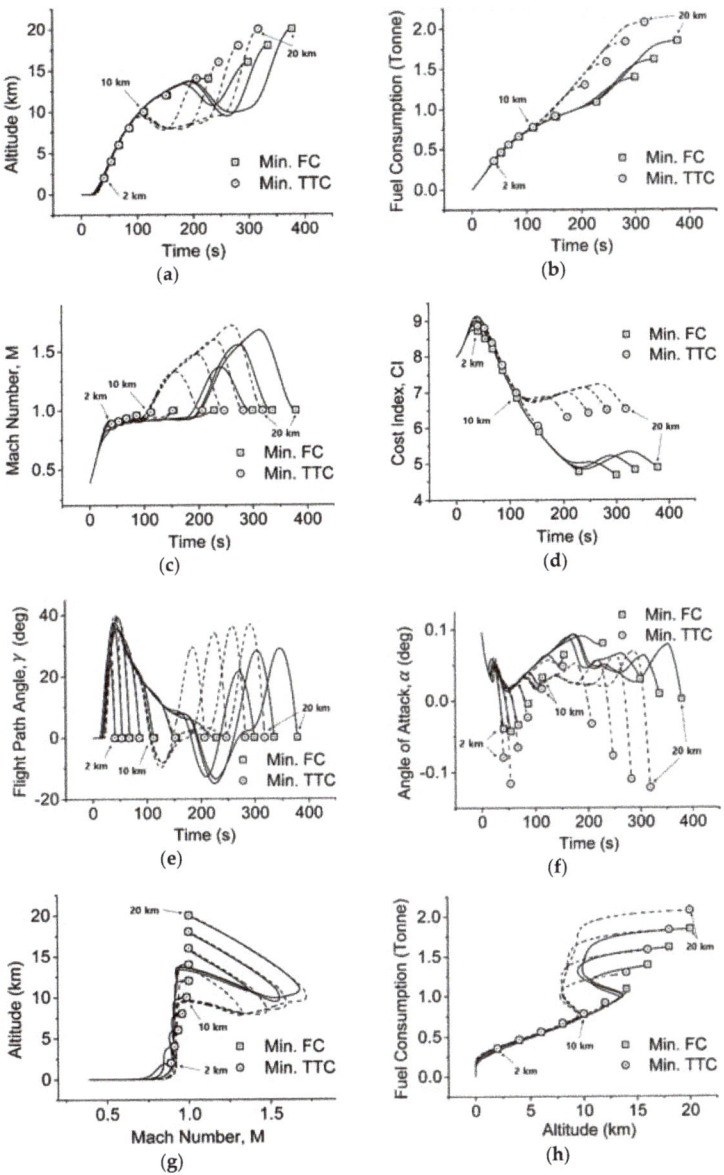

Figure 6. Climb path trajectories of: altitude z, fuel consumption (FC), Mach Number M, cost index (CI), flight path angle γ, angle of attack α in terms of time t, M, and z for Aircraft 1 at the altitudes of interval of 2 km from 2 to 20 km. (**a**) Time against z; (**b**) time against FC; (**c**) time against M; (**d**) time against CI; (**e**) time against γ; (**f**) time against α; (**g**) M against z; (**h**) z against FC.

3.4. Optimized Trajectory to MOCA

As shown in Figure 7, the altitude-specific, optimized trajectories at the minimum FC and TTC demonstrated similar patterns of overlapping line segments during the initial stage of the climb. Hence, the MOCA specific, minimum FC climb paths were plotted in Figure 7, where MOCA was approximated to be 11 km (which was 10.91 km). Similar to Figure 7, the properties of altitude z, fuel consumption (FC), Mach number M, cost index

(CI), flight path angle γ, and angle of attack α were plotted. Specifically, z continuously increased until it reached the desired MOCA (Figure 7a), and FC showed an identical trend in Figure 7b,h. M increased steeply until a value of 0.9 in Figure 7c, which then increased to 1.0 at the end of the trajectory. Figure 7g had a similarity with Figure 7c for having M in relation. γ and α varied continuously over the course of the trajectory in Figure 7e,f. CI in Figure 7d also changed over time.

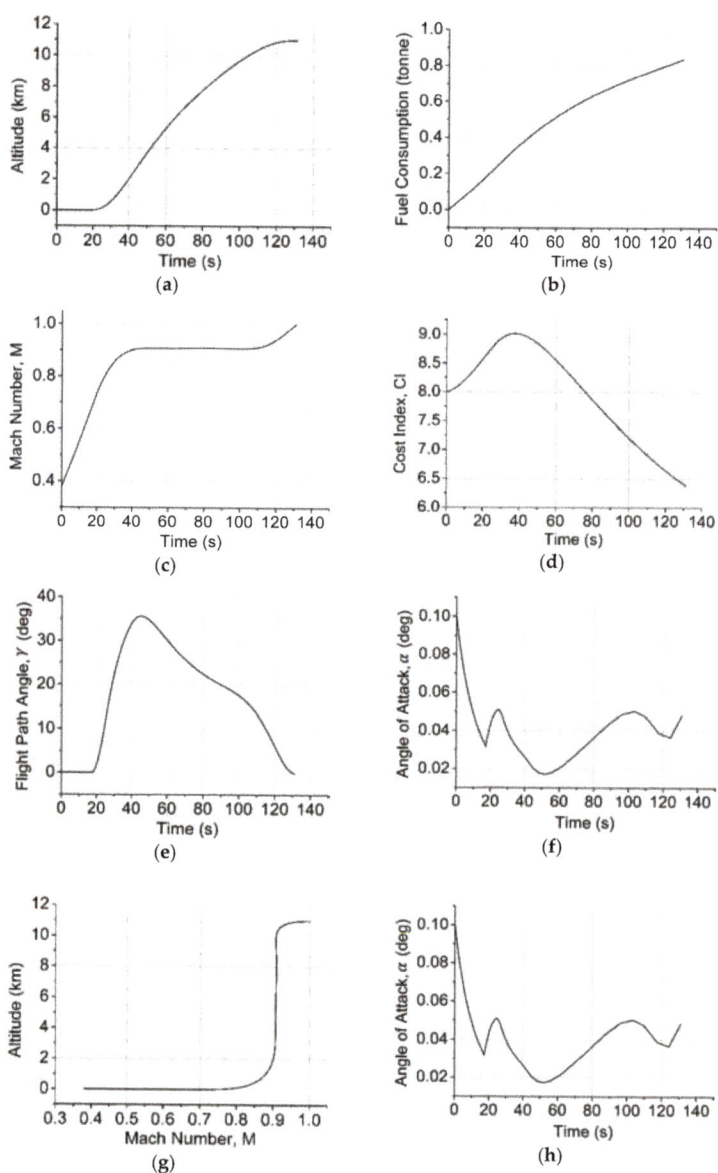

Figure 7. Climb path trajectories of: altitude z, fuel consumption (FC), Mach Number M, cost index (CI), flight path angle γ, angle of attack α in terms of time t, M, and z for Aircraft 1 at the multi-objective cruise altitude (MOCA) of 11 km. (**a**) Time against z; (**b**) time against FC; (**c**) time against M; (**d**) time against CI; (**e**) time against γ; (**f**) time against α; (**g**) M against z; (**h**) z against FC.

4. Discussion

4.1. Effectiveness of Terminal Residual Analysis (TRA) and Modified TRA

The minimum FC and TTC trajectory solutions were generated, and the values at t_E were plotted to show the gradual increase in minimum FC and TTC to the given altitude (Figure 3a). The trends in the ascension were similar in both trajectories but diverged from a certain altitude (Figures 3 and 6). Using the novel residual method proposed in this study, TRA, an acceptable error level σ was selected to be 1, which acted like a terminal residual criterion (Figure 4). In this specific application on the determination of the multi-objective cruise altitude, the residuals abruptly increase from $\max(\omega(z))$, which gave TRA-determined altitude, MOCA, satisfying the given multi-objectives. TRA initially failed to provide feasible solutions at lower σ values, and a modified m-σ TRA was proposed, which linearly extrapolated the optimized solution curve towards $\sigma = 0$. It took account of only the points before the first local maximum of $d^2\nu/d\sigma^2$ near the turning point, where ν represents the MOCA. Substituting the fuel-saving cruise altitude value obtained from m-σ TRA to the original TRA argument yielded the desired MOCA values denoting the required additional fuel usage from the minimum FC cruise altitude to achieve the minimum FC and TTC cruise altitude.

The extrapolated value from m-σ TRA was 10.91 km at $\sigma = 0$, and this value was almost equivalent to the cruise altitude of 10.64 km for absolute maximum specific ground range (SGR) from Ref. [9]. The authors believe that this was a very close estimation with computational simulation as the value from Ref. [9] was based on the accumulated, regressed statistical result of more than 200,000 actual flight data. Hence, TRA had proven its value in finding the optimum aircraft cruise altitude achieving minimum FC, which then could be implemented to acquire the minimum FC and TTC MOCA value with the supplemental method, m-σ TRA.

The continuous climb path in the present results was similar to the fuel consumption model in Ref. [21]. Ref. [21] adopted a CCO model to optimize fuel consumption during the climb, which introduced external influences such as crosswind. The accuracy was 96%, and the maximum amount of fuel saved was 12%. The current study also had a continuous climb path with reliable atmospheric model and aircraft aerodynamic data showing a difference of 2.5% in the altitude solution. While the model in Ref. [21] considered various external factors such as crosswind, our study did not employ any external factors in the calculation. Hence, the present study is significant in two aspects: first, the proposed novel methods, TRA and m-σ TRA, were robust even without the consideration of external factors such as crosswind, and second, such external factors were not significant in the determination of the optimized performance of long-range flight. The latter, especially, would be valid as Ref. [9] already suggested that 46.8% of 200,000 flight data already had less than 1% benefit from the optimization.

Additionally, the MOCA value within $\sigma = 2.0$ in the m-σ TRA line, fell below 12 km (Figure 5a), leading to the MOCA range of 10.91 km to 11.97 km. The optimum cruise altitudes determined in the statistical results of Refs. [9,51] were 10.64, 11.19, 11.67, and 11.46 km, which implies that the altitude range calculated in the present study was appropriate.

4.2. Difference in TRA and Modified TRA

A newly proposed method, TRA, was modified into m-σ TRA as the original TRA with σ below a certain level changed abruptly. Such abrupt change indicated that σ was critical in determining the multi-objective goal in this study. The residual value below the MOCA of 11.5 km was nearly constant, making a low σ criterion below 1.4% ineffective. The linearly extrapolated value of minimum FC at $\sigma = 0$ was validated with Ref. [9], as discussed in Section 4.1. Hence m-σ could also be an alternative approach for such cases of significantly low σ.

4.3. Optimum Altitude MOCA with Minimum FC and TTC

The altitude determined with the m-σ TRA method at $\sigma = 0$ was the MOCA only with minimum FC objective. As described in Section 3.2, this approach incorporated with the TRA could be utilized to determine the MOCA. MOCA satisfied both the minimum FC and TTC at any altitude with the additional fuel usage denoted as σ. The initial selection of $\sigma = 1$ was to find the cruise altitude showing the potential fuel reduction within 1%, which eventually was determined to be 10.43 km, and its difference to the altitude 11.68 km, selected on m-σ TRA, seemed relatively large concerning only a difference of 0.4 in the value of σ. Interestingly, the difference in altitudes became minimum with increasing σ (Figure 5). This result emphasized the difficulties in achieving both objectives simultaneously for an optimal altitude, which varies even with the slightest changes in σ.

Another interesting point here is that the minimum FC altitude, 10.91 km, was about 0.8 km lower than the MOCA at $\sigma = 1.4$ (11.68 km). As mentioned in Section 1, a climb to a higher altitude takes advantage of lower air density for reduced aerodynamic drag but consequently consumes extra fuel for ascension. Hence, cruising at the MOCA at $\sigma = 1.4$ may enjoy the minimum FC travel with considerably reduced travel time (TT) due to reduced aerodynamic drag, but the same at the MOCA at the adjusted $\sigma = 1.1$ may enjoy the fuel-economy even with higher aerodynamic drag than the former. In detail, the MOCA with the minimum FC and TTC could be achieved at an altitude of 10.91 km with an additional fuel usage of 1.1% of the fuel used for minimum fuel climb to an altitude of 10.91 km. It implies the potential for flight mode transition between the single-objective, minimum FC, and the multi-objective, minimum FC and TTC. With this argument, an aircraft cruising at an altitude of 10.91 km could switch the flight mode from the minimum FC to the minimum FC and TTC with the extra fuel usage of 1.1%. Likewise, minimum FC and TTC cruise mode could be achieved at an altitude of 11.68 km with the additional fuel usage within 1.4% of the minimum FC cruise mode. The MOCA results of different s values provided quantitative explanations for the theory and the transition criterion for the flight mode between single-objective, the minimum FC, and the multi-objective, the minimum FC and TTC. Additionally, it also demonstrated that the fuel-economy may not always be achieved just by cruising at higher altitudes.

4.4. Individual Trajectory and Variable Cost Index (CI)

The trajectory solution at the determined MOCA of 10.91 km shows the trajectory of a supersonic jet (Figure 7). These graphs clearly show the relationships between the variables and parameters in this study, as described in Section 3.4. In the initial segment of the climb path, the aircraft accelerated by increasing T, which increased both V and M (Figure 7c). Throughout the whole trajectory, a fluctuated to make stable transitions in g, which resulted in a continuous and smooth climb path of the aircraft in terms of z (Figure 7a), FC (Figure 7b,h), M (Figure 7c), and CI (Figure 7d).

Interestingly, CI in (Figure 7d) varied over time. As mentioned in Section 1, numerous literatures targeting for multi-objective optimization of TOC usually employ a model with constant CI throughout a flight trajectory [4,6,8]. It may be due to the difficulties involved in the estimation of actual CI or an attempt to reduce the number of variables to solve in complex total cost equation such as:

$$C_{TOT} = \int_0^{T_{flight}} fbr(t)\, dt + 60 \cdot CI \cdot T_{flight}, \qquad (54)$$

where C_{TOT} is the total cost, CI is the cost index as defined in this study, T_{flight} is the cruise segment flight time, and $fbr(t)$ is the time-dependent aircraft fuel-burn rate function. This equation was re-written from Equation (6) of Ref. [52] by Dancila, B. D., et al. [8]. Given the number of variables involved, a range of constant CI values definitely would have made the problems more manageable. On the contrary, CI was not consistent throughout a flight path in this study (Figure 7d). Considering the realistic aspect of an actual flight path, the variation in CI during a flight is a more convincing phenomenon. This variation implies that the present approach was able to describe more pragmatic phenomena in aircraft motion.

4.5. Limitations

The first limitation is the use of aerodynamic data of a supersonic aircraft from Ref. [12] to find a multi-objective optimum cruise altitude. The selection of the optimum cruise altitude is a necessity in the economic management of airliners, and in most cases, the aircraft is subsonic. The use of a supersonic aircraft model may not be suitable to determine general cruise altitude, but as seen in Figure 7d, M remains subsonic before reaching the cruise altitude. Such probably may have made the MOCA result in the present study similar to the statistically optimum cruise altitude of commercial airplanes. There also was lack of available data for subsonic aircraft in terms of thrust, Mach number and altitude. Such data was necessary for climb trajectory optimization. With the inadequacy in the available data, the authors tried to maintain the feasibility of the approach and obtained confident results for cruise altitude determination of a long-range flight. Overall, a more refined study could be conducted using the aerodynamic data of a subsonic aircraft.

The second limitation may be the applicability of the present methodology to determine the optimum cruise altitude for all-range flight. As mentioned in Section 1, Aircraft 1 and 2 are different in their specifications, especially in their initial weight and maximum thrust. This difference made the authors speculate Aircraft 1 to be the heavy, long-range plane for the former and lightweight, short-range plane for the latter. The results in Figure 8 clearly shows the difference where the present approach failed to achieve reasonable cruise altitude but only showed the potential to indicate the costs for the transition between two objectives. The present methodology may only apply to the cruise altitude determination of long-range flight, but since such flights require cruise altitude optimization the most [9], our approach could be a feasible solution for a cost-minimizing strategy.

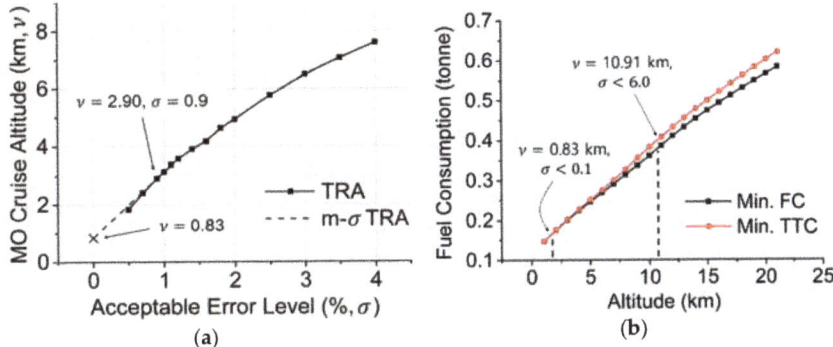

Figure 8. The effect of different σ on the MOCA (a) and the corresponding MOCA for the minimum TTC objective (b) of Aircraft 2. (a) The AEL and the m-σ TRA line segment were plotted against the MOCA, and along the TRA line, respectively. The AEL, σ, is plotted against multi-objective cruise altitude (MOCA) ν, and linearly extrapolated m-σ TRA is also drawn along the terminal residual analysis (TRA) line. (b) The altitude 10.91 km was the predicted minimum FC cruise altitude with m-σ TRA of Aircraft 1, and the corresponding altitude along the minimum TTC trajectory in Aircraft 2 is achieved with σ < 6.0. AEL, acceptable error level, σ; MOCA, multi-objective cruise altitude; TRA, terminal residual analysis.

This study proposed TRA and a modified TRA (m-σ TRA) to determine an optimum cruise altitude satisfying minimum FC and TTC. Despite the promising results for long-range flight, further investigation would be necessary whether this approach is also suitable in other multi-objective optimization problems. The multi-objective optimization of aircraft trajectory problems may be the only problem solvable using the methods proposed in this study. Hence, this approach may need further validation in various optimization problems in future studies.

5. Conclusions

Multi-objective determination of cruise altitude of a supersonic aircraft was conducted using the Hermite-Simpson method. Individual optimum climb trajectories were generated in the discretized problem domain. A novel approach, TRA, was proposed with modified TRA, m-σ TRA, to select the optimum cruise altitude with the minimum FC and TTC, where σ also worked as a criterion to represent the magnitude of the transition between the single- or multi-objective flight mode. As a result, a multi-objective cruise altitude (MOCA) was determined to be 10.91 km for aircraft 1, which was validated with the statistical results of subsonic, commercial airliner data. The method presented in this study was found reliable in the multi-objective problem to determine the optimum cruise altitude of a long-range flight, achieving both minimum FC and TTC. Although various assumptions were used, this approach confirmed the complex multi-objective solution in aircraft cruise altitude problems in an environment formulated as realistic as possible. Further studies would be needed to verify this technique in other applications to validate its utility in other fields requiring multi-objective optimum solutions.

Author Contributions: Conceptualization, methodology, software, validation, formal analysis, investigation, resources, data acquisition, visualization, conclusion, and writing–original draft preparation, T.K.; writing–review, and editing, T.K., J.R.; supervision, project administration, and funding acquisition, J.R. All authors have read and agreed to the published version of the manuscript.

Funding: This research was supported by the National Research Foundation of Korea(NRF) grant funded by the Korean government(MEST)(No. 2019R1A2C1087763), and the Chung-Ang University Research Scholarship Grants in 2019.

Conflicts of Interest: The authors declare no conflict of interest.

Abbreviations

The following abbreviations are used in this manuscript:

FC	Fuel consumption
TT	Travel time
TTC	Time-to-climb
CI	Cost index
TOC	Total operating cost
OCP	Optimal control problem
NLP	Nonlinear programming
EOM	Equation of motion
SET	Specific excess thrust
TRA	Terminal residual analysis
AEL	Acceptable error level
MOCA	Multi-objective cruise altitude

References

1. IATA. Economic Performance of the Airline Industry: 2017 End-Year Report 2018. 2020. Available online: https://www.iata.org/en/iata-repository/publications/economic-reports/airline-industry-economic-performance-june-2020-report (accessed on 6 September 2020).
2. Bureau of Transportation Statistics. First Quarter 2020 U.S. Airline Financial Data. 2020. Available online: https://www.bts.gov/newsroom/first-quarter-2020-us-airline-financial-data (accessed on 15 September 2020).
3. Bureau of Transportation Statistics. U.S. Airlines Report Second Quarter 2020 Losses. 2020. Available online: https://www.bts.gov/newsroom/us-airlines-report-second-quarter-2020-losses (accessed on 15 September 2020).
4. Dai, R.; Cochran, J.E. Three-Dimensional Trajectory Optimization in Constrained Airspace. *J. Aircr.* **2009**, *46*, 627–634. [CrossRef]
5. Vavrina, M.; Howell, K. Multiobjective Optimization of Low-Thrust Trajectories Using a Genetic Algorithm Hybrid. In Proceedings of the AAS/AIAA Space Flight Mechanics Meeting, Pittsburgh, PA, USA, 9–13 August 2009.

6. Franco, A.; Rivas, D. Minimum-Cost Cruise at Constant Altitude of Commercial Aircraft Including Wind Effects. *J. Guid. Control Dyn.* **2011**, *34*, 1253–1260. [CrossRef]
7. Lovegren, J.; Hansman, R.J. *Estimation of Potential Aircraft Fuel Burn Reduction in Cruise via Speed and Altitude Optimization Strategies*; MIT International Center for Air Transport (ICAT): Cambridge, MA, USA, 2011.
8. Dancila, B.; Botez, R.; Labour, D. Altitude Optimization Algorithm for Cruise, Constant Speed and Level Flight Segments. In Proceedings of the AIAA Guidance, Navigation, and Control Conference, Minneapolis, MI, USA, 13–16 August 2012.
9. Jensen, L.; Hansman, R.J.; Venuti, J.C.; Reynolds, T. Commercial airline altitude optimization strategies for reduced cruise fuel consumption. In Proceedings of the 14th AIAA Aviation Technology, Integration, and Operations Conference, Atlanta, GA, USA, 16–20 June 2014.
10. Antonakis, A.; Nikolaidis, T.; Pilidis, P. Multi-Objective Climb Path Optimization for Aircraft/Engine Integration Using Particle Swarm Optimization. *Appl. Sci.* **2017**, *7*, 469. [CrossRef]
11. Rutowski, E.S. Energy approach to the general aircraft performance problem. *J. Aeronaut. Sci.* **1954**, *21*, 187–195. [CrossRef]
12. Bryson, A.E., Jr.; Desai, M.N.; Hoffman, W. Energy-state approximation in performance optimization of supersonicaircraft. *J. Aircr.* **1969**, *6*, 481–488.
13. Ong, S.Y. A model comparison of a supersonic aircraft minimum time-to-climb problem. In Proceedings of the 25th AIAA Aerospace Sciences Meeting, Reno, NV, USA, 12–15 January 1986.
14. Pierson, B.L.; Ong, S.Y. Minimum-Fuel Aircraft Transition Trajectories. *Math. Comput. Model.* **1989**, *12*, 925–934. [CrossRef]
15. Morimoto, H.; Chuang, J. Minimum-fuel trajectory along entire flight profile for a hypersonic vehicle with constraint. In Proceedings of the Guidance, Navigation, and Control Conference and Exhibit, Boston, MA, USA, 10–12 August 1998.
16. Benson, D. *A Gauss Pseudospectral Transcription for Optimal Control*; Massachusetts Institute of Technology: Cambridge, MA, USA, 2005.
17. Von Stryk, O.; Bulirsch, R. Direct and Indirect Methods for Trajectory Optimization. *Ann. Oper. Res.* **1992**, *37*, 357–373. [CrossRef]
18. Hargraves, C.R.; Paris, S.W. Direct Trajectory Optimization Using Nonlinear-Programming and Collocation. *J. Guid. Control Dyn.* **1987**, *10*, 338–342. [CrossRef]
19. Bayen, A.M.; Mitchell, I.M.; Oishi, M.M.K.; Tomlin, C.J. Aircraft autolander safety analysis through optimal control-based reach set computation. *J. Guid. Control Dyn.* **2007**, *30*, 68–77. [CrossRef]
20. Tsiotras, P.; Bakolas, E.; Zhao, Y. Initial guess generation for aircraft landing trajectory optimization. In Proceedings of the AIAA Guidance, Navigation, and Control Conference, Portland, OR, USA, 8–11 August 2011.
21. Roach, K.; Robinson, J. A terminal area analysis of continuous ascent departure fuel use at Dallas/Fort Worth international airport. In Proceedings of the 10th AIAA Aviation Technology, Integration, and Operations (ATIO) Conference, Fort Worth, TX, USA, 13–15 September 2010.
22. Zhang, M.; Huang, Q.; Liu, S.; Zhang, Y. Fuel Consumption Model of the Climbing Phase of Departure Aircraft Based on Flight Data Analysis. *Sustainability* **2019**, *11*, 4362. [CrossRef]
23. Pargett, D.M.; Ardema, M.D. Flight path optimization at constant altitude. *J. Guid. Control Dyn.* **2007**, *30*, 1197–1201. [CrossRef]
24. NOAA; NASA; USAF. *US Standard Atmosphere, 1976*; National Oceanic and Atmospheric Administration: Washington, DC, USA, 1976; Volume 76.
25. Herman, A.L.; Conway, A.C. Direct optimization using collocation based on high-order Gauss-Lobatto quadrature rules. *J. Guid. Control Dyn.* **1996**, *19*, 592–599. [CrossRef]
26. Bryson, A.E.; Denham, W.F. A steepest-ascent method for solving optimum programming problems. *J. Appl. Mech.* **1962**, *29*, 247. [CrossRef]
27. Gelfand, I.M.; Silverman, R.A. *Calculus of Variations*; Dover Publications: Mineola, NY, USA, 2000.
28. Wan, F. *Introduction to the Calculus of Variations and Its Applications*; Routledge: London, UK, 2017.
29. Leitmann, G. *The Calculus of Variations and Optimal Control: An Introduction*; Springer Science & Business Media: New York, NY, USA, 2013; Volume 24.
30. Pontryagin, L.S. *Mathematical Theory of Optimal Processes*; Routledge: New York, NY, USA, 2018.
31. Bryson, A.E. *Applied Optimal Control: Optimization, Estimation and Control*; CRC Press: Boca Raton, FL, USA, 1975.
32. Thompson, G.L. *Optimal Control Theory: Applications to Management Science and Economics*; Springer: New York, NY, USA, 2006.
33. Topputo, F.; Zhang, C. Survey of direct transcription for low-thrust space trajectory optimization with applications. In *Abstract and Applied Analysis*; Hindawi: London, UK, 2014.
34. Nie, Y.; Faqir, O.; Kerrigan, E.C. ICLOCS2: Try this optimal control problem solver before you try the rest. In Proceedings of the 2018 UKACC 12th International Conference on Control (CONTROL), Sheffield, UK, 5–7 September 2018.
35. Wachter, A.; Biegler, L.T. On the implementation of an interior-point filter line-search algorithm for large-scale nonlinear programming. *Math. Program.* **2006**, *106*, 25–57. [CrossRef]
36. Barnes, W.; McCormick, W. *Aerodynamics Aeronautics and Flight Mechanics*; Wiley: New York, NY, USA, 1995.
37. Daniel, D.; Livescu, D.; Ryu, J. Reaction analogy based forcing for incompressible scalar turbulence. *Phys. Rev. Fluids* **2018**, *3*, 094602. [CrossRef]
38. Ryu, J.; Livescu, D. Turbulence structure behind the shock in canonical shock–vortical turbulence interaction. *J. Fluid Mech.* **2014**, *756*. [CrossRef]
39. Ryu, J.; Lele, S.K.; Viswanathan, K. Study of supersonic wave components in high-speed turbulent jets using an LES database. *J. Sound Vib.* **2014**, *333*, 6900–6923. [CrossRef]

40. Denham, W.F. *Steepest-Ascent Solution of Optimal Programming Problems*; Harvard University: Cambridge, MA, USA, 1963.
41. Patterson, M.A.; Rao, A.V. GPOPS-II: A *MATLAB* software for solving multiple-phase optimal control problems using hp-adaptive Gaussian quadrature collocation methods and sparse nonlinear programming. *ACM Trans. Math. Softw. (TOMS)* **2014**, *41*, 1–37. [CrossRef]
42. Prautzsch, H.; Boehm, W.; Paluszny, M. *Bézier and B-Spline Techniques*; Springer Science & Business Media: New York, NY, USA, 2002.
43. Pareto, V. *Manual of Political Economy: A Critical and Variorum Edition*; OUP Oxford: New York, NY, USA, 2014.
44. Deb, K. *Multi-Objective Optimization Using Evolutionary Algorithms*; John Wiley & Sons: Chichester, UK, 2001; Volume 16.
45. Zadeh, L. Optimality and non-scalar-valued performance criteria. *IEEE Trans. Autom. Control* **1963**, *8*, 59–60. [CrossRef]
46. Bureau of Transportation Statistics. 2016 Annual and 4th Quarter Airline Finalcial Data. Available online: https://www.bts.gov/newsroom/2016-annual-and-4th-quarter-airline-financial-data (accessed on 12 September 2020).
47. Bureau of Transportation Statistics. 2017 Annual and 4th Quarter U.S. Airline Financial Data. Available online: https://www.bts.gov/newsroom/2017-annual-and-4th-quarter-us-airline-financial-data (accessed on 12 September 2020).
48. Bureau of Transportation Statistics. 2018 Annual and 4th Quarter U.S. Airline Financial Data. Available online: https://www.bts.gov/newsroom/2018-annual-and-4th-quarter-us-airline-financial-data (accessed on 12 September 2020).
49. Bureau of Transportation Statistics. 2019 Annual and 4th Quarter U.S. Airline Financial Data. Available online: https://www.bts.gov/newsroom/2019-annual-and-4th-quarter-us-airline-financial-data (accessed on 12 September 2020).
50. Roberson, B. Fuel Conservation Strategies: Cost index explained. *AERO* **2007**, *2*, 26–28.
51. Jensen, L.; Hansman, R.J.; Venuti, J.; Reynolds, T. Commercial airline speed optimization strategies for reduced cruise fuel consumption. In Proceedings of the 2013 Aviation Technology, Integration, and Operations Conference, Los Angeles, CA, USA, 12–14 August 2013.
52. Liden, S. Optimum cruise profiles in the presence of winds. In Proceedings of the IEEE/AIAA 11th Digital Avionics Systems Conference, Seattle, WA, USA, 5–8 October 1992.

Article

Mathematical Approach to Improve the Thermoeconomics of a Humidification Dehumidification Solar Desalination System

Rasikh Tariq [1,*], Jacinto Torres Jimenez [2], Nadeem Ahmed Sheikh [3] and Sohail Khan [4]

1 Facultad de Ingeniería, Universidad Autónoma de Yucatán, Av. Industrias No Contaminantes por Anillo Periférico Norte, Apdo. Postal 150, Cordemex, Mérida 97310, Yucatán, Mexico
2 Ingeniería Eléctrica-Maestría en Tecnologías de la Información, Tecnológico Nacional de México/Instituto Tecnológico Superior de Huauchinango, Av. Tecnológico No. 80, Col. 5 de Octubre, Huauchinango 73173, Puebla, Mexico; jacinto.torres@huauchinango.tecnm.mx
3 Department of Mechanical Engineering, Faculty of Engineering and Technology, International Islamic University, Islamabad 44000, Pakistan; ndahmed@gmail.com
4 Programa de Maestría y Doctorado en Ingeniería, Especialidad en Sistemas Eléctricos de Potencia, Universidad Nacional Autónoma de México, Ciudad de México 04510, Mexico; sohailmomand6@gmail.com
* Correspondence: rasikhtariq@gmail.com or rasikhtariq@alumnos.uady.mx; Tel.: +52-1-999-174-1935

Citation: Tariq, R.; Jimenez, J.T.; Ahmed Sheikh, N.; Khan, S. Mathematical Approach to Improve the Thermoeconomics of a Humidification Dehumidification Solar Desalination System. *Mathematics* 2021, 9, 33. https://dx.doi.org/10.3390/math9010033

Received: 24 June 2020
Accepted: 15 October 2020
Published: 25 December 2020

Publisher's Note: MDPI stays neutral with regard to jurisdictional claims in published maps and institutional affiliations.

Copyright: © 2020 by the authors. Licensee MDPI, Basel, Switzerland. This article is an open access article distributed under the terms and conditions of the Creative Commons Attribution (CC BY) license (https://creativecommons.org/licenses/by/4.0/).

Abstract: Water desalination presents a need to address the growing water-energy nexus. In this work, a literature survey is carried out, along an application of a mathematical model is presented to enhance the freshwater productivity rate of a solar-assisted humidification-dehumidification (HDH) type of desalination system. The prime novelty of this work is to recover the waste heat by reusing the feedwater at the exit of the condenser in the brackish water storage tank and to carry out the analysis of its effectiveness in terms of the system's yearly thermoeconomics. The developed mathematical model for each of the components of the plant is solved through an iterative procedure. In a parametric study, the influence of mass flow rates (MFRs) of inlet air, saline water, feedwater, and air temperature on the freshwater productivity is shown with and without the waste heat recovery from the condensing coil. It is reported that the production rate of water is increased to a maximum of 15% by recovering the waste heat. Furthermore, yearly analysis has shown that the production rate of water is increased to a maximum of 16% for June in the location of Taxila, Pakistan. An analysis is also carried out on the economics of the proposed modification, which shows that the cost per litre of the desalinated water is reduced by ~13%. It is concluded that the water productivity of an HDH solar desalination plant can be significantly increased by recovering the waste heat from the condensing coil.

Keywords: desalination; humidification-dehumidification; waste heat recovery; mathematical model; yearly analysis; thermo-economics

1. Introduction

Water covers almost 71% of the total Earth's surface [1]. Sea contains 97% of the total water of the Earth [2] and the remaining 3% is stored in the form of rivers, glaciers, underground water storage, and lakes, etc. The freshwater is not evenly distributed in the world, as some geography near the equator has less availability of freshwater. The seawater contains a large number of salts. Therefore, it is not feasible to be used for household, agricultural, or commercial purposes [3]. Considering this aspect, desalination is an important need of the human being. In this regard, Manju et al. [4] provided an extensive review of the need for a desalination system to overcome future freshwater demand for India. Among many desalination processes [5], some can be energy inefficient, costly, and/or can have environmental impacts (CO$_2$ emissions and other dangerous byproducts as referenced by [6]) depending on the design parameters [7].

The use of solar radiation is undergoing intensive research for the desalination process [5] since solar energy is a low-grade heat source considering the exergetic usefulness [8]. In this aspect, Reif et al. [9] focused on the desalination system powered by solar energy and reviewed its potential and challenges. Considering the need of the modern world, Giwa et al. [10] proposed recent advances in the solar-assisted humidification-dehumidification (HDH) type desalination system in terms of improved design and productivity. Afterwards, Kabeel et al. [11] and Hamed et al. [12] developed an experimental setup for the solar power HDH type of desalination system. Kabeel et al. [11] concluded that the condenser with a cylindrical shell and corrugated fins led to an increase in the rate of heat transfer, the cellulose of 5 mm gives higher productivity as compared to the usage of cellulose of 7 mm under both natural and forced flow circulation. Whereas, Hamed et al. [12] concluded that average productivity of the desalination unit is 22 litres per day with an estimated cost of 0.0578 USD per litre, the best operating time during the day is between 1 and 5 pm, and the productivity of the unit increases as the temperature of the water which is entering the humidifier is increased. Balaji et al. [13] carried out the numerical analysis of HDH desalination system by developing code in 'C' language to study the performance for various operating conditions. Their results have indicated that increasing the mass flow rate of air increases the gain-output-ratio and they reported that the economic feasibility of the system is in the lower range because no external sources were used.

Zhani et al. [14] also developed a prototype of a solar-assisted HDH desalination system and tested it for the weather conditions of Tunisia during the summer season (June, July, and August). They concluded that their proposed system is quite efficient technically, however it lacks in economic efficiency. Owing to the requirement of a large surface area for solar energy collection, Elminshawy et al. [15] proposed to run the desalination system using two energy sources i.e., solar as well as a low-grade heat source. Elminshawy et al. [15] also developed an analytical model and the results were compared with the experimental results. It was concluded that the cost of the water is 0.014 USD per litre, and this corresponds to a fuel-saving equivalent to 1844 kg per hour.

Narayan et al. [16] evaluated the potential of a solar-driven humidification- dehumidification desalination plant for small-scale decentralized water production, presented [17] the thermodynamic analysis of desalination cycles, introduced [18] an experimental investigation on the thermal design of humidification dehumidification desalination system, and also presented [19] a thermodynamic balancing of HDH desalination by mass extraction and injection. The authors, in their reference work [17], concluded that the air-heated cycles reported in the literature are insufficient, a dehumidifier is more vital than the humidifier to the performance of a conventional water-heated cycle, and the varied pressure systems can have a better performance than a single pressure system. The authors, in their reference work [18], summarized that, for a water-heated closed-air-open-water humidification, a dehumidification system without any mass extraction represents a maximum gained-output-ratio. Similarly, the authors in their reference work [19] concluded that the uncertainty of the final results with the approximation of the air being saturated to all the points in the humidification and dehumidification process seems to be reasonably small based on the boundary layer data from Thiel and Lienhard [20].

Summers et al. [21] presented a comparison of the energy efficiency of a single-stage membrane distillation desalination cycles in various configurations and reported that the rate-limiting processes and their impact on gained-output-ratio can be determined from the development model, and a single-stage vacuum membrane distillation is inherently limited by the low temperature of condensation which results from the reduction in pressure. McGovern [22] presented the performance limits of zero and single extraction humidification-dehumidification desalination systems and reported that the usage of an ideal gas model for water vapour and the air is highly accurate to model HDH systems, the influence of salinity at 35,000 parts per million is to reduce the change in moist air humidity ratio and enthalpy by approximating 1% and 3%, respectively, for a feed temperature of 25 °C and a top air temperature of 70 °C.

Mistry et al. [23] carried out an entropy generation analysis of desalination technologies, including multiple effect distillation, HDH, reverse osmosis, mechanical vapour compression, membrane distillation, and multistage flash, and also presented [24] the effect of entropy generation rate on the performance indicators, i.e., gained output ratio, of HDH desalination cycles. In summary, the authors in reference work [24] has reported that for any given cycle, there is a specific mass flow rate ratio that simultaneously minimizes the entropy generation rate and maximizes the gained-output-ratio; in other words, it corresponds that the minimization of specific irreversibility leads to peak performance.

Sharqaw et al. [25] presented the exergy calculations of seawater with applications in desalination systems and reported that the ideal mixture models give flow exergy values that are far from the actual ones and the exergetic efficiency can differ by 80% for some cases, and in another article, Sharqaw et al. [26] also presented the optimum thermal design of such desalination systems and reported that the optimum mass flow rate ratio is always greater than unity, as increasing the effectiveness of the humidifier and dehumidifier increases the recovery ratio almost linearly, and the higher maximum temperature can yield a higher gained-output-ratio.

Some authors like Khalifa et al. [27] carried out experimental and theoretical research on water desalination using a direct contact membrane distillation. The authors developed an analytical model based on heat and mass transfer equations and utilized it to predict the temperature difference across the membrane surfaces and then calculating the vapour pressure difference leading to the permeate flux. It was noted that the productivity of the system is very promising since a permeate flux of 100 kg/m^2·h was achieved at 90 °C for hot feed side and 5 °C for cold side steam.

Several researchers also carried out investigations to integrate the desalination unit with thermal energy storage. In this case, Summers et al. [28] proposed the design and optimization of an air heating solar collector with phase changing material integrated with an HDH desalination plant. A two-dimensional transient finite element method was developed, and it was reported that a layer of phase-changing material of 8 cm below the absorber plate is sufficient to produce a consistent output temperature yielding a thermal efficiency of 35%. Moreover, the phase changing material can produce consistent air outlet temperature throughout the day or night.

Several other researchers integrated the desalination unit with other types of energy systems. In a work of Sulaiman et al. [29], the authors integrated the desalination plant with a parabolic trough solar air collector and evaluated two configurations of open-water open-air desalination units with collector installed before the humidifier or between the humidifier and dehumidifier. It was reported that the second configuration with the collector between the humidifier and the dehumidifier has much more advantages than the other configuration. Whereas the gained-output-ratio of the first and second configuration were 1.5 and 4.7, respectively. In a work of Lawal et al. [30], the authors integrated the HDH desalination system with a heat pump and concluded that the maximum gain-output-ratio of 8.88 and 7.63 is obtained at 80% components effectiveness using a mass flow rate ratio of 0.63 and 1.3 for modified air heated and water heated cycle, respectively.

Gabrielli and Mazzotti [31] presented a solar-driven HDH process for water desalination analyzed and optimized via an equilibrium theory and concluded it as an immediate tool for easily determining the optimal system operation. Gabra et al. [32] presented the mathematical models for the components constructed using CARNOT toolbox in a MATLAB environment and the results have shown that FOPID (fractional-proportional–integral–derivative) controlled offers a superior dynamic and static performance and reported that it can be automatically adjusted to compensate the weather changes.

As noted, significant research has been conducted on solar-assisted HDH desalination systems focusing on various design improvements with an objective to enhance the thermoeconomics of the system. One of the works in the performance and cost-effectiveness of a solar-driven humidification-dehumidification desalination system is presented by Zubair et al. [33] in which the capital cost. including supply well, equipment costs, and building costs, were considered. A multi-location (six locations in Saudi Arabia) analysis has concluded that the highest annual output is noted for Sharurah and lowest for Dhahran. Similarly, Jamil et al. [34] also reported the thermoeconomics of desalination system and concluded that the levelized cost of water production can be variable for various type of desalination system, for example, the production cost of reverse osmosis, mechanical vapour compression, multi-effect evaporation/desalination, multistage flash, and thermal vapour compression are $0.9 \pm 0.3\$/m^3$, $1.0 \pm 0.5\$/m^3$, $1.5 \pm 0.5\$/m^3$, $2.0 \pm 0.5\$/m^3$, and $2.7 \pm 0.8\$/m^3$, respectively. Likewise, Jamil et al. [34] presented that the hybrid desalination plants like energy recovery devices along with the reverse osmosis can have the lowest water production cost at $0.7 \pm 0.2\$/m^3$, leading us to the conclusion that further system improvement is needed to cope with the uncertain water security situation in the future.

One of the improvement can be related with the internal heat recovery mechanisms within the same desalination system and Xu et al. [35] has emphasized that the ongoing research on HDH desalination system has demonstrated that the internal heat recovery is a significant and potential method for improving the system performance and reducing the freshwater cost. Strictly speaking, conventionally, the HDH systems are driven by a low-grade heat source in the form of a waste heat recovery from another energy system. However, in this work, the authors are emphasizing the waste heat recovery option within the processing circuit of the desalination plant. Although, in literature, some research is available on different types of waste heat recovery procedures either with integration with another external energy resource or from an internal resource. However, this area needs more research to fully understand the potential and benefits of internal waste heat recovery.

Therefore, in this work, an opportunity of waste heat recovery is identified in HDH desalination and the system behaviour is reported with and without this waste heat recovery. The waste heat is recovered from the condenser coil by supplying it back to the hot water tank. Although, a variety of research is available on different strategies and methodologies of waste heat recovery within the system and/or integration with other energy systems; nevertheless, no research is focused on the practical thermoeconomic benefits of the HDH desalination plant with waste heat recovery from the condenser coil. The summary of various waste heat recovery in desalination plants is reported in Table 1 along with the identified gaps with the literature, thus highlighting the novelty of the work. Additionally, the waste heat recovery from the system would influence the system performance. However, it would also be a subject to the local conditions, either in terms of climate to influence the thermal indicators or in terms of economic conditions to influence the levelized cost of water production. Therefore, it is very important to realize the analysis of the desalination plant considering the local climatic and economic conditions. This aspect is still missing in the literature and needs more research. Another pivot point of the analysis is that a single-day demonstration of the waste heat recovery in the desalination plant might not be enough (see Table 1 as most of the analysis is based on a selected duration), because the solar integration makes the performance transient. Therefore, a yearly analysis demonstrating the pros and cons of the internal waste heat recovery has quite a significance. Therefore, in conclusion from this discussion, and based on the identified literature gaps, there is still a need to strengthen the research area of the yearly demonstration of waste heat recovery considering the local climatic and economic conditions to fully understand the gain of the system. A checklist is included in Figure 1, highlighting various literature gaps in the literature along with the novelty of the work.

Literature gaps	Novelty of the work
✗ Little or inadequate published work on humidification-dehumidification desalination unit with **internal waste heat recovery**[1].	✓ Considered humidification-dehumidification desalination unit with **internal waste heat recovery**[1] from the condenser coil.
✗ Little or inadequate technical and quantitative demonstration of the benefits of internal waste heat recovery in a humidification dehumidification desalination unit.	✓ Demonstrated thermoeconomic benefits of the internal waste heat recovery in a humidification dehumidification desalination unit.
✗ No available analysis on the yearly technical exploitation of internal waste heat recovery in a humidification dehumidification desalination unit.	✓ Demonstrated yearly technical exploitation of internal waste heat recovery in a humidification dehumidification desalination unit.
✗ No analysis is available on the behavior of internal waste heat recovery in a humidification dehumidification desalination unit for Pakistani climatic and economical conditions.	✓ Demonstrated the benefits of internal waste heat recovery in a humidification-dehumidification desalination unit for the climatic and economical conditions of Taxila, Pakistan.

Figure 1. Checklist of various identified literature gaps which are addressed in this work contributing towards its novelty.

In this work, using a mathematical model and simulations for actual solar irradiance, assessment of improvements in the productivity of the proposed system is presented. The improvement in the HDH system is proposed in terms of recovering waste energy from the feed water at the exit of the dehumidification part of the system. Usually, water leaving the condenser has relatively higher enthalpy. The prime novelty of this work is to present a yearly thermoeconomic analysis of the solar-assisted humidification dehumidification desalination plant by recovering the energy from the wasted feed water enthalpy while considering the local conditions.

Here, the proposed solar-assisted HDH desalination system utilizes the feed water at the exit of the condenser/dehumidification coil by reusing it in the brackish water storage tank. This allows the waste heat recovery (WHR) for the HDH desalination system and lessens the requirement of secondary energy sources. The governing mathematical model for the flat-plate solar collector, humidification chamber, and dehumidification chamber is solved through an iterative procedure [36]. The mass and energy balance on the brackish water storage is also solved including waste heat recovery from the condensing coil. A parametric study in which the influence of mass flow rate (MFR) of inlet air, saline water, feedwater, and temperature of the air on the freshwater productivity with and without waste heat recovery is conducted. Moreover, a yearly assessment of the proposed HDH system is also carried out to study the impact of waste energy recovery. In this regard, the data of the Taxila city (Pakistan) are used for solar irradiance over the complete year. A comparison of solar-assisted HDH desalination system with and without waste heat recovery is presented for the complete year; thereby indicating the impact of energy recovery. An economic analysis is also presented to reflect the advantages of waste heat recovery in terms of the cost of the desalinated water.

Table 1. Summary of different available waste heat recovery options in desalination systems and their gaps with the proposed scope of work.

Reference	Description	Findings	Location of the Analysis	Demonstration Dates of the Analysis	Gaps with the Current Work
[37]	Performance of a humidification dehumidification desalination technology is experimentally investigated in which new corrugated packing aluminium sheets were used in the humidifier.	The production of freshwater using the system can be increased by raising the inlet temperature, the mass flow rate of water entering the humidifier, and the rate of cooling water in the dehumidifier. The authors also concluded that the inlet temperature of the humidifier has a small effect on the productivity of the system as compared to other parameters. The authors also concluded that the waste heat from any industrial source as to be used to run the desalination unit and the total cost per litre of fresh water can be around $0.01 considering hot water as an input source driven from the gas turbine.	–	–	The focus of the work is on the HDH desalination system driven by a heat source supplied from a waste heat recovery of another energy system and the focus is given to the waste heat recovery on the gas turbines. Additionally, the presented configuration does not utilize the waste heat from the condenser coil.
[38]	The authors carried out a theoretical investigation of a humidification-dehumidification desalination unit for the climatological conditions of Antalya, Turkey.	The authors have concluded that water heating has major importance on clean water production because the heat capacity of water is higher than that of air. Therefore, the solar air heater doesn't lead to any significant improvement as compared to the usage of water solar collectors in the energy system. The annual clean water production can be around 12 tons given the specified variables.	Antalya, Turkey	August 15, 2011	The design of the study is completely different in which the focus is on the attainable benefits from the usage of solar air heaters or solar water collectors for the location of Antalya, Turkey. Additionally, the outlet of the cooling water circuit is discharged.
[39]	The authors experimentally investigated a solar energy-driven humidification-dehumidification desalination unit constructed by the Chinese Academy of Sciences having a capacity of 1000 litres per day of the water production rate.	The authors have concluded that the outlet temperature can rise to 118 °C when the solar radiation reaches 760 W/m^2 for parallel field configuration of the collectors. The water production can reach until 1200 litres per day when the average intensity of solar radiation is around 550 W/m^2 and the water production cost can be RMB 19.2 Yuan per m^3.	China	October 27 and November 10.	The design of the study is to develop and experimentally test a huge scale desalination plant. In the work, a cooling water pond is considered to extract the water to supply in the condenser coil and re-supplied to the same water pond without considering any waste heat recovery.

Table 1. *Cont.*

Reference	Description	Findings	Location of the Analysis	Demonstration Dates of the Analysis	Gaps with the Current Work
[40]	A humidification-dehumidification desalination plant is designed and tested for the actual conditions with the following dimensions of analysis: energy, exergy, economic and environmental.	Overall energy and exergy efficiencies of the system can be ranging from 4.1 to 31.54% and 0.03 to 1.867% respectively. The average water production rate can be 10.87 litres per day. The cost of the desalination can be 0.0981 USD per litre and finally, it was concluded that the productivity of the unit can increase with the increase in the temperature of the water and the air in the humidifier.	Karabuk city (longitude: 32.37°E, latitude: 41.12°N), Turkey	Starting from July 2015 for six days from 9:00 am to 6:00 pm	The configuration of the desalination unit is different in which the solar air heater is also considered. The cooling water from the condenser coil is supplied back to the salty water tank; however, no demonstration is carried out to signify its benefits.
[41]	A high capacity wind turbine is integrated with the multi-effect desalination system to recover the waste heat from the high capacity wind turbine.	The results have shown that the waste heat in a 7580 kW wind turbine is 231 kW at 140 °C for a wind speed of 11 m/s. The steam produced at 100 °C and 101.3 kPa can be enough to produce 45.069 m^3 per day of distilled water which can be sufficient to 4507 people with their daily consumption. The reported rate of return is 6.76% and the payback period is 6.33 years.	-		Although, the integration is very interesting considering its benefits as the wind turbines are installed near a huge source of the water body, and the waste heat of wind turbines can be used to desalinate the saline water. Nevertheless, the concept of waste heat recovery in this article is quite different than the one proposed in this work.
[35]	The authors have presented a novel heat pump with internal heat recovery desalination system based on the humidification-dehumidification process.	The freshwater cost of the two-stage HDH desalination system is reduced by 17.36%, the gained-output-ratio is increased by 55.64%, and the system productivity is increased by 15.51% as compared to a single-stage configuration. The authors also concluded that the productivity of the system is also vulnerably affected by the flow rates of cooling seawater and the working air.	China	-	The focus of the work is to recover the waste heat by using two stages of humidifier called low temperature and high-temperature humidifier. Although, the work is dedicated to the significance of the waste heat, yet the configuration and the type of waste heat recovery are different than the one proposed in this work. Additionally, the findings of this work are very different than the current one.

Table 1. *Cont.*

Reference	Description	Findings	Location of the Analysis	Demonstration Dates of the Analysis	Gaps with the Current Work
[42]	An integrated trigeneration system for electricity, hydrogen, and freshwater production using waste heat from a glass melting furnace are analyzed. It consisted of a flue gas emitted from a glass melting furnace which is used as a heat source of the system. Afterwards, it consists of a Rankine cycle, a thermochemical Cu-Cl cycle and a reverse osmosis desalination unit.	The Rankine cycle produces 1.9 MW of electricity, its energetic efficiency is 28.9%, and the exergetic efficiency is 31.2%. The energy efficiency of reverse osmosis desalination unit is 62.8% and its exergetic efficiency is 29.6%. Finally, the overall energetic and exergetic efficiencies of the trigeneration integrated system are 39.3% and 40.8% respectively.	The standard reference conditions of temperature and pressure of 20 °C and 1 atm, respectively, are used irrespective to any time or regional dynamics.		The trigeneration system is driven by a waste heat recovery unit which is quite different than the concept of internal waste heat recovery presented in this article.
[43]	It is a review article dedicated to the utilization of waste heat in desalination processes and its advances are presented. It is concluded that waste heat has successfully been used to drive different desalination processes and has proven to be a significant economic and environmental benefit.	(The cell is left blank intentionally because these columns are not applicable for reference [43])			The work does not quantitively present a case study of energetic and economic benefits of a waste heat recovery unit in desalination, as outlined in this work.
[44]	The authors presented multipurpose desalination, cooling, and air conditioning system powered by waste heat recovery from diesel exhaust fumes and cooling water.	The authors have concluded that the effect of water temperature on the mass flux through the membrane is higher than the hot water mass flow rate. The Coefficient-of-Performance within the range of 0.83–0.88 can be achieved when the heat transfer coefficient is 0.45 while the exhaust gas mass ratio is between 0.37 and 0.53.	-		The configuration, the mode of waste heat recovery, and the analysis in the proposed configuration are different than reference [44].

Table 1. *Cont.*

Reference	Description	Findings	Location of the Analysis	Demonstration Dates of the Analysis	Gaps with the Current Work
[45]	The authors have mentioned that the seawater in copper tubes is usually used in condenser but owing to the drawbacks of pipe erosion; the usage of a water-cooled condenser can be a suitable option. However, the prime challenge is to attain sufficient enough heat flux in air-cooled condensers owing to the poor thermal conductivity of air. Therefore, in the work, the performance of a humidification-dehumidification desalination plant is presented with integration with an air-cooling condenser and cellulose evaporative pad.	Maximum productivity of 120 kg per day was achieved using the humidifier of cellulose pad operating on a water temperature of 49.5 °C. The maximum gained-output-ratio was 0.53 with a maximum coefficient-of-performance of 20.7.	Taiwan	The configuration is not integrated with a solar-driven energy resource; therefore, the time dependency does not have significance.	The configuration is not integrated with a solar-driven source; instead, it utilized an electric heater in the water storage tank. Additionally, the dehumidifier design is different than the one proposed in this work, as the dehumidifier of reference [45] is an air-cooled condenser, and the one presented in the work is the conventional water-cooled condenser.
[46]	The authors have presented a configuration in which the heat required to drive the HDH solar desalination plant is recovered from a source (there can be many of them depending on its type). The seawater is appointed to recover the waste heat of any process and based on the mass and energy conservation principles; a mathematical model is derived to simulation the process. Similar work is also presented by the authors in reference [47].	The results of the simulation have shown that the maximum value of the freshwater at 99.05 kg per hour and of gained-output-ratio at 1.51 is obtained when the balanced condition of the dehumidifier appears at the design conditions. It was also concluded that reducing the seawater spraying temperature and evaluating its humidification effectiveness is beneficial for its thermoeconomic performance.	China		There are many fundamental differences between this and the current work. The HDH system of reference [46] is driven from a waste heat whereas the system in the current work is driven from a solar energy source. Moreover, the concept of waste heat recovery is quite different in both works; where the current work is dedicated to the internal waste heat recovery; whereas, the work of reference [46] focuses on the desalination system delivered from a waste heat recovery unit.

Table 1. Cont.

Reference	Description	Findings	Location of the Analysis	Demonstration Dates of the Analysis	Gaps with the Current Work
[48]	A hybrid humidification-dehumidification desalination system operated through a waste heat recovered from a vapour compression refrigeration based on a household air-conditioning unit is presented. The heat rejected from the condenser in the form of heated air is utilized to drive the desalination unit. The heat and mass transfer characteristics of various components were simulated using TRNSYS and experimental validation is also carried out.	It is concluded that the average maximum freshwater produced using the waste heat from the vapour-compression refrigeration system for hot and medium (pre-monsoon) climatic conditions were 4.63 kg per hour and 4.13 kg per hour, respectively. The freshwater production rate is higher during the summer season as the ambient air has lower relative humidity and higher temperature. Finally, the economic analysis has indicated that the cost of freshwater produced from the integration is around $0.1658 per kilogram.	Chennai city (12.98° N latitude, 80.17° E longitude), Southern India	Months of March and May.	The configuration, research design, analysis procedure, findings, and conclusion of this work are completely different from the one proposed by Santosh et al. [48]. The recovered waste heat in the work of reference [48] is attained to drive the solar desalination unit from a vapour-compression refrigeration cycle. However, the emphasis of the waste heat recovery is quite different in the case of the current work.
[49]	The authors have presented a review article focusing that desalination market has greatly expanded in recent decades and expected to continue growing in the coming years. This study reviews some of the most promising desalination techniques including humidification-dehumidification desalination powered by a solar energy resource. The review focuses on water sources, demand, availability of potable water, and purification method.	It is concluded by the authors that desalination seems to be a reasonable and technically attractive option towards the emerging water-energy scarcity problems. The authors have mentioned that cheap freshwater can be produced from brackish, sea, and ocean water by using solar panels, wind turbines, along with other emerging renewable energy technologies. The authors have mentioned that the HDH system has some advantages for a small-scale decentralized water production which includes simpler brine pretreatment, disposal requirements, and simplified operation and maintenance. It is recommended that a multi-effect closed-air-open-water-heated system is the most energy efficiency. The authors have also proposed several methods to improve the performance such as water heating and innovative ways of atomization (misting) of the hot water.			The work of reference [49] is a review work and doesn't quantitatively describe the performance indicators of any desalination plant which is making the current work unique in its novelty.

Table 1. Cont.

Reference	Description	Findings	Location of the Analysis	Demonstration Dates of the Analysis	Gaps with the Current Work
[50]	The work presents an experimental and a theoretical model for the utilization of Fresnel lens in solar water desalination system working on the humidification-dehumidification principle. The thermodynamic analysis considering the mass and energy balance equations are developed for the following processes: water heater, humidifier, and other cycle components. The models were solved numerically, and the validation process has shown that the model outcomes are 25% higher than the experimental data owing to some energy losses.	The results have concluded that the Fresnel lens has a good efficiency in the range of ~70% for the clear days. The authors have also concluded that at an inlet water temperature of 90 °C, the flow rates were 27 and 40.8 litre/h/m^3 of feed saline water for open and closed systems, respectively.	Egypt	-	The components of the work of reference [50] are different from the current work because the authors of reference [50] have utilized a Fresnel lens; however, the current work utilizes a flat-plate solar collector. Owing to many other basic differences along with problem formulation, research design, and methodology, the outcomes/findings and conclusion of both of the works are quite different.
[51]	A humidification-dehumidification desalination unit having bubbler humidifier and thermoelectric cooler for the dehumidification purposes is presented with a theoretical and an experimental justification to evaluate the influence of temperature of air and water, the diameter of the hole on the periphery of circling tube, the height of hot water column, and the air mass flow rate in the performance indicator i.e., production of freshwater.	It is concluded that the achieved daily distillate production was in the range of 7 to 13 litres per day for different operational conditions; whereas the maximum productivity of the system was 12.96 litre per day for a hole diameter of 2mm, the mass flow rate of air of 0.016 kg/s, the water temperature of 60 °C, air temperature of 27 °C, and column height of 7cm.	India	-	The equipment utilized for the humidification and the dehumidification of the work of reference [51] is different from the one employed in this work.

2. System Description

A schematic diagram of the desalination system working on humidification and dehumidification along with waste heat recovery from the condenser/dehumidification section is presented in Figure 2. Saline water is pumped into the solar collector (state 2) from brackish water storage (state 1). Water is heated in the solar collector and re-enters the heated brackish water storage at state 3.

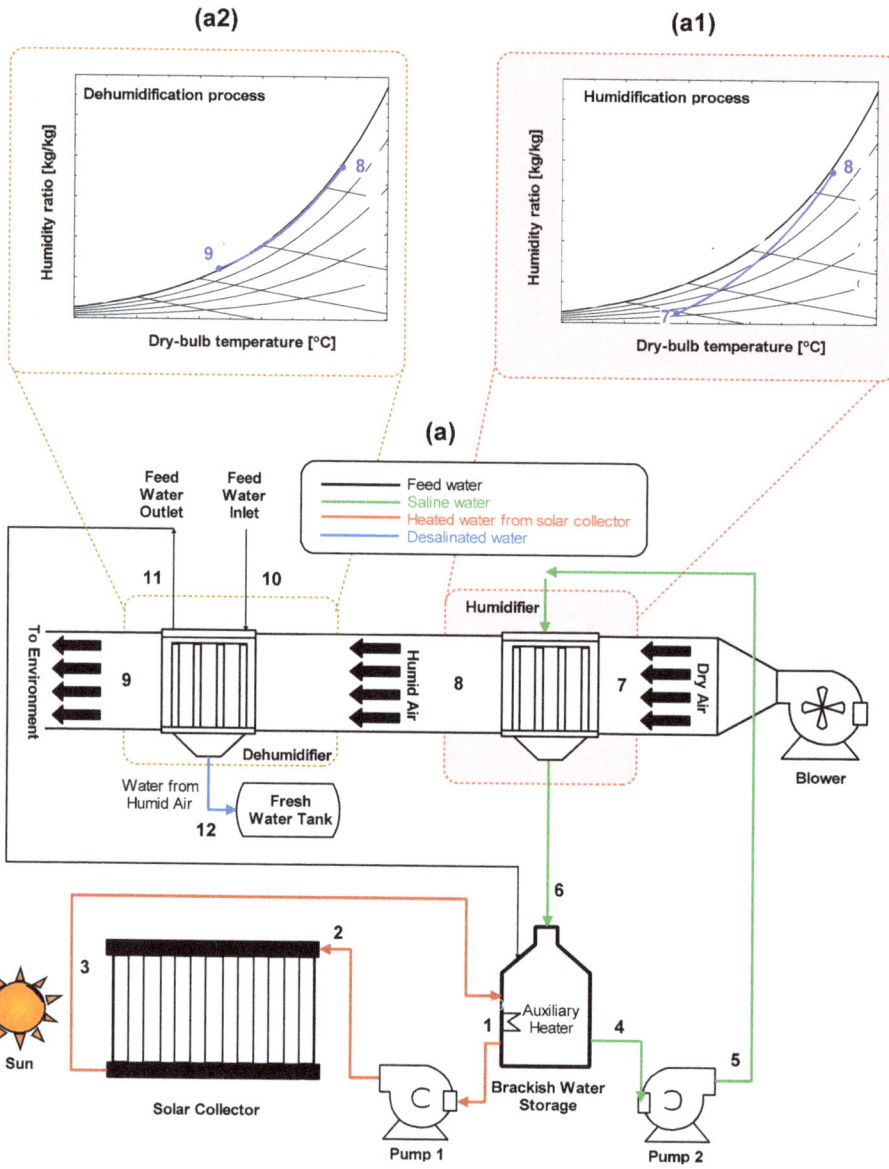

Figure 2. Desalination system scheme with waste heat recovery process (the description of the state numbers 1–12 is discussed in Section 2).

Afterwards, the salty water is pumped to the humidification section from state 4 to 5 where it comes in direct contact with the ambient air entering at state 7. Depending on the climatic conditions, ambient air blown through the humidification chamber at state 7 where it absorbs moisture from the falling film of water. The saturated moist air leaves the humidification section at state 8, whereas the remaining salty water is re-circulated to the brackish water storage. The corresponding psychometric description of this process is presented in Figure 2a1 in which the air leaves at the saturation conditions [49].

The moist air at state 8 enters the dehumidification section where feed/cooling water is circulated at a relatively high flow rate to facilitate dehumidification. As a result, fresh condensed water is collected in a tank at state 12, and the air leaves the system to the environment. The corresponding dehumidification process is described on a psychometric chart as in Figure 2a2 [49].

The feed water at exit 11 of the dehumidification section has relatively high enthalpy because it receives heat from the humid air. In this work, the heat from this feed water is recovered by supplying it back to the brackish water storage tank.

There can be many practical ways to collect the condensed water from the dehumidification section. For instance, in other configurations of thermal desalination, Patel et al. [52] and Nayi et al. [53] have shown that an outlet pipe supported with a trough placed inside assembly of a solar still can be used and finally the condensed water can be collected in a beaker outside. In a theoretical and experimental study of seawater desalination based on humidification-dehumidification technique, Mohamed et al. [54] have demonstrated that the freshwater can be collected at the bottom of the condensation coil for a vertical section. In another work, Rajaseenivasan et al. [55] presented an experimentally verified HDH system with a dual-purpose collector and employed a horizontal flow shell-and-tube heat exchanger with the condensed water collected at the bottom from the end of the dehumidifier. Another configuration is demonstrated by Xu et al. [56] in which a novel enhanced HDH method with weakly compressed air and internal heat recovery based on traditional mechanical vapour compression is developed in which the moist air is used as a working fluid instead of a vapour. In this experimentally developed configuration [57], the freshwater is extracted from the evaporator-condenser assembly having airflow in the horizontal direction.

3. Mathematical Model

The thermal performance of the solar-assisted desalination plant is evaluated by developing a mathematical [58] expression of the components involved such as solar collector, brackish water storage, humidifier [59], and dehumidifier.

3.1. Flat Plate Solar Collector

Ioan Sarbu and Calin Sebarchievici [60] have reported that the flat-plate collectors are the heart of any solar energy collection system designed for operation in the low-temperature range (less than 60 °C) [60] or the medium temperature range (less than 100 °C) [60]. It is used to absorbed solar energy, convert it into heat, and then to transfer that heat to a stream of liquid (as in this case). They use both direct and diffuse solar radiations [60], do not require tracking of the sun [60], and require little maintenance [60]. They are mechanically simpler than concentrating collectors [60]. The major applications of these units are in solar water heating, building heating, air conditioning, and industrial process heating [60]. For quasi-steady-state conditions [61] at a given solar time, along with other standard assumptions [62–64], the following energy balance equation can be written

$$Q_u^s = A_c^s \left[S^s - U_L^s \left(T_p^s - T_{amb} \right) \right] = A_c^s F_R^s [S^s - U_L^s (T_2^s - T_{amb})] = m_{w,1}^s c_{p,w} (T_3^s - T_2^s) \quad (1)$$

where:

$$F_R^s = \frac{m_{w,1}^s c_{p,w}}{A_c^s U_L^s} \left[1 - \exp\left(\frac{A_c^s U_L^s F'^s}{m_{w,1}^s c_{p,w}} \right) \right] \quad (2)$$

$$Fr^s = \left[W^s U_L^s \left(\frac{1}{U_L^s[D^s + (M^s - D^s)F^s]} + \frac{1}{\pi D_i^s h_{c,w}^s} \right) \right]^{-1} \tag{3}$$

$$F^s = \frac{\tanh[\zeta^s(M^s - D^s)/2]}{\zeta^s(M^s - D^s/2)} \tag{4}$$

$$\zeta^s = \sqrt{\frac{U_L^s}{k_p^s t_p^s}} \tag{5}$$

The temperature of the absorber plate is calculated through an iterative method by the solution of Equations (1)–(5) coupled with:

$$T_p^s = T_2^s + \frac{Q_u^s/A_c^s}{F_R^s U_L^s}(1 - F_R^s) \tag{6}$$

The total absorbed solar radiation (S) is evaluated by considering the beam and diffused components of incident solar radiation along with the incorporation of optical losses (i.e., transmittance-absorptance product).

$$S^s = I_b^s(\tau\alpha)_b^s + I_d^s(\tau\alpha)_d^s \tag{7}$$

The transmittance-absorptance product is calculated separately for beam and diffused radiation.

$$(\tau\alpha)_{b,d}^s = \frac{\tau_{b,d}^s \alpha_p^s}{1 - \left(1 - \alpha_p^s\right)\Re^s} \tag{8}$$

Fresnel's expressions [61] are derived for the reflection of unpolarized radiation passing from medium 1 to medium 2 with different refractive indexes given the angle of incidence and refraction for the parallel and perpendicular components of the beam and diffused radiation. This procedure gives two Fresnel's expressions of the beam and diffused radiation with averaged parallel and the perpendicular component of each one, and finally, it is used to calculate the transmissivity of solar radiation from the glass cover to the absorber plate.

Along with that, the incident angle for direct radiation is the angle of incidence, given by:

$$\cos\theta_b^s = \sin\delta^s \sin\phi^s + \cos\delta^s \cos\phi^s \cos\omega^s \tag{9}$$

Afterwards, Snell's law [61] is utilized to obtain the refractive angles from the glazing.

By taking into account the top heat loss coefficient only, the overall heat loss coefficient U_{top}^s is computed as follows.

$$U_{top}^s = \left[\frac{1}{h_{c,g-a}^s + h_{r,g-a}^s} + \frac{1}{h_{c,p-g}^s + h_{r,p-g}^s} \right]^{-1} \tag{10}$$

The bottom and side heat loss coefficients are considered negligible. Practically, it can be made possible through the usage of insulation materials which can be glass mineral roll [65], rice husk and sunflower stalks [66], petiole piece, fibres and gypsum [67], and mineral wool [68]. The convective heat transfer coefficient is calculated using [61,69]:

$$h_a^s = 5.7 + 3.8 v_a^s \tag{11}$$

The method proposed in [70] is used to compute the Nusselt number between the absorber plate and the glass cover.

$$Nu_{p-g}^s = 1 + \left[\left(\frac{Ra_{p-g}^s}{5830} \right)^{1/3} - 1 \right]^+ + 1.44 \left[1 - \frac{1708}{Ra_{p-g}^s} \right]^+ \tag{12}$$

The radiative heat transfer coefficient is given by:

$$h^s_{r,p-g} = \frac{\sigma \left(T^s_p + T^s_g\right)\left(T^{s^2}_p + T^{s^2}_g\right)}{\left(1/\varepsilon^s_p\right) + \left(1/\varepsilon^s_g\right) - 1} \tag{13}$$

3.2. Water Storage Tank

In this research study, the feed water of the dehumidifier is being re-used in the brackish water storage, therefore, a mass and energy balance is of key importance. A mass and energy balance applied on the storage tank yields:

$$m_3 + m_6 + m_{11} - m_1 - m_4 = 0 \tag{14}$$

$$m_3 h_3 + m_6 h_6 + m_{11} h_{11} - m_1 h_1 - m_4 h_4 = 0 \tag{15}$$

3.3. Humidifier

The humidification section consists of a falling film of water with accompanying airflow. It is assumed that the process is quasi-steady with negligible heat loss. The falling film is an assumed laminar with a smooth liquid-gas interface. Figure 3 shows a typical element in the humidifier section with accompanying zones. The mean velocity of a falling film is calculated using the following equation [71]:

$$u^h_w = \frac{\rho_w g \delta^2_w}{3 \mu_w} \tag{16}$$

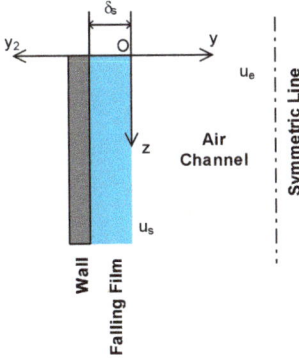

Figure 3. Schematic diagram of control volume analysis of the humidification section.

With the thickness of the falling film, given by [72]:

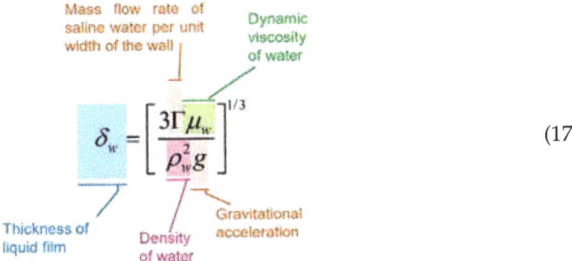

$$\delta_w = \left[\frac{3 \Gamma \mu_w}{\rho^2_w g}\right]^{1/3} \tag{17}$$

The energy balance equation is [72,73]:

$$u_w^h \frac{\partial T_w^h}{\partial z} = \alpha_w \frac{\partial^2 T_w^h}{\partial y_2^2} \tag{18}$$

The solution of Equation (18) requires the following boundary condition [73]:

$$\alpha_w \frac{\partial T}{\partial y}\bigg|_{y=0} = -\alpha_s \frac{\partial T_w}{\partial y_2}\bigg|_{y_2=0} + \frac{\lambda W_j}{\rho_s c_{p,a}} \tag{19}$$

Eams [74] predicted the water evaporation rate as shown in Equation (20).

$$W_j = \varepsilon * (P_s - P_v) \sqrt{M/2\pi R T_n} \tag{20}$$

Knudson constant of evaporation [75] $\varepsilon*$ is written in terms of the coefficient of evaporation:

$$\varepsilon* = \frac{2\varepsilon}{2 - \varepsilon} \tag{21}$$

whereas the coefficient of evaporation is calculated using [74]:

$$\varepsilon = h * \sqrt{\frac{2\pi R T_s}{M}} \frac{T_s}{\rho_v \lambda^2} \tag{22}$$

Considering the evaporation affecting the heat transfer coefficient $h*$, the concept of wet bulb coefficient of heat transfer as proposed by MacLaine-Cross and Banks [76] can be used to correct the heat transfer coefficient. It can be written as:

$$h* = h\left[1 + \frac{e\lambda}{C_{pa}}\right] \tag{23}$$

where e is derived from the wet-bulb coefficient, given by [76]:

$$e = \frac{\omega_{max} - \omega_{min}}{T_{max} - T_{min}} \tag{24}$$

3.4. Dehumidifier/Condenser Section

The efficacy of the condenser, modelled as a counter-flow concentric heat exchanger, is computed using NTU (Number of Transfer Units) method [77–79] which is used to calculate the rate of heat transfer in heat exchangers when there is an insufficient data to evaluate the Log-Mean-Temperature-Difference [77–79].

Therefore, in NTU method, the effectiveness of the heat exchanger is calculated which can be converted into the actual heat transfer of the heat exchanger. It also involves the calculation of minimum heat capacity (C_{min}) and maximum heat capacity (C_{max}) which can be either heat capacity of the cold fluid or the hot fluid depending on whichever can be minimum or maximum. The evaluation of U is carried out using the thermal resistance diagram by computing the convective heat transfer coefficient of the humidifier and feed water. Once the actual heat transfer is calculated, it can be used to evaluate the outlet temperatures of both steam and finally, the mass balance can give the quantity of the condensed fresh water. This methodological framework is presented in a stepwise format as in Figure 4.

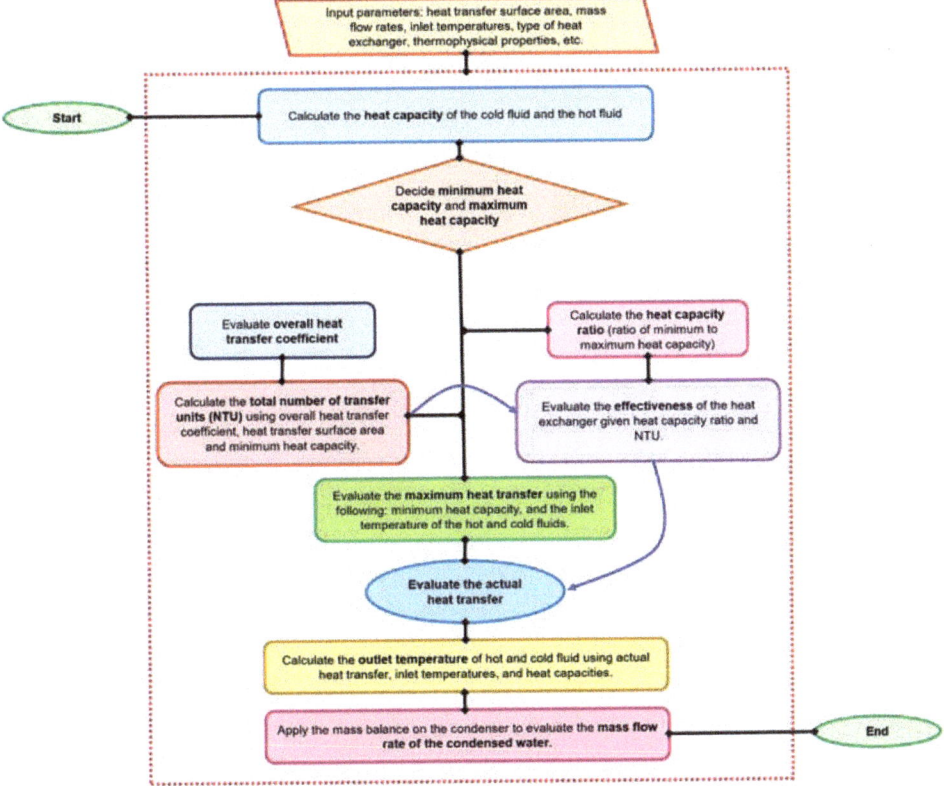

Figure 4. The methodological framework for the analysis of the dehumidifier.

3.5. Auxilary Equations

The auxiliary equations aids in the extraction of the temperature-dependent thermophysical properties of different fluids. These are mostly empirical equations and their usage in this work was practised with precaution in which the limit of each model is verified with the database of Engineering Equation Solver [80] before the final implementation.

The thermal conductivity of the dry air is given by [81]:

$$k_a = -4.937787 \times 10^{-4} + 1.018087 \times 10^{-4} T_a - 4.627937 \times 10^{-8} T_a^2 + 1.250603 \times 10^{-11} T_a^3 \quad (25)$$

The thermal conductivity of the water vapour is given by [81]:

$$k_v = 1.3 - 46 \times 10^{-2} - 3.756191 \times 10^{-5} T_a + 2.217964 \times 10^{-7} T_a^2 + 1.111562 \times 10^{-14} T_a^3 \quad (26)$$

The thermal conductivity of the humid air is given by [81]:

$$k_{humid-air} = \frac{\left(\frac{1}{1+1.608\omega}\right) K_a M_a^{1/3} + \left(\frac{\omega}{\omega+0.622}\right) K_v M_v^{1/3}}{\left(\frac{1}{1+1.608\omega}\right) M_a^{1/3} + \left(\frac{\omega}{\omega+0.622}\right) M_v^{1/3}} \quad (27)$$

The latent heat of vaporization is given by [82]:

$$i_v = 3483181.4 - 5862.7703T + 12.139568T^2 - 0.0140290431T^3 \quad (28)$$

The specific heat of dry air is given by [81]:

$$c_a = 1045.356_a + 0.0007083814 T_a^2 - 2.705209 \times 10^{-7} T_a^3 - 0.3161783 T \tag{29}$$

The specific heat of water vapour is given by [81]:

$$c_v = 1360.5 + 2.31334 T_a - 2.46 \times 10^{-10} T_a^5 + 5.9 \times 10^{-13} T_a^6 \tag{30}$$

Finally, the specific heat of humid air is calculated using [81]:

$$c_{humid-air} = c_a + \omega c_v \tag{31}$$

The dynamic viscosity of dry air in terms of the temperature is given by [81]:

$$\mu_a = 2.287973 \times 10^{-6} + 6.259793 \times 10^{-8} T_a - 3.131956 \times 10^{-11} T_a^2 + 8.15038 \times 10^{-15} T_a^3 \tag{32}$$

The dynamic viscosity of water vapour is given by [81]:

$$\mu_v = 2.562435 \times 10^{-6} + 1.816683 \times 10^{-8} T_a - 2.579066 \times 10^{-11} T_a^2 - 1.067299 \times 10^{-14} T_a^3 \tag{33}$$

Finally, for the humid air, the dynamic viscosity is [81]:

$$\mu_{humid-air} = \frac{\left(\frac{1}{1+1.608\omega}\right) \mu_a M_a^{0.5} + \left(\frac{\omega}{\omega+0.622}\right) \mu_v M_v^{0.5}}{\left(\frac{1}{1+1.608\omega}\right) M_a^{0.5} + \left(\frac{\omega}{\omega+0.622}\right) M_v^{0.5}} \tag{34}$$

4. Benchmarking of Simulation Results

The benchmarking of the current simulation results is accomplished by comparison with the experimental and the simulation results of Dai et al. [75] without considering the waste heat recovery of the feedwater from the condensing coil. A solar desalination study having humidification and dehumidification processes, both mathematically and experimentally, was presented. All of the conditions considered by Dai et al. [75] are codified in MATLAB simulation and the desired results are obtained. The initial conditions for this benchmarking are: inlet temperature of the air is 35 °C, inlet relative humidity is 40%, MFR of saline water is 1500 kg/h, the NTUs of condensing coil are 4, the MFR of cooling coil is 2500 kg/h, and the incident solar radiation is 700 W/m². The humidifier is 0.6m long and has a cross-sectional area of 0.56m². Considering these input parameters in the developed code, the compliance between the current simulation results and the simulation results of Dai et al. [75] are shown in Figure 5 in which the variation in water productivity is shown with changing MFR of working air along with different water film temperature. A maximum discrepancy of 2.33% is observed between both simulation results. The discrepancy is caused by the advanced modelling of solar collector adapted by the authors as compared to the relatively simplified mathematical model of [75]. It is to be stressed here that the feed water recirculation (waste heat recovery) [83] is not included for this comparison to fully replicate the conditions of [75].

The results of the simulation study are also benchmarked by comparison with the experimental results of Dai et al. [75]. A comparison is carried out for two cases. For the case, I, the MFR of air is 615.6 kg/h, and the ambient relative humidity is 54%. For case II: the MFR of air is 661.8 kg/h, and relative humidity is 49%. For both cases: the ambient air temperature is ~22 °C, the feedwater temperature is ~19 °C, the MFR of saline water is 2310 kg/h and the MFR of feed water in condensing coil is 3780 kg/h. Here it is stressed that the authors of [75] replaced the solar collector by a boiler to obtain quick lab results during experimentation. The authors of the current work carried out analysis by considering the solar collector. Subsequently, an analysis is also carried out by solving the boiler as a heat input. The analysis considering boiler and solar collector along with experimental data of [75] for each of the case I and case II is shown in Figure 6a,b. It can be observed here that the water productivity level for the solar collector is lower than the experimental data points because the heat loss coefficient (U_L) for the solar collector case is contributing significantly which was absent in the experimental setup of [75]. However, the water productivity level for the case in which the authors simulated the boiler is higher than the experimental data points because of the assumptions considered in Section 3.3. Based on this discussion, the maximum discrepancy observed for this benchmarking is ~8% for both cases as reported in Figure 6a,b which is a plot between the temperature of saline water and water productivity level.

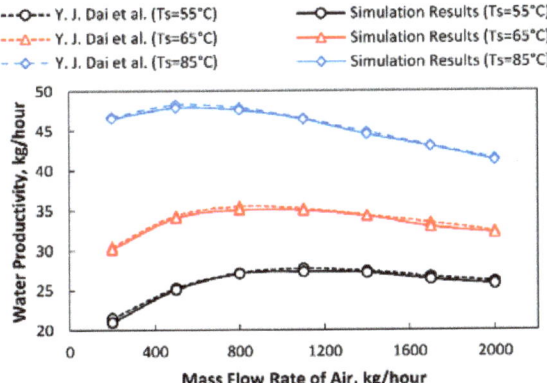

Figure 5. Benchmarking of simulation results by comparison with simulation results of Dai et al. [75] without waste heat recovery.

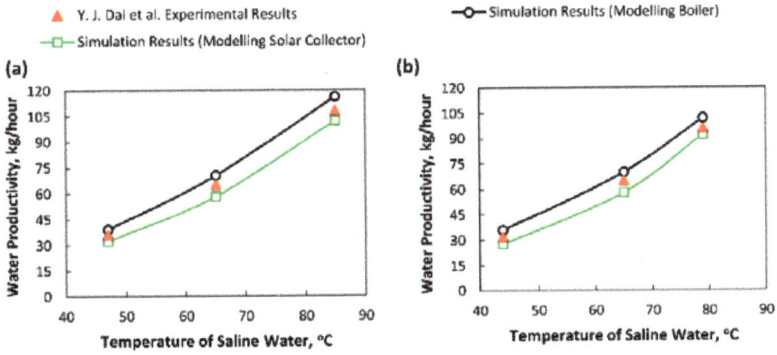

Figure 6. Comparison of simulation results with the experimental results of Dai et al. [75] for (**a**) case I, (**b**) case II, without waste heat recovery.

5. Results and Discussion

This work aims to recover the heat from the condensing coil. Therefore, in this section, different parameters are varied to observe their effect on freshwater productivity with and without the waste heat recovery.

5.1. Mass Flow Rate (MFR) of Inlet Air Effect on Freshwater Production

The intake of MFR of air is an important parameter in a desalination unit because it determines the blowing power. Therefore, simulations are carried out to observe the influence of MFR of air on water productivity with and without waste heat recovery from the condensing coil and it is presented in Figure 7.

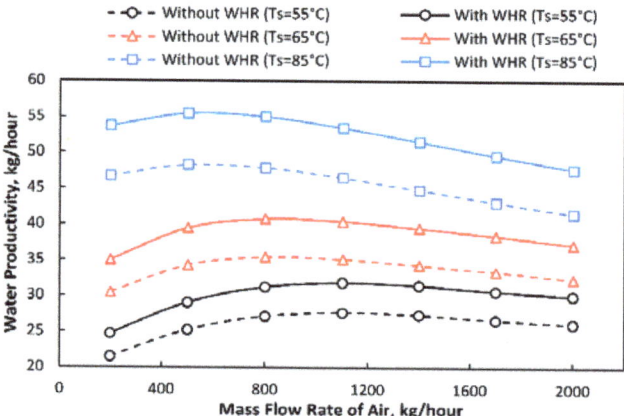

Figure 7. Effect of MFR of air on freshwater productivity with and without waste heat recovery.

It is observed that the water productivity is increased with the increase in MFR of air because the tendency of moisture transfer increases; therefore, water productivity attains a maximum value at 500–800 kg/h of MFR of air. However, water productivity decreases significantly after 1100 kg/h of MFR of air. It is because very high values of MFR of air do not give sufficient time to an air in the humidification section to cause effective evaporation; therefore, as a result, the water productivity decreases. Dai et al. [75] recommended a range of 500–800 kg/h of the MFR of air for the HDH desalination unit, which can be also observed here. It is also observed that elevating the temperature of the saline water also increases the productivity of freshwater. However, it is deduced from Figure 5 that recovering the heat from the condensing coil by reusing the feedwater in the brackish water storage significantly increases the freshwater productivity. It is reported that ~14% of freshwater productivity is increased for an MFR of 800 kg/h by waste heat recovered from the condensing coil at a saline water temperature of 85 °C.

5.2. Mass Flow Rate (MFR) of Saline Water Effect on Freshwater Production

Several simulations are carried out to observe the effect of MFR of saline water on freshwater productivity from the dehumidifier/condenser. It can be observed from Figure 8 that an increase in the MFR of the saline water increases the water production rate. It is because a high MFR of saline water contributes to a high Reynolds number, thus increasing the heat transfer coefficient between the water and air mixture in the humidification section, which yields higher values of water productivity. It can also observe that increasing the saline water temperature increases the freshwater productivity. Since the mass transfer coefficient is a strong function of temperature, therefore, it contributes towards a more effective mass transfer of water vapours to air. Furthermore, it is observed that the freshwater productivity also increases by waste heat recovered. Here, an increase of ~15% in water productivity is reported by waste heat recovery from the condensing coil for a saline water MFR of 1800 kg/h at a water film temperature of 85 °C.

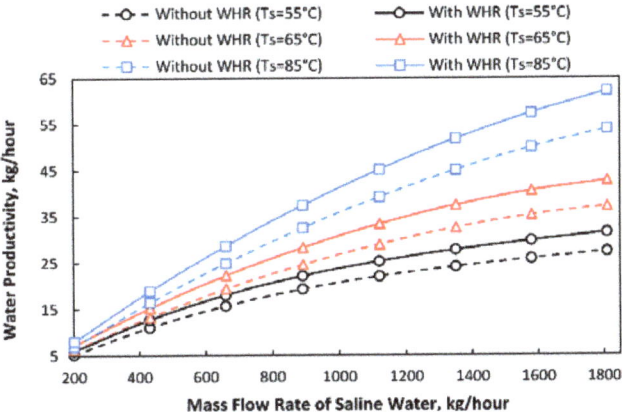

Figure 8. Effect of MFR of saline water on water production with and without waste heat recovery.

5.3. Mass Flow Rate (MFR) of Feedwater Effect on Freshwater Productivity

The MFR of cooling water in the condensing coil is changed from 1200 kg/h to 4000 kg/h and its influence on freshwater productivity is studied. The simulation was carried out on three different values (1, 2, and 4) of NTU of the condensing coil. The influence of waste heat recovery is also presented here.

It can be witnessed in Figure 9 that the hourly freshwater production is directly proportional to the MFR of the cooling water because its increment decreases the surface temperature of the condenser which can increase condensation rate. Furthermore, water productivity is increasing with the increase in the NTU of the condensing coil. A higher value of NTU corresponds to a higher value of the overall heat transfer coefficient in the condensing coil. Therefore, it increases the actual heat transfer; and as a result, the hourly freshwater production is increased. Furthermore, here it is shown that the hourly production rate of freshwater is increased to 10% by waste heat recovered from the condensing coil at an MFR of cooling water of 4000 kg/h.

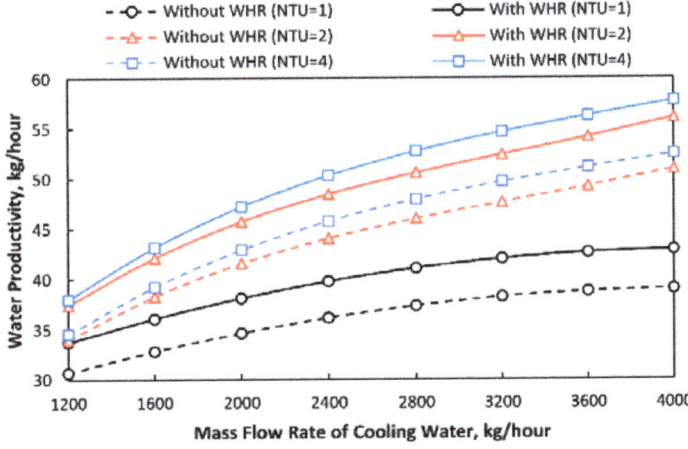

Figure 9. Effect of MFR of cooling water on freshwater productivity with and without waste heat recovery.

5.4. Effect of the Temperature of the Air on Freshwater Productivity

In this section, a simulation is carried out to study the effect of the ambient air temperature on the freshwater productivity. These results are important as they will govern the applicability of the desalination unit. Since this desalination unit is working on humidification principle; therefore, the ambient air conditions are crucial. The behaviour is reported in Figure 10.

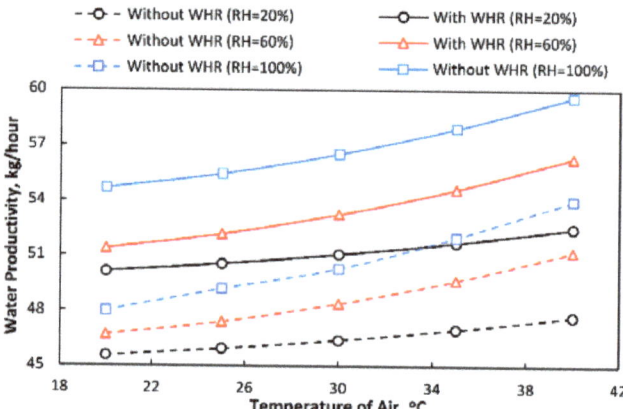

Figure 10. Influence of temperature of the air on freshwater productivity with and without waste heat recovery.

It is observed that the hourly freshwater production rate is directly proportional to the ambient temperature of air because the increment in it enhances the heat and mass transfer characteristics in the humidification chamber. Furthermore, it is reported that the freshwater production rate is also directly proportional to the relative humidity at the exit of the humidification chamber (state 8). This is because higher relative humidity at the exit of the humidification section indicates a small difference between the dry bulb and the dew point temperature of the air which is an assertive parameter towards condensation. It is also reported here that the hourly freshwater production rate is increased to ~11% by waste heat recovered from the condensing coil at an ambient air temperature of 45 °C at full saturation conditions at the exit of the humidification chamber.

5.5. Yearly Analysis of Freshwater Productivity with and without Waste Heat Recovery

In this section, an analysis is carried out to observe the freshwater productivity rate yearly. As the plant uses solar radiation for heating, therefore its actual output will be important for different days of the year. For this purpose, the selected day of each month is used for the calculation of flat-plate solar collectors. Table 2 shows the representative day of each month [84] that is selected to carry out an analysis. Selected data such as ambient temperature, ambient relative humidity, and velocity for analysis purposes are reported in Table 2. The solar radiation data, including total incident, beam, and diffuse radiation for Taxila, is presented in Figure 11.

Table 2. Parameters for the calculation of water productivity.

Month	Representative Day [84]	Temperature (°C)	RH (%)	Wind Velocity (m/s)
January	17	11.1	84	3.5
February	16	12.3	85	2.2
March	16	18.7	84	4.7
April	15	26.8	56	10.0
May	15	28.5	22	9.4
June	11	28.8	77	4.6
July	17	32.4	81	2.7
August	16	26.7	82	3.2
September	15	23.7	45	4.1
October	15	20.8	39	2.4
November	14	15.9	100	1.0
December	10	9.1	81	1.5

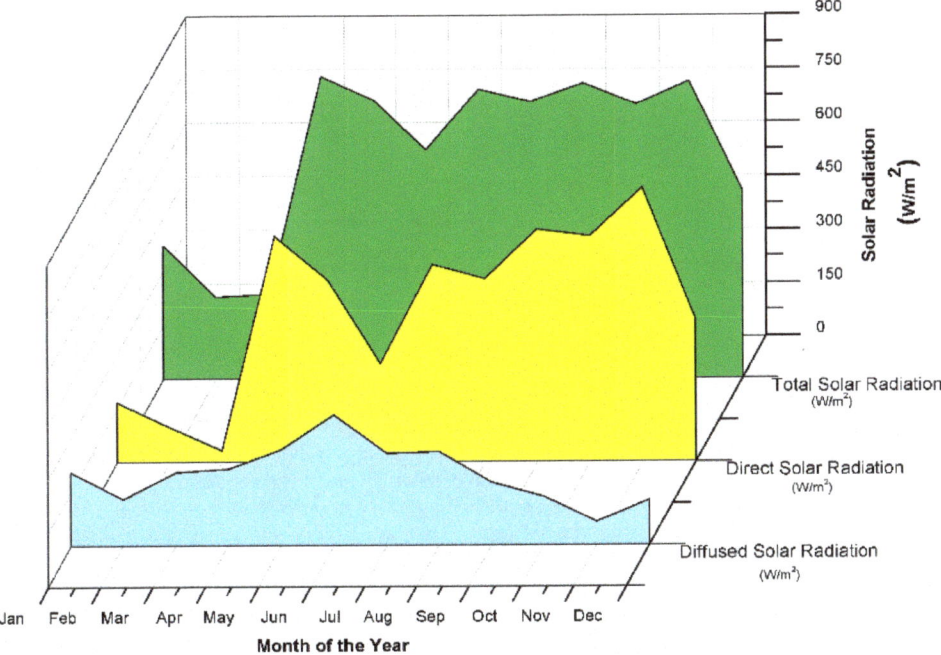

Figure 11. Distribution of total, beam/direct, and diffused solar radiation.

Figure 12 shows the yearly freshwater productivity from the solar-assisted desalination HDH plant with and without waste heat recovery from the condensing coil. The lower freshwater production rate is observed for April and May. Although a significant amount of solar radiation is experienced in these months (Figure 11) but the region of Taxila (33.745833, 72.7875) experiences high wind velocities in these months. Therefore, it increases the overall heat transfer coefficient of the solar collector, which yields lower water production rates. The highest freshwater production rate is observed for July where the incident radiation is maximum, and the wind velocities are low. The water production rate in November, December, January, and February are relatively high. Even though low ambient temperatures are observed in these months, but relatively higher values of incident radiation and lower wind velocities are observed; therefore, the water production rate is also relatively high. The water production rate in August is low because Taxila experiences very high humidity this month; therefore, it decreases the potential of the humidification process in the desalination unit. One important phenomenon can be observed from Figure 12 that the water production is high for each month of the year if the waste heat from the condensing coil is recovered. It is reported here that the freshwater productivity is increased by ~16% for June by recovering the waste heat from the condensing coil.

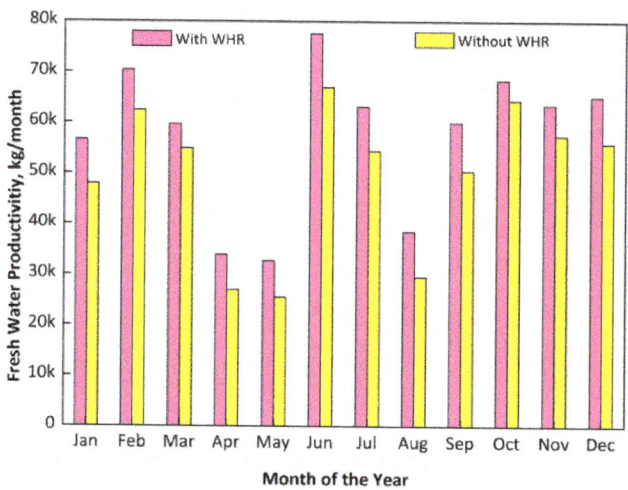

Figure 12. Yearly freshwater production rate with and without waste heat recovery.

6. Economic Assessment

Kaya et al. [6] carried out a levelized cost analysis for solar-energy-powered seawater desalination in the Emirate of Abu Dubai and considered multi-stage flash and multi-effect distillation coupled with thermal power plants while mentioning that these thermal desalination methods are responsible for more than 90% of the desalination capacity in the Emirate. The findings of the article [6] suggest that the analysis considering the levelized cost of water for a combination of solar PV and a reverse osmosis system is technologically well-positioned in terms of cheap and clean desalination systems. The authors of reference [6] considered total capital cost, annual energy consumption with a required solar energy system to balance a zero in the grid along with the chemical costs and overhead cost yielding a final levelized cost of water. A similar approach is adopted here in which the economic assessment involves an estimation of the levelized cost of water production using the total cost of ownership method, which involves capital cost, maintenance cost, energy cost, and salvage value. This process is depicted in Figure 13.

The following variables are considered for the economical assessment. Solar collector: The thickness of the glass cover is 3 mm, the emittance of the glass is 0.88, the refractive index of glass is 1.526, the refractive index of air is 1, the absorptance factor of the absorber plate is 0.94, the extinction coefficient of the glass is 32 m^{-1}, the area of absorber plate is 1.5 m^2, the number of tubes are 20, having an effective length of 20 m, while the spacing between the tubes is 50 mm, with each tube thickness of 1 mm, having an outdoor diameter of 10 mm, and the mass flow rate of water at the inlet of the solar collector is 0.005 kg/s. Humidifier: The length of the section is 1 m with a width of 600 mm and a height of 20 mm, the total mass flow rate is 0.05 kg/s. Dehumidifier: The total heat transfer surface area is 1.5 m^2, the temperature of water at the inlet of the condenser is 20 °C, the feed water mass flow rate is 0.01 kg/s, the inner diameter of tubes is 5 mm, the total number of tubes are 20 having an effective length of 10 m, and the shell diameter is 200 mm. Turbomachinery: The isentropic efficiency of pumps and blowers is 0.85, the mechanical efficiency of pumps is 0.96, and the mechanical efficiency of blowers is 0.90. Based on these data, the total initial cost of the system including the land and the equipment cost is estimated from the local manufacturers and comes out to be 18,000 PKR.

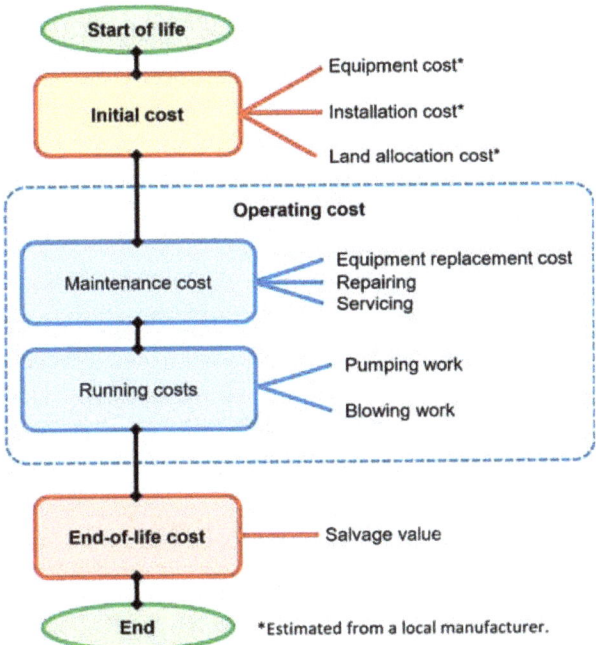

Figure 13. A framework of life-cycle-cost analysis adapted for the economic assessment of the desalination plant.

For an interest rate of 4%, and a life expectancy of 10 years, the capital recovery factor is given by:

$$CRF = \frac{i(1+i)^n}{(1+i)^n - 1} \tag{35}$$

The first annual cost can be calculated using the initial cost, and the capital recovery factor, given by:

$$First\ Annual\ Cost = 180,000 PKR \times CRF \tag{36}$$

The maintenance cost is considered 5% of the first annual cost. The total running cost of the equipment involves the annual energy consumption given by:

$$Annual\ Energy\ Consumption = 365 \times (W_{pump} + W_{blower}) \times t \tag{37}$$

The energy consumption of the pump (W_{pump}) is calculated by considering the enthalpy difference times mass flow rate corrected by its mechanical efficiency, and it is given by [85]:

$$W_{pump} = \frac{m(H_{out} - H_{in})}{\eta_m} \tag{38}$$

where m is the mass flow rate (kg/s) of water in the section, H is the enthalpy (kJ/kg) and η_m is the mechanical efficiency of pump which is considered 0.96 [86]. The blower W_{blower} is calculated by considering the pressure drop and it is given by [87]:

$$W_{blower} = \frac{Q \times \Delta P}{\eta_0 \times \eta_1} \times \mathbb{C} \qquad (39)$$

The details of the parameters used in Equation (39) is given as follows:

1. The pressure drop is calculated by considering the hydraulic calculation formulas, and it is given by:

$$\Delta P = \frac{ff}{Re} \cdot \frac{L}{d_e} \cdot \frac{\rho u^2}{2} \qquad (40)$$

where ff is the friction factor, Re is the non-dimensional Reynolds number, L is the effective length of the channel, d_e is the hydraulic diameter, ρ is the density of air, and u is the averaged velocity of the channel calculated using the ratio of volume flow rate per unit cross-sectional area. These parameters are calculated for humidifier and dehumidifier section differently and finally, the aggregation of both pressure differences gives the total pressure difference. The friction factor (ff) is given by the empirical formula:

$$ff = 96\left[1 - 1.3553AR + 1.9467AR^2 - 1.7012AR^3 + 0.9564AR^4 - 0.2537AR^5\right] \qquad (41)$$

The aspect ratio is a comparison between the shorter and the longer side of the cross-section of the air channel given by:

$$AR = \frac{\text{Shorter dimension of the cross} - \text{section of air channel}}{\text{Longer dimension of the cross} - \text{section of air channel}} \qquad (42)$$

2. η_0 is the internal efficiency of the fan which is considered 0.75 [88].
3. η_1 is the mechanical efficiency of the fan which is considered 0.9 [89].
4. $\mathbb{C}$ is the motor capacity coefficient which is 1.1 [90].

For an industrial rate of the cost of electricity at 12 PKR/kWh, the annual running cost is 8,176,000 PKR. The salvage value is calculated using the sinking fund factor, given by:

$$SRF = \frac{i}{(1+i)^n - 1} \qquad (43)$$

Finally, the total annual cost is calculated using the balance of life cycle costs, given by:

$$\begin{aligned}\text{Total Annual cost} &= \text{First Annual Cost} + \text{Annual Maintenance Cost} \\ &+ \text{Annual Running Cost} - \text{Annual Salvage Value}\end{aligned} \qquad (44)$$

With this methodology, the production cost of desalinated water is 17.16 PKR/litre (0.15 USD/litre) without waste heat recovery and it is 14.85 PKR/litre (0.13 USD/litre) with waste heat recovery. Therefore, the production cost of the water is decreased by ~13% by recovering the waste heat from the condensing coil.

7. Conclusions

In this research paper, a simulation is carried out for the humidification-dehumidification (HDH) type of desalination system in which the emphasis is given to the performance of the plant by recovering the heat from the condensing coil by reusing the feed water in the brackish water storage tank. The feedwater is recirculated because it has relatively high enthalpy as it absorbs energy in the dehumidification process of air. The mathematical model of the system is simulated, and the results are validated by comparing them with the previously published experimental and numerical results without waste heat recovery. Afterwards, an extensive parametric study is conducted in which MFR of inlet air, saline water, feedwater, and the temperature of ambient air are varied. It is observed that; for all cases, the freshwater productivity rate is increased to a maximum of 15% by recovering the heat from the condensing coil. The yearly analysis of freshwater was carried out and it is observed that the freshwater productivity is increased to a maximum of ~16% in June. Furthermore, economic analysis has demonstrated that the cost of desalinated water is decreased by ~13% by recovering the heat from the condensing coil and it is 0.13 USD/litre. It is concluded that the proposed system enhances the performance of an existing HDH desalination system.

However, this work is a theoretical analysis of the waste heat recovery from the condenser of the desalination system, which is finally a physical connection, in the form of piping and tubing, from the dehumidifier to the storage tank. This piping can contribute to some practical limitation in the cycle and other studies considering the 'design for manufacturing and assembly' are suggested before the mass-scale production.

Author Contributions: Conceptualization, R.T., and N.A.S.; methodology, R.T., and N.A.S.; software, R.T.; validation, R.T. and N.A.S.; formal analysis, R.T., J.T.J., N.A.S., and S.K.; investigation, R.T., J.T.J., N.A.S., and S.K.; resources, J.T.J.; data curation, R.T.; writing—original draft preparation, R.T.; writing—review and editing, R.T., J.T.J., N.A.S., and S.K.; visualization, R.T.; supervision, N.A.S.; project administration, S.K.; funding acquisition, J.T.T. All authors have read and agreed to the published version of the manuscript.

Funding: Tecnológico Nacional de México / Instituto Tecnológico Superior de Huauchinango, Av. Tecnológico No. 80, Col. 5 de Octubre, Huauchinango 73173, Puebla, México.

Acknowledgments: The first author, Rasikh Tariq, is thankful for the financial support granted by CONACYT ('*Consejo Nacional de Ciencia y Tecnología*' as in Spanish); with the following details CVU no. 949314, scholarship no. 730315, to pursue a postgraduate degree in *Universidad Autónoma de Yucatán*, Mexico. The authors are thankful for the financial support of *Instituto Tecnológico Superior de Huauchinango*. The second author, Jacinto Torres Jimenez, is thankful to CONACYT for the SNI ('*Sistema Nacional de Investigadores*' as in Spanish) system.

Conflicts of Interest: The authors declare no conflict of interest. The funders had no role in the design of the study; in the collection, analyses, or interpretation of data; in the writing of the manuscript, or in the decision to publish the results.

Nomenclature

Letters

A_c	Collector Area, m^2
c_p	Specific Heat of Air, J/kg. K
d	Thickness of Film, m
D^s	Diameter of collector tubes, m
E	Effectiveness of dehumidifier
F^s	Fin efficiency
F_R^s	Collector heat removal factor
$F\prime^s$	Collector efficiency factor
g	Gravitational acceleration, m/s^2
h	Convective Heat Transfer Coefficient, W/m^2. K
h_{number}	Enthalpy, J/kg
H	Enthalpy, kJ/kg
I	Radiation, W/m^2
m	Mass flow rate, kg/s
M	Molecular Weight of Water
HDH	Humidification-Dehumidification
P	Pressure, Pa

Q_u	Useful heat gain, W
R	General Gas Constant, J/mol. K
Ra^s_{p-g}	Rayleigh number
S	Absorbed Solar Radiation, W/m^2
T	Temperature, K
U	Heat Transfer Coefficient, W/m^2. K
u, v	Velocity, m/s
M^s	Spacing between the collector tubes, m
W_j	Water Evaporation Rate, kg/m^2
WER	Waste Heat Recovery
y, z	Coordinate system, m

Subscript

a	Air
s	Saturated
v	Vapor
w	Water
d	Diffused Radiation

Symbols

α	Thermal Diffusivity, m^2/s
α^s_p	Thermal absorptivity of absorber plate of the solar collector
β	Title Angle, degree
$(\tau\alpha)$	Transmittance-absorptance product
ε	Knudson Coefficient of Evaporation
ε^s_p	Emissivity of absorber plate
ϕ^s	Latitude angle
$\Re^s$	Reflectance of a single cover
μ_w	Dynamic viscosity of water, kg/m.s
Γ	Mass flow rate of saline water per unit width of the wall, kg/s/m
δ^s	Earth's declination angle
δ_s	Falling Film Water Thickness, m
λ	Latent Heat of Water, kJ/kg
ρ	Density, kg/m^3
θ_e	Angle of Incidence, Degree
ω	Absolute Humidity of Air, kg/kg
ω^s	Hour Angle

Superscript

s	Related to Solar Collector
h	Related to Humidifier

References

1. Sharshir, S.W.; Peng, G.; Yang, N.; El-Samadony, M.O.A.; Kabeel, A.E. A continuous desalination system using humidification–Dehumidification and a solar still with an evacuated solar water heater. *Appl. Therm. Eng.* **2016**, *104*, 734–742. [CrossRef]
2. Zhang, L.Z.; Li, G.P. Energy and economic analysis of a hollow fiber membrane-based desalination system driven by solar energy. *Desalination* **2017**, *404*, 200–214. [CrossRef]
3. Predescu, A.; Truică, C.-O.; Apostol, E.-S.; Mocanu, M.; Lupu, C. An Advanced Learning-Based Multiple Model Control Supervisor for Pumping Stations in a Smart Water Distribution System. *Mathematics* **2020**, *8*, 887. [CrossRef]
4. Manju, S.; Sagar, N. Renewable energy integrated desalination: A sustainable solution to overcome future fresh-water scarcity in India. *Renew. Sustain. Energy Rev.* **2017**, *73*, 594–609. [CrossRef]
5. Chandrashekara, C.; Yadav, A. Water desalination system using solar heat: A review. *Renew. Sustain. Energy Rev.* **2017**, *67*, 1308–1330.
6. Kaya, A.; Tok, M.; Koc, M. A Levelized Cost Analysis for Solar-Energy-Powered Sea Water Desalination in The Emirate of Abu Dhabi. *Sustainability* **2019**, *11*, 1691. [CrossRef]
7. Abdelkareem, M.A.; Assad, M.E.H.; Sayed, E.T.; Soudan, B. Recent progress in the use of renewable energy sources to power water desalination plants. *Desalination* **2018**, *435*, 97–113. [CrossRef]
8. Rabbani, M.; Ratlamwala, T.A.H.; Dincer, I. Transient energy and exergy analyses of a solar based integrated system. *J. Sol. Energy Eng. Trans. ASME* **2014**, *137*, 011010. [CrossRef]
9. Reif, J.H.; Alhalabi, W. Solar-thermal powered desalination: Its significant challenges and potential. *Renew. Sustain. Energy Rev.* **2015**, *48*, 152–165. [CrossRef]

10. Giwa, A.; Akther, N.; Al Housani, A.; Haris, S.; Hasan, S.W. Recent advances in humidification dehumidification (HDH) desalination processes: Improved designs and productivity. *Renew. Sustain. Energy Rev.* **2016**, *57*, 929–944. [CrossRef]
11. Kabeel, A.E.; Hamed, M.H.; Omara, Z.M.; Sharshir, S.W. Experimental study of a humidification-dehumidification solar technique by natural and forced air circulation. *Energy* **2014**, *68*, 218–228. [CrossRef]
12. Hamed, M.H.; Kabeel, A.E.; Omara, Z.M.; Sharshir, S.W. Mathematical and experimental investigation of a solar humidification-dehumidification desalination unit. *Desalination* **2015**, *358*, 9–17. [CrossRef]
13. Bakthavatchalam, B.; Rajasekar, K.; Habib, K.; Saidur, R.; Basrawi, F. Numerical analysis of humidification dehumidification desalination system. *Evergreen* **2019**, *6*, 9–17. [CrossRef]
14. Zhani, K.; Ben Bacha, H. Experimental investigation of a new solar desalination prototype using the humidification dehumidification principle. *Renew. Energy* **2010**, *35*, 2610–2617. [CrossRef]
15. Elminshawy, N.A.S.; Siddiqui, F.R.; Sultan, G.I. Development of a desalination system driven by solar energy and low grade waste heat. *Energy Convers. Manag.* **2015**, *103*, 28–35. [CrossRef]
16. Narayan, G.P.; Sharqawy, M.H.; Summers, E.K.; Lienhard, J.H.; Zubair, S.M.; Antar, M.A. The potential of solar-driven humidification-dehumidification desalination for small-scale decentralized water production. *Renew. Sustain. Energy Rev.* **2010**, *14*, 1187–1201. [CrossRef]
17. Narayan, G.P.; Sharqawy, M.H.; Lienhard, V.J.H.; Zubair, S.M. Thermodynamic analysis of humidification dehumidification desalination cycles. *Desalin. Water Treat.* **2010**, *16*, 339–353. [CrossRef]
18. Narayan, G.P.; John, M.G.S.; Zubair, S.M.; Lienhard, J.H. Thermal design of the humidification dehumidification desalination system: An experimental investigation. *Int. J. Heat Mass Transf.* **2013**, *58*, 740–748. [CrossRef]
19. Narayan, G.P.; Chehayeb, K.M.; McGovern, R.K.; Thiel, G.P.; Zubair, S.M.; Lienhard, V.J.H. Thermodynamic balancing of the humidification dehumidification desalination system by mass extraction and injection. *Int. J. Heat Mass Transf.* **2013**, *57*, 756–770. [CrossRef]
20. Thiel, G.P.; Lienhard, J.H. Entropy generation in condensation in the presence of high concentrations of noncondensable gases. *Int. J. Heat Mass Transf.* **2012**, *55*, 5133–5147. [CrossRef]
21. Summers, E.K.; Arafat, H.A.; Lienhard, V.J.H. Energy efficiency comparison of single-stage membrane distillation (MD) desalination cycles in different configurations. *Desalination* **2012**, *290*, 54–66. [CrossRef]
22. McGovern, R.K.; Thiel, G.P.; Narayan, G.P.; Zubair, S.M.; Lienhard, V.J.H. Performance limits of zero and single extraction humidification-dehumidification desalination systems. *Appl. Energy* **2013**, *102*, 1081–1090. [CrossRef]
23. Mistry, K.H.; McGovern, R.K.; Thiel, G.P.; Summers, E.K.; Zubair, S.M.; Lienhard, J.H. Entropy Generation Analysis of Desalination Technologies. *Entropy* **2011**, *13*, 1829–1864. [CrossRef]
24. Mistry, K.H.; Lienhard, J.H.; Zubair, S.M. Effect of entropy generation on the performance of humidification-dehumidification desalination cycles. *Int. J. Therm. Sci.* **2010**, *49*, 1837–1847. [CrossRef]
25. Sharqawy, M.H.; Lienhard, V.J.H.; Zubair, S.M. On exergy calculations of seawater with applications in desalination systems. *Int. J. Therm. Sci.* **2011**, *50*, 187–196. [CrossRef]
26. Sharqawy, M.H.; Antar, M.A.; Zubair, S.M.; Elbashir, A.M. Optimum thermal design of humidification dehumidification desalination systems. *Desalination* **2014**, *349*, 10–21. [CrossRef]
27. Khalifa, A.; Ahmad, H.; Antar, M.; Laoui, T.; Khayet, M. Experimental and theoretical investigations on water desalination using direct contact membrane distillation. *Desalination* **2017**, *404*, 22–34. [CrossRef]
28. Summers, E.K.; Antar, M.A.; Lienhard, J.H. Design and optimization of an air heating solar collector with integrated phase change material energy storage for use in humidification-dehumidification desalination. *Sol. Energy* **2012**, *86*, 3417–3429. [CrossRef]
29. Al-Sulaiman, F.A.; Zubair, M.I.; Atif, M.; Gandhidasan, P.; Al-Dini, S.A.; Antar, M.A. Humidification dehumidification desalination system using parabolic trough solar air collector. *Appl. Therm. Eng.* **2015**, *75*, 809–816. [CrossRef]
30. Lawal, D.; Antar, M.; Khalifa, A.; Zubair, S.; Al-Sulaiman, F. Humidification-dehumidification desalination system operated by a heat pump. *Energy Convers. Manag.* **2018**, *161*, 128–140. [CrossRef]
31. Gabrielli, P.; Mazzotti, M. Solar-Driven Humidification-Dehumidification Process for Water Desalination Analyzed and Optimized via Equilibrium Theory. *Ind. Eng. Chem. Res.* **2019**, *58*, 15244–15261. [CrossRef]
32. Gabra, B.; Rady, M.; Ghany, A.M.A.; Shamseldin, M.A. Modelling and control of solar-driven humidification–dehumidification desalination plant. *J. Electr. Syst. Inf. Technol.* **2019**, *6*, 7. [CrossRef]
33. Zubair, M.I.; Al-Sulaiman, F.A.; Antar, M.A.; Al-Dini, S.A.; Ibrahim, N.I. Performance and cost assessment of solar driven humidification dehumidification desalination system. *Energy Convers. Manag.* **2017**, *132*, 28–39. [CrossRef]
34. Jamil, M.A.; Shahzad, M.W.; Zubair, S.M. A comprehensive framework for thermoeconomic analysis of desalination systems. *Energy Convers. Manag.* **2020**, *222*, 113188. [CrossRef] [PubMed]
35. Xu, H.; Zhao, Y.; Dai, Y.J. Experimental study on a solar assisted heat pump desalination unit with internal heat recovery based on humidification-dehumidification process. *Desalination* **2019**, *452*, 247–257. [CrossRef]
36. Tariq, R.; Sheikh, N.A.; Xamán, J.; Bassam, A. An innovative air saturator for humidification-dehumidification desalination application. *Appl. Energy* **2018**, *228*, 789–807. [CrossRef]
37. Ahmed, H.A.; Ismail, I.M.; Saleh, W.F.; Ahmed, M. Experimental investigation of humidification-dehumidification desalination system with corrugated packing in the humidifier. *Desalination* **2017**, *410*, 19–29. [CrossRef]

38. Yildirim, C.; Solmuş, I. A parametric study on a humidification-dehumidification (HDH) desalination unit powered by solar air and water heaters. *Energy Convers. Manag.* **2014**, *86*, 568–575. [CrossRef]
39. Yuan, G.; Wang, Z.; Li, H.; Li, X. Experimental study of a solar desalination system based on humidification-dehumidification process. *Desalination* **2011**, *277*, 92–98. [CrossRef]
40. Deniz, E.; Çınar, S. Energy, exergy, economic and environmental (4E) analysis of a solar desalination system with humidification-dehumidification. *Energy Convers. Manag.* **2016**, *126*, 12–19. [CrossRef]
41. Khalilzadeh, S.; Hossein Nezhad, A. Utilization of waste heat of a high-capacity wind turbine in multi effect distillation desalination: Energy, exergy and thermoeconomic analysis. *Desalination* **2018**, *439*, 119–137. [CrossRef]
42. Ishaq, H.; Dincer, I.; Naterer, G.F. New trigeneration system integrated with desalination and industrial waste heat recovery for hydrogen production. *Appl. Therm. Eng.* **2018**, *142*, 767–778. [CrossRef]
43. Elsaid, K.; Taha Sayed, E.; Yousef, B.A.A.; Kamal Hussien Rabaia, M.; Ali Abdelkareem, M.; Olabi, A.G. Recent progress on the utilization of waste heat for desalination: A review. *Energy Convers. Manag.* **2020**, *221*, 113105. [CrossRef]
44. Shafieian, A.; Khiadani, M. A multipurpose desalination, cooling, and air-conditioning system powered by waste heat recovery from diesel exhaust fumes and cooling water. *Case Stud. Therm. Eng.* **2020**, *21*, 100702. [CrossRef]
45. Xu, L.; Chen, Y.-P.; Wu, P.-H.; Huang, B.-J. Humidification–Dehumidification (HDH) Desalination System with Air-Cooling Condenser and Cellulose Evaporative Pad. *Water* **2020**, *12*, 142. [CrossRef]
46. He, W.F.; Han, D.; Zhu, W.P.; Ji, C. Thermo-economic analysis of a water-heated humidification-dehumidification desalination system with waste heat recovery. *Energy Convers. Manag.* **2018**, *160*, 182–190. [CrossRef]
47. He, W.F.; Xu, L.N.; Han, D. Parametric analysis of an air-heated humidification-dehumidification (HDH) desalination system with waste heat recovery. *Desalination* **2016**, *398*, 30–38. [CrossRef]
48. Santosh, R.; Kumaresan, G.; Selvaraj, S.; Arunkumar, T.; Velraj, R. Investigation of humidification-dehumidification desalination system through waste heat recovery from household air conditioning unit. *Desalination* **2019**, *467*, 1–11. [CrossRef]
49. Abdelmoez, W.; Mahmoud, M.S.; Farrag, T.E. Water desalination using humidification/dehumidification (HDH) technique powered by solar energy: A detailed review. *Desalin. Water Treat.* **2014**, *52*, 4622–4640. [CrossRef]
50. Mahmoud, M.S.; Farrag, T.E.; Mohamed, W.A. Experimental and Theoretical Model for Water Desalination by Humidification–dehumidification (HDH). *Procedia Environ. Sci.* **2013**, *17*, 503–512. [CrossRef]
51. Patel, V.; Patel, R.; Patel, J. Theoretical and experimental investigation of bubble column humidification and thermoelectric cooler dehumidification water desalination system. *Int. J. Energy Res.* **2020**, *44*, 890–901. [CrossRef]
52. Patel, S.K.; Modi, K.V. Techniques to improve the performance of enhanced condensation area solar still: A critical review. *J. Clean. Prod.* **2020**, *268*, 122260. [CrossRef]
53. Nayi, K.H.; Modi, K.V. Pyramid solar still: A comprehensive review. *Renew. Sustain. Energy Rev.* **2018**, *81*, 136–148. [CrossRef]
54. Mohamed, A.S.A.; Ahmed, M.S.; Shahdy, A.G. Theoretical and experimental study of a seawater desalination system based on humidification-dehumidification technique. *Renew. Energy* **2020**, *152*, 823–834. [CrossRef]
55. Rajaseenivasan, T.; Srithar, K. Potential of a dual purpose solar collector on humidification dehumidification desalination system. *Desalination* **2017**, *404*, 35–40. [CrossRef]
56. Xu, H.; Sun, X.Y.; Dai, Y.J. Thermodynamic study on an enhanced humidification-dehumidification solar desalination system with weakly compressed air and internal heat recovery. *Energy Convers. Manag.* **2019**, *181*, 68–79. [CrossRef]
57. Xu, H.; Dai, Y.J. Experimental investigation on an enhanced humidification-dehumidification solar desalination system with weakly compressed air and internal heat recovery. In Proceedings of the ISES Solar World Congress 2019 and IEA SHC International Conference on Solar Heating and Cooling for Buildings and Industry 2019, Santiago, Chile, 4–7 November 2019.
58. Tariq, R.; Hussain, Y.; Sheikh, N.A.; Afaq, K.; Ali, H.M. Regression-Based Empirical Modeling of Thermal Conductivity of CuO-Water Nanofluid using Data-Driven Techniques. *Int. J. Thermophys.* **2020**, *41*, 1–28. [CrossRef]
59. Tariq, R.; Zhan, C.; Ahmed Sheikh, N.; Zhao, X. Thermal Performance Enhancement of a Cross-Flow-Type Maisotsenko Heat and Mass Exchanger Using Various Nanofluids. *Energies* **2018**, *11*, 2656. [CrossRef]
60. Sarbu, I.; Sebarchievici, C. *Solar Heating and Cooling Systems: Fundamentals, Experiments and Applications*; Elsevier Inc.: London, UK, 2016; ISBN 9780128116630.
61. Duffie, J.A.; Beckman, W.A. *Solar Engineering of Thermal Processes*, 4th ed.; John Wiley & Sons: New York, NY, USA, 2013; ISBN 9780470873663.
62. Noam, L. Thermal Theory and Modeling of Solar Collectors. In *Solar Collectors, Energy Storage, and Materials*; de Winter, F., Ed.; The MIT Press: Cambridge, MA, USA, 1990; Chapter 4; pp. 99–182. ISBN 9780262041041.
63. Blaine, F. Parker Design Equations for Solar Air Heaters. *Trans. ASAE* **1980**, *23*, 1494–1499. [CrossRef]
64. Norton, B. *Solar Energy Thermal Technology*, 1st ed.; Norton, B., Ed.; Springer: London, UK, 1992; Volume 1.
65. Knauf Insulation TSP SOLAR ROLL (TSP SR)–Glass Mineral Wool Insulation Mats for Thermal Solar Collectors. Available online: https://www.oem.knaufinsulation.com/products/tsp-solar-roll (accessed on 22 September 2020).
66. Da Rosa, L.C.; Santor, C.G.; Lovato, A.; Da Rosa, C.S.; Güths, S. Use of rice husk and sunflower stalk as a substitute for glass wool in thermal insulation of solar collector. *J. Clean. Prod.* **2015**, *104*, 90–97. [CrossRef]
67. Nadir, N.; Bouguettaia, H.; Boughali, S.; Bechki, D. Use of a new agricultural product as thermal insulation for solar collector. *Renew. Energy* **2019**, *134*, 569–578. [CrossRef]

68. Beikircher, T.; Osgyan, P.; Reuss, M.; Streib, G. Flat plate collector for process heat with full surface aluminium absorber, vacuum super insulation and front foil. *Energy Procedia* **2014**, *48*, 9–17. [CrossRef]
69. Florez, F.; Fernández de Cordoba, P.; Taborda, J.; Polo, M.; Castro-Palacio, J.C.; Pérez-Quiles, M.J. Sliding Modes Control for Heat Transfer in Geodesic Domes. *Mathematics* **2020**, *8*, 902. [CrossRef]
70. Hollands, K.G.T.; Unny, T.E.; Raithby, G.D.; Konicek, L. Free Convective Heat Transfer Across Inclined Air Layers. *J. Heat Transf.* **1976**, *98*, 189. [CrossRef]
71. Portalski, S. Velocities in film flow of liquids on vertical plates. *Chem. Eng. Sci.* **1964**, *19*, 575–582. [CrossRef]
72. Lamb, S.H. *Hydrodynamics*, 1st ed.; Courier Corporation: Chelmsford, MA, USA, 1945; Volume 1, ISBN 0486602567.
73. Anderson, J.D. Mathematical properties of the fluid dynamic equations. In *Computational Fluid Dynamics*; Springer: Berlin/Heidelberg, Germany, 2009; pp. 77–86. ISBN 9783540850557.
74. Eames, I.W.; Marr, N.J.; Sabir, H. The evaporation coefficient of water: A review. *Int. J. Heat Mass Transf.* **1997**, *40*, 2963–2973. [CrossRef]
75. Dai, Y.J.; Wang, R.Z.; Zhang, H.F. Parametric analysis to improve the performance of a solar desalination unit with humidification and dehumidification. *Desalination* **2002**, *142*, 107–118. [CrossRef]
76. Maclaine-cross, I.L.; Banks, P.J. A General Theory of Wet Surface Heat Exchangers and its Application to Regenerative Evaporative Cooling. *J. Heat Transfer* **1981**, *103*, 579–585. [CrossRef]
77. Incropera, F.P.; DeWitt, D.P.; Bergman, T.L.; Lavine, A.S. *Fundamentals of Heat and Mass Transfer*; John Wiley & Sons: Hoboken, NJ, USA, 2007; Volume 7, ISBN 9780471457282.
78. Tariq, R.; Sheikh, N.A. Numerical heat transfer analysis of Maisotsenko Humid Air Bottoming Cycle—A study towards the optimization of the air-water mixture at bottoming turbine inlet. *Appl. Therm. Eng.* **2018**, *133*, 49–60. [CrossRef]
79. Tariq, R.; Sheikh, N.A.; Bassam, A.; Xamán, J. Analysis of Maisotsenko humid air bottoming cycle employing mixed flow air saturator. *Heat Mass Transf.* **2018**, *55*, 1477–1489. [CrossRef]
80. Klein, S.; Alvarado, F. *EES, Engineering Equation Solver*; F-Chart Software: Madison, WI, USA, 2015.
81. Kroger, G. Air Cooled Heat Exchangers and Cooling Towers: Thermal Flow Performance Evaluation and Design. Ph.D. Thesis, Stellenbosch University, Stellenbosch, South Africa, 2004.
82. Majchrzak, E.; Mochnacki, B. Second-Order Dual Phase Lag Equation. Modeling of Melting and Resolidification of Thin Metal Film Subjected to A Laser Pulse. *Mathematics* **2020**, *8*, 999. [CrossRef]
83. Tariq, R.; Sohani, A.; Xamán, J.; Sayyaadi, H.; Bassam, A.; Tzuc, O.M. Multi-objective optimization for the best possible thermal, electrical and overall energy performance of a novel perforated-type regenerative evaporative humidifier. *Energy Convers. Manag.* **2019**, *198*, 111802. [CrossRef]
84. Klein, S.A. Calculation of monthly average insolation on tilted surfaces. *Sol. Energy* **1977**, *19*, 325–329. [CrossRef]
85. Moran, M.J.; Shapiro, H.N. Fundamentals of Engineering Thermodynamics, Second Edition. *Eur. J. Eng. Educ.* **1993**, *18*, 215. [CrossRef]
86. Sohani, A.; Sayyaadi, H. Thermal comfort based resources consumption and economic analysis of a two-stage direct-indirect evaporative cooler with diverse water to electricity tari ff conditions. *Energy Convers. Manag.* **2018**, *172*, 248–264. [CrossRef]
87. White, F.M. Fluid Mechanics. *Book* **2009**, *17*, 864. [CrossRef]
88. European Commission. On internal Specific Fan Power, SFPint and draft transitional methods. In *Technical Assistance Study for the Ventilation Units Product Group*; Preliminary DRAFT prepared for the first stakeholder meeting of the Technical Assistance Study; Danish Technological Institute: Taastrup, Denmark, 2016.
89. Chen, Y.; Luo, Y.; Yang, H. A simplified analytical model for indirect evaporative cooling considering condensation from fresh air: Development and application. *Energy Build.* **2015**, *108*, 387–400. [CrossRef]
90. Toliyat, H.A.; Kliman, G.B. *Handbook of Electric Motors*; Marcel Dekker: New York, NY, USA, 2004; ISBN 0824741056.

Article

A Novel Comparative Statistical and Experimental Modeling of Pressure Field in Free Jumps along the Apron of USBR Type I and II Dissipation Basins

Seyed Nasrollah Mousavi [1] and Daniele Bocchiola [2,*]

1. Department of Water Engineering, University of Tabriz, Tabriz 5166616471, Iran; s.n.mousavi@tabrizu.ac.ir
2. Department of Civil and Environmental Engineering, Politecnico di Milano, L. da Vinci, 32, 20133 Milano, Italy
* Correspondence: daniele.bocchiola@polimi.it

Received: 7 September 2020; Accepted: 29 November 2020; Published: 3 December 2020

Abstract: Dissipation basins are usually constructed downstream of spillways to dissipate energy, causing large pressure fluctuations underneath hydraulic jumps. Little systematic experimental investigation seems available for the pressure parameters on the bed of the US Department of the Interior, Bureau of Reclamation (USBR) Type II dissipation basins in the literature. We present the results of laboratory-scale experiments, focusing on the statistical modeling of the pressure field at the centerline of the apron along the USBR Type I and II basins. The accuracy of the pressure transducers was ±0.5%. The presence of accessories within $basin_{II}$ reduced the maximum pressure fluctuations by about 45% compared to $basin_I$. Accordingly, in some points, the bottom of $basin_{II}$ did not collide directly with the jet due to the hydraulic jump. As a result, the values of pressure and pressure fluctuations decreased mainly therein. New original best-fit relationships were proposed for the mean pressure, the statistical coefficient of the probability distribution, and the standard deviation of pressure fluctuations to estimate the pressures with different probabilities of occurrence in $basin_I$ and $basin_{II}$. The results could be useful for a more accurate, safe design of the slab thickness, and reduce the operation and maintenance costs of dissipation basins.

Keywords: $basin_I$; $basin_{II}$; mean pressure head; pressure head with different probabilities of occurrence; standard deviation of the pressure fluctuations; statistical modeling; USBR

---

1. Introduction

Hydraulic jump with the turbulent entrainment process is a function of time and position. This phenomenon is a complex and stochastic process, so that hydrodynamic pressure fluctuations can be analyzed using statistical methods. Energy dissipation through the hydraulic jumps with the conversion of energy downstream of spillways is usually confined within the dissipation basins [1]. This type of hydraulic structure protects the soil against flow erosion, which can affect the dam's safety. Due to the large heads upstream of spillways, dissipation basins may be subjected to enormous instantaneous pressure and velocity fluctuations, causing significant stresses in such energy dissipators. This may cause the uplift of a basin lining, making it necessary to provide this structure with sufficient weight or anchorage. Through the analysis of collected data, it is possible to characterize the forces under a hydraulic jump according to the values of mean pressures, pressure fluctuations, and extreme pressures [2].

US Department of the Interior, Bureau of Reclamation (USBR) Type II dissipation basins [3] are designed to reduce excess kinetic energy downstream of the spillway [4–10], reduce largely (≈30%) the required length compared to smooth basins [11], and help in reducing the costs of the structure [12]. Furthermore, knowledge of the geometric characteristics of the hydraulic jump is fundamental for the design of the dissipation structures. Measurement of fluctuating pressures or forces may be difficult to carry out in the field or at the scale of real structures. Overall, there is a lack of information concerning the hydrodynamic loading on the bottom slabs. Little systematic experimental investigation seems available for Type II dissipation basins, and only a general understanding of the hydraulic behavior is attained [13]. A better understanding of the distribution of pressure fluctuations may lead to a more economical design with high safety of energy dissipation structures.

Toso and Bowers [14] stated that the peak value of the pressure fluctuations intensity coefficient (C'_P) varies up to 60% when comparing results from different works. It seems that these differences were related to the degree of development (larger or smaller) of the flow boundary layer. Accordingly, the fully developed flows show lower values of C'_P than undeveloped ones. Endres [15] developed a real-time acquisition system and treatment of data representative of the instantaneous pressure fields in a hydraulic jump by analyzing instantaneous pressures downstream of a spillway. Pinheiro [16] measured the pressure fields inside the hydraulic jump downstream of a spillway. He concluded that the pressure fields near the bottom and along the hydraulic jump are lower than the corresponding depth of the mean flow. Marques et al. [17] measured pressures within a dissipation basin with the smooth bed downstream of a spillway. They proposed a dimensionless methodology that groups the fluctuating pressures with different incident Froude numbers (Fr_1) in a single trend, being a function of the jump position. The values of Fr_1 used by Endres [15], Pinheiro [16], and Marquez et al. [17] were in the range of 4.2 to 8.6, 6 to 10, and 5 to 8, respectively. Based on the pressure data by Enders [15], Teixeira [18] proposed second-order polynomial relationships for estimating different pressure parameters in smooth dissipation basins.

According to Alves [2], the measurement of fluctuating pressures is highly influenced by laboratory conditions. This may include the Reynolds number of flow, transducer accuracy, transducer installation method, hose length, pressure point diameter, channel width, model roughness, etc. Farhoudi et al. [19] studied the pressure fluctuations around some chute blocks in a St. Anthony Fall (SAF) type dissipation basin. Novakoski et al. [20] showed that the negative pressures in the zone near the spillway toe represent the risk of cavitation in the dissipation basin. They concluded that the extreme pressures with the probabilities of occurrence equal to 0.1% and 1% require careful assessment. Macián-Pérez et al. [21] used a numerical model to analyze pressure distributions in a USBR Type II dissipation basin. Hampe et al. [22] estimated extreme pressures in hydraulic jumps with low Froude numbers. Samadi et al. [23] used some explicit data-driven approaches to estimate the C'_P coefficient underneath hydraulic jumps on a sloping channel.

Mousavi et al. [24] focused on the minimal and maximal pressures, the pressure coefficients, the power spectral density (PSD), the probability density function (PDF), and the uncertainty analysis of the pressures along a USBR Type I basin (basin$_I$). Mousavi et al. [25] assessed the statistical parameters of free jumps, including mean pressure (P^*_m), the standard deviation of pressure fluctuations (σ^*_X), the probability distribution coefficient ($N_{K\%}$), and the pressures with different probabilities ($P^*_{K\%}$) along basin$_I$. Mousavi et al. [26] evaluated artificial intelligence models to estimate the C'_P coefficient for the free and submerged jumps at the bottom of a USBR Type II basin (basin$_{II}$). The results showed the deep learning model could estimate the C'_P coefficient more accurately.

However, pressure patterns on the apron of basin$_{II}$ have not been widely investigated in the literature. We designed and pursued experiments to obtain some information about the effect of chute blocks and dentated end sill on the free jumps' characteristics and pressure fluctuation. Experiments were conducted in the centerline of the apron along basin$_I$ and basin$_{II}$ with the incident Froude numbers (Fr_1) in the range of 6.14 to 8.29. In summary, the differences between the previous works and the present paper are explained as follows:

i. Analysis of the minimal and maximal values of pressures along the free jumps within basin$_I$ and basin$_{II}$. These parameters for basin$_{II}$ have not been investigated in the literature.

ii. Evaluation of the PSD analysis to determine the dominant frequency of fluctuating pressures in the free jumps for basin$_I$ and basin$_{II}$. In addition, assessment of the PDF histograms for

iii. the fluctuating pressures at different pressure points and investigation of the skewness and kurtosis coefficients, P^*_m, extreme pressures (P^*_{min} and P^*_{max}), σ^*_X, $N_{K\%}$, and $P^*_{K\%}$ along basin$_I$ and basin$_{II}$. For reference, we benchmarked and compared our findings with previous similar results of other authors focusing on hydraulic jumps we could retrieve in the present literature.

iii. Proposition of some new original best-fit relationships to estimate the dimensionless forms of statistical parameters including P^*_m, σ^*_X, $N_{K\%}$, and $P^*_{K\%}$ for the free jumps as a function of the dimensionless position along basin$_I$ and basin$_{II}$.

iv. Proposition of the hydraulic jump length (L_j) as a scaling factor for the dimensionless position from the toe of the spillway (X^*). Marques al. [17] proposed the dimensionless adjustments for the pressure parameters. Due to the presence of significant air bubbles at the beginning of the jump, it is difficult to measure the initial depth of the jump (Y_1) with great accuracy. It seems that the expression of Y_2-Y_1 (conjugated depths of hydraulic jumps) is not appropriate as a scaling factor. In this case, the X^* parameter was defined as X/L_j, where L_j is the length of hydraulic jump. In addition, the values of Y_1 were calculated using the well-known equation of Bélanger [27].

2. Materials and Methods

2.1. Experimental Setup

In this research, the pressure field of free jumps was investigated in the hydraulic laboratory, University of Tabriz, Iran (see Figure 1). The laboratory flume used had a length of 10 m, a width of 0.51 m, and a height of 0.5 m. The channel's bed was considered in the form of a horizontal line in all experiments. An Ogee spillway of 70 cm in height (H) was equipped with two different configurations of the dissipation basins, designed according to the USBR criteria [3]. In addition, the accessories of basin$_{II}$, including eight chute blocks (3.2 cm width, 3 cm height, and 7.94 cm length) and a dentated end sill with 6 cm height, were designed based on the maximum flow discharge. The spillway was installed at a distance of 260 cm from the entrance head tank of the flume.

We performed some experiments on basin$_I$ and basin$_{II}$ with different flow discharges, ranging from 33 to 60.4 L/s, and supercritical Froude numbers (Fr$_1$) between 6.14 and 8.29. According to the USBR recommendation, the lengths of basin$_I$ (L_I) and basin$_{II}$ (L_{II}) were 200 and 125 cm, respectively. The width of the basins was considered equal to the width of the flume (see Figure 2). At the end of the flume, a hinged weir was used to create and stabilize the free jump position. Therefore, the hydraulic jump was positioned at the basins' beginning and contained within the basins (i.e., jump type A-jump [28]).

The subcritical flow depth (Y_2) at the endpoint of the jumps was measured along the flume's centerline. To do this, we used a Data logic ultrasonic sensor device model US30, made in Italy with a nominal accuracy of 1 mm. The discharge in the flume (Q) was measured with a transit-time clamp-on ultrasonic flow meter. The values of supercritical flow depth (Y_1) were calculated using the Bélanger's equation [27,29,30], which is defined as follows:

$$\frac{Y_1}{Y_2} = \frac{1}{2}(-1 + \sqrt{1 + 8\,\text{Fr}_2^2}) \tag{1}$$

$$\text{Fr}_2 = \frac{V_2}{\sqrt{g \times Y_2}} \tag{2}$$

where V_2 is the mean subcritical velocity, calculated using the continuity law, Fr$_2$ is the subcritical Froude number, and g is the gravitational acceleration.

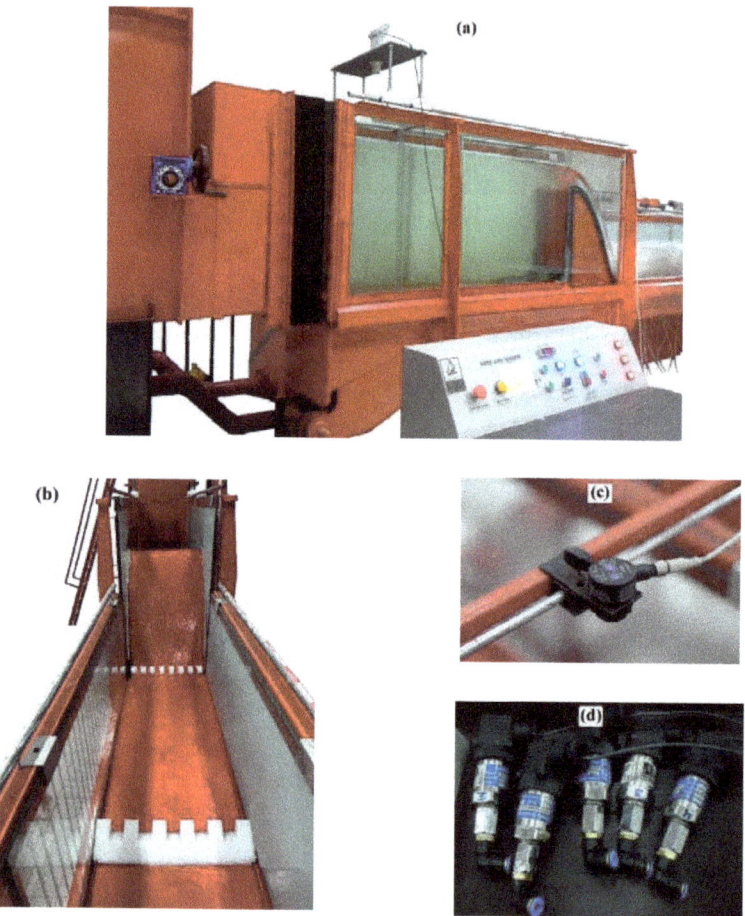

Figure 1. Laboratory flume and experimental setup. (**a**) Hydraulic jump during an experiment, (**b**) basin$_{II}$ with chute blocks and dentated end sill, (**c**) Data logic ultrasonic sensor device model US30, and (**d**) pressure transducers (Atek BCT 110 series with 100 mbar-A-G1/4 model).

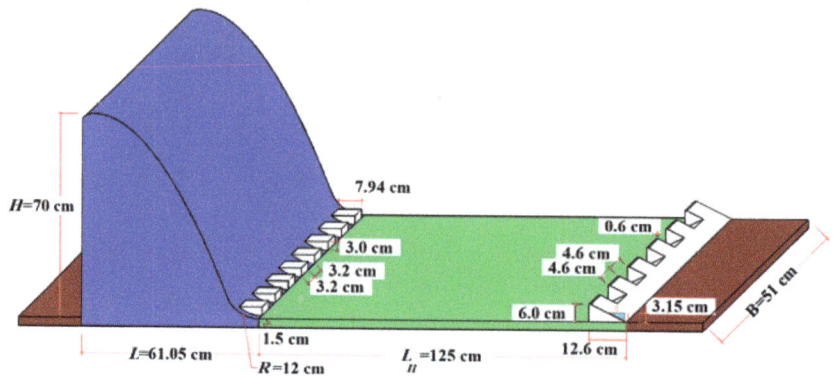

Figure 2. Dimensions of spillway and accessories installed in basin$_{II}$.

To measure the (dynamic) pressure fluctuations, 25 measurement points were considered along the centerline of the apron inside and outside the basin$_{II}$ (see Figure 3). Pressure therein was measured by way of piezometers installed at the centerline of the apron along the basins. The position of the pressure points (X) was from 2.5 cm (piezometer No. 1) to 189 cm (piezometer No. 25). The instantaneous pressures were measured with pressure transducers (Atek BCT 110 series with 100 mbar-A-G1/4 model). The pressure transducers used a 6-channel digital board and have an accuracy of ±0.5% within the range of −1.0 to 1.0 m [24–26].

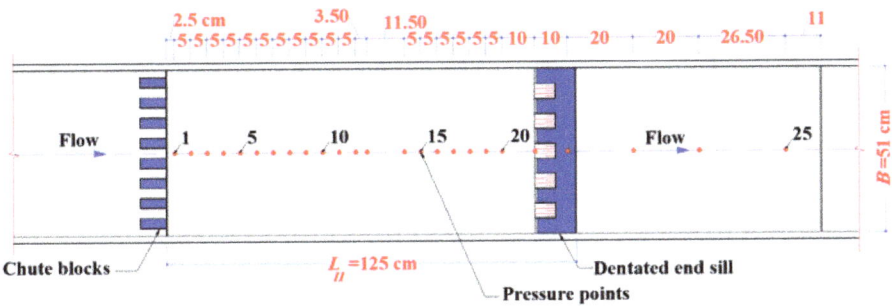

Figure 3. Distribution of the pressure points at the centerline of the apron along basin$_{II}$.

Pressure transducers were calibrated before the experiments using a static pressure gauge in the laboratory. Therefore, the mean fluctuating pressures were approximately equal to the static pressures. The transducers were mounted on a support plate, placed under the bottom of the flume. Thus, it was possible to eliminate possible distortion effects in the pressure signal due to the connection with rubber hoses. The transparent plastic hoses used here had an internal diameter of 3 mm and were 200 cm in length. Hydrodynamic pressure data were measured in time series. Accordingly, some statistical methods were used to analyze the collected pressure data.

2.2. Statistical Parameters

Investigation of the pressure head (cm) parameter is a first step to describe the pressure field in the hydraulic jump. The pressure parameters at each point (P_X) include the minimum pressure (P_{min}), the mean pressure (P_m), the maximum pressure (P_{max}), and the pressure with a certain probability of occurrence ($P_{K\%}$). Marques al. [17] proposed $P^*_X = (P_X - Y_1)/(Y_2 - Y_1)$ for the dimensionless form of pressure parameters as a function of the dimensionless position of each point (X^*), defined as $X^* = X/(Y_2 - Y_1)$, where X is the longitudinal position of each point inside the hydraulic jump. As the upstream part of the jump exhibited significant air bubbles, it seems that the scaling of $Y_2 - Y_1$ is not appropriate for the non-dimensional position of the pressure point. As a result, in the present study, the X^* parameter was defined as X/L_j, where L_j is the hydraulic jump length.

Knowledge of the extreme pressure heads in the dissipation basins helps to understand the energy dissipation of the hydraulic jumps. In the present study, the extreme pressure heads (P^*_{min} and P^*_{max}) were investigated in detail. Marques et al. [17] proposed $(\sigma^*_X/E_l) \times (Y_2/Y_1)$ to analyze the dimensionless standard deviation of the pressure fluctuations at point X. There, E_l is energy head loss (cm) along the hydraulic jump. The experimental values of $P_{K\%}$ were achieved using the pressure time series data collected at each pressure point. The statistical coefficient of the probability distribution ($N_{K\%}$) can be varied at different points of the dissipation basins. Therefore, it is necessary to determine the longitudinal distribution of $N_{K\%}$ to estimate the $P^*_{K\%}$ parameter with a probability to be less than or equal to a certain value (K) along basin$_I$ and basin$_{II}$. As the estimated values of P_m, σ_X, and $N_{K\%}$ were determined at each point inside the basins, the values of $P_{K\%}$ can be estimated using equation $P_{K\%} = P_m + N_{K\%} \times \sigma^*_X$ [17].

In this study, a new statistical methodology was proposed to estimate the values of $P^*_{K\%}$ in basin$_I$ and basin$_{II}$. To evaluate the estimated values of pressure parameters, some statistical performance criteria were determined [31–34]. The PDF function of the normalized pressures along the hydraulic jumps was calculated according to $P^*(Z) = (1/\sqrt{2\pi}) \times Exp(-Z^2/2)$. The normalized pressure variable (Z) was defined as $(P[X,t]) - P_m/\sigma_X$, where $P[X,t]$ is the instantaneous pressure [35]. We pursued an analysis of the skewness and kurtosis coefficients of pressure fluctuations [36]. Due to the high variation in S and K coefficients, it is difficult to define a single statistical distribution to describe the overall behavior along the jump.

3. Results and Discussion

3.1. Flow Characteristics

Table 1 presents some experimental and calculated parameters of the flow downstream of the spillway in two dissipation basins under different free jump conditions.

Table 1. Experimental parameters in two dissipation basins.

Q (L/s)	V_1 (m/s)	Fr_1	Re_1	Y_1 (cm)	Y_2 (cm)		L_j (cm)	
					basin$_I$	basin$_{II}$	basin$_I$	basin$_{II}$
33.0	3.52	8.29	58,200	1.84	20.65	19.69	142.50	102.50
43.0	3.59	7.48	74,400	2.35	23.70	22.44	162.50	112.50
47.5	3.60	7.14	81,500	2.59	24.87	23.57	189.00	122.50
52.7	3.58	6.72	89,500	2.89	26.05	24.70	189.00	122.50
55.0	3.56	6.52	92,900	3.03	26.49	25.33	189.00	122.50
60.4	3.53	6.14	100,900	3.36	27.55	26.60	189.00	122.50

$_1$ Supercritical flow, $_2$ Subcritical flow.

Re_1 is the Reynolds number for the supercritical flow of the hydraulic jump. The mean velocity of the incoming flow to the dissipation basins (V_1) was computed using the continuity law. According to Table 1, the Y_2 parameter in basin$_{II}$ decreases compared to those in the basin$_I$ case (classical hydraulic jump). As Q increases, Y_1 increases faster than V_1. Accordingly, the Froude number reduces when increasing flow discharges. The flow conditions downstream of the spillways are different compared to the sluice gates. The Y_1 parameter has an essential role in determining the values of Fr_1. Reducing Fr_1 with increasing Q for the free jumps downstream of the spillway has been confirmed in previous similar results we were able to retrieve in the present literature [13,17,37,38].

3.2. Power Spectral Density Analysis

The power spectral density (PSD) analysis of the pressure data demonstrates the variation of the PSD parameter in a wide range of frequencies. According to Figure 4, the maximum values of the amplitude corresponding to the dominant frequency decrease by increasing distance from the jump toe. The results indicated that the maximum variation of the PSD parameter in free jumps within basin$_{II}$ was achieved at frequencies less than 5 Hz. It should be noted that the PSD analysis of the fluctuating pressures for different points of basin$_I$ with free jumps has been studied by Mousavi et al. [24]. The minimum frequency of pressure transducers is considered to be almost twice the dominant frequency of the signal in the literature [39]. In the present study, a pressure data collection frequency of 20 Hz for 90 s was used for each pressure point. The maximum amplitude at low frequencies along the free jumps indicates large-scale vortices, which is due to the dominance of gravitational forces [40]. Therefore, the Froude law is valid for modeling fluctuating pressures in free jumps.

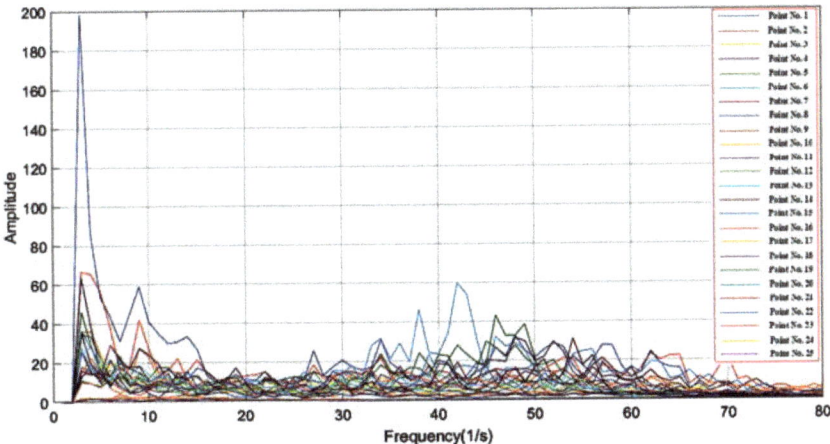

Figure 4. Power spectral density (PSD) analysis of fluctuating pressures for basin$_{II}$ (Fr$_1$ = 6.14).

3.3. Probability Density Function

In this section, the $P^*(Z)$ parameter is plotted as a function of the normalized pressure level (Z). Furthermore, the appropriate probability distributions for each pressure point are compared with the normal distribution. The PDF histograms of pressure fluctuations at some points on the bed of basin$_{II}$ in free jumps are shown in Figure 5. In addition, the PDF function of the fluctuating pressures for different points of basin$_I$ with the free jumps has been previously investigated by Mousavi et al. [24]. The results indicate that the PDF function does not follow the normal distribution at different points along the free jumps, especially for the initial points of basin$_{II}$. In other words, the S and K coefficients do not match with the normal distribution values.

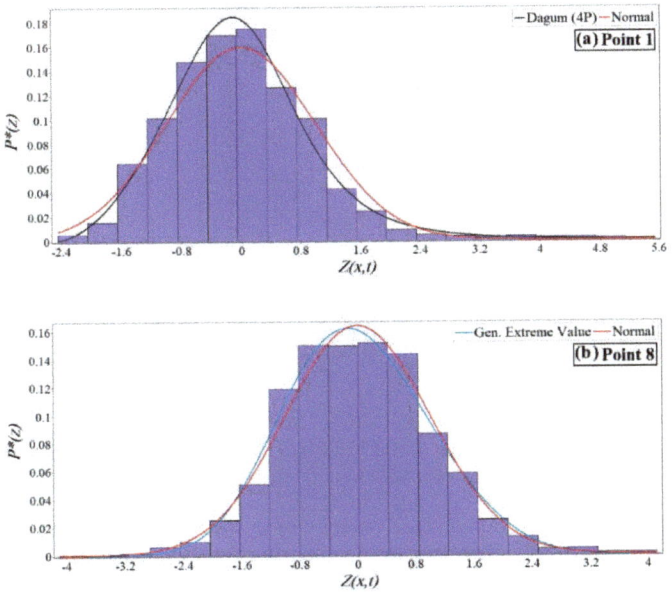

Figure 5. *Cont.*

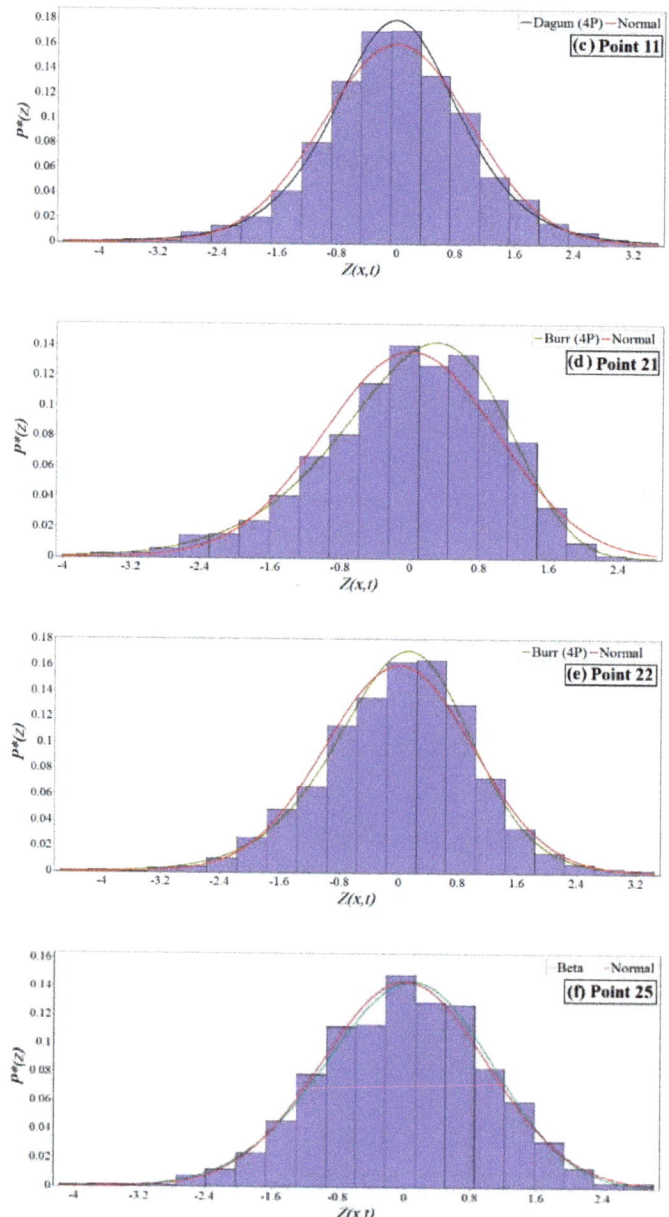

Figure 5. Probability density function (PDF) histograms of fluctuating pressures at some points of basin$_{II}$ (Fr$_1$ = 6.14): (**a**) point 1; (**b**) point 8; (**c**) point 11; (**d**) point 21; (**e**) point 22; (**f**) point 25.

According to Fiorotto and Rinaldo [35], positive pressure fluctuations at the beginning of the basins have a relatively high frequency compared to negative pressure fluctuations. In this zone, the S coefficient has positive and maximum values, and the PDF curve tail is drawn to the right. The K values are more than the normal distribution, and the PDF curve is drawn upwards at these points.

At the characteristic point of X^*_d (point of expected flow detachment), the frequency of positive and negative pressure fluctuations is almost identical, and $S \approx 0$ (pressure point No. 11). At this point, pressure values distribution is somewhat similar to the normal distribution. For point No. 21, located at X^*_r (endpoint of the roller), the S coefficient has negative values. Also, the K coefficient is greater than the normal distribution value. For points located at X^*_j (endpoint of the hydraulic jump), the S values tend to move towards zero, and the data somewhat follow the normal distribution. Pressure point No. 25 is outside basin$_{II}$, and it has a normal distribution. The flow energy of the incoming jet is dissipated after passing through the roller point of the jump, and the uniform flow is established almost downstream of the basin. Due to the presence of accessories in basin$_{II}$, all the considered characteristic points are closer than in the absence of such structures.

3.4. Extreme Pressures

Figure 6 represents variations of the dimensionless minimum, mean, and maximum (scaled) pressures (P^*_{min}, P^*_m, and P^*_{max}) as a function of X^* in basin$_{II}$ and basin$_{I}$ for different values of Fr_1. It is observed that the P^*_{min} data reach lower values and have more significant fluctuations concerning the P^*_m data at the position nearest to the spillway toe (probably due to the incidence of flow in the basin).

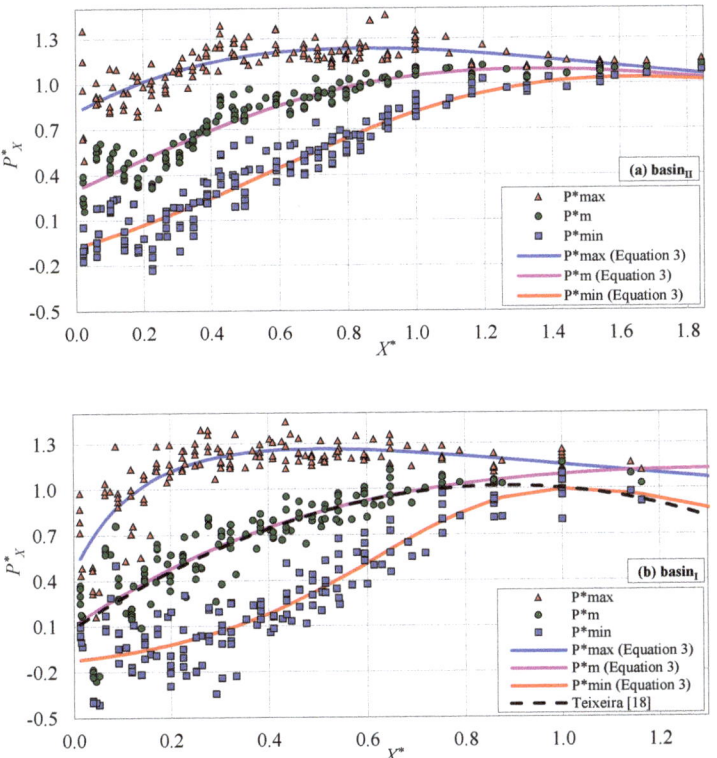

Figure 6. Distribution of the experimental and estimated values of P^*_X: (**a**) basin$_{II}$; (**b**) basin$_{I}$.

The P^*_{min} data reach negative values around −0.2 approximately at $X^* \leq 0.20$ for basin$_{II}$, and of −0.4 at $X^* \leq 0.30$ for basin$_{I}$, increasing with oscillations after that. This may indicate zones subject to low pressure, which may be associated with erosion or cavitation processes. Therefore, basin$_{II}$ is more reliable than basin$_{I}$ in terms of the possibility of cavitation. At the position of X^*_j, P^*_{min} data begin to oscillate near the value 1.0 and slightly lower. Concerning the values of P^*_{max}, the higher and more disparate values versus the P^*_m values occur near the spillway, caused by the direct impact of the flow jet on the dissipation basins. The values of extreme pressures in basin$_{II}$ are lower than those for basin$_I$. There is a narrower pressure range in basin$_{II}$ compared to basin$_I$. The results indicate that P^*_{max} seemingly decreases with the increasing Froude number, with P^*_m and P^*_{min} somewhat constant, in the explored range. Using the results obtained in the present study by adjusting the values of P^*_X, including P^*_{min}, P^*_m, and P^*_{max}, a new second-order rational expression was developed for basin$_{II}$ and basin$_I$. Equation (3) is valid for $0 < X^* \leq 1.85$ in basin$_{II}$ and $0 < X^* \leq 1.30$ in basin$_I$. According to Figure 6, one can estimate P^*_X using Equation (3). The values of α, β, γ, and δ to estimate P^*_X for basin$_{II}$ and basin$_I$ are provided in Table 2.

$$P^*_X = \frac{\alpha + \beta X^*}{1 + \gamma X^* + \delta X^{*2}} \quad (3)$$

Table 2. Coefficients of α, β, γ, δ, and the statistical performance criteria to estimate P^*_X.

Basin	P^*_X	A	B	γ	δ	R	RMSE	MAE
basin$_{II}$	P^*_{min}	−0.0758	0.6885	−0.6537	0.4041	0.950	0.110	0.082
	P^*_m	0.3057	0.9186	−0.2466	0.4086	0.944	0.085	0.063
	P^*_{max}	0.8171	1.5498	0.4397	0.4879	0.753	0.100	0.072
basin$_I$	P^*_{min}	−0.1220	0.5397	−1.6625	1.0825	0.909	0.155	0.122
	P^*_m	0.1094	2.2112	0.6233	0.4925	0.882	0.150	0.105
	P^*_{max}	0.4690	8.5806	4.2451	2.5554	0.789	0.145	0.099

* Dimensionless value.

3.5. Standard Deviation of Fluctuating Pressures

The σ^*_X parameter is a function of the flow discharge and the pressure point position relative to the beginning of the jump. In Figure 7, increasing flow discharge (i.e., decreasing Froude number) results in σ^*_X increasing. As Q increases, the dynamic energy increases, and the fluctuating component of pressure (P') increases as well, indicating the turbulence intensity of the flow. Along the jump, σ^*_X increases to a maximum value, in the range of $X^* \leq 0.33$ for basin$_I$ and basin$_{II}$, and decreases after that. It seems that the main factors for the fluctuations of pressures along the jump are turbulent flow, eddies formation, and their movement during the jump. Therefore, in some positions, the interaction of eddies and the basin bed causes a sudden increase in the bed pressure.

Figure 7b shows a comparison of the σ^*_X values in the case of basin$_I$ with the results obtained by Pinheiro [16] and Marques et al. [17] for free jumps. It is seen that our study displays similar patterns to their work. However, in the downstream zone of the basin, σ^*_X values are relatively higher than the results obtained by others. This is likely to be linked to the determination of Y_1 and Y_2 and identification of the initial position of the hydraulic jump. The σ^*_X values for the smooth bed (basin$_I$) are greater than in basin$_{II}$ with blocks. Accordingly, the presence of accessories within the hydraulic jumps significantly decreases σ^*_X. Figure 7 demonstrates the values of $\sigma^*_{X\,max}$ for different Froude numbers in basin$_{II}$ and basin$_I$. A high value of $\sigma^*_{X\,max}$ may indicate a considerable variation of the dynamic pressures on the bottom slab, damaging the structure. According to Figure 7, as the Froude number (Fr_1) increases, the intensity of pressure fluctuations decreases. According to Teixeira [18], the average value of $\sigma^*_{X\,max}$ in a smooth basin was about 0.7.

As seen in Table 3, one has $\sigma^*_{X\,max} \approx 0.50\sim0.68$ for basin$_{II}$, and $\sigma^*_{X\,max} \approx 1.02\sim1.20$ for basin$_I$, similarly for all Froude numbers. Accordingly, the values of $\sigma^*_{X\,max}$ in basin$_{II}$ decreased down to about −45% compared with basin$_I$ for the free jumps. The $X^*_{\sigma max}$ position in the presence of the blocks and end sill is closer to the spillway toe. The accessories on the bed of basin$_{II}$ may cause the jet to be spread or submerged. Due to the presence of chute blocks at some points, the bottom of basin$_{II}$ does not collide directly with the jet due to the hydraulic jump. Consequently, the values of pressure and pressure fluctuations decrease mainly therein. Figure 7 illustrates the σ^*_X values for different Froude numbers in basin$_I$ and basin$_{II}$.

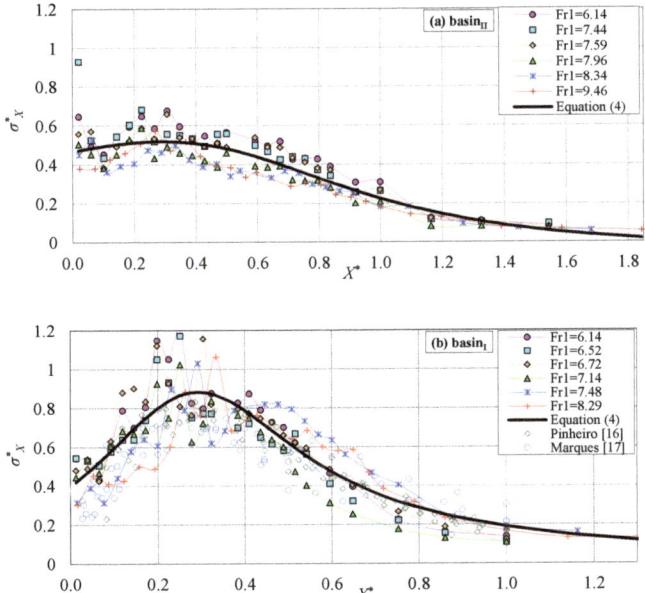

Figure 7. Distribution of the experimental and estimated values of σ^*_X: (**a**) basin$_{II}$; (**b**) basin$_I$.

Table 3. Range of $\sigma^*_{X\,max}$ values and the position of $X^*_{\sigma max}$.

Results	$\sigma^*_{X\,max}$	$X^*_{\sigma max}$
basin$_{II}$	0.50~0.68	0.07~0.33
basin$_I$ [25]	1.02~1.20	0.25~0.33
Endres [15]	0.65~0.77	0.03~0.18
Pinheiro [16]	0.73~0.83	0.25~0.33
Marques [17]	0.69~0.76	0.22~0.40

* Dimensionless value.

We optimized Teixeira's method [18] to assess σ^*_X for basin$_{II}$ and basin$_I$. A new second-order rational expression was developed in the range of $0 < X^* \leq 1.85$ for basin$_{II}$ and $0 < X^* \leq 1.30$ for basin$_I$. According to Figure 7, one can estimate σ^*_X using Equation (4). The values of a, b, c, and d to determine σ^*_X are provided in Table 4.

$$\sigma^*_X = \frac{a + b\,X^*}{1 + c\,X^* + d\,X^{*2}} \tag{4}$$

Table 4. Coefficients of a, b, c, d, and the statistical performance criteria to estimate σ^*_X.

Results	a	B	c	d	R	RMSE	MAE
basin$_{II}$	0.4661	−0.2218	−1.1229	1.2068	0.910	0.065	0.053
basin$_I$	0.3975	0.3735	−3.3347	6.4248	0.872	0.120	0.095

* Dimensionless value.

Therefore, the new adjustment can estimate σ^*_X very well with a correlation coefficient (R) equal to 0.910 and 0.872 for basin$_{II}$ and basin$_I$, respectively.

3.6. Statistical Coefficient of the Probability Distribution

Figure 8 presents the distribution of the experimental values of the $N_{K\%}$ coefficient obtained from the pressure data along basin$_{II}$ for different probabilities from 0.1% to 99.9% with different flow conditions in free jumps. The distribution of the $N_{K\%}$ coefficient along basin$_{I}$ with the free jumps has been previously investigated by Mousavi et al. [25].

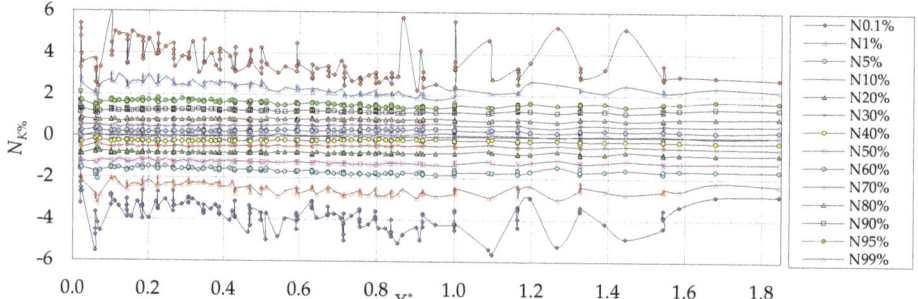

Figure 8. Distribution of the experimental values of the $N_{K\%}$ coefficient along basin$_{II}$.

From Figure 8, one can verify the dispersion of the $N_{K\%}$ coefficient with the minimum and maximum extreme pressures in the initial zone of the jumps. It is observed that for probabilities greater than 50%, the $N_{K\%}$ coefficient has positive values, and for probabilities less than 50%, it has negative values. At the beginning area of the basins, the values of $N_{0.1\%}$ are approximately −3, and for positions $X^* \geq 0.40$, it has values less than −4. In addition, the $N_{99.9\%}$ coefficient at the beginning of basin$_{II}$ has values around 4 to 6. At the downstream of the basins, the $N_{K\%}$ values are slightly stabilized and vary in the range of 2 to 4. The results show that the variation rate of the $N_{K\%}$ coefficient along basin$_{II}$ has decreased somewhat compared to basin$_{I}$.

Teixeira [18] demonstrated that in free jumps, the longitudinal distribution of the $N_{K\%}$ coefficient follows a second-order polynomial relationship. In the present study, the results show that the $N_{K\%}$ coefficient has relatively constant values along the jumps, mainly for the probabilities from 5% to 95%. Accordingly, depending on the probability to be identified, the $N_{K\%}$ coefficient shows a trend more or less close to a single (average) value for each probability, regardless of Fr_1 values. Table 5 displays the average experimental values of the $N_{K\%}$ coefficient with different probabilities along the basins.

Table 5. Average experimental values of $N_{K\%}$ coefficient in the dissipation basins.

$N_{K\%}$	$N_{5\%}$	$N_{10\%}$	$N_{20\%}$	$N_{30\%}$	$N_{40\%}$	$N_{50\%}$	$N_{60\%}$	$N_{70\%}$	$N_{80\%}$	$N_{90\%}$	$N_{95\%}$
basin$_{II}$	−1.66	−1.25	−0.80	−0.48	−0.22	0.02	0.252	0.51	0.80	1.23	1.60
basin$_{I}$	−1.62	−1.25	−0.82	−0.50	−0.24	0.00	0.242	0.50	0.81	1.25	1.63

To develop a method for estimating the $P^*_{K\%}$ parameter in the case of basin$_{II}$ and basin$_{I}$, we identified variations in the $N_{K\%}$ coefficient as a function of probability. Therefore, it was decided to use the average value of $N_{K\%}$ for each probability. According to Wiest [41], there is little effect of Fr_1 on $N_{K\%}$, and the latter remains constant along the dissipation basin. Accordingly, $N_{K\%}$ follows a specific curve acceptably well, making it possible to establish a new adjustment for $N_{K\%}$ as a function of the probability of occurrence (K). Therefore, we propose a second-order rational relationship to estimate $N_{K\%}$:

$$N_{K\%} = \frac{\alpha + \beta K}{1 + \gamma K + \delta K^2} \quad (5)$$

Here α, β, γ, and δ are the coefficients of Equation (5), and K is the value of the probability in decimal. The values of coefficients in Equation (5) are shown in Table 6. The residual of the experimental and estimated data set of the $N_{K\%}$ coefficient for different probabilities in basin$_{II}$ and basin$_I$ is plotted in Figure 9. This parameter is defined as the difference between the experimental and estimated values of $N_{K\%}$.

Table 6. Coefficients of Equation (5) for estimating $N_{K\%}$ coefficient in dissipation basins.

Results	α	β	γ	δ
basin$_{II}$	−2.1625	4.3873	3.8320	−3.7389
basin$_I$	−2.0752	4.1402	3.3326	−3.3448

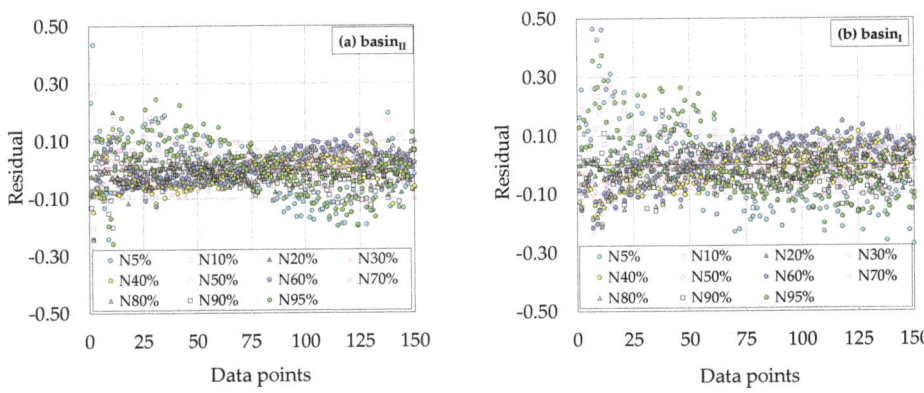

Figure 9. Residual plots of the experimental and estimated data set of the $N_{K\%}$ coefficient for different probabilities: (**a**) basin$_{II}$ and (**b**) basin$_I$.

3.7. Estimation of Pressures with Different Probabilities of Occurrence

In this study, new original adjustments were proposed for P^*_m (Equation (3)), σ^*_X (Equation (4)), and $N_{K\%}$ (Equation (5)) to estimate the pressure values with different probabilities of occurrence ($P_{K\%}$). Therefore, the estimated values of $P_{\alpha\%}$ were determined using Equation $P_{\alpha\%} = P_m + N_{K\%} \times \sigma_X$.

Some statistical criteria for the estimated values of the $P^*_{K\%}$ parameter in basin$_{II}$ and basin$_I$ are presented in Table 7. For instance, the longitudinal distribution of the experimental and estimated data of the $P^*_{K\%}$ parameter with different probabilities along basin$_{II}$ is shown in Figure 10. The distribution of the $P^*_{K\%}$ parameter for different probabilities of occurrence along basin$_I$ with the free jumps has been previously investigated by Mousavi et al. [25].

Table 7. Statistical criteria to estimate $P^*_{K\%}$ with different probabilities of occurrence.

$P^*_{K\%}$	basin$_{II}$				basin$_I$			
	R	RMSE	MAE	WI	R	RMSE	MAE	WI
$P^*_{5\%}$	0.948	0.096	0.073	0.973	0.880	0.166	0.122	0.934
$P^*_{10\%}$	0.946	0.094	0.071	0.972	0.879	0.164	0.120	0.933
$P^*_{20\%}$	0.944	0.092	0.069	0.971	0.879	0.161	0.116	0.932
$P^*_{30\%}$	0.944	0.090	0.067	0.970	0.880	0.158	0.112	0.932
$P^*_{40\%}$	0.943	0.088	0.065	0.970	0.882	0.155	0.109	0.933
$P^*_{50\%}$	0.943	0.087	0.063	0.970	0.884	0.152	0.106	0.934
$P^*_{60\%}$	0.940	0.085	0.062	0.969	0.884	0.150	0.103	0.934
$P^*_{70\%}$	0.942	0.082	0.060	0.969	0.884	0.147	0.100	0.934
$P^*_{80\%}$	0.941	0.080	0.059	0.969	0.884	0.145	0.097	0.934
$P^*_{90\%}$	0.939	0.077	0.057	0.968	0.881	0.141	0.093	0.934
$P^*_{95\%}$	0.929	0.078	0.057	0.963	0.848	0.138	0.093	0.933

* Dimensionless value.

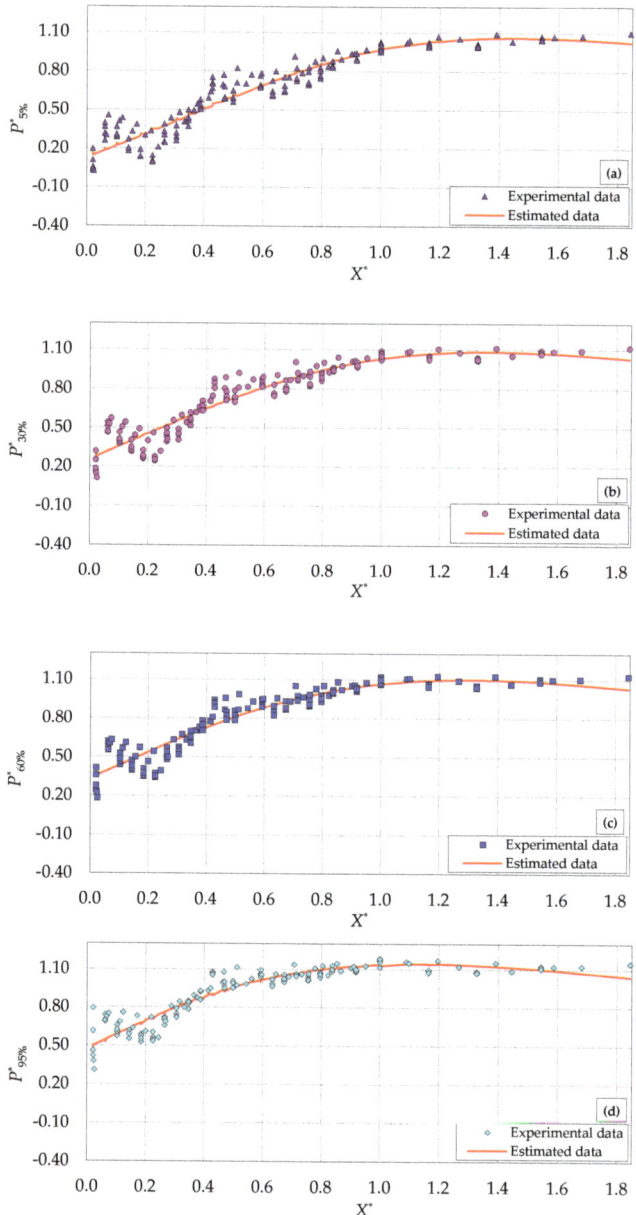

Figure 10. Longitudinal distribution of the experimental and estimated data of $P^*_{K\%}$ parameter with different probabilities along basin$_{II}$: (**a**) $P^*_{5\%}$; (**b**) $P^*_{5\%}$; (**c**) $P^*_{60\%}$; (**d**) $P^*_{95\%}$.

4. Conclusions

In this study, a lab-scale model of an Ogee spillway, either equipped with the USBR Type I and II dissipation basins was installed downstream of an Ogee spillway, based on the USBR criteria, to investigate pressure fields therein. The present study aimed to measure and provide useful insights about the pressure fluctuations at the bottom of basin$_{II}$. We can provide here some conclusions from our research, covering the (different) patterns of pressures along the free hydraulic jumps, as follows:

(i) For the first time to our knowledge, our results allow calculation of the statistics and extreme values of the pressure field occurring on the bed of the dissipation basins, and demonstrate the advantage of using a USBR Type II basin in terms of reduced stress over the basin's bed.

(ii) The Y_2 parameter in basin$_{II}$ was decreased against that in basin$_{I}$. In addition, with increasing flow discharge (Q), supercritical flow depth (Y_1) increased more than velocity (V_1). As a result, Fr$_1$ reduced with higher Q values.

(iii) The P^*_{min} data reached negative values of around −0.2 approximately at $X^* \leq 0.2$ for basin$_{II}$, and of −0.4 at $X^* \leq 0.3$ for basin$_I$ (i.e., very close to spillway toe). Therefore, basin$_{II}$ was more reliable than basin$_I$ in terms of the possibility of cavitation. More fluctuating values of P^*_{max} against the mean values occurred near the spillway, justified by the direct impact of the flow jet on the dissipation basin.

(iv) Analysis of σ^*_X showed that the dimensionless position of $X^*_{\sigma max}$ is close to 0.20 and 0.29 for basin$_{II}$ and basin$_I$, respectively, with pressure fluctuations decreasing after that. Accordingly, the position of $X^*_{\sigma max}$ was closer to the spillway toe for basin$_{II}$. With increasing flow discharge, the pressure fluctuations increased. The pressure fluctuations range on the basin bed was visibly narrower for basin$_{II}$ than for basin$_I$. For basin$_{II}$, σ^*_{Xmax} values along the free jumps were reduced by −40% compared to basin$_I$.

(v) Based on the methodologies proposed by Marques et al. [17] and Teixeira [18], new original best-fit adjustments were proposed here for the P^*_m, σ^*_X, and $N_{K\%}$ parameters to estimate the $P^*_{K\%}$ parameter in the case of basin$_I$ and basin$_{II}$. In addition, we originally displayed that $N_{K\%}$ values show a trend towards a single average value independently of the Froude number, and we proposed an adjustment for $N_{K\%}$ as a function of probability.

(vi) Some effort may be devoted to investigating the statistical distribution of pressures on the basin bed. As observed, a deviation of the skewness from the $S = 0$ value for normal distribution in the beginning area of the basins indicates a different and asymmetric distribution. Positively skewed distributions indicate the potential for more (than normally expected) frequent outbursts of large flow pressure, possibly requiring the increase of the structural resistance of the basin apron.

(vii) The laboratory-scale models presented herein have several limitations that should guide further research on the topic. It should be noted that there is a potential error in scaling the pressure heads. Therefore, just indicating the dimensionless terms may be misleading.

(viii) The results of this work contribute to the present debate about the use of dissipation basins, and especially of USBR Type II ones for spillway flow calming, providing a quantitative assessment of some main features of the hydraulic jump within the dissipation basin, and the modified (reduced) maximum pressure on the basin apron, and are potentially useful for designing dissipation basins in real-world applications.

Author Contributions: Methodology, S.N.M. and D.B.; validation, D.B.; writing—original draft, S.N.M.; writing—review and editing, D.B. Both authors have read and agreed to the published version of the manuscript.

Funding: This research received no external funding.

Acknowledgments: All co-authors would like to express their gratitude to the editors and reviewers for their time spent reviewing our manuscript and for helping to improve the manuscript. Daniele Bocchiola acknowledges the support from the Climate-Lab of the Polytechnic University of Milan (https://www.climatelab.polimi.it/en/), an interdepartmental laboratory on climate change at the Polytechnic University of Milan.

Conflicts of Interest: The authors declare no conflict of interest.

Abbreviation

The following symbols are used in this paper:

B	Basin width (L)
$basin_I$	USBR Type I dissipation basin
$basin_{II}$	USBR Type II dissipation basin
C'_p	Pressure fluctuations intensity coefficient
E_l	Energy head loss along the hydraulic jump (L)
Fr_1	Supercritical Froude number
Fr_2	Subcritical Froude number
g	Gravitational acceleration (LT^{-2})
H	Ogee spillway height
K	Kurtosis coefficient
L_I	Length of $basin_I$ (L)
L_{II}	Length of $basin_{II}$ (L)
L_j	Length of hydraulic jump (L)
MAE	Mean absolute error
$N_{K\%}$	Statistical coefficient of the probability distribution
$P_{K\%}$	Pressure head with a certain probability of occurrence (L)
P_{min}	Minimum extreme pressure (L)
P_m	Mean pressure head at each pressure point (L)
P_{max}	Maximum extreme pressure (L)
PSD	Power spectral density of the pressure data
$P(X,t)$	Instantaneous pressure (L)
P^*_Z	Probability density function (PDF) of the normalized fluctuating pressures
P'	Fluctuating component of pressure (L)
Q	Flow discharge (L^3T^{-1})
R	Correlation coefficient
Re_1	Reynolds number for the supercritical flow of the hydraulic jump
RMSE	Root mean squared error
S	Skewness coefficient
USBR	US Department of the Interior, Bureau of Reclamation
V_1	Mean velocity of the coming flow to the dissipation basin (LT^{-1})
V_2	Mean subcritical velocity (LT^{-1})
WI	Willmott's index of agreement
X	Longitudinal position of each point inside the hydraulic jump (L)
X^*	Dimensionless position of each point (X/L_j)
X^*_d	Characteristic point of the expected flow detachment
X^*_r	Characteristic endpoint of the roller
X^*_j	Characteristic endpoint of the hydraulic jump
Y_1	Supercritical flow depth at the jump toe (L)
Y_2	Subcritical flow depth at the end of the jump (L)
Z	Normalized pressure variable
σ_X	Standard deviation of the pressure fluctuations at point X (L)
σ^*_X	Dimensionless standard deviation of the pressure fluctuations at point X (L)
1	Supercritical flow
2	Subcritical flow
m	Mean value
max	Maximum value
min	Minimum value
*	Dimensionless value

References

1. Khatsuria, R.M. *Hydraulics of Spillways and Energy Dissipators*; Marcel Dekker: New York, NY, USA, 2005.
2. Alves, A.A.M. *Characterization of Hydrodynamic Forces in Dissipation Basins under Hydraulic Jumps with Low Froude Number*; Universidade Federal do Rio Grande do Sul: Porto Alegre, Brazil, 2008. (In Portuguese)
3. USBR. Spillways. In *Design of Small Dams*, 3rd ed.; US Department of the Interior, Bureau of Reclamation: Washington, DC, USA, 1987; pp. 339–437.
4. Mohamed Ali, H. Effect of roughened-bed stilling basin on length of rectangular hydraulic jump. *J. Hydraul. Eng.* **1991**, *117*, 83–93. [CrossRef]
5. Verma, D.; Goel, A. Stilling basins for pipe outlets using wedge-shaped splitter block. *J. Irrig. Drain. Eng.* **2000**, *126*, 179–184. [CrossRef]
6. Alikhani, A.; Behrozi-Rad, R.; Fathi-Moghadam, M. Hydraulic jump in stilling basin with vertical end sill. *Int. J. Phys. Sci.* **2010**, *5*, 25–29.
7. Tiwari, H.; Goel, A.; Gahlot, V. Experimental Study of effect of end sill on stilling basin performance. *Int. J. Eng. Sci. Technol.* **2011**, *3*, 3134–3140.
8. Cancian Putton, V.; Marson, C.; Fiorotto, V.; Caroni, E. Supercritical flow over a dentated sill. *J. Hydraul. Eng.* **2011**, *137*, 1019–1026. [CrossRef]
9. Tiwari, H.; Goel, A. Effect of end sill in the performance of stilling basin models. *Am. J. Civil Eng. Archit.* **2014**, *2*, 60–63. [CrossRef]
10. Fecarotta, O.; Carravetta, A.; Del Giudice, G.; Padulano, R.; Brasca, A.; Pontillo, M. Experimental results on the physical model of an USBR type II stilling basin. In Proceedings of the River Flow 2016, International Conference on Fluvial Hydraulics, St. Louis, MI, USA, 10–14 July 2016.
11. Chanson, H.; Carvalho, R. Hydraulic jumps and stilling basins. In *Energy Dissipation in Hydraulic Structures*; Chanson, H., Ed.; CRC Press: Leiden, The Netherlands, 2015; pp. 65–104.
12. Vischer, D.; Hager, W.H. *Dam Hydraulics*; Wiley: Ürich, Switzerland, 1998; Volume 2.
13. Padulano, R.; Fecarotta, O.; Del Giudice, G.; Carravetta, A. Hydraulic design of a USBR Type II stilling basin. *J. Irrig. Drain. Eng.* **2017**, *143*, 1–9. [CrossRef]
14. Toso, J.W.; Bowers, C.E. Extreme pressures in hydraulic-jump stilling basins. *J. Hydraul. Eng.* **1988**, *114*, 829–843. [CrossRef]
15. Endres, L.A.M. *Contribution to the Development of a System for the Acquisition and Processing of Instantaneous Pressures Data in the Laboratory*; Federal University of Rio Grande do Sul: Porto Alegre, Brazil, 1990. (In Portuguese)
16. Pinheiro, A.A.d.N. *Hydrodynamic Actions in Thresholds for Energy Dissipation Basin by Hydraulic Jumps*; Universidade Técnica de Lisboa: Lisbon, Portugal, 1995. (In Portuguese)
17. Marques, M.G.; Drapeau, J.; Verrette, J.-L. Pressure fluctuation coefficient in a hydraulic jump. *Braz. J. Water Resour. (Rbrh)* **1997**, *2*, 45–52. (In Portuguese)
18. Teixeira, E.D. *Scale Effect on Estimating Extreme Pressure Values on the Bed of the Hydraulic Dissipation Basins*; Universidade Federal do Rio Grande do Sul: Porto Alegre, Brazil, 2008. (In Portuguese)
19. Farhoudi, J.; Sadat-Helbar, S.; Aziz, N.I. Pressure fluctuation around chute blocks of SAF stilling basins. *J. Agric. Sci. Technol.* **2010**, *12*, 203–212.
20. Novakoski, C.K.; Hampe, R.F.; Conterato, E.; Marques, M.G.; Teixeira, E.D. Longitudinal distribution of extreme pressures in a hydraulic jump downstream of a stepped spillway. *Braz. J. Water Resour. (Rbrh)* **2017**, *22*. [CrossRef]
21. Macián-Pérez, J.F.; García-Bartual, R.; Huber, B.; Bayon, A.; Vallés-Morán, F.J. Analysis of the flow in a typified USBR II stilling basin through a numerical and physical modeling approach. *Water* **2020**, *12*, 227. [CrossRef]
22. Hampe, R.F.; Steinke Júnior, R.; Prá, M.D.; Marques, M.G.; Teixeira, E.D. Extreme pressure forecasting methodology for the hydraulic jump downstream of a low head spillway. *Braz. J. Water Resour. (Rbrh)* **2020**, *25*, 1–10. [CrossRef]
23. Samadi, M.; Sarkardeh, H.; Jabbari, E. Explicit data-driven models for prediction of pressure fluctuations occur during turbulent flows on sloping channels. *Stoch. Environ. Res. Risk Assess.* **2020**, *34*, 691–707. [CrossRef]

24. Mousavi, S.N.; Farsadizadeh, D.; Salmasi, F.; Dalir, A.H.; Bocchiola, D. Analysis of minimal and maximal pressures, uncertainty and spectral density of fluctuating pressures beneath classical hydraulic jumps. *Water Supply* **2020**, *20*, 1909–1921. [CrossRef]
25. Mousavi, S.N.; Júnior, R.S.; Teixeira, E.D.; Bocchiola, D.; Nabipour, N.; Mosavi, A.; Shamshirband, S. Predictive modeling the free hydraulic jumps pressure through advanced statistical methods. *Mathematics* **2020**, *8*, 323. [CrossRef]
26. Mousavi, S.N.; Farsadizadeh, D.; Salmasi, F.; Hosseinzadeh Dalir, A. Evaluation of pressure fluctuations coefficient along the USBR Type II stilling basin using experimental results and AI models. *ISH J. Hydraul. Eng.* **2020**. [CrossRef]
27. Belanger, J.B. *Essay on the Numerical Solution of Some Problems Related to the Constant Motion of Water Flow*; Carilian-Goeury: Paris, France, 1828. (In French)
28. Hager, W.H.; Li, D. Sill-controlled energy dissipator. *J. Hydraul. Res.* **1992**, *30*, 165–181. [CrossRef]
29. Chaudhry, M.H. *Open-Channel Flow*, 2nd ed.; Springer Science & Business Media: New York, NY, USA, 2008.
30. Chanson, H. Development of the Bélanger equation and backwater equation by Jean-Baptiste Bélanger (1828). *J. Hydraul. Eng.* **2009**, *135*, 159–163. [CrossRef]
31. Bennett, N.D.; Croke, B.F.; Guariso, G.; Guillaume, J.H.; Hamilton, S.H.; Jakeman, A.J.; Marsili-Libelli, S.; Newham, L.T.; Norton, J.P.; Perrin, C. Characterising performance of environmental models. *Environ. Model. Softw.* **2013**, *40*, 1–20. [CrossRef]
32. Willmott, C.J.; Matsuura, K. Advantages of the mean absolute error (MAE) over the root mean square error (RMSE) in assessing average model performance. *Clim. Res.* **2005**, *30*, 79–82. [CrossRef]
33. Chai, T.; Draxler, R.R. Root mean square error (RMSE) or mean absolute error (MAE)?–Arguments against avoiding RMSE in the literature. *Geosci. Model Dev.* **2014**, *7*, 1247–1250. [CrossRef]
34. Willmott, C.J.; Robeson, S.M.; Matsuura, K. A refined index of model performance. *Int. J. Climatol.* **2012**, *32*, 2088–2094. [CrossRef]
35. Fiorotto, V.; Rinaldo, A. Turbulent pressure fluctuations under hydraulic jumps. *J. Hydraul. Res.* **1992**, *30*, 499–520. [CrossRef]
36. Sharma, C.; Ojha, C. Statistical parameters of hydrometeorological variables: Standard deviation, SNR, skewness and kurtosis. In *Advances in Water Resources Engineering and Management*; Springer: Singapore, 2020; Volume 39, pp. 59–70.
37. Novakoski, C.K.; Conterato, E.; Marques, M.; Teixeira, E.D.; Lima, G.A.; Mees, A. Macro-turbulent characteristcs of pressures in hydraulic jump formed downstream of a stepped spillway. *Braz. J. Water Resour. (Rbrh)* **2017**, *22*. [CrossRef]
38. Prá, M.D.; Priebe, P.d.S.; Teixeira, E.D.; Marques, M.G. Evaluation of pressure fluctuation in hydraulic jump by dissociation of hydraulic forces. *Braz. J. Water Resour. (Rbrh)* **2016**, *21*, 221–231. (In Portuguese)
39. Yan, Z.-M.; Zhou, C.-T.; Lu, S.-Q. Pressure fluctuations beneath spatial hydraulic jumps. *J. Hydrodyn.* **2006**, *18*, 723–726. [CrossRef]
40. Pei-Qing, L.; Ai-Hua, L. Model discussion of pressure fluctuations propagation within lining slab joints in stilling basins. *J. Hydraul. Eng.* **2007**, *133*, 618–624. [CrossRef]
41. Wiest, R.A. *Evaluation of the Pressure Field in Hydraulic Jump Formed Downstream of a Spillway with Different Submergence Degrees*; Universidade Federal do Rio Grande do Sul: Porto Alegre, Brazil, 2008. (In Portuguese)

Publisher's Note: MDPI stays neutral with regard to jurisdictional claims in published maps and institutional affiliations.

© 2020 by the authors. Licensee MDPI, Basel, Switzerland. This article is an open access article distributed under the terms and conditions of the Creative Commons Attribution (CC BY) license (http://creativecommons.org/licenses/by/4.0/).

Article

Experimental Validation of a Sliding Mode Control for a Stewart Platform Used in Aerospace Inspection Applications

Javier Velasco [1,*], Isidro Calvo [2], Oscar Barambones [2], Pablo Venegas [1] and Cristian Napole [2]

1. Fundación Centro de Tecnologías Aeronáuticas (CTA), Juan de la Cierva 1, 01510 Miñano, Spain; pablo.venegas@cta.aero
2. Department of System Engineering and Automation, Faculty of Engineering Vitoria-Gasteiz, University of the Basque Country (UPV/EHU), Nieves Cano 12, 01006 Vitoria-Gasteiz, Spain; isidro.calvo@ehu.eus (I.C.); oscar.barambones@ehu.eus (O.B.); cristianmario.napole@ehu.eus (C.N.)
* Correspondence: javier.velasco@cta.aero

Received: 20 October 2020; Accepted: 16 November 2020; Published: 17 November 2020

Abstract: The authors introduce a new controller, aimed at industrial domains, that improves the performance and accuracy of positioning systems based on Stewart platforms. More specifically, this paper presents, and validates experimentally, a sliding mode control for precisely positioning a Stewart platform used as a mobile platform in non-destructive inspection (NDI) applications. The NDI application involves exploring the specimen surface of aeronautical coupons at different heights. In order to avoid defocusing and blurred images, the platform must be positioned accurately to keep a uniform distance between the camera and the surface of the specimen. This operation requires the coordinated control of the six electro mechanic actuators (EMAs). The platform trajectory and the EMA lengths can be calculated by means of the forward and inverse kinematics of the Stewart platform. Typically, a proportional integral (PI) control approach is used for this purpose but unfortunately this control scheme is unable to position the platform accurately enough. For this reason, a sliding mode control (SMC) strategy is proposed. The SMC requires: (1) a priori knowledge of the bounds on system uncertainties, and (2) the analysis of the system stability in order to ensure that the strategy executes adequately. The results of this work show a higher performance of the SMC when compared with the PI control strategy: the average absolute error is reduced from 3.45 mm in PI to 0.78 mm in the SMC. Additionally, the duty cycle analysis shows that although PI control demands a smoother actuator response, the power consumption is similar.

Keywords: automatic optical inspection; kinetic theory; parallel robots; robust control; sliding mode control

1. Introduction

Contemporary markets tend to constantly increase the number of product variants, and product life cycles are also changing. Demand for product families in small batch sizes is increasing, therefore reconfigurable tooling is becoming a key technology for fulfilling production requirements [1–3]. For that reason, Stewart platforms, which provide precise motion in six degrees of freedom, are being introduced in production scenarios where continuous positioning and orientating is required. In addition, parallel robot architectures provide high rigidity, high payload-to-weight-ratio, high positioning accuracy, and low inertia of moving parts subjected to high loads. Compared to anthropomorphic robotic arms, they do not have such a wide range of displacements, but they present higher stiffness and precision as well as a simpler solution to the inverse kinematics equations [4,5]. Previous benefits, in addition to the reduction in hexapod production costs, justify the use of Stewart platforms in industrial applications.

Stewart platforms can be found in different domains, such as machinery [1,6], test beds [7], real-time simulators for vehicles [2,8], and antenna/solar orientation platforms [9,10]. Typically, every arm is controlled by an electro mechanical actuator (EMA), but hydraulic actuators are more suitable for large loads [2,11], and piezoelectric actuators are commonly used in micro- or nano-scale platforms (found in biomedical science, optics, and microscopical devices) [12]. Stewart platforms are frequently used as camera levelling bases in a wide range of applications. These devices can be found in rescue operations, traffic control, identification, surveillance of frontiers, agriculture control and fire detection in forest areas, among others. The use of Stewart platforms for levelling cameras has been introduced in automated non-destructive inspections (NDI) in order to reduce labour costs [13,14]. Thermographic cameras, sensitive to radiation in the infrared spectral range, are frequently used in combination with external thermal stimulation systems. Thermographic inspection is normally conducted statically by setting the infrared camera at a fixed distance from the object [15]. However, this approach is subject to the curvature changes in the object inspected so that the spatial resolution is reduced due to the difference in distance between the camera and the target. A different approach in thermographic NDIs consists of applying the inspection dynamically by moving the sensor over the surface of interest. This approach improves the inspection quality, because it permits the detection of smaller defects, but requires higher performance in positioning control systems. Figure 1 illustrates the influence of using static positioning by means of two infrared radiation (IR) images, only one being properly focused.

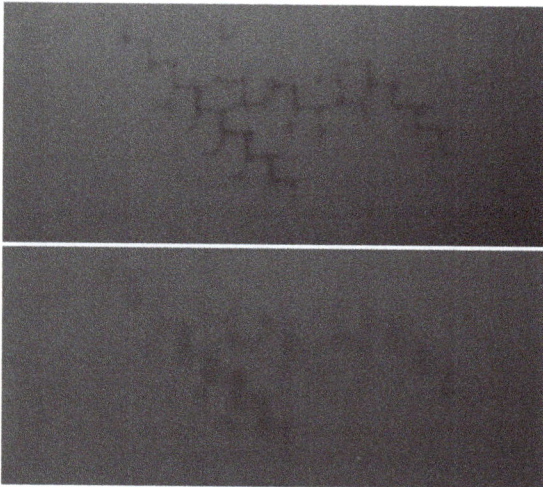

Figure 1. Comparison between a corrected IR image (**above**) and a defocused IR image (**below**).

The control of Stewart platforms must deal with a complex kinetics calculation. It is a well-known fact that, for serial robot manipulators, the solution of a forward kinematics problem is easier than an inverse kinematics problem. However, in parallel robots the situation is just the opposite [4]. Several solutions for the forward and the inverse kinematics are proposed in the literature. The inverse kinematic problem, i.e., obtaining the joint space position or the six link lengths given the position/orientation of the platform or the Cartesian space position, is straightforward to calculate at Stewart Platforms. On the contrary, the forward kinematics problem, viz. The determination of the Cartesian space position for a given joint space position, is more demanding computationally [16]. The most common approach is to calculate the closed-form solution to the inverse kinematic transformation, and then estimate the forward kinematic transformation using the Newton–Raphson method [7]. A different approach for solving the algebraic system of equations utilises the reduced Gröbner basis form of the system of equations under total degree term ordering of its monomials and Sylvester's Dialytic elimination method [17]. Particle swamp optimisation is another powerful tool that can be used for analysing different robot configurations [18].

In some cases, the kinematic nonlinearity of robotic actuators is a source of error which requires special consideration [3,6]. The use of appropriate algorithms considerably reduces the calculation time of the kinematics, and the use of simulation software is often extended to dynamics modelling (considering the mass and the stiffness of the moving platform and legs) [19,20]. By using a combination of design and finite elements, an integrated approach can be developed that simulates the machine dynamics and solves the kinematics, taking into account rigid body dynamics, vibration and strength [21].

The control strategy must command the actuators to achieve precise positioning, minimise the error, and deal with system uncertainties. A proportional–integral–derivative controller (PID) is a simple closed-loop method to correct uncertainties, the proportional integral (PI) controller version being one of the most popular control approaches in the market. Several control strategies were proposed in the literature to improve the performance of actuators in parallel robot architectures, nevertheless, most of them were validated only through simulations. Among other schemes, the use of inverse models, which represent the behaviour of a system mathematically, may become a simple approach to improve the controller performance when compared to PI controllers. This strategy corrects known nonlinearities in open loop control, even though errors caused by uncertainties and external perturbations are not corrected. The combination of inverse models with feedback controllers [8], such as H-infinity [11] and genetic algorithms [22], enables us to achieve higher accuracy. Another approach is the use of inverse models to condition the closed loop feedback signal [9]. These schemes were validated by simulations.

Sliding mode control (SMC) is a nonlinear control approach that drives the state trajectory of the system onto a specified sliding surface and maintains the trajectory on that surface for the subsequent time under system uncertainties and perturbations. However, in conventional SMC design, a priori knowledge of the bounds on system uncertainties must be acquired [23–25]. Several SMC-based strategies to control Stewart platforms are proposed and verified by simulations: SMC with perturbation estimation [26], integral SMC [5], continuous higher order SMC [27], and SMC with fuzzy tuning design [28]. The combination of SMC with estimation techniques, such as state observers [29], allows the conditioning of high frequency input signals [13], and estimating variables when not all states are measured directly, by using super-twisting algorithms [2] and adaptive super-twisting algorithms [10].

This paper presents the development of an SMC aimed at positioning an inspection camera precisely over a Stewart platform. This device is designed to perform inspections on aeronautical coupons, which must fulfil severe quality standards, so that any positioning enhancements leads to substantial improvements in the defect detection ratio. In the literature, the application of SMC in Stewart platforms is validated mainly by simulations. In this study, the proposed SMC strategy is validated experimentally as an alternative to the vendor-provided PI controller, which was producing improperly focused images for the inspection application because it was not accurate enough. The experimental results show that the overall platform performance can be improved by using this control scheme. To develop the SMC strategy, in the first place the inspection platform is analysed and modelled in Section 2, where the inverse and forward kinetics are calculated. Additionally, the system dynamics are modelled in this section, which serve as basis to develop the SMC, in Section 3. The performance of the proposed control scheme is evaluated and compared with the vendor-provided PI controller in Section 4. As a final point, discussion conclusions are presented in Section 5.

2. Stewart Platform Mathematical Modelling

The Stewart platform presented in this paper must be controlled to ensure the position and orientation of an inspection camera. The camera must check different aeronautical samples: several sweep scans at a distance of 40 mm must be performed to ensure the quality of the specimen at different heights. The NDI system is shown in Figure 2a and the trajectory to follow when inspecting the composite panel is shown in Figure 2b.

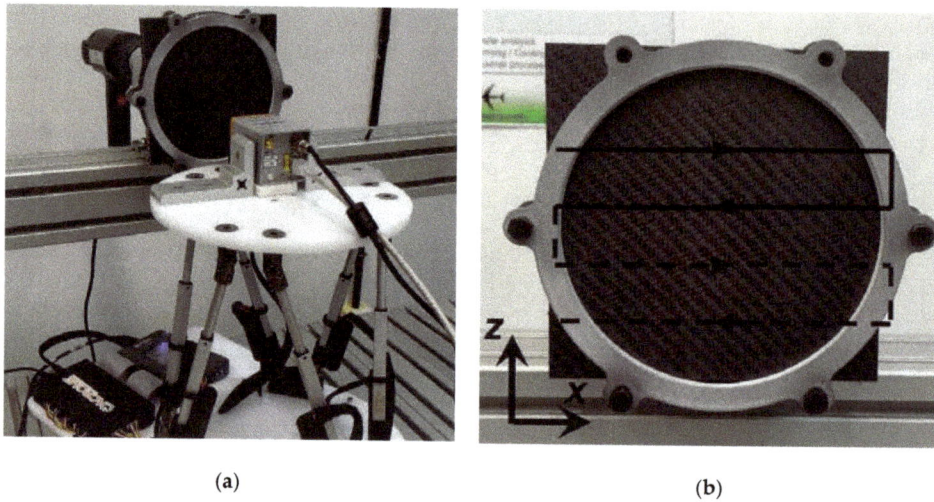

(a)   (b)

Figure 2. Inspection system prototype. (**a**) Test setup; (**b**) Trajectory scheme.

While recording images, it becomes essential to control the position with the greatest accuracy possible in order to ensure that the camera is on the target, and it is vitally important that the camera remains at a specific distance to the target to avoid defocusing, as shown in Figure 1. In this kind of mechatronic system, an effective control strategy is mandatory to achieve successful positioning, and consequently improve the image quality. To do this, firstly, it is necessary to perform an analysis of the system, and create a mathematical model that represents it, as accurately as possible. Creating the model according to the real needs (open or closed loop control, lifecycle, system instabilities, etc.) is vitally important, so the design is optimized and accurate. In the following subsections, the kinetic modelling and dynamic modelling are presented.

2.1. Platform Kinematics

In order to describe the Stewart platform position and orientation, six coordinates are needed. Three of these coordinates, described by d, are positional displacements that locate the position of a reference point in the moving platform with reference to a fixed coordinate system, selected as the base B. The other three coordinates are angular displacements that describe the orientation of the moving platform, represented as e, again with reference to a nonrotating coordinate system located in the base.

$$^B d = \begin{bmatrix} x & y & z \end{bmatrix}^T \tag{1}$$

$$^B e = \begin{bmatrix} \alpha & \beta & \gamma \end{bmatrix}^T \tag{2}$$

The platform position and orientation are controlled by the length of each actuator l_i. The calculation of the actuator lengths is required to drive the Stewart platform over a certain trajectory, this operation is done by using the Stewart platform kinematic transformations. Additionally, to develop a successful control strategy, a correct modelling of the system is required. The following subsections describe the inverse and forward kinematics [7,16] as well as the platform model.

2.1.1. Inverse Kinematics

The inverse kinematics determine the required actuator length for achieving a certain position and orientation of the moving platform. Frame assignment for the robot wrist is illustrated in Figure 3, where two coordinate frames, p and B, are assigned to the payload and base platforms, respectively. The Cartesian variables are chosen to be the relative position and orientation of Frame p with respect to Frame B, where the position of Frame p is specified by the position of its origin with respect to Frame B. The position vectors of the centres of spherical joints in frame B and p can be expressed as:

$$^P p_i = \begin{bmatrix} r_P \cos(\lambda_i) & r_P \sin(\lambda_i) & 0 \end{bmatrix}^T \tag{3}$$

$$^B b_i = \begin{bmatrix} r_B \cos(\Lambda_i) & r_B \sin(\Lambda_i) & 0 \end{bmatrix}^T \tag{4}$$

$$\Lambda_i = \frac{\pi}{3} - \frac{\theta_B}{2}; \; \lambda_i = \frac{\pi}{3} - \frac{\theta_P}{2}, \text{ for } i = 1, 3, 5 \tag{5}$$

$$\Lambda_i = \Lambda_{i-1} + \theta_B; \; \lambda_i = \lambda_{i-1} + \theta_P, \text{ for } i = 2, 4, 6 \tag{6}$$

where r_B and r_P are the radius of the lower and upper platforms, $\theta_P = 23°$ and $\theta_B = 96°$ are the platform and base angles between the first and second joints.

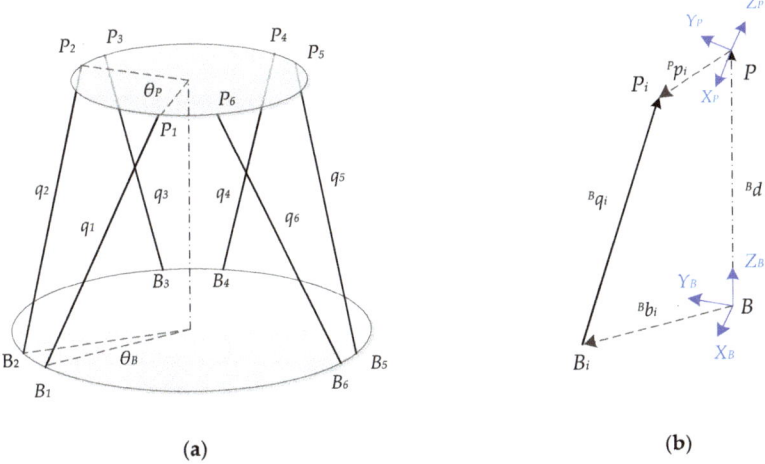

Figure 3. Stewart platform schemes: (**a**) Stewart platform frame assignment; (**b**) vector diagram for the ith actuator.

The leg vector $^B q_i$ pointing from attachment point B_i to point P_i can be expressed as follows:

$$^B q_i = {}^B_P R\, ^P p_i + {}^B d - {}^B b_i \tag{7}$$

where $^B_P R$ is the orientation matrix required to calculate the platform spherical joint vectors in the base coordinate system used in the rest equation terms.

$$^B_P R = \begin{bmatrix} \cos\alpha\cos\beta & \cos\alpha\sin\beta\sin\gamma - \sin\alpha\cos\gamma & \cos\alpha\sin\beta\cos\gamma + \sin\alpha\sin\gamma \\ \sin\alpha\cos\beta & \sin\alpha\sin\beta\sin\gamma + \cos\alpha\cos\gamma & \sin\alpha\sin\beta\cos\gamma - \cos\alpha\sin\gamma \\ \sin\beta & \cos\beta\sin\gamma & \cos\beta\cos\gamma \end{bmatrix} \tag{8}$$

The length of each actuator can be calculated from the modules of the length vectors ^Bq_i:

$$l_i = \left|^Bq_i\right| \tag{9}$$

2.1.2. Forward Kinematics

The forward kinematics allows calculating the platform position and orientation given specific actuator lengths. Equation (9) represents a set of six highly nonlinear simultaneous equations, therefore the Newton–Raphson method can be used to solve the kinematic problem. The objective is to calculate the estimated lengths l_i^* on an iterative process given an initial position estimation a_{j-1}, until the value of a_j, which minimizes the $f_i(d,e)$ scalar function, is found.

$$f_i(d,e) = l_i - l_i^* \tag{10}$$

The Newton–Raphson procedure for this case is:

1. Select an initial guess $(j-1)$:

$$a_{j-1} = \begin{bmatrix} d^T & e^T \end{bmatrix}^T \tag{11}$$

2. Calculate the rotation matrix ^B_PR, with Equation (8), as the initial guess.
3. Calculate ^Bq_i with Equation (7) and $l^*_{i\,j-1}$ with Equation (9).
4. Calculate the unit vector $^B\hat{q}_i$:

$$^B\hat{q}_i = {^Bq_i}/\left|^Bq_i\right| \tag{12}$$

5. Compute the inverse Jacobian matrix for the initial guess J_{j-1}^{-1}:

$$J_{j-1}^{-1} = \begin{bmatrix} ^B\hat{q}_1^T & \left(^B_PRa_1 \times {^B\hat{q}_1}\right)^T \\ \vdots & \vdots \\ ^B\hat{q}_6^T & \left(^B_PRa_6 \times {^B\hat{q}_6}\right)^T \end{bmatrix} \tag{13}$$

6. Compute the estimation a_j:

$$a_j = a_{j-1} - J_{j-1}\left(l^*_{i\,j-1} - l_i\right) \tag{14}$$

7. If $l^*_{i\,j-1} - l_i$ is acceptable, take a_j as the solution, otherwise repeat the procedure with the last a_j estimation.

2.2. Stewart Platform Dynamics

The Stewart platform is moved by means of six EMAs, attached in pairs to three positions on the baseplate of the platform, crossing over to three mounting points on a top plate. The whole system can be divided in six subsystems, each composed of one EMA. In these actuators, the rotary motion of the motor is converted into linear displacements. If the interactions between actuators and external disturbances are considered as perturbations, each EMA can be modelled as a second-order model with a single element, as shown in Figure 4. p_{est} represents the external forces, m_i the mass attached to the actuator, b_i the friction and k_i the actuator stiffness.

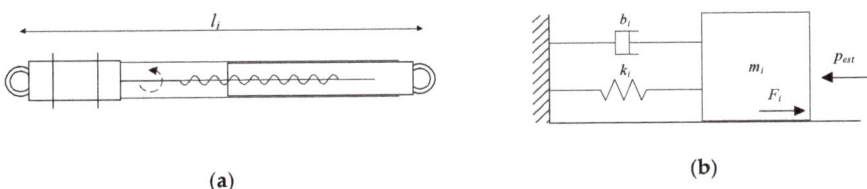

Figure 4. Actuator models. (**a**) Actuator scheme, (**b**) actuator mechanical model.

With the proposed scheme, the equation that models the actuator dynamics results is as follows:

$$k_i l_i(t) + b_i \dot{l}_i(t) + m_i \ddot{l}_i(t) = F_i(t) + p_{est}(t) \tag{15}$$

where t denotes the time variable. The force applied by the actuator F_i is proportional to the maximum force F_{max} and the pulse-width modulation (PMW) duty cycle d_{ci}.

$$F_i(t) = d_{ci}(t) F_{max} \tag{16}$$

3. Sliding Mode Control Calculation

The control strategy must command each actuator in order to achieve the desired trajectory with a precise platform positioning and orientation, minimizing error and dealing with system uncertainties. The proposed control scheme is shown in Figure 5. The platform setpoint is given in terms of position and orientation (a), so at first the trajectory must be transformed to the equivalent actuator length setpoint (b) by using the inverse kinematics explained in Section 2.1. Then, an SMC closed loop control estimates the voltage outputs (c) that are needed to achieve each actuator length, based on the EMA setpoints (b) and the measured lengths (e). The Stewart platform EMAs are controlled by the SMC voltage input (c) and affected by possible disturbances (e.g., external forces, non-linearities) (d). Finally, the EMA lengths (e) are measured by the encoders and transformed to the equivalent platform position (f), which is done using the forward kinematics shown in Section 2.2. The inverse and forward kinematics have been studied previously, therefore in this section the SMC strategy is developed.

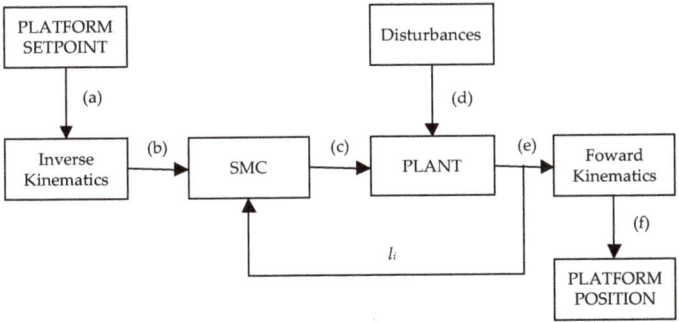

Figure 5. Stewart platform control scheme. The inverse and forward kinematics are used to calculate the actuator length for a determined platform position, and vice versa. The sliding mode control (SMC) generates the control action to bring the Stewart platform to the desired position.

The Stewart platform native closed loop control was a PI control strategy that provided a stable response. The major drawback of this control was that it did not provide an accurate platform control, therefore the NDI camera produced a low-quality IR image. To improve the positioning accuracy of the platform, an SMC strategy was chosen, because it showed accurate results in previous studies. SMC is a nonlinear control approach that drives the state trajectory of the system onto a specified sliding surface and maintains the trajectory on that surface for the subsequent time. However, in conventional SMC design, a priori knowledge of the bounds on system uncertainties must be acquired in order to calculate the sliding gain value that can surmount these uncertainties [23]. Perturbation estimation strategies were studied in the literature [23,30]: one common approach to estimating perturbation p_{est} is as in Equation (15). Then, the system model becomes the following:

$$k_i l_i(t) + b_i \dot{l}_i(t) + m_i \ddot{l}_i(t) = F_{max} d_{ci}(t) + p_{est}(t) + \widetilde{p}(t) \quad (17)$$

where $\widetilde{p}(t) = p(t) - p_{est}(t)$ represents the error between the real and estimated perturbations at the system. To design the SMC controller, the position error is defined as follows:

$$e(t) = l_i(t) - l_{di}(t) \quad (18)$$

where l_{di} represents the desired length. In the rest of this section, the time indices have been omitted for the sake of brevity. Because the dynamic system of the EMA is a second-order system, a second-order PID sliding surface was selected:

$$s = \dot{e} + \lambda_P e + \lambda_I \int e \, dt \quad (19)$$

where λ_P is the proportional sliding gain and λ_I is the integral sliding gain. Studies show that the use of output integral sliding mode control considerably improves the stabilisation of the desired position of platform p as well as the velocity with which it stabilises [13].

For the system represented with Equation (17), with sliding surface defined in Equation (19) and the position error given by Equation (18) which satisfies $\lim_{t \to 0} e(t) = 0$, the control law results as follows, as proved in [30]:

$$d_{ci} = \frac{m_i}{F_{max}}\left(\frac{b_i}{m_i} - \lambda_P\right)\dot{l}_i + \frac{m_i}{F_{max}}\left(\frac{k_i}{m_i} - \lambda_I\right)l_i - \frac{1}{F_{max}}p_{est} + \frac{m_i}{F_{max}}\left(\ddot{l}_{di} + \lambda_P \dot{l}_{di} + \lambda_I l_{di}\right) - \eta \operatorname{sgn}(s) \quad (20)$$

where η is a positive switching gain and $sgn(s)$ represents the signum function, as defined in the following expression:

$$\operatorname{sgn}(s) = \begin{cases} -1, & \text{for } s < 0 \\ 0, & \text{for } s = 0 \\ 1, & \text{for } s > 0 \end{cases} \quad (21)$$

The following candidate Lyapunov function was selected for analysing the stability of the system. Its first derivative can be obtained as:

$$\dot{V} = m_i s \dot{s} \quad (22)$$

By taking the time derivative of both sides of Equation (19), the following sliding dynamics can be generated:

$$\begin{aligned}\dot{s} = \ddot{e} + \lambda_P \dot{e} + \lambda_I e &= -\left(\ddot{l}_i - \ddot{l}_{di}\right) - \lambda_P\left(\dot{l}_i - \dot{l}_{di}\right) - \lambda_I(l_i - l_{di}) \\ &= -\left(\frac{b_i}{m_i} - \lambda_P\right)\dot{l}_i - \left(\frac{k_i}{m_i} - \lambda_I\right)l_i + \frac{d_{ci} F_{max}}{m_i} + \frac{1}{m_i}(p_{est} + \widetilde{p}) - \ddot{l}_{di} - \lambda_P \dot{l}_{di} - \lambda_I l_{di}\end{aligned} \quad (23)$$

Then, substituting Equation (23) into Equation (22) with Equation (20) also taken into account yields:

$$\begin{aligned}\dot{V} = &-m_i\left(\frac{b_i}{m_i} - \lambda_P\right)\dot{l}_i s - m_i\left(\frac{k_i}{m_i} - \lambda_I\right)l_i s + d_{ci} F_{max} s + (p_{est} + \widetilde{p})s \\ &-m_i\left(\ddot{l}_{di} + \lambda_P \dot{l}_{di} + \lambda_I l_{di}\right)s = -(F_{max}\eta \operatorname{sgn}(s) - \widetilde{p})s = -(F_{max}\eta|s| - \widetilde{p}s)\end{aligned} \quad (24)$$

If the gain is designed to meet the condition

$$\eta > \frac{|\vec{p}|}{F_{max}} - \frac{\varepsilon}{F_{max}} \quad (25)$$

where $\varepsilon > 0$ is an arbitrary constant, then it follows that for $|s| \neq 0$,

$$\eta|s| > \frac{|\vec{p}|}{F_{max}}|s| + \frac{\varepsilon}{F_{max}}|s| \quad (26)$$

$$\eta|s| - \frac{\widetilde{p}}{F_{max}}s > \frac{\varepsilon}{F_{max}}|s| \quad (27)$$

$$F_{max}\eta|s| - \widetilde{p}s > \varepsilon|s| \quad (28)$$

$$-(F_{max}\eta|s| - \widetilde{p}s) < -\varepsilon|s| \quad (29)$$

Hence, considering Equations (24) and (29), one can derive that

$$\dot{V} < -\varepsilon|s| \quad (30)$$

It can be concluded that the states can reach the switching surface $s = 0$ in finite time. Equation (30) also ensures that the states will be confined to the surface $s = 0$ for all future time, because leaving the surface requires $\dot{V}$ to be positive, which is impossible as the above inequality implies. Thus, the switching variable $s \to 0$ as $t \to \infty$. According to the definition of s, we can conclude that the tracking error satisfies $\lim_{t \to \infty} e(t) = 0$ and $\lim_{t \to \infty} \dot{e}(t) = 0$ and that $l_i \to l_{di}$ and $\dot{l}_i \to \dot{l}_{di}$ as $t \to \infty$. Therefore, the SMC controller guarantees a zero steady-state tracking error [30].

Due to the discontinuity of the sign function, the control input may produce chattering. To reduce this phenomenon, the boundary layer technique was used by replacing the signum function by hyperbolic tangent function. Hence, the proposed control law in Equation (20) using a hyperbolic function and the proposed dynamic correction gives the following:

$$d_{ci} = \frac{m_i}{F_{max}}\left(\frac{b_i}{m_i} - \lambda_P\right)\dot{l}_i + \frac{m_i}{F_{max}}\left(\frac{k_i}{m_i} - \lambda_I\right)l_i - \frac{1}{F_{max}}p_{est} + \frac{m_i}{F_{max}}\left(\ddot{l}_{di} + \lambda_P\dot{l}_{di} + \lambda_I l_{di}\right) \\ -\eta \tanh\left(\dot{e} + \lambda_P e + \lambda_I \int e\, dt\right) \quad (31)$$

The equivalent control scheme for Equation (31) is presented in Figure 6.

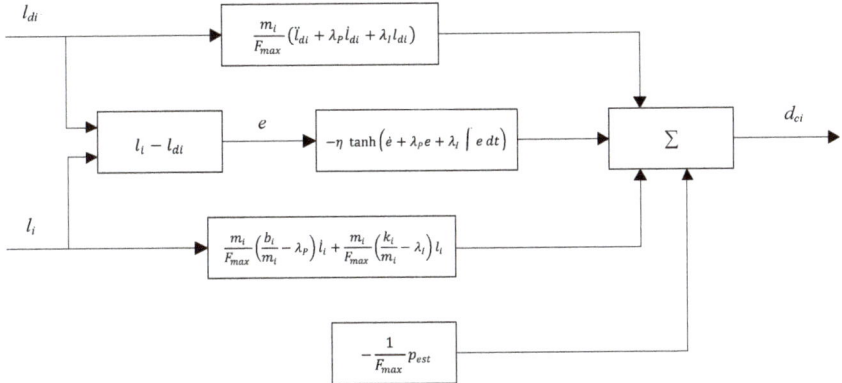

Figure 6. SMC scheme. The PMW outputs (d_{ci}) are calculated with the actuators position feedbacks (l_i) and the desired positions (l_{di}) as inputs.

4. Experimental Results

In this section, the performance of the proposed SMC strategy is analysed experimentally. In order to evaluate the SMC strategy, both accuracy and power consumption were compared with the native PI control. For each analysis, graphs and average data were used to analyse the system behaviour. NI MyRIO hardware system and NI LabVIEW 2017 software (National Instruments, Austin, TX, USA) were used to control the Acrome Stewart platform (Acrome, Istanbul, Turkey). Six linear DC actuators drove the platform and provided length feedbacks analogically. The MyRIO PWM output was applied as a voltage to the DC motor via dual H-bridge motor drivers, which actuated as current amplifiers.

The monitoring and control loops were executed in the Xilinx Z-7010 RT processor at 1 kHz. The PWM output, actuator lengths and setpoint data were registered at 50 Hz. The SMC gains were adjusted experimentally: several trials were carried out to tune the control gains adequately and obtain the optimal response of the controller. Low gain values produced high error and a slow system response. On the other hand, it was observed that using control gains that were too high led to chattering and an unstable system response. The PI original gains and the SMC-adjusted gains are displayed in Table 1.

Table 1. Control gains.

PI	SMC
$P = 0.04$	$\lambda_P = 0.4$
$I = 0.02$	$\lambda_I = 0.02$

In order to perform the surface inspection, the platform must follow the fixed sweep path schematically marked in black in Figure 2b, which corresponds to the trajectory shown in Figure 7a. In this study, the first two sweeps were analysed to limit the amount of data. The trajectory firstly requires a displacement along the x axis at constant speed, and then, a second displacement in the opposite x direction at a different z level. As mentioned in Section 2, the positioning must be accurate, and particularly, so must the distance between the surface and the camera lens (z axis). Variations lower than +/−1 mm were required to obtain valid inspection images. By using the inverse kinematics, the required leg length setpoint profile was calculated from the inspection trajectory, which is shown in Figure 7b. Because the platform moved along the y axis, each pair of legs moved with the same profile with the following combinations: l_1–l_6, l_2–l_5 and l_3–l_4.

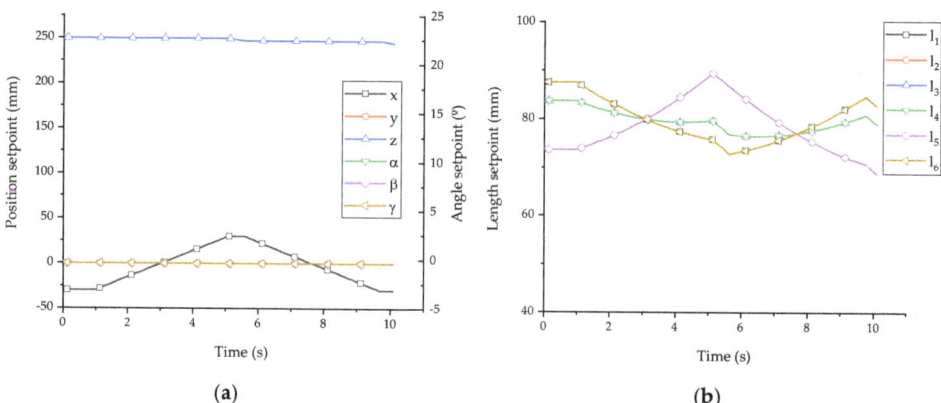

Figure 7. Trajectories graphs: (a) Stewart platform absolute position and orientation (b) Actuator lengths.

To evaluate the positioning accuracy of each control strategy, both the trajectory error and length error were analysed. The absolute position and angles measurements were not available, therefore the forward kinetic transformation was used to calculate these data from the known actuator lengths. Figure 8a,b shows the length errors in PI and SMC strategies. The PI control produced a smooth actuator response which caused length deviations up to +/− 2.0 mm, as can be observed in Figure 8a. In this figure we can observe that the error in each pair of EMAs l_1–l_6, l_2–l_5 and l_3–l_4 is similar. The SMC provided a better performance and reduced the maximum actuator length error to less than +/− 0.5 mm, as depicted in Figure 8b.

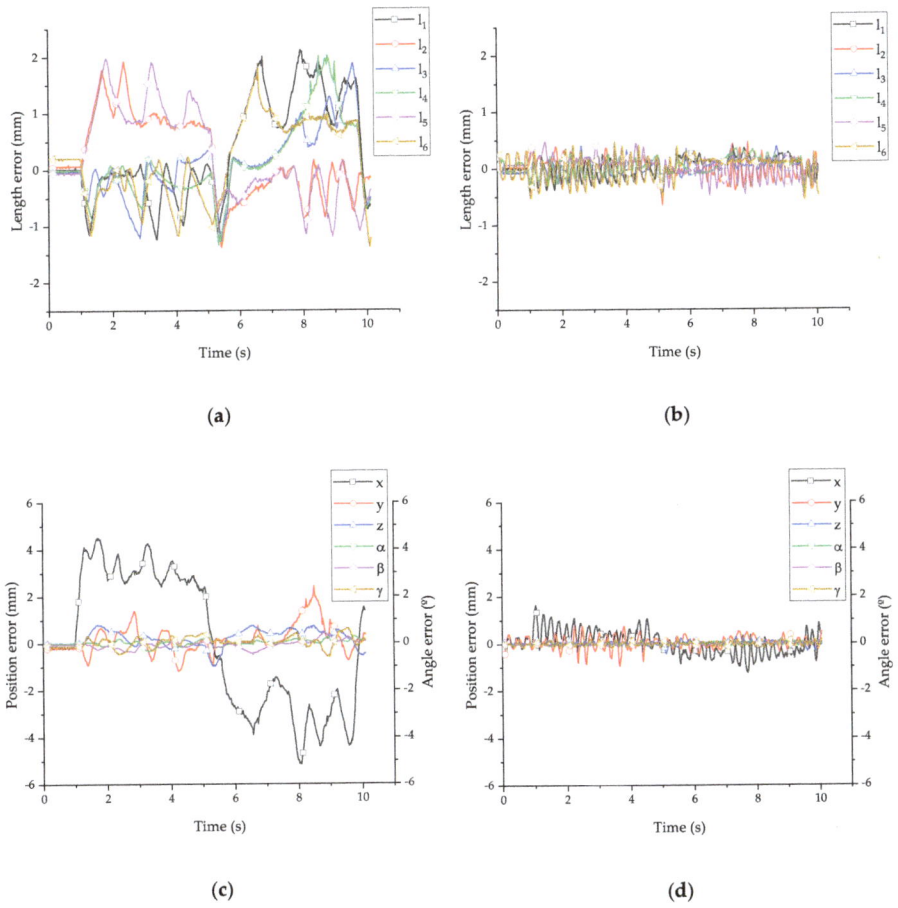

Figure 8. Positioning error graphs: (**a**) proportional integral (PI) length errors; (**b**) SMC length errors graph; (**c**) PI trajectory errors graph; (**d**) SMC trajectory errors graph.

PI compared with the SMC shows a similar response during the first 0.5 s, when a continuous setpoint is required. Under those conditions, the integral gain in both control strategies reduced the error successfully. PI command was stable due to the low proportional gain, whereas SMC corrected the actuator positions continuously. Nevertheless, the accumulated error in both strategies was practically the same during this fixed period of time, as shown in Figure 9. Afterwards, as the setpoint commands changed and EMA lengths variations were required, SMC effectively controlled the EMA lengths within an error of less than 0.5 mm, proving the great response of this control strategy. On the contrary, using the PI EMAs errors increased up to 2 mm: the PI control does not provide a proper PWM command to achieve the EMAs setpoints. The PI control showed a slight difference between the upper error (+2 mm) and lower error (−1.5 mm). This deviation was caused by the platform weight, which helps to shorten the EMAs downwards, but impedes their extension upwards.

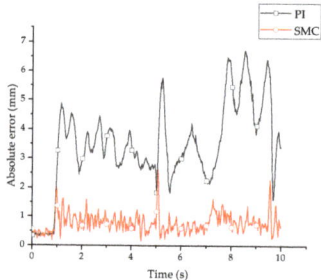

Figure 9. Absolute error comparison between the two control schemes graph.

The errors in the EMAs mentioned above caused deviations in the positioning of the Stewart platform. Figure 8c,d shows that both control strategies followed the desired trajectory. However, only SMC achieved a positioning y error of between +/− 1 mm. Additionally, the positioning error along the x and z axes was considerably higher when using the PI control. Hence, the PI control, which is not accurate enough, caused most of the recorded images to be defocused. In both strategies, the x and z axis deviation was remarkably smaller than in the y axis, mainly because the movement was in the y direction in the plane x/y. Because each pair of actuators was symmetrically placed along the plane x/y, the control in the z axis was more precise and deviations were compensated between the pairs of actuators. Although SMC presented deviations higher than 1 mm in the x axis, the NDI exhibited a sharp image quality.

Another approach to evaluate the system accuracy is to compare the accumulated absolute EMA error, which is the sum of the absolute error of each actuator. Figure 8 represents the absolute error comparison between PI and SMC. Whereas the PI average error is 3.45 mm, the SMC average error is 0.78 mm. Only when no dynamic response of the EMAs was required, were the PI and SMC absolute errors practically the same. Hence, it may be concluded that the accuracy provided by the SMC strategy is remarkably higher.

To analyse the power consumption, its value was estimated from the duty cycle commands. Once the duty cycle and the actuator maximum consumption were known, the power of each actuator was calculated. Figure 10 shows that the PI control required a smooth power variation. The SMC, as a nonlinear control, required a more demanding peak–valley consumption. In any case, the average power consumption for both control strategies were quite similar: PI = 9.41 watts and SMC = 9.62 watts.

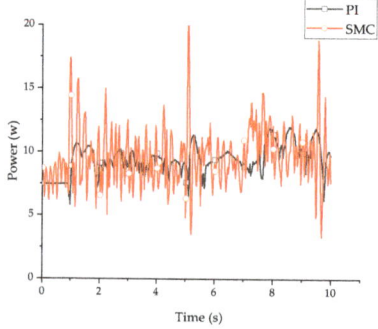

Figure 10. Power consumption comparison between the two control schemes graph.

5. Conclusions

The use of Stewart platforms as NDI orientation apparatus offers six degrees of freedom, high rigidity, a high payload-to-weight-ratio, and low moving inertia. The aeronautical inspection application presented in this paper requires highly precise positioning, otherwise, non-desirable blurred imaging is obtained affecting the quality of the inspection system. A new control strategy approach, based on the system model analysis, was proposed and validated experimentally as a solution to overcome positioning errors.

The Stewart platform is driven by six electro mechanic actuators that must be coordinated to achieve the desired trajectory and position. The calculation of the kinematics of the mechanism is necessary to obtain the relationship between the position/orientation of the platform and the EMA lengths. Whereas the inverse kinematics can be calculated directly analytically, the forward kinematics require the use of the Newton–Raphson method to find a solution. As an alternative to the native proportional integral (PI) control, which was unable to position the platform accurately enough, a sliding mode control (SMC) system was proposed. The SMC strategy was calculated based on the actuator second-order model with a single element. Interactions between actuators and external disturbances were modelled as perturbations.

The validation, which was performed in the physical technology demonstrator, shows that SMC can be a successful solution to control Stewart platform devices. In comparison with the vendor-provided PI control strategy, the SMC achieves higher performance while executing the actuators commands and desired platform positioning: the average absolute error is reduced from 3.45 mm in PI to 0.78 mm when using the SMC approach presented here. In this range of error, the NDI system succeeds in obtaining sharp images. Although the PI control demands a smoother actuator response, the experimental results prove that the power consumption is similar in both control approaches.

The proposed device was constructed to perform analyses of small plain coupons; however this control scheme can be implemented in bigger parallel platforms and anthropomorphic robotic arms aimed at structure diagnoses of bigger and more complex structures, e.g., aircraft components, car parts, etc. Future research on the topic should be focused on the comparison of SMC with other robust control schemes, such as fuzzy logic controllers, H-infinite and artificial neural networks, and on the development of a hybrid SMC, e.g., fuzzy logic SMC and artificial neural network SMC, to verify whether there is room for improvement.

Author Contributions: Conceptualization, O.B. and C.N.; investigation, J.V. and C.N.; software, J.V. and P.V.; validation, J.V. and P.V.; supervision, I.C. and O.B.; writing—original draft, J.V.; writing—review and editing, J.V., I.C. All authors have read and agreed to the published version of the manuscript.

Funding: This research was funded by the Basque Government through the project SMAR3NAK (ELKARTEK KK-2019/00051), by the Ministerio de Economía y Competitividad (RTI2018-094669-B-C31) and by Aernnova and the Diputación Foral de Álava (DFA) through the project CONAVAUTIN 2 (Collaboration Agreement).

Acknowledgments: The authors wish to express their gratitude to AERNNOVA, the DFA, the Basque Government and UPV/EHU for supporting this work.

Conflicts of Interest: The authors declare no conflict of interest.

References

1. Matar, G. Hexapod initiative—Configurable manufacturing. *Comput. Ind.* **1999**, *39*, 71–78. [CrossRef]
2. Olma, S.; Kohlstedt, A.; Traphöner, P.; Jäker, K.P.; Trächtler, A. Observer-based nonlinear control strategies for Hardware-in-the-Loop simulations of multiaxial suspension test rigs. *Mechatronics* **2018**, *50*, 212–224. [CrossRef]
3. Zhang, T.; Zhang, A. Robust finite-time tracking control for robotic manipulators with time delay estimation. *Mathematics* **2020**, *8*, 165. [CrossRef]
4. Kizir, S.; Bingul, Z. Position Control and Trajectory Tracking of the Stewart Platform. *Ser. Parallel Robot Manip. Kinemat. Dyn. Control Optim.* **2012**, *1965*. [CrossRef]

5. Kumar, P.R.; Chalanga, A.; Bandyopadhyay, B. Smooth integral sliding mode controller for the position control of Stewart platform. *ISA Trans.* **2015**, *58*, 543–551. [CrossRef]
6. Karimi, D.; Nategh, M.J. Kinematic nonlinearity analysis in hexapod machine tools: Symmetry and regional accuracy of workspace. *Mech. Mach. Theory* **2014**, *71*, 115–125. [CrossRef]
7. Nguyen, C.; Antrazi, S.; Zhou, Z.L. Analysis and design of a six-degree-of-freedom Stewart platform-based robotic wrist. *Comput. Electr. Eng.* **1991**, *17*, 191–203. [CrossRef]
8. Huang, Y.; Pool, D.M.; Stroosma, O.; Chu, Q.P. Incremental Nonlinear Dynamic Inversion Control for Hydraulic Hexapod Flight Simulator Motion Systems. *IFAC-PapersOnLine* **2017**, *50*, 4294–4299. [CrossRef]
9. Lin, H.; McInroy, J.E. Disturbance attenuation in precise hexapod pointing using positive force feedback. *Control Eng. Pract.* **2006**, *14*, 1377–1386. [CrossRef]
10. Keshtkar, S.; Keshtkar, J.; Poznyak, A. Adaptive sliding mode control for solar tracker orientation. *Proc. Am. Control Conf.* **2016**, *2016*, 6543–6548. [CrossRef]
11. Davliakos, I.; Papadopoulos, E. Model-based control of a 6-dof electrohydraulic Stewart-Gough platform. *Mech. Mach. Theory* **2008**, *43*, 1385–1400. [CrossRef]
12. Ting, Y.; Li, C.C.; Jar, H.C.; Kang, Y. Task-space measurement and control for a 6DOF Stewart nanoscale platform. In Proceedings of the 2007 IEEE International Conference on Mechatronics and Automation ICMA 2007, Harbin, China, 5–8 August 2007; pp. 174–179. [CrossRef]
13. Fraguela, L.; Fridman, L.; Alexandrov, V.V. Output integral sliding mode control to stabilize position of a Stewart platform. *J. Frankl. Inst.* **2012**, *349*, 1526–1542. [CrossRef]
14. Chen, W.; Xu, T.; Liu, J.; Wang, M.; Zhao, D. Picking Robot Visual Servo Control Based on Modified Fuzzy Neural Network Sliding Mode Algorithms. *Electronics* **2019**, *8*, 605. [CrossRef]
15. Usamentiaga, R.; Venegas, P.; Guerediaga, J.; Vega, L.; Molleda, J.; Bulnes, F.G. Infrared thermography for temperature measurement and non-destructive testing. *Sensors* **2014**, *14*, 12305–12348. [CrossRef] [PubMed]
16. Harib, K.; Srinivasan, K. Kinematic and dynamic analysis of Stewart platform-based machine tool structures. *Robotica* **2003**, *21*, 541–554. [CrossRef]
17. Dhingra, A.K.; Almadi, A.N.; Kohli, D. A gröbner-sylvester hybrid method for closed-form displacement analysis of mechanisms. *J. Mech. Des. Trans.* **2000**, *122*, 431–438. [CrossRef]
18. Umar, A.; Shi, Z.; Khlil, A.; Farouk, Z.I.B. Developing a new robust swarm-based algorithm for robot analysis. *Mathematics* **2020**, *8*, 158. [CrossRef]
19. Xi, F.; Sinatra, R. Inverse dynamics of hexapods using the natural orthogonal complement method. *J. Manuf. Syst.* **2002**, *21*, 73–82. [CrossRef]
20. Chen, J.; Lan, F. Instantaneous stiffness analysis and simulation for hexapod machines. *Simul. Model. Pract. Theory* **2008**, *16*, 419–428. [CrossRef]
21. Akdağ, M.; Karagülle, H.; Malgaca, L. An integrated approach for simulation of mechatronic systems applied to a hexapod robot. *Math. Comput. Simul.* **2012**, *82*, 818–835. [CrossRef]
22. Omran, A.; El-Bayiumi, G.; Bayoumi, M.; Kassem, A. Genetic algorithm based optimal control for a 6-dof non redundant stewart manipulator. *Int. J.* **2008**, *2*, 73–79.
23. Velasco, J.; Barambones, O.; Calvo, I.; Zubia, J.; de Ocariz, I.S.; Chouza, A.; Saez de Ocariz, I.; Chouza, A. Sliding Mode Control with Dynamical Correction for Time-Delay Piezoelectric Actuator Systems. *Materials* **2019**, *13*, 132. [CrossRef] [PubMed]
24. Le Nhu Ngoc Thanh, H.; Vu, M.T.; Mung, N.X.; Nguyen, N.P.; Phuong, N.T. Perturbation observer-based robust control using a multiple sliding surfaces for nonlinear systems with influences of matched and unmatched uncertainties. *Mathematics* **2020**, *8*, 1371. [CrossRef]
25. Florez, F.; de Córdoba, P.F.; Higón, J.L.; Olivar, G.; Taborda, J. Modeling, simulation, and temperature control of a thermal zone with sliding modes strategy. *Mathematics* **2019**, *7*, 503. [CrossRef]
26. Kim, N.I.; Lee, C.W. High speed tracking control of Stewart platform manipulator via enhanced sliding mode control. In Proceedings of the 1998 IEEE International Conference on Robotics and Automation (Cat. No.98CH36146), Leuven, Belgium, 16–20 May 1998; Volume 3, pp. 2716–2721.
27. Kumar, P.R.; Chalanga, A.; Bandyopadhyay, B. Position control of Stewart platform using continuous higher order sliding mode control. In Proceedings of the 2015 10th Asian Control Conference (ASCC), Kota Kinabalu, Malaysia, 31 May–3 June 2015; pp. 1–6. [CrossRef]
28. Flottmeier, S.; Olma, S.; Trächtler, A. Sliding mode and continuous estimation techniques for the realization of advanced control strategies for parallel kinematics. *IFAC Proc. Vol.* **2014**, *19*, 182–190. [CrossRef]

29. Bo, W.; Yanlang, D.; Shenglin, W.; Dongguang, X.; Keding, Z. An integral variable structure controller with fuzzy tuning design for electro-hydraulic driving Stewart Platform. In Proceedings of the 2006 1st International Symposium on Systems and Control in Aerospace and Astronautics, Harbin, China, 19–21 January 2006; pp. 941–945. [CrossRef]
30. Li, Y.; Xu, Q. Adaptive Sliding Mode Control With Perturbation Estimation and PID Sliding Surface for Motion Tracking of a Piezo-Driven Micromanipulator. *IEEE Trans. Control Syst. Technol.* **2010**, *18*, 798–810. [CrossRef]

Publisher's Note: MDPI stays neutral with regard to jurisdictional claims in published maps and institutional affiliations.

© 2020 by the authors. Licensee MDPI, Basel, Switzerland. This article is an open access article distributed under the terms and conditions of the Creative Commons Attribution (CC BY) license (http://creativecommons.org/licenses/by/4.0/).

Article

An Efficient Design and Implementation of a Quadrotor Unmanned Aerial Vehicle Using Quaternion-Based Estimator

Eva H. Dulf [1,2], Mihnea Saila [3], Cristina I. Muresan [1,*] and Liviu C. Miclea [1]

[1] Department of Automation, Faculty of Automation and Computer Science, Technical University of Cluj-Napoca, Memorandumului Str. 28, 400014 Cluj-Napoca, Romania; Eva.Dulf@aut.utcluj.ro (E.H.D.); Liviu.Miclea@aut.utcluj.ro (L.C.M.)
[2] Physiological Controls Research Center, Óbuda University, H-1034 Budapest, Hungary
[3] Huisman Equipment BV, Admiraal Trompstraat, 3115HH Schiedam, The Netherlands; msaila@huisman-nl.com
* Correspondence: Cristina.Muresan@aut.utcluj.ro

Received: 21 August 2020; Accepted: 14 October 2020; Published: 18 October 2020

Abstract: The main goal of the research is to design a low-cost, performing quadrotor unmaned aerial vehicle (UAV) system. Because of low cost limits, the performance must be ensured by other ways. The present proposal is a quaternion-based estimator used in the control loop. In order to make the proposed solution easy to be reproduced by the reader, step-by-step instructions are given, including component choices, design, and implementation. Throughout the article, detailed description of the system model is given. The efficacy of the suggested quaternion-based predictive control is evaluated by extended experimental results.

Keywords: unmanned aerial vehicle (UAV); quaternion-based estimator; low-cost design

1. Introduction

Unmanned aerial vehicles (UAV)s have fascinated many researchers and engineers, as they turned out to be accessible in a large variety of applications, not just for costly military operations. Nowadays, UAVs have a broad range of applications, such as: image capturing, aerial recording, military operations, operations in hard-to-reach areas, etc., [1–7]. Along with the development of wireless communications, the control of UAVs has become extremely precise, robust, and even predictive. New research results in the design of UAVs and new application areas include advanced and complex control techniques like robust and adaptive control, algorithms for different flight conditions, fault tolerance, disturbance rejection, etc., [8–13]. All these methods increase the complexity and the cost of the UAV. Because of the extremely alert technological progress registered in the past two decades, the global industrialization and the minimization of the costs of electronic components, countless researchers have shown a high interest in the development of various devices helpful for the society.

One key issue regarding the control of UAVs resides in the estimation of their position. Various methods have been proposed. An efficient method is presented in [14], both from the point of view of the algorithm's performance and from the point of view of using the processing capacity of a microcontroller. Several estimation algorithms are compared in [15], with the results showing that the extended Kalman algorithm is slower in terms of processing time than Madgwick algorithm [15].

The approach of estimating the pitch and roll coordinates presented in [16] constitutes a reference that fits perfectly in the context of the present paper. For the application of the algorithm proposed in the paper, a method of fusion of the data received from an accelerometer, a gyroscope, and a magnetometer was used to estimate, accurately, the position of a flight apparatus. A combination of the extension of the classical Kalman filtering algorithm and the sequential geometric correction is proposed, completely eliminating the magnetic distortions captured by the sensor. In addition, the paper offers a clear and concise comparison between certain popular approaches to the problem of estimating the coordinates of a flight apparatus and the method proposed by the author. Both the improvements and the problems that arise in the implementation of this method are presented.

From all the knowledge resulting from this state-of-art, it can be concluded that UAVs can be made at a relatively lower cost. Simple transducers can be used, as long as this is compensated by a high performance and optimized estimation algorithm. The authors already designed a cheap and easy to use two-rotor equipment, in order to be multiplied for laboratory works [17,18].

Quaternion framework is widely used today to avoid locks and to ensure better computational efficiency [19,20]. The field of application is large, ranging from mechanical systems [21] and medical robots [22] to neural networks [23] and human activities and postures recognition [24], all research papers reporting remarkable results. Quaternions are also used in UAV control with great success. In [25], the authors developed a nonlinear state space model using the quaternion and angular velocity as state variables, which simplifies the system dynamics. The main focus of the research is directed toward the feedback linearization of the model. The simulation results are presented solely for the attitude stabilization task of the quadcopter. A quaternion representation of the attitude of a quadrotor is also used in [26], where various control methods are discussed and compared, such as the PD, LQR, and backstepping methods. Various case scenarios are discussed including noisy data, actuator restrictions, external disturbances. The attitude control of a quadrotor is designed in a quaternion framework in [27], to avoid gimbal lock and for better computational efficiency. The controllers are tuned based on third-order sliding mode control, with a low-pass filter to reduce chattering and a disturbance observer to cover disturbance estimation problems. To ensure the robustness, a disturbance compensation term is also included in the control law. The simulation results show that the proposed method is efficient. In [28], two variants of adaptive state space controllers for attitude stabilization and self-tuning of a quadrotor are proposed. The effectiveness of the approach is demonstrated through simulations that use a quaternion-based nonlinear dynamic model of a quadrotor. A quaternion representation of the attitude of a quadrotor is also used in [29], where a quaternion-energy-based control law is defined as a Lyapunov function, with the control laws described with unit quaternions and their axis-angle representation. Various simulation and experimental results are presented. Unit quaternions are also used in [30] to describe a simple yet complete dynamic model for the rotational and translational dynamics of unmanned aerial vehicles, whereas dual quaternions are explained and used for robotic systems with multiple rotations and translations. An unmanned aerial vehicle described with unit quaternions is presented in [31]. In this case, a quaternion-passivity-based control is derived. The experimental results and numerical simulations validate the results. Intermediary quaternions are used in the design of a backstepping control technique with integral properties in [32]. Compared to classical quaternions, the proposed approach has also the advantage that one specific orientation corresponds to only one intermediary quaternion, which helps coping with the unwinding phenomenon. Numerical simulations, as well as experimental tests, are presented. The robustness of the algorithm is also tested during the numerical simulations only. In [33], a quaternion-based guidance law is proposed which feeds into an attitude control system based on a PD+ control law. A quaternion control scheme for a quadrotor is also proposed in [34]. An attitude control algorithm is developed to stabilize the vehicle's heading and an additional position control law for stabilization of the vehicle in all states. In this case also, numerical and experimental results are presented to validate the approach. An advanced control scheme, also based on quaternions, is presented in [35] for the attitude control of a quadrotor. Here, both the model and the proportional squared control algorithm are implemented in the quaternion space. Extended simulation results are included to demonstrate the efficacy of the suggested novel approach. Quaternions for attitude control are also used in [36], where a quaternion multiplicative formula is proposed to obtain the change of the attitude angle of a quadrotor. Only some practical solutions are presented.

Other recent significant results in UAV control includes more complex structures or calculus. In [37], a control structure based on a hierarchical scheme is proposed, consisting of an energy-based control to stabilize the vehicle translational dynamics and to attenuate the payload oscillation and a nonlinear state feedback controller based on a linear matrix inequality (LMI) to control the quadrotor rotational dynamics. The authors of [38] propose a neuroadaptive integral robust controller, while [39] discusses the dynamic motion planning and control of an UAV using Direct and the Second Method of Lyapunov. An interesting, but complex solution is proposed in [40], where the dynamic system is divided in two subsystems driven by the translational and the rotational dynamics, based on a linear parameter-varying model.

All these approaches have in common the use of advanced control algorithms, with the major drawback of requiring expensive hardware for implementation purposes. Thus, the main objective of the present work is to design and implement a low-cost, easy to use quadrotor UAV, accessible for any user. A quaternion-based estimator is proposed, similarly to existing research studies. However, in terms of the proposed control strategy, the classical PID controller is used, instead of advanced control algorithms. In this way, the implementation of the control strategy is simplified, which triggers the possibility of using low-cost devices for measurement and control. The final control structure includes four controllers, one for each direction of movement. Step by step design and implementation details are presented in order to be easily reproduced by the reader. Using simulation and experimental data, the proposed method is validated. The results show that similar closed loop performance can be achieved using our proposed approach, compared to other more advanced control strategies. The major advantage is that using our proposed method, these results are achieved using a low-cost UAV with a simple, yet efficient control strategy. The novelty of this work consists, thus, in a quaternion-based estimator and classical Proportional-Integral-Derivative (PID) control strategy, implemented using low-cost microcontroller and sensors. For the proposed remote control, performances are imposed in terms of rejecting a moderate range of disturbances and filtering sensor noisy signals.

The rest of this paper is organized as follows. The materials and methods used are presented in the next section. The resulting quadrotor UAV prototype, along with experimental data, is detailed in Section 3. Finally, conclusions are presented in Section 4.

2. Materials and Methods

From construction point of view, the system includes the following elements: plastic skeleton for the flight apparatus; support for the electrical circuit of the remote control; electrical circuits; ATmega32U4 and ATmega 328 microcontrollers; four DC motors; four electronic velocity controllers; four propellers; two wireless remote communication modules and a position detection module.

The main aspects of the flight apparatus described in this work are defined by: the number of engines, the position of the support arms, the mass and the center of gravity of the whole assembly. The arms are mounted in "X," to allow easy change of direction, and the center of gravity is fixed at the intersection of the axes of the arms. The change of direction is facilitated by the control of the angular velocity of the engines. The motors are positioned as follows: two motors on one diagonal are rotated in the same direction, while the remaining two motors on the other diagonal rotate in the opposite direction. Viewed as a whole, the system is composed of a four-arm flight apparatus mounted in "X" and a remote control that provides references to the control circuit located on the quadrotor. They communicate via the UART protocol, using two RF transmission and reception modules. Two-way data exchanges are made between the quadrotor and the remote control, so both items send data and await receipt.

Regarding the mechanical design of the system, a variety of computer-aided design environments could be used to create 3D drawings and model the parts necessary for the physical realization of the system. In the present work AutoCAD and SolidWorks were adopted. In addition, Ultimaker Cura–a G code generator and a 3D printer that could correctly interpret the generated code—was operated to create the remote control.

After choosing the components, measuring their dimensions and making the connections, an electrical scheme could be conceived. The present practice used a CAD/CAM environment provided by Autodesk, called Eagle.

The device is designed in such way that the center of gravity coincides with the geometric one, also serving as the center of the coordinate system attached to the quadrotor. This coordinate system describes the relative movements of the flight apparatus to a fixed coordinate system, with an axis perpendicular to the earth's surface. The other two axes of the fixed coordinate system can be chosen so as to coincide with cardinal points whose axes are perpendicular (for example north-east or south-west).

Like any aerial vehicle, this system has also six degrees of freedom, meaning three movements of translation and three of rotation. All these movement possibilities are strongly dependent on the velocity and implicitly the angular velocity of the four engines. Therefore, depending on these aspects, the following kinetic forces and moments developed and applied to the quadrotor can be distinguished: the altitude advance, the gyroscopic effect, the yaw moment, the pitch moment, the roll moment and, of course, the force of gravitational attraction. The increase or decrease of altitude is possible by simultaneously increasing or decreasing the velocity of all engines. In order to maintain a constant altitude it is necessary to drive the engines at the same velocity, each developing the same angular velocity. Unlike the altitude movement, the kinetic yaw, pitch, and roll momenta are obtained by differentiating the engine velocity. The yaw moment, or rotation around the vertical axis, is obtained by simultaneously increasing the velocity of two motors rotating in the same direction. Depending on the chosen engine group, the flight apparatus will rotate clockwise or trigonometrically.

2.1. Quaternion-Based Estimator

In order to obtain the orientation angles and to facilitate the calculus, two representations can be used, namely: Euler angles and quaternions.

Quaternions are used to express the orientation of a coordinate system to a reference system [41]. Given an angle of rotation Ψ about the axis $\hat{r}$, an orientation of the coordinate system B can be represented with respect to the system A as follows [41]:

$$^A_B\hat{q} = [q_1 q_2 q_3 q_4] = \left[\cos\frac{\Psi}{2} - \hat{r}_x\sin\frac{\Psi}{2} - \hat{r}_y\sin\frac{\Psi}{2} - \hat{r}_z\sin\frac{\Psi}{2}\right] \tag{1}$$

The terms $\hat{r}_x, \hat{r}_y, \hat{r}_z$ represent the components of the unity vector $\hat{r}$ of the reference system A.

A very important advantage presented by this angle expression method is that the product of two quaternions $^C_D\hat{q}$ and $^D_E\hat{q}$ represents the orientation of the system E with respect to the reference system C.

Moreover, the orientation described by a quaternion $^A_B\hat{q} = [q_1\ q_2\ q_3\ q_4]$ can be expressed by the rotation matrix ^A_BR, representing the rotation of the coordinate system B with respect to the reference system A. The dependence between the quaternion terms and the rotation matrix is presented in Equation (2) [16,20].

$$^A_BR = \begin{bmatrix} 2q_1^2 + 2q_2^2 - 1 & 2(q_1q_4 + q_2q_3) & 2(q_2q_4 - q_1q_3) \\ 2(q_2q_3 - q_1q_4) & 2q_1^2 + 2q_3^2 - 1 & 2(q_1q_2 + q_3q_4) \\ 2(q_1q_3 + q_2q_4) & 2(q_3q_4 - q_1q_2) & 2q_1^2 + 2q_4^2 - 1 \end{bmatrix} \tag{2}$$

Although, from a computational point of view, obtaining orientation using quaternions is more efficient, they are hard to interpret physically. Thus, in order to have a clear picture of the real movement, the orientations expressed by quaternions are transformed into representations using Euler angles. To carry out these transformations, Equations (3)–(5) could be used [21].

$$\Psi = \operatorname{atan}(2q_2q_3 - 2q_1q_4,\ 2q_1^2 + 2q_2^2 - 1) \tag{3}$$

$$\theta = \arcsin(2q_1q_3 + 2q_2q_4) \tag{4}$$

$$\Phi = \operatorname{atan}(2q_3q_4 - 2q_1q_2,\ 2q_1^2 + 2q_4^2 - 1) \tag{5}$$

In order to obtain the real values of the angles, a sensor with 9 degrees of freedom was used, consisting of an accelerometer, a gyroscope, and a magnetometer. The sensor used is from the MPU9250 family and communicates with the microcontroller via the I2C interface, at a frequency of 400 kHz. The I2C protocol is a very popular data transmission protocol, due to the multitude of advantages it presents [42]. For data filtering and estimating the orientation of the aerial vehicle, the quaternion representation described above was used. With a physical interpretation much closer to reality, the data provided by the gyroscope are filtered and estimated easily. Thus, the angular positions on the X, Y, and Z axes are arranged in a vector W as described in Equation (6). In addition to these three elements, on the first position in the vector is inserted the term 0 in order to be able to perform quaternion products.

$$W = \begin{bmatrix} 0 & w_x & w_y & w_z \end{bmatrix} \tag{6}$$

$${}^{\text{Ref}}_{\text{Sensor}}\dot{q} = {}^{\text{Ref}}_{\text{Sensor}}q \otimes W \tag{7}$$

With the angular position arranged in the vector W it is possible to compute the orientation change of the coordinate system given by the earth to the coordinate system attached to the UAV. This calculus is represented in Equation (7).

Where the term ${}^{\text{Ref}}_{\text{Sensor}}q$ represents the current orientation of the coordinate system given by the earth to the quadrotor coordinate system.

In order to obtain an orientation of the coordinate system attached to the quadrotor with respect to the reference one, at a time t it is necessary to perform the mathematical operations detailed in Equations (8) and (9).

$$ {}^{\text{Sensor}}_{\text{Ref}}\dot{q}_{\text{gyro},k} = \frac{1}{2}\, {}^{\text{Sensor}}_{\text{Ref}}q_{\text{est},\,k-1} \otimes W \tag{8}$$

$$ {}^{\text{Sensor}}_{\text{Ref}}q_{\text{gyro},k} = {}^{\text{Sensor}}_{\text{Ref}}q_{\text{est},\,k-1} + {}^{\text{Sensor}}_{\text{Ref}}\dot{q}_{\text{gyro},k} T_s \tag{9}$$

where T_S represents the sampling time, and $t = k \cdot T_S$.

Because of the nature of the data from the accelerometer, an optimization problem can be formulated in which the orientation of the sensor ${}^{\text{Sensor}}_{\text{Ref}}\hat{q}$ and, implicitly of the flight system, is given by minimizing the difference between the orientation of the reference system of the earth d_{ref} and that of the sensor, d_{sensor}. The objective function to be minimized of is described by Equations (10) and (11), with the components detailed in (12)–(14).

$$of\left({}^{\text{Sensor}}_{\text{Ref}}\hat{q},\ d_{\text{ref}},\ d_{\text{sensor}}\right) = {}^{\text{Sensor}}_{\text{Ref}}\hat{q}^{*} \otimes d_{\text{ref}} \otimes {}^{\text{Sensor}}_{\text{Ref}}\hat{q} - d_{\text{sensor}} \tag{10}$$

$$of\left({}^{\text{Sensor}}_{\text{Ref}}\hat{q},\ d_{\text{ref}},\ d_{\text{sensor}}\right) = \begin{bmatrix} 2d_{rx}\left(\frac{1}{2} - q_3^2 - q_4^2\right) + 2d_{ry}(q_1q_4 + q_2q_3) + 2d_{rz}(q_2q_4 - q_1q_3) - d_{sx} \\ 2d_{rx}(q_2q_3 - q_1q_4) + 2d_{ry}\left(\frac{1}{2} - q_2^2 - q_4^2\right) + 2d_{rz}(q_1q_2 + q_3q_4) - d_{sy} \\ 2d_{rx}(q_1q_3 + q_2q_4) + 2d_{ry}(q_3q_4 - q_1q_2) + 2d_{rz}\left(\frac{1}{2} - q_2^2 - q_3^2\right) - d_{sz} \end{bmatrix} \tag{11}$$

$${}^{\text{Sensor}}_{\text{Ref}}\hat{q} = \begin{bmatrix} q_1 & q_2 & q_3 & q_4 \end{bmatrix} \tag{12}$$

$$d_{\text{ref}} = \begin{bmatrix} 0 & d_{rx} & d_{ry} & d_{rz} \end{bmatrix} \tag{13}$$

$$d_{\text{sensor}} = \begin{bmatrix} 0 & d_{sx} & d_{sy} & d_{sz} \end{bmatrix} \tag{14}$$

In order to solve this optimization problem, the conjugate gradient method is used, a simple, efficient method that requires a relatively low computing power [43]. However, the conjugate gradient method presents a number of disadvantages related to the algorithm step, μ and the initial point $^{Sensor}_{Ref}\hat{q}_0$. Equations (15) and (16) describe the estimation of future orientation $^{Sensor}_{Ref}\hat{q}_{k+1}$.

$$^{Sensor}_{Ref}\hat{q}_{k+1} = {}^{Sensor}_{Ref}\hat{q}_k - \mu \frac{F}{\|F\|} \quad (15)$$

$$F = \frac{\partial f\left(^{Sensor}_{Ref}\hat{q}, d_{ref}, d_{sensor}\right)}{\partial \left(^{Sensor}_{Ref}\hat{q}, d_{ref}\right)} of\left(^{Sensor}_{Ref}\hat{q}, d_{ref}, d_{sensor}\right) \quad (16)$$

The general cost function of given in (10) can be simplified to be easy to implement even in a low-cost microcontroller. Because of the fact that by convention gravitational acceleration determines only the Z axis of the reference system, this objective function can be expressed as in Equation (17), while the vectors d_{ref} and d_{sensor} are given in (18) and (19).

$$of\left(^{Sensor}_{Ref}\hat{q}, d_{ref}, d_{sensor}\right) = \begin{bmatrix} 2(q_2 q_4 - q_1 q_3) - d_x \\ 2(q_1 q_2 + q_3 q_4) - d_y \\ 2\left(\frac{1}{2} - q_2^2 - q_3^2\right) - d_z \end{bmatrix} \quad (17)$$

$$d_{ref} = [0\ 0\ 0\ 1] \quad (18)$$

$$d_{sensor} = [0\ d_x\ d_y\ d_z] \quad (19)$$

The data obtained from the magnetometer will be processed in the same way as the data obtained from the accelerometer, but with a more laborious processing given by the decomposition of the earth's magnetic field in both a component on the X axis and one on the Z axis. To obtain the next orientations $^{Sensor}_{Ref}\hat{q}_{k+1}$, the same conjugate gradient algorithm will be used. Equation (20) describes the objective function, with the terms detailed in (21) and (22), while Equation (23) presents the gradient of the objective function.

$$of\left(^{Sensor}_{Ref}\hat{q}, m_{ref}, m_{sensor}\right) = \begin{bmatrix} 2m_{rx}\left(\frac{1}{2} - q_3^2 - q_4^2\right) + 2m_{rz}(q_2 q_4 - q_1 q_3) - m_{sx} \\ 2m_{rx}(q_2 q_3 - q_1 q_4) + 2m_{rz}(q_1 q_2 + q_3 q_4) - m_{sy} \\ 2m_{rx}(q_1 q_3 + q_2 q_4) + 2m_{rz}\left(\frac{1}{2} - q_2^2 - q_3^2\right) - m_{sz} \end{bmatrix} \quad (20)$$

$$m_{ref} = [0\ m_{rx}\ 0\ m_{rz}] \quad (21)$$

$$m_{sensor} = [0\ m_{sx}\ m_{sy}\ m_{sz}] \quad (22)$$

$$F = \frac{\partial of\left(^{Sensor}_{Ref}\hat{q}, m_{ref}, m_{sensor}\right)}{\partial \left(^{Sensor}_{Ref}\hat{q}, m_{ref}\right)} of\left(^{Sensor}_{Ref}\hat{q}, m_{ref}, m_{sensor}\right) \quad (23)$$

In order to obtain both a measurement and an accurate estimation of the orientation of the quadrotor, it is necessary to compose the two objective functions presented in Equations (10) (or the simplified form in (17)) and (20). Also, the gradient of both functions will be used to implement the conjugate gradient algorithm for the combination of functions. The composition will be noted with f_{com} and the gradient of this compound function will be noted by F_{com}. In addition, to make the algorithm more efficient, the step μ will be variable and recomputed at each iteration, as shown in Equation (24). The algorithm and the gradient of the new objective function are presented in Equations (25) and (26):

$$\mu_t = \alpha \|^{Actual}_{Ref}\dot{q}_{gyro,k}\| T_S \quad (24)$$

$$^{Sensor}_{Ref}\hat{q}_{com,k} = {}^{Sensor}_{Ref}\hat{q}_{est,k-1} - \mu_t \frac{F_{com}}{\|F_{com}\|} \quad (25)$$

$$F_{com} = \begin{bmatrix} \dfrac{\partial f\left(^{Sensor}_{Ref}\hat{q}_{est,k-1}, d_{Sensor}\right)}{\partial \left(^{Sensor}_{Ref}\hat{q}_{est,k-1}\right)} f\left(^{Sensor}_{Ref}\hat{q}_{est,k-1}, d_{Sensor}\right) \\ \dfrac{\partial f\left(^{Sensor}_{Ref}\hat{q}_{est,k-1}, m_{ref}, m_{Sensor}\right)}{\partial \left(^{Sensor}_{Ref}\hat{q}_{est,k-1}, m_{ref}\right)} f\left(^{Sensor}_{Ref}\hat{q}_{est,k-1}, m_{ref}, m_{Sensor}\right) \end{bmatrix} \tag{26}$$

where α is a constant chosen experimentally to minimize the measurements noise from the accelerometer and magnetometer, T_S is the sampling period, $^{Actual}_{Ref}\dot{q}_{gyro,k}$, represents the orientation given by the gyroscope, computed using Equation (8).

Because of the fusion of measurements from the gyroscope, $^{Sensor}_{Ref}q_{gyro,k}$ and those from the accelerometer and magnetometer $^{Sensor}_{Ref}q_{com,k}$, a weighted, very accurate estimate is obtained, as presented in Equation (27). The weight P_k will be computed at each iteration based on the step μ_t, a control constant β, and the sampling period T_S, as in (28).

$$^{Sensor}_{Ref}\hat{q}_{est,k} = P_k \, ^{Sensor}_{Ref}q_{com,k} + (1 - P_k) \, ^{Sensor}_{Ref}q_{gyro,k} \tag{27}$$

$$P_k = \dfrac{\beta}{\dfrac{\mu_t}{T_E} + \beta} \tag{28}$$

The proposed filter in (27) and (28) ensures an accurate estimation such that $^{Sensor}_{Ref}\hat{q}_{est,k} \to \, ^{Sensor}_{Ref}q_k$ as $k \to \infty$. This can be easily proved using the classical Lyapunov function.

At each iteration, after obtaining the current estimate, Equations (3)–(5) are used to express the Euler angle orientation, which gives a much easier to understand perspective on the movement of the quadrotor.

After obtaining the orientation angles and converting them from quaternions to Euler angles, at each iteration the rotation matrices $R_x(\Phi)$, $R_y(\theta)$, and $R_z(\Psi)$ will be constructed. With these matrices, the rotation matrix of the entire system $R_{xyz}(\Psi, \theta, \Phi)$ is computed, as described in Equations (29)–(32).

$$R_x(\Phi) = \begin{bmatrix} 1 & 0 & 0 \\ 0 & c(\Phi) & -s(\Phi) \\ 0 & s(\Phi) & c(\Phi) \end{bmatrix} \tag{29}$$

$$R_y(\theta) = \begin{bmatrix} c(\theta) & 0 & s(\theta) \\ 0 & 1 & 0 \\ -s(\theta) & 0 & c(\theta) \end{bmatrix} \tag{30}$$

$$R_z(\Psi) = \begin{bmatrix} c(\Psi) & -s(\Psi) & 0 \\ s(\Psi) & c(\Psi) & 0 \\ 0 & 0 & 1 \end{bmatrix} \tag{31}$$

$$R_{xyz}(\Phi, \theta, \Psi) = R_x(\Psi) \cdot R_y(\theta) \cdot R_z(\Phi) =$$
$$\begin{bmatrix} c(\theta)c(\Psi) & s(\Phi)s(\theta)c(\Psi) - c(\Phi)s(\Psi) & c(\Phi)s(\theta)c(\Psi) + s(\Phi)s(\Psi) \\ c(\theta)s(\Psi) & s(\Phi)s(\theta)s(\Psi) + c(\Phi)c(\Psi) & c(\Phi)s(\theta)s(\Psi) - s(\Phi)c(\Psi) \\ -s(\theta) & s(\Phi)c(\theta) & c(\Phi)c(\theta) \end{bmatrix} \tag{32}$$

where $c(\Psi) = \cos(\Psi)$, $s(\Psi) = \sin(\Psi)$, $c(\theta) = \cos(\theta)$, $s(\theta) = \sin(\theta)$, $c(\Phi) = \cos(\Phi)$, $s(\Phi) = \sin(\Phi)$.

2.2. Quadrotor Kinematic and Dynamic Model

In order to establish an efficient mathematical model, as close as possible to the reality, which ensures greater system controllability, it is necessary to use the Equations of Newton classical mechanics and of Euler for angular motions. It is also necessary to take into account both the relative movements of the fixed coordinate system (in this case, the earth), as well as the relative dynamics of the coordinate system attached to the quadrotor. Thus, two vectors, P_p and P_a, will be used, described by Equations (33) and (34). P_p is the vector of the linear and angular positions of the flight system relative to earth, while P_a is the vector of the linear and angular velocities of the quadrotor.

$$P_p = \begin{bmatrix} x & y & z & \Phi & \theta & \Psi \end{bmatrix}^T \tag{33}$$

$$P_a = \begin{bmatrix} u & v & w & p & q & r \end{bmatrix}^T \tag{34}$$

To link these two vectors, the rotation matrix $R_{xyz}(\Phi, \theta, \Psi)$ and a matrix of angular velocity transformations, $T_v(\Phi, \theta)$ is used, derived from the inverse of the derivative of the Euler angle change rate. Thus in Equations (35)–(40) the dependencies between the vectors P_p and P_a are detailed.

$$v_p = \begin{bmatrix} \dot{x} & \dot{y} & \dot{z} \end{bmatrix}^T \tag{35}$$

$$\omega_p = \begin{bmatrix} \dot{\Phi} & \dot{\theta} & \dot{\Psi} \end{bmatrix}^T \tag{36}$$

$$v_a = \begin{bmatrix} u & v & w \end{bmatrix}^T \tag{37}$$

$$\omega_a = \begin{bmatrix} p & q & r \end{bmatrix}^T \tag{38}$$

$$v_p = R_{xyz}(\Phi, \theta, \Psi) \cdot v_a \tag{39}$$

$$\omega_p = T_V(\Phi, \theta) \cdot \omega_a \tag{40}$$

The vectors v_p and ω_p are the derivatives of the linear and angular positions of P_p, while v_a and ω_a are the linear and angular velocities of the vector P_a. The matrix of angular velocity transformations $T_V(\Phi, \theta)$ is constructed as described by Equation (41).

$$T_V(\Phi, \theta) = \begin{bmatrix} 1 & c(\Phi)\tan(\theta) & c(\Phi)\tan(\theta) \\ 0 & c(\Phi) & -s(\Phi) \\ 0 & \frac{s(\Phi)}{c(\theta)} & \frac{c(\Phi)}{c(\theta)} \end{bmatrix} \tag{41}$$

Performing the multiplications leads to the kinematic model:

$$\begin{cases} \dot{x} = uc(\Psi)c(\theta) - v[c(\Phi)s(\Psi) - c(\Psi)s(\Phi)s(\theta)] + w[s(\Phi)s(\Psi) + c(\Phi)s(\Psi)s(\theta)] \\ \dot{y} = uc(\theta)c(\Psi) + v[c(\Phi)c(\Psi) + s(\Phi)s(\theta)s(\Psi)] - w[c(\Psi)s(\Phi) - c(\Phi)s(\Psi)s(\theta)] \\ \dot{z} = -uc(\theta) + vc(\theta)s(\Phi) + wc(\Phi)c(\theta) \\ \dot{\Phi} = p + qs(\Phi)t(\theta) + rc(\Phi)t(\theta) \\ \dot{\theta} = qc(\Phi) - rs(\Phi) \\ \dot{\Psi} = q\frac{s(\Phi)}{c(\theta)} + q\frac{c(\Phi)}{c(\theta)} \end{cases} \tag{42}$$

From Newton's laws, the forces acting on the quadrotor can be determined. These will be denoted with vector F_a and calculated as described in Equations (43) and (44).

$$F_a = m_q(\omega_a \times v_a + \dot{v}_a) \tag{43}$$

$$F_a = \begin{bmatrix} f_x & f_y & f_z \end{bmatrix}^T \tag{44}$$

where m_q denotes the mass of the quadrotor, "×" is the vector product of the linear and angular velocity relative to the quadrotor coordinate system, while $\dot{v}_a$ is the linear acceleration.

Similar to the computation of the force, the angular velocity applied to the quadrotor will also be determined from Euler's Equation. These velocities will be noted with M_a, and are strongly dependent on the inertia matrix I, as it is presented in Equations (45)–(47).

$$M_a = I \cdot \dot{\omega}_a + \omega_a \times (I \cdot \omega_a) \tag{45}$$

$$M_a = \begin{bmatrix} m_x & m_y & m_z \end{bmatrix}^T \tag{46}$$

$$I = \begin{bmatrix} I_x & 0 & 0 \\ 0 & I_y & 0 \\ 0 & 0 & I_z \end{bmatrix} \tag{47}$$

Combining Equations (44) and (46), the dynamic model of the quadrotor relative to its own coordinate system can be expressed as:

$$\begin{cases} f_x = m(\dot{u} + qw - rv) \\ f_y = m(\dot{v} - pw + ru) \\ f_z = m(\dot{w} + pv - qu) \\ m_x = \dot{p}I_x - qrI_y + qrI_z \\ m_y = prI_x + \dot{q}I_y - prI_z \\ m_z = -pqI_x + pqI_y + \dot{r}I_z \end{cases} \tag{48}$$

The forces and velocities described above can also be expressed by Equations (49) and (50).

$$F_a = m_q g R_{xyz}(\Phi, \theta, \Psi)^T \cdot \hat{e}_z - f_p \cdot \hat{e}_3 + f_v \tag{49}$$

$$M_a = \tau_a - g_a + \tau_v \tag{50}$$

In the above expression m_q means the total mass of the quadrotor, g is the gravitational acceleration, $\hat{e}_z$ and $\hat{e}_3$ are the unit vectors on the Z axis of the reference coordinate system, respectively of the coordinate system attached to the quadrotor. The element f_p represents the total propulsion force developed by the engines, and $f_v = \begin{bmatrix} f_{vx} & f_{vy} & f_{vz} \end{bmatrix}^T$ represents the disturbances or forces that are opposed to the rotation of each engine, caused by air currents. τ_a represents the angular velocity generated by the velocity differences of the four motors, while τ_v stands for the angular velocities produced by air currents on each motor, detailed in Equations (51) and (52). g_a are the gyroscope moments caused by the combined velocities of the four motors. Given the fact that the inertia of the motors is negligible compared to the developed force, the gyroscopic moments may be neglected from Equation (50).

$$\tau_a = \begin{bmatrix} \tau_x & \tau_y & \tau_z \end{bmatrix}^T \tag{51}$$

$$\tau_v = \begin{bmatrix} \tau_{vx} & \tau_{vy} & \tau_{vz} \end{bmatrix}^T \tag{52}$$

Replacing these new Equations for forces and velocities, a new dynamic model is obtained:

$$\begin{cases} -m_q gs(\theta) + f_{vx} = m_q(\dot{u} + qw - rv) \\ m_q gc(\theta)s(\Phi) + f_{vy} = m_q(\dot{v} - pw + ru) \\ m_q gc(\theta)c(\Phi) + f_{vz} - f_p = m_q(\dot{w} + pv - qu) \\ \tau_x + \tau_{vx} = \dot{p}I_x - qrI_y + qrI_z \\ \tau_y + \tau_{vy} = prI_x + \dot{q}I_y - prI_z \\ \tau_z + \tau_{vz} = -pqI_x + pqI_y + \dot{r}I_z \end{cases} \tag{53}$$

In order to control the quadrotor, the dependence between the propulsion force f_p, velocity τ_a, and the motor's angular velocities $\Omega_a = \begin{bmatrix} \Omega_1 & \Omega_2 & \Omega_3 & \Omega_4 \end{bmatrix}$ needs to be introduced in the model, using Equation (54).

$$\begin{cases} f_p = b\left(\Omega_1^2 + \Omega_2^2 + \Omega_3^2 + \Omega_4^2\right) \\ \tau_x = b \cdot l\left(\Omega_1^2 + \Omega_2^2 - \Omega_3^2 - \Omega_4^2\right) \\ \tau_y = b \cdot l\left(\Omega_1^2 + \Omega_4^2 - \Omega_2^2 - \Omega_3^2\right) \\ \tau_z = d\left(\Omega_1^2 + \Omega_3^2 - \Omega_2^2 - \Omega_4^2\right) \end{cases} \tag{54}$$

where b is a propulsion coefficient and d is the aerodynamic resistance coefficient. The term l represents the distance from the center of gravity of the quadrotor to the center of rotation of the engine. This term is equal for all four arms of the quadrotor. In addition, replacing the terms obtained from Equation (54) in (53) leads to a new expression of the dynamic model, given by:

$$\begin{cases} -m_q g\, s(\theta) + f_{vx} = m_q\left(\dot{u} + qw - rv\right) \\ m_q g\, c(\theta)s(\Phi) + f_{vy} = m_q\left(\dot{v} - pw + ru\right) \\ m_q g c(\theta)c(\Phi) + f_{vz} - b\left(\Omega_1^2 + \Omega_2^2 + \Omega_3^2 + \Omega_4^2\right) = m_q\left(\dot{w} + pv - qu\right) \\ b \cdot l\left(\Omega_1^2 + \Omega_2^2 - \Omega_3^2 - \Omega_4^2\right) + \tau_{vx} = \dot{p}I_x - qrI_y + qrI_z \\ b \cdot l\left(\Omega_1^2 + \Omega_4^2 - \Omega_2^2 - \Omega_3^3\right) + \tau_{vy} = prI_x + \dot{q}I_y - prI_z \\ d\left(\Omega_1^2 + \Omega_3^2 - \Omega_2^2 - \Omega_4^2\right) + \tau_{vz} = -pqI_x + pqI_y + \dot{r}I_z \end{cases} \tag{55}$$

This model will be used as predictor in the control structure.

2.3. Quadrotor State Space Model Used for Controller Design

The next step consists in the model design in a state space form, in order to easily apply the controller design methods. Therefore, the state vector $\mathbb{X}$, the input vector u, and the output vector y will be chosen as it is presented below:

$$\mathbb{X} = \begin{bmatrix} \Phi & \theta & \Psi & p & q & r & u & v & w & x & y & z \end{bmatrix}^T \tag{56}$$

$$u = \begin{bmatrix} f_p & \tau_x & \tau_y & \tau_z \end{bmatrix}^T \tag{57}$$

$$y = \begin{bmatrix} \Phi & \theta & \Psi & f_p \end{bmatrix}^T \tag{58}$$

Using this state vector $\mathbb{X}$ and Equations (42) and (55), one can determine the derivative of this state vector, $\dot{\mathbb{X}}$:

$$\begin{cases} \dot{\Phi} = p + qs(\Phi)t(\theta) + rc(\Phi)t(\theta) \\ \dot{\theta} = qc(\Phi) - rs(\Phi) \\ \dot{\Psi} = q\frac{s(\Phi)}{c(\theta)} + r\frac{c(\Phi)}{c(\theta)} \\ \dot{p} = \frac{I_y - I_z}{I_x}qr + \frac{\tau_x + \tau_{vx}}{I_x} \\ \dot{q} = \frac{I_z - I_x}{I_y}pr + \frac{\tau_y + \tau_{vy}}{I_y} \\ \dot{r} = \frac{I_x - I_y}{I_z}pq + \frac{\tau_z + \tau_{vz}}{I_z} \\ \dot{u} = rv - qw - gs(\theta) + \frac{f_{vx}}{m} \\ \dot{v} = pw - ru + gs(\Phi)c(\theta) + \frac{f_{vy}}{m} \\ \dot{w} = qu - pv + gc(\Phi)c(\theta) + \frac{f_{vz} - f_p}{m} \\ \dot{x} = uc(\Psi)c(\theta) - v[c(\Phi)s(\Psi) - c(\Psi)s(\Phi)s(\theta)] + w[s(\Phi)s(\Psi) + c(\Phi)c(\Psi)s(\theta)] \\ \dot{y} = uc(\theta)s(\Psi) + v[c(\Phi)c(\Psi) + s(\Phi)s(\theta)s(\Psi)] - w[c(\Psi)s(\Phi) - c(\Phi)s(\Psi)s(\theta)] \\ \dot{z} = -us(\theta) + vc(\theta)s(\Phi) + wc(\Phi)c(\theta) \end{cases} \tag{59}$$

As can be seen from Equation (59), the system is strongly nonlinear, presenting major problems in the design of a control system based on models. In order to be linearized, a Jacobian matrix is used, at certain chosen equilibrium points. Given that it is desired that in the absence of a command the system be maintained at a fixed point at a predetermined altitude, the equilibrium points are chosen as described below.

$$\mathbb{X}_e = \begin{bmatrix} 0 & 0 & 0 & 0 & 0 & 0 & 0 & 0 & 0 & x_e & y_e & z_e \end{bmatrix}^T \tag{60}$$

$$u_e = \begin{bmatrix} m_q \cdot g & 0 & 0 & 0 \end{bmatrix}^T \tag{61}$$

where $g = 9.8 \, \frac{m}{s^2}$ is the gravitational acceleration and m_q is the total mass of the quadrotor.

Also, since the trigonometric dependencies between the system states do not disappear even after the linearization by the Jacobian method, a preliminary simplification is made. Thus, in order to eliminate the trigonometric functions from the system model, all the values of the sine functions are approximated with their argument, respectively the cosine functions with 1. The approximate model, resulting from the simplification, has the form as described in (62).

$$\begin{cases} \dot{\Phi} = p + q\Phi\theta + r\theta \\ \dot{\theta} = q - r\Phi \\ \dot{\Psi} = q\Phi + r \\ \dot{p} = \frac{I_y - I_z}{I_x} qr + \frac{\tau_x + \tau_{vx}}{I_x} \\ \dot{q} = \frac{I_z - I_x}{I_y} pr + \frac{\tau_y + \tau_{vy}}{I_y} \\ \dot{r} = \frac{I_x - I_y}{I_z} pq + \frac{\tau_z + \tau_{vz}}{I_z} \\ \dot{u} = rv - qw - g\theta + \frac{f_{vx}}{m_q} \\ \dot{v} = pw - ru + g\Phi + \frac{f_{vy}}{m_q} \\ \dot{w} = qu - pv + g + \frac{f_{vz} - f_p}{m_q} \\ \dot{x} = u - v(\Psi - \Phi\theta) + w(\Phi\Psi + \theta) \\ \dot{y} = u\Psi + v(1 + \Phi\theta\Psi) - w(\Phi - \Psi\theta) \\ \dot{z} = -u\theta + v\Phi + w \end{cases} \tag{62}$$

In the state space form, the system is $\dot{\mathbb{X}} = h(\mathbb{X}, u)$. Applying the linearization by the Jacobian matrix method and using the equilibrium points expressed in (60) and (61), the linearized state space system became:

$$\begin{cases} \dot{\mathbb{X}} = A_e \cdot \mathbb{X} + B_e \cdot u \\ y = C \cdot \mathbb{X} \end{cases} \tag{63}$$

with

$$A_e = \left. \frac{\partial h(\mathbb{X}, u)}{\partial \mathbb{X}} \right|_{\substack{\mathbb{X} = \mathbb{X}_e \\ u = u_e}} = \begin{bmatrix} 0 & 0 & 0 & 1 & 0 & 0 & 0 & 0 & 0 & 0 & 0 & 0 \\ 0 & 0 & 0 & 0 & 1 & 0 & 0 & 0 & 0 & 0 & 0 & 0 \\ 0 & 0 & 0 & 0 & 0 & 1 & 0 & 0 & 0 & 0 & 0 & 0 \\ 0 & 0 & 0 & 0 & 0 & 0 & 0 & 0 & 0 & 0 & 0 & 0 \\ 0 & 0 & 0 & 0 & 0 & 0 & 0 & 0 & 0 & 0 & 0 & 0 \\ 0 & 0 & 0 & 0 & 0 & 0 & 0 & 0 & 0 & 0 & 0 & 0 \\ 0 & -g & 0 & 0 & 0 & 0 & 0 & 0 & 0 & 0 & 0 & 0 \\ g & 0 & 0 & 0 & 0 & 0 & 0 & 0 & 0 & 0 & 0 & 0 \\ 0 & 0 & 0 & 0 & 0 & 0 & 0 & 0 & 0 & 0 & 0 & 0 \\ 0 & 0 & 0 & 0 & 0 & 0 & 1 & 0 & 0 & 0 & 0 & 0 \\ 0 & 0 & 0 & 0 & 0 & 0 & 0 & 1 & 0 & 0 & 0 & 0 \\ 0 & 0 & 0 & 0 & 0 & 0 & 0 & 0 & 1 & 0 & 0 & 0 \end{bmatrix} \tag{64}$$

$$B_e = \left.\frac{\partial h(\mathbb{X}, u)}{\partial u}\right|_{\substack{\mathbb{X}=\mathbb{X}_e \\ u=u_e}} = \begin{bmatrix} 0 & 0 & 0 & 1 \\ 0 & 0 & 0 & 0 \\ 0 & 0 & 0 & 0 \\ 0 & \frac{1}{I_x} & 0 & 0 \\ 0 & 0 & \frac{1}{I_y} & 0 \\ 0 & 0 & 0 & \frac{1}{I_z} \\ 0 & 0 & 0 & 0 \\ 0 & 0 & 0 & 0 \\ \frac{1}{m_q} & 0 & 0 & 0 \\ 0 & 0 & 0 & 0 \\ 0 & 0 & 0 & 0 \\ 0 & 0 & 0 & 0 \end{bmatrix} \qquad (65)$$

$$C = \begin{bmatrix} 0 & 0 & 0 & 0 & 0 & 0 & 0 & g & 0 & 0 & 0 & 0 \\ 0 & 0 & 0 & 0 & 0 & 0 & -g & 0 & 0 & 0 & 0 & 0 \\ 0 & 0 & 0 & 0 & 0 & 0 & 0 & 0 & 0 & 0 & 0 & 0 \\ 0 & 0 & 0 & 0 & 0 & 0 & 0 & 0 & -1/m_q & 0 & 0 & 0 \end{bmatrix} \qquad (66)$$

The same model can be described as a system of Equations as indicated in (67).

$$\begin{cases} \dot{\Phi} = p \\ \dot{\theta} = q \\ \dot{\Psi} = r \\ \dot{p} = \frac{\tau_x + \tau_{vx}}{I_x} \\ \dot{q} = \frac{\tau_y + \tau_{vy}}{I_y} \\ \dot{r} = \frac{\tau_z + \tau_{vz}}{I_z} \\ \dot{u} = -g\theta + \frac{f_{vx}}{m_q} \\ \dot{v} = g\Phi + \frac{f_{vy}}{m_q} \\ \dot{w} = \frac{f_{vz} - f_p}{m_q} \\ \dot{x} = u \\ \dot{y} = v \\ \dot{z} = w \end{cases} \qquad (67)$$

2.4. Controller Design

The controller design method uses the linear model of the system (63). Considering that references for the orientation of the quadrotor and for the altitude of flight will be transmitted from the remote controller, and the system inputs depend on the angular velocities of the four engines, a number of four controllers will be implemented for each direction of movement. Each controller can be designed with any tuning algorithm, ensuring the cancellation of the steady state error and a short settling time. An interesting choice is presented for example in [44]. If advanced controller tuning methods are used, the performances could be increased. The idea of the present work is to implement a very low-cost quadrotor, with the simplest control algorithm, but with results comparable with advanced control methods. With this regard a simple PID controller is designed for each rotor, using the classical root locus method [45]. For this method, given the characteristic polynomial of the closed-loop system, the parameters of controller are chosen depending on the location of the poles of the system. Overshoot, settling time and steady state error cancellation are imposed for each controller. Figure 1 illustrates the block diagram of the control strategy chosen for this quadrotor, with the PID blocks detailed in Figure 2. The proposed feedback control requires feedback signals and disturbance identification. To obtain these signals, a sensor with 9 degrees of freedom, consisting of an accelerometer, a gyroscope, and a magnetometer is used. Signals from this sensor must be processed because they suffer from noise disturbance and other drawbacks. For example the gyroscope has a flowing bias. This inconvenient is mitigated by the estimator. Both data filtering and estimating the orientation of the aerial vehicle are realized with the quaternion representation of the estimator (27,28). The nonlinear dynamic model (55) is used as predictor in the control structure.

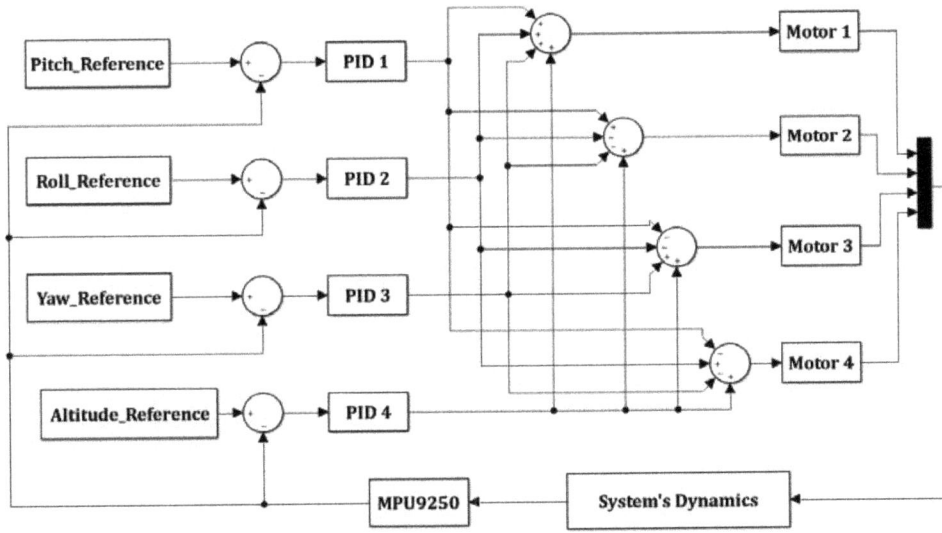

Figure 1. Block diagram of the three-dimensional space orientation control.

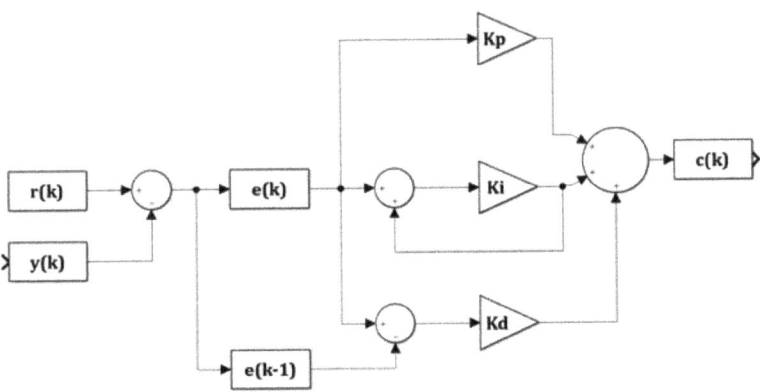

Figure 2. Block diagram of each PID controller.

In Figure 2 the signals are denoted as follows: r(k) is the reference signal at current iteration k; c(k) represents the control signal at this current iteration k; y(k) is the output of the system, measured by sensors at iteration k; e(k) is the error signal at iteration k; K_p, K_i, and K_d represent the proportionality, integration, and derivation constant, respectively. Regarding the angular velocities of the motors, it is necessary to ensure that they behave according to the control signals received from the angular position and altitude controllers. In view of the microcontroller's processing capacity and the relatively large dimensions of the program used to obtain the inclination angles and the control law previously determined, four electronic speed control modules (ESCs) will be used to control the angular velocities of the motors.

The verification of the designed controllers was first performed through a numerical simulation. The non-linear model from Equation (55) was used to carry out all simulations. Several simulation scenarios were adjusted in order to set the simulation closer to reality. Furthermore, some restrictions related to the actuators were applied based on real data measurements. The delay of the actuators was implemented because of the use of the electronic speed controller (ESC). Moreover, sensor noise was implemented to the measured feedback signals. The evaluation of the designed controller was done both in disturbance free, constant disturbance, and real disturbance conditions. In each case the quadrotor has to follow the same trajectory, including takes off, flying from point A to B, and rotation around the Z axis. As quality indicators chosen to discuss the efficiency of the proposed algorithm are the steady state position error, overshoot, and settling time. In all cases the proposed simple control structure exhibit very similar behavior to advanced, expensive solutions.

3. Results

Figure 3 presents the resulted low-cost quadrotor UAV. It has four motors controlled by electronic speed controllers rotating as described in Figure 4. Each motor is mounted on a plastic arm, which in turn is attached to the carbon fiber central structure. All pieces were chosen so that the assembly has the lowest weight and, at the same time, to maintain the condition of the center of gravity described in the previous section.

Figure 3. The unmaned aerial vehicle (UAV) prototype.

Figure 4. Rotational directions.

In each step of the design, the aim was to understand the functionality of each component of the system and to describe the relationships between them using block diagrams. In this regard Figures 5 and 6 detail the block diagrams of each subsystem, highlighting the type of data provided by/for each element.

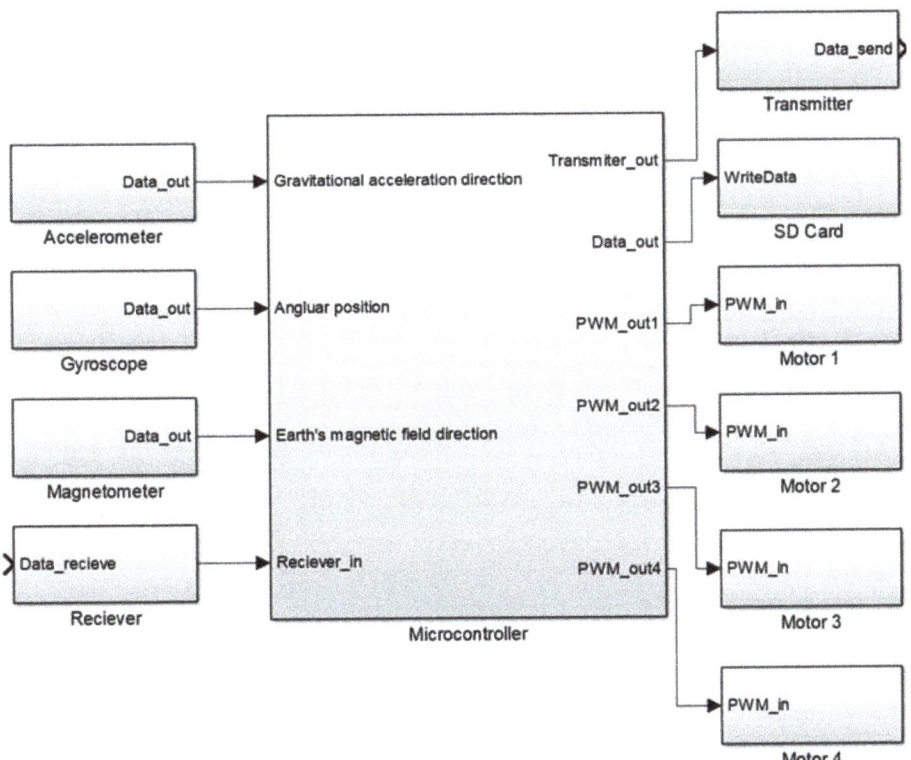

Figure 5. Block scheme of the quadrotor UAV.

Mathematics **2020**, *8*, 1829

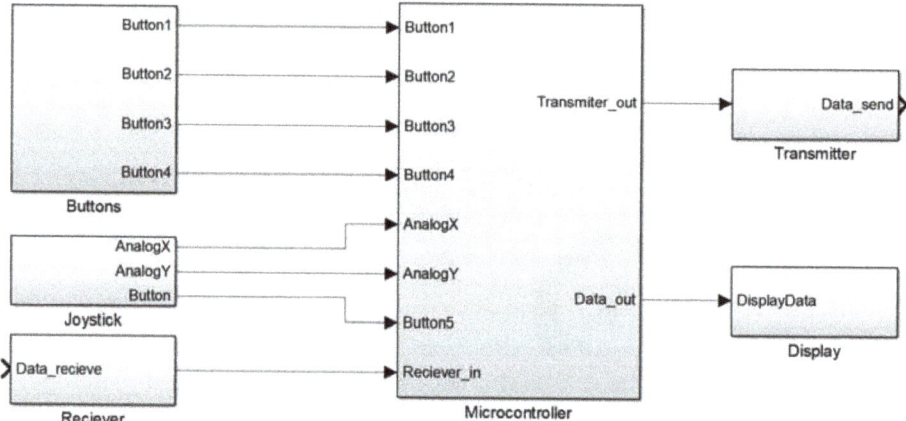

Figure 6. Block scheme of the remote controller.

In accordance with these block schemes and the dimensions imposed by the mechanical elements, a series of electrical components were chosen. These have been selected so that they can achieve the specifications of the desired control, allow flexibility in resolving errors and have low cost. Also, from the point of view of the processing capacity and the number of input and output signals, respectively, a microcontroller was chosen that satisfies these conditions.

The wiring diagram and the implemented UAV system are presented in Figure 7. The corresponding remote controller schemes are in Figure 8, where (a) represents the wiring diagram designed in Eagle, while (b) is the implemented circuit.

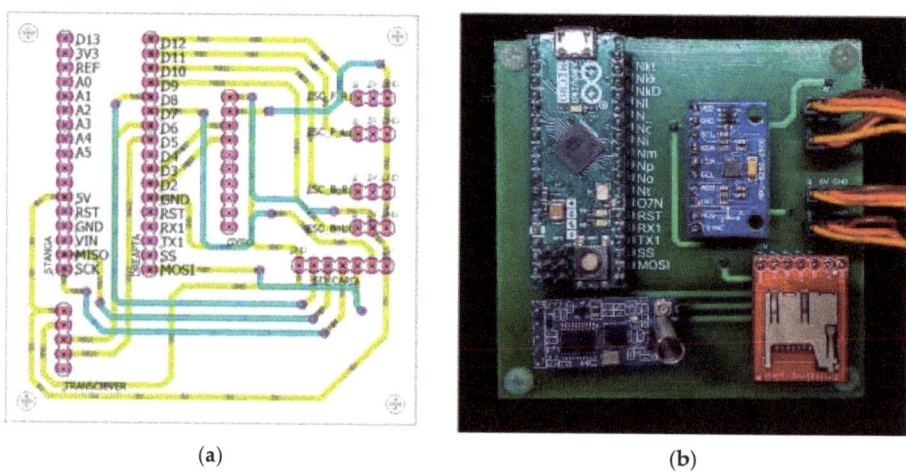

(a)   (b)

Figure 7. The designed (**a**) and implemented (**b**) electronic circuit of UAV.

Measurements were realized without using the developed estimator. Figure 9 presents the raw results of the gyroscope, accelerometer and magnetometer for a linear movement. It can be concluded that in the case of noisy signals, such an approach is not usable in a feedback control structure. It is obvious the necessity of the estimator.

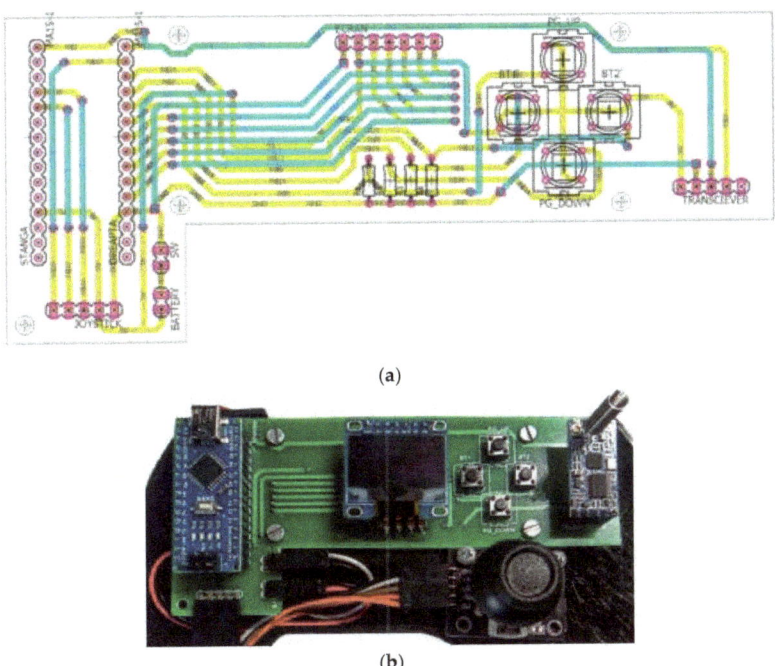

Figure 8. The remote controller: (**a**) the designed and (**b**) the implemented electronic circuit.

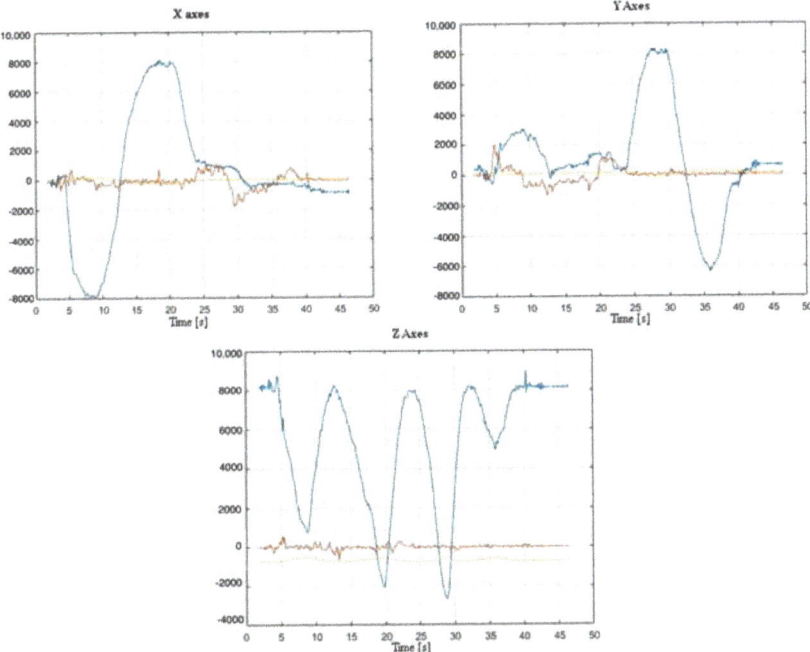

Figure 9. Raw measured output for the three degree of freedom.

The quaternion-based estimation algorithm described in the previous section was implemented on the microcontroller. In order to test the obtained system, a reference sequence of the form: 0, maximum value to the right, maximum value to the left was applied. The obtained results are plotted in Figure 10.

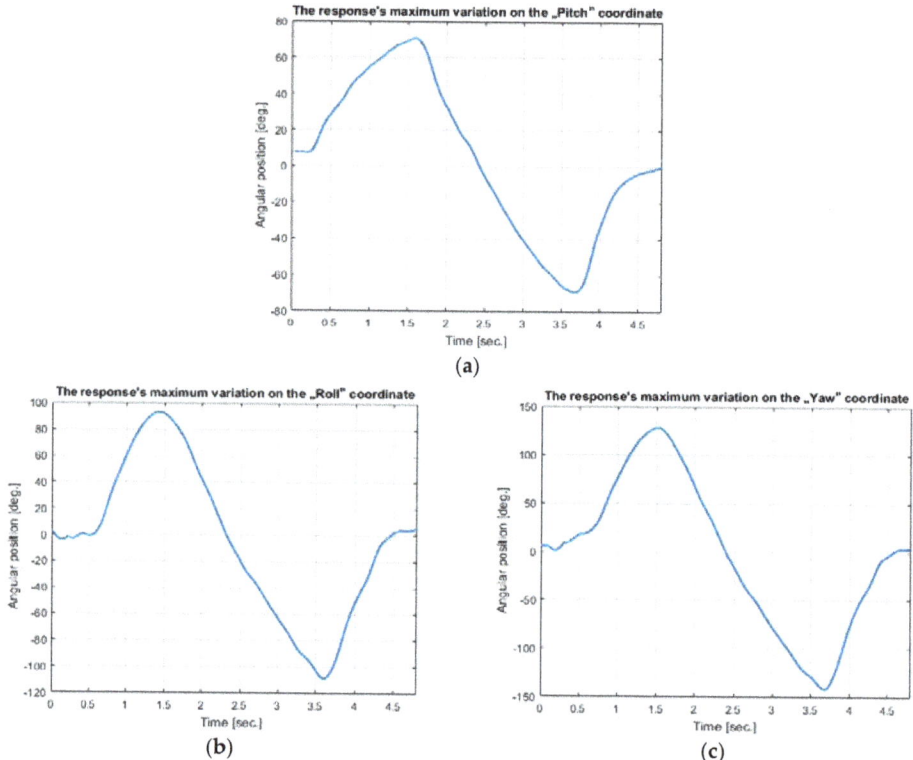

Figure 10. Measured output for the three degree of freedom: (**a**) pitch; (**b**) roll; (**c**) yaw.

The designed nonlinear model was tested, obtaining the results from Figure 11.

The designed control algorithms were also implemented on the microcontroller. The obtained results in the worst-case scenario, windy conditions, are plotted on Figure 12, presenting the response of the closed loop system to a step reference on each of the four directions of movement.

The performance was measured for different operation scenarios, including different step inputs on each axis, wind-free and windy conditions. The results for one of the "classical" scenarios—16° step input for angular position on X and Y axis and 45° on Z axis, 5 m altitude, with relatively high wind speed—are presented in Table 1, highlighting good performances. All these results are comparable with the results of advanced control algorithms in [25–36], without needing expensive hardware equipment. In [26], where the studied quadrotor is similar with our prototype, a LQR controller is used for altitude and a PD controller for position, resulting in a settling time for a step input between 2 and 3.7 s and overshoot 13–20%. Using a backstepping controller combined with the PD, the settling time varies between 2.10 and 3.70 s and the overshoot between 12 and 14%. The LQR controller used both for altitude and position, the settling time are 2.7–3.35 s, while the overshoot is 19–25%. The combination of the backstepping controller with LQR leads to values varying between 2.7–3.3 s, overshoot 15–25%. In our experiments the overshot does not exceed 13.75% neither in worst case and the largest settling time is 1.2 s. The model identification adaptive control (MIAC) used in [28] leads to settling time between 0.8 and 1.4 s, while with the Model Reference Adaptive Control (MRAC) from the same study, the achieved settling time is of 0.75–2 s, very close to our values. The main advantages of these two (MIAC and MRAC) controllers are the overshoot cancellation, but the cost is the control effort. Analyzing the active disturbance rejection controller designed in [11], the presented settling times are 0.85–1.5 s for a 20° step input, overshoot is 7–25%, comparable with our results. The advantage of the high-order sliding mode-based fixed-time active disturbance rejection control from [11] is that it tracks the unknown disturbances in about 3 s.

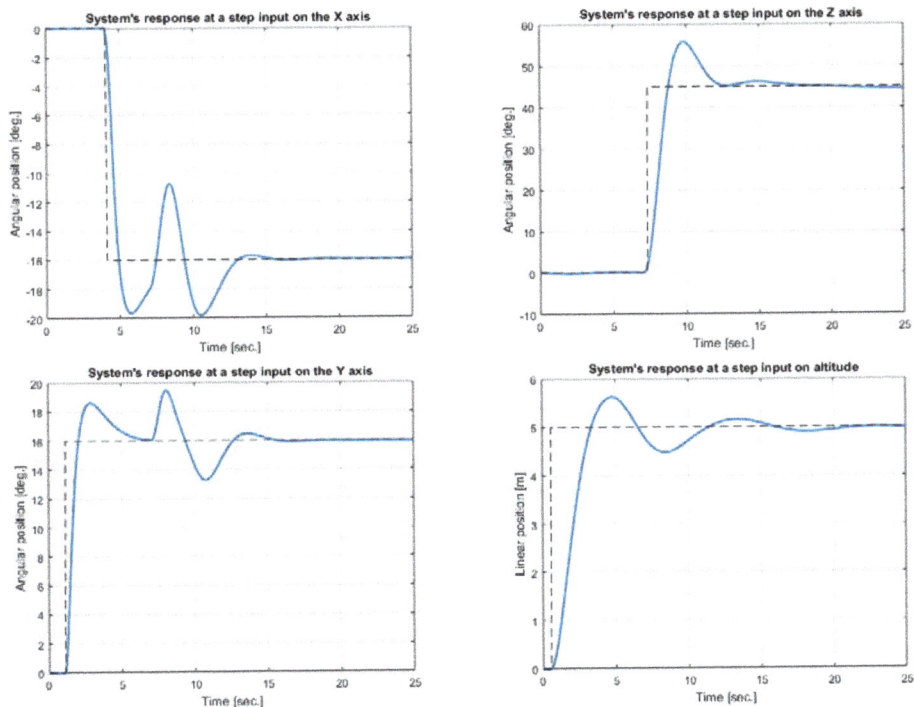

Figure 11. The nonlinear model output for step input on each axis.

Table 1. The obtained performance measures.

Movement	Overshoot [%]	Settling Time [sec]	Steady State Position Error
Front-back	13.75	0.65	0
Left-right	12.5	0.51	0
Rotation around the central axis	0	1.2	0
Up-down	2	0.8	0

Comparing our results with the results of the bioinspired controller from [10], the present results are still competitive. Moreover, imposing different settling time and overshoot in the design stage, it is possible to set a different transient response. Reducing the overshot will increase the settling time and vice versa. Obviously, designing an advanced controller could increase the performances, but the idea of present work was to analyze the most simple algorithm, a PID controller.

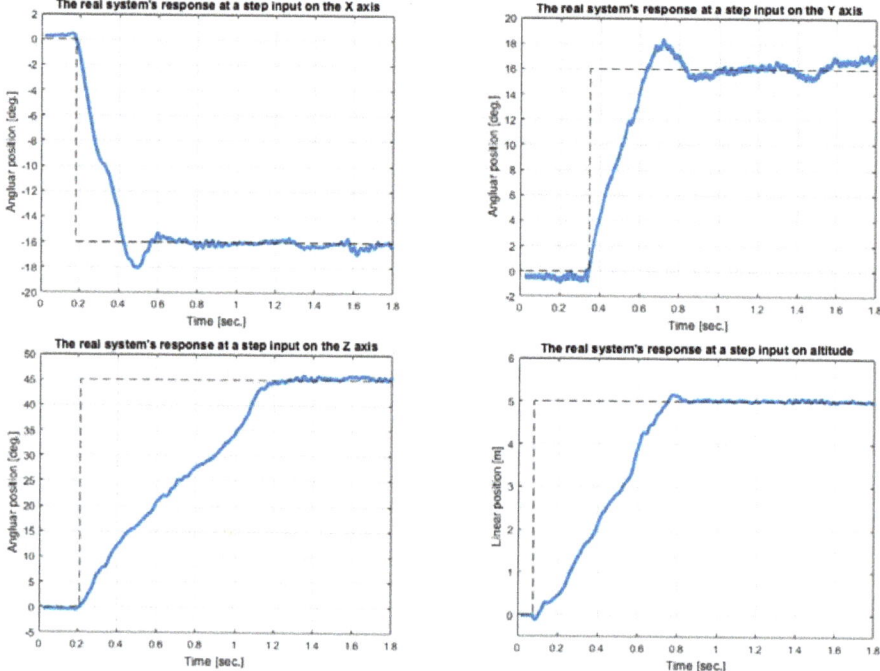

Figure 12. The closed loop system output for step input on each axis.

4. Conclusions

The present research is focused on a low-cost, but performing UAV system design. Taken as a whole, such a flight system presents great difficulties in obtaining positioning data, in particular due to their complex determination or estimation algorithms. In addition to the high complexity of the estimation algorithms, the problem of measurement errors and the resolution of the sensors must often be taken into consideration, so that the control structures are provided with the most accurate data. The offered solution is the quaternion-based estimation. In addition, the tuning of the proportional, integrative, and derivative terms of the control laws is another major problem of the UAV system. Also, the nonlinearities present in such a system introduce challenging problems.

The prototype described in the previous sections offers solution for all these problems.

As future works a global positioning system (GPS) would be added to the equipment model in order to acquire more functionalities.

Author Contributions: Conceptualization and methodology, M.S.; validation, E.H.D.; formal analysis, C.I.M. and L.C.M.; writing—review and editing, C.I.M. and E.H.D.; supervision, E.H.D.; funding acquisition, L.C.M. All authors have read and agreed to the published version of the manuscript.

Funding: This work was supported by a grant of the Romanian Ministry of Research and Innovation, CCCDI–UEFISCDI, project number PN-III-P1-1.2-PCCDI-2017-0734/ROBIN—"Roboții și Societatea: Sisteme Cognitive pentru Roboți Personali și Vehicule Autonome," within PNCDI III. E.H.D. was funded by Hungarian Academy of Science, Janos Bolyai Grant (BO/00313/17) and the ÚNKP-19-4-OE-64 New National Excellence Program of the Ministry for Innovation and Technology. Cristina Muresan is financed by a grant of the Romanian National Authority for Scientific Research and Innovation, CNCS/CCCDI-UEFISCDI, project number P PN-III-P1-1.1-TE-2019-0745.

Conflicts of Interest: The authors declare no conflict of interest.

Abbreviations

Notations Used

A_e, B_e, C	matrices of the state space representation
b	propulsion coefficient
c(k)	the control signal at this current iteration k
$c(\Psi)$	$\cos(\Psi)$
$s(\Psi)$	$\sin(\Psi)$
$c(\theta)$	$\cos(\theta)$
$s(\theta)$	$\sin(\theta)$
$c(\Phi)$	$\cos(\Phi)$
$s(\Phi)$	$\sin(\Phi)$
d	aerodynamic resistance coefficient
d_{sensor}	orientation of the sensor
d_{ref}	orientation of the reference system
e(k)	the error signal at iteration k
$\hat{e}_z$	the unit vector on the Z axis of the reference coordinate system
$\hat{e}_3$	unit vector on the Z axis of the coordinate system attached to the quadrotor
f_{com}	composition of functions
f_x, f_y, f_z	forces acting on the quadrotor on the X, Y and Z axes
F	gradient of function f
F_a	vector of forces acting on the quadrotor
f_p	total propulsion force developed by the engines
$f_v = \begin{bmatrix} f_{vx} & f_{vy} & f_{vz} \end{bmatrix}^T$	disturbances or forces that are opposed to the rotation of each engine on the X, Y and Z axes
g	the gravitational acceleration
g_a	the gyroscope moments caused by the combined velocities of the four motors
I	inertia matrix, with components I_x, I_y, I_z on each axis
K_p, K_i, K_d	the proportionality, integration, and derivative constant
l	the distance from the center of gravity of the quadrotor to the center of rotation of the engine
m	magnetic field
M_a	angular velocities vector applied to the quadrotor
m_q	mass of the quadrotor
of	objective function
P_a	vector of the linear and angular velocities of the quadrotor
P_k	weighting factor
P_p	vector of the linear and angular positions of the quadrotor
R	rotation matrix
$\hat{q}$	quaternion
$^D_E\hat{q}$	quaternion of system E with respect to the reference system D
^A_BR	rotation matrix of the coordinate system B with respect to the reference system A
$\hat{q}_0$	initial point
$\hat{q}_{k+1}$	future orientation
$q_1\ q_2\ q_3\ q_4$	components of the quaternion q
r(k)	the reference signal at current iteration k
$\hat{r}_x, \hat{r}_y, \hat{r}_z$	components of the unity vector $\hat{r}$
T_S	sampling period
$T_v(\Phi, \theta)$	matrix of angular velocities transformations
u	input vector
u_e	input vector at equilibrium point
v_p	vector of the derivatives of the linear positions of P_p
v_a	vector of the linear velocities of the vector P_a
$\dot{v}_a$	linear acceleration
$w_x\ w_y\ w_z$	angular positions on X, Y and Z axes
W	angular positions vector
$\mathbb{X}$	state vector
$\mathbb{X}_e$	state vector of equilibrium point

x_e, y_e, z_e	equilibrium points on the X, Y and Z axes
y	output vector
$y(k)$	the output of the system, measured by sensors at iteration k
Greek Letters	
α, β	constants
Ψ, θ, Φ	Euler angles
τ_a	vector of angular velocities generated by the velocity differences of the four motors
$\tau_x\ \tau_y\ \tau_z$	angular velocities on the X, Y and Z axes generated by the velocity differences of the four motors
τ_v	vector of angular velocities produced by air currents on each motor
$\tau_{vx}\ \tau_{vy}\ \tau_{vz}$	angular velocities produced by air currents on each motor on the X, Y and Z axes
μ	algorithm step
μ_t	variable step at time t
ω_a	vector of the angular velocities of the vector P_a
ω_p	vector of the derivatives of the angular positions of P_p
$\Omega_a = \begin{bmatrix} \Omega_1 & \Omega_2 & \Omega_3 & \Omega_4 \end{bmatrix}$	vector of angular velocities of the four motors

References

1. Shafi, U.; Mumtaz, R.; García-Nieto, J.; Hassan, S.A.; Zaidi, S.A.R.; Iqbal, N. Precision Agriculture Techniques and Practices: From Considerations to Applications. *Sensors* **2019**, *19*, 3796. [CrossRef]
2. Avanzato, R.; Beritelli, F. An Innovative Technique for Identification of Missing Persons in Natural Disaster Based on Drone-Femtocell Systems. *Sensors* **2019**, *19*, 4547. [CrossRef] [PubMed]
3. Rosa, R.; Wehrmeister, M.A.; Brito, T.; Lima, J.L.; Pereira, A.I.P.N. Honeycomb Map: A Bioinspired Topological Map for Indoor Search and Rescue Unmanned Aerial Vehicles. *Sensors* **2020**, *20*, 907. [CrossRef] [PubMed]
4. Hinas, A.; Ragel, R.; Roberts, J.M.; Gonzalez, F. A Framework for Multiple Ground Target Finding and Inspection Using a Multirotor UAS. *Sensors* **2020**, *20*, 272. [CrossRef]
5. Hu, Z.; Wan, K.; Gao, X.; Zhai, Y.; Wang, Q. Deep Reinforcement Learning Approach with Multiple Experience Pools for UAV's Autonomous Motion Planning in Complex Unknown Environments. *Sensors* **2020**, *20*, 1890. [CrossRef] [PubMed]
6. Fabra, F.; Zamora, W.; Sanguesa, J.A.; Calafate, C.T.; Cano, J.C.; Manzoni, P. A Distributed Approach for Collision Avoidance between Multirotor UAVs Following Planned Missions. *Sensors* **2019**, *19*, 2404. [CrossRef] [PubMed]
7. González-Rocha, J.; De Wekker, S.F.J.; Ross, S.D.; Woolsey, C.A. Wind Profiling in the Lower Atmosphere from Wind-Induced Perturbations to Multirotor UAS. *Sensors* **2020**, *20*, 1341. [CrossRef] [PubMed]
8. Zhang, R.; Zhang, J.; Yu, H. Review of modeling and control in UAV autonomous maneuvering flight. In Proceedings of the 2018 IEEE International Conference on Mechatronics and Automation (ICMA), Changchun, China, 5–8 August 2018; pp. 1920–1925.
9. Garcia-Nieto, S.; Velasco-Carrau, J.; Paredes-Valles, F.; Salcedo, J.V.; Simarro, R. Motion Equations and Attitude Control in the Vertical Flight of a VTOL Bi-Rotor UAV. *Electronics* **2019**, *8*, 208. [CrossRef]
10. Armendariz, S.; Becerra, V.; Bausch, N. Bio-inspired Autonomous Visual Vertical and Horizontal Control of a Quadrotor Unmanned Aerial Vehicle. *Electronics* **2019**, *8*, 184. [CrossRef]
11. Song, C.; Wei, C.; Yang, F.; Cui, N. High-Order Sliding Mode-Based Fixed-Time Active Disturbance Rejection Control for Quadrotor Attitude System. *Electronics* **2018**, *7*, 357. [CrossRef]
12. Ferrero, A. Control of a Supersonic Inlet in Off-Design Conditions with Plasma Actuators and Bleed. *Aerospace* **2020**, *7*, 32. [CrossRef]
13. Qin, Z.; Dong, C.; Wang, H.; Li, A.; Dai, H.; Sun, W.S.; Xu, Z. Trajectory Planning for Data Collection of Energy-Constrained Heterogeneous UAVs. *Sensors* **2019**, *19*, 4884. [CrossRef] [PubMed]
14. Madgwick, S.O.H. An Efficient Orientation Filter for Inertial and Inertial/Magnetic Sensor Arrays, Internal_Report. 2010. Available online: https://www.samba.org/tridge/UAV/madgwick_internal_report.pdf (accessed on 15 March 2020).
15. Cavallo, A.; Cirillo, A.; Cirillo, P.; De Maria, G.; Falco, P.; Natale, C.; Pirozzi, S. Experimental Comparison of Sensor Fusion Algorithms for Attitude Estimation. In *IFAC World Congress*; IFAC: Cape Town, South Africa, 2014; Volume 19, pp. 7585–7591.

16. Feng, K.; Li, J.; Zhang, X.; Shen, C.; Bi, Y.; Zheng, T.; Liu, J. A New Quaternion-Based Kalman Filter for Real-Time Attitude Estimation Using the Two-Step Geometrically-Intuitive Correction Algorithm. *Sensors* **2017**, *17*, 2146. [CrossRef]
17. Danku, A.; Kovari, A.; Miclea, L.C.; Dulf, E.-H. Intelligent Control of an Aerodynamical System. In Proceedings of the 2019 IEEE 15th International Conference on Intelligent Computer Communication and Processing (ICCP), Cluj-Napoca, Romania, 5–7 September 2019; pp. 49–52.
18. Dulf, E.-H.; Timis, D.D.; Szekely, L.; Miclea, L.C. Adaptive Fractional Order Control Applied to a Multi-Rotor System. In Proceedings of the 2019 22nd International Conference on Control Systems and Computer Science (CSCS), Bucharest, Romania, 28–30 May 2019; pp. 696–699.
19. Humphries, U.; Rajchakit, G.; Kaewmesri, P.; Chanthorn, P.; Sriraman, R.; Samidurai, R.; Lim, C.P. Stochastic Memristive Quaternion-Valued Neural Networks with Time Delays: An Analysis on Mean Square Exponential Input-to-State Stability. *Mathematics* **2020**, *8*, 815. [CrossRef]
20. Kizilateş, C.; Catarino, P.; Tuglu, N. On the Bicomplex Generalized Tribonacci Quaternions. *Mathematics* **2019**, *7*, 80. [CrossRef]
21. Cabarbaye, A.; Lozano, R.; Estrada, M.B. Adaptive quaternion control of a 3-DOF inertial stabilised platforms. *Int. J. Control* **2018**, *93*, 473–482. [CrossRef]
22. Birlescu, I.; Husty, M.; Calin, V.; Gherman, B.; Tucan, P.; Pisla, D. Joint-Space Characterization of a Medical Parallel Robot Based on a Dual Quaternion Representation of SE(3). *Mathematics* **2020**, *8*, 1086. [CrossRef]
23. Wei, R.; Cao, J. Synchronization control of quaternion-valued memristive neural networks with and without event-triggered scheme. *Cogn. Neurodynamics* **2019**, *13*, 489–502. [CrossRef]
24. Zmitri, M.; Fourati, H.; Vuillerme, N. Human Activities and Postures Recognition: From Inertial Measurements to Quaternion-Based Approaches. *Sensors* **2019**, *19*, 4058. [CrossRef]
25. Long, Y.; Lyttle, S.; Pagano, N.; Cappelleri, D.J. Design and Quaternion-Based Attitude Control of the Omnicopter MAV Using Feedback Linearization. In Proceedings of the ASME 2012 International Design Engineering Technical Conferences & Computers and Information in Engineering Conference IDETC/CIE 2012, Chicago, IL, USA, 12–15 August 2012.
26. Chovancová, A.; Fico, T.; Hubinský, P.; Duchoň, F. Comparison of various quaternion-based control methods applied to quadrotor with disturbance observer and position estimator. *Robot. Auton. Syst.* **2016**, *79*, 87–98. [CrossRef]
27. Sanwale, J.; Trivedi, P.; Kothari, M.; Malagaudanavar, A. Quaternion-based position control of a quadrotor unmanned aerial vehicle using robust nonlinear third-order sliding mode control with disturbance cancellation. *Proc. Inst. Mech. Eng. Part G J. Aerosp. Eng.* **2019**, *234*, 997–1013. [CrossRef]
28. Schreier, M. Quaternion-based adaptive attitude control schemes for quadrotor systems. *Int. J. Mechatron. Autom.* **2013**, *3*, 217. [CrossRef]
29. Sanchez, M.E.G.; Abaunza, H.; Castillo, P.; Lozano, R.; García-Beltrán, C.D. Quadrotor Energy-Based Control Laws: A Unit-Quaternion Approach. *J. Intell. Robot. Syst.* **2017**, *88*, 347–377. [CrossRef]
30. Abaunza, H.; Castillo, P.; Lozano, R. Quaternion Modeling and Control Approaches. In *Handbook of Unmanned Aerial Vehicles*; Vachtsevanos, G., Ed.; Springer: Cham, Switzerland, 2018; pp. 1–29.
31. Sanchez, M.E.G.; Abaunza, H.; Castillo, P.; Lozano, R.; García-Beltrán, C.; Rodriguez-Palacios, A. Passivity-Based Control for a Micro Air Vehicle Using Unit Quaternions. *Appl. Sci.* **2016**, *7*, 13. [CrossRef]
32. Colmenares-Vazquez, J.; Marchand, N.; Castillo, P.; Gomez-Balderas, J.E. An intermediary quaternion-based control for trajectory following using a quadrotor. In Proceedings of the 2017 IEEE/RSJ International Conference on Intelligent Robots and Systems (IROS), Vancouver, BC, Canada, 24–28 September 2017; pp. 5965–5970.
33. Andersen, T.S.; Kristiansen, R. Quaternion guidance and control of quadrotor. In Proceedings of the 2017 International Conference on Unmanned Aircraft Systems (ICUAS), Miami, FL, USA, 13–16 June 2017; pp. 1567–1601.
34. Carino, J.; Abaunza, H.; Castillo, P. Quadrotor quaternion control. In Proceedings of the 2015 International Conference on Unmanned Aircraft Systems (ICUAS), Denver, CO, USA, 9–12 June 2015; pp. 825–831.
35. Fresk, E.; Nikolakopoulos, G. Full quaternion based attitude control for a quadrotor. In Proceedings of the 2013 European Control Conference (ECC), Zurich, Switzerland, 17–19 July 2013; pp. 3864–3869.

36. Ji, X. Partial study of quadrotor based on quaternions. In Proceedings of the 6th International Conference on Computer-Aided Design, Manufacturing, Modeling and Simulation (CDMMS 2018), Busan, South Korea, 14–15 April 2018.
37. Sanchez, M.E.G.; Gonzalez, O.H.; Lozano, R.; García-Beltrán, C.; Valencia-Palomo, G.; López-Estrada, F.-R. Energy-Based Control and LMI-Based Control for a Quadrotor Transporting a Payload. *Mathematics* **2019**, *7*, 1090. [CrossRef]
38. Shao, X.; Liu, N.; Wang, Z.; Zhang, W.; Yang, W. Neuroadaptive integral robust control of visual quadrotor for tracking a moving object. *Mech. Syst. Signal Process.* **2020**, *136*, 106513. [CrossRef]
39. Raj, J.; Raghuwaiya, K.S.; Vanualailai, J. Novel Lyapunov-Based Autonomous Controllers for Quadrotors. *IEEE Access* **2020**, *8*, 47393–47406. [CrossRef]
40. Guzmán-Rabasa, J.A.; López-Estrada, F.R.; González-Contreras, B.M.; Valencia-Palomo, G.; Chadli, M.; Pérez-Patricio, M. Actuator fault detection and isolation on a quadrotor unmanned aerial vehicle modeled as a linear parameter-varying system. *Meas. Control.* **2019**, *52*, 1228–1239. [CrossRef]
41. Chiella, A.C.B.; Teixeira, B.O.S.; Pereira, G.A.S. Quaternion-Based Robust Attitude Estimation Using an Adaptive Unscented Kalman Filter. *Sensors* **2019**, *19*, 2372. [CrossRef]
42. Naharro, R.J.; Gómez-Bravo, F.; Garcia, J.M.; Sánchez-Raya, M.; Gómez-Galán, J.A. A Smart Sensor for Defending against Clock Glitching Attacks on the I2C Protocol in Robotic Applications. *Sensors* **2017**, *17*, 677. [CrossRef]
43. Abubakar, A.B.; Kumam, P.; Mohammad, H.; Awwal, A.M. An Efficient Conjugate Gradient Method for Convex Constrained Monotone Nonlinear Equations with Applications. *Mathematics* **2019**, *7*, 767. [CrossRef]
44. Waliszkiewicz, M.; Wojtowicz, K.; Rochala, Z.; Balestrieri, E. The Design and Implementation of a Custom Platform for the Experimental Tuning of a Quadcopter Controller. *Sensors* **2020**, *20*, 1940. [CrossRef] [PubMed]
45. Bavafa-Toosi, Y. *Introduction to Linear Control Systems*; Elsevier: Amsterdam, The Netherlands, 2019. [CrossRef]

Publisher's Note: MDPI stays neutral with regard to jurisdictional claims in published maps and institutional affiliations.

© 2020 by the authors. Licensee MDPI, Basel, Switzerland. This article is an open access article distributed under the terms and conditions of the Creative Commons Attribution (CC BY) license (http://creativecommons.org/licenses/by/4.0/).

Article

Solitary Wave Solutions of the Generalized Rosenau-KdV-RLW Equation

Zakieh Avazzadeh [1,2], Omid Nikan [3] and José A. Tenreiro Machado [4,*]

[1] Institute of Research and Development, Duy Tan University, Da Nang 550000, Vietnam
[2] Faculty of Natural Sciences, Duy Tan University, Da Nang 550000, Vietnam; zakiehavazzadeh@duytan.edu.vn
[3] School of Mathematics, Iran University of Science and Technology, Narmak, Tehran 16846-13114, Iran; omidnikan77@yahoo.com
[4] Department of Electrical Engineering, Institute of Engineering, Polytechnic of Porto, 4249-015 Porto, Portugal
* Correspondence: jtm@isep.ipp.pt

Received: 21 August 2020; Accepted: 9 September 2020; Published: 17 September 2020

Abstract: This paper investigates the solitary wave solutions of the generalized Rosenau–Korteweg-de Vries-regularized-long wave equation. This model is obtained by coupling the Rosenau–Korteweg-de Vries and Rosenau-regularized-long wave equations. The solution of the equation is approximated by a local meshless technique called radial basis function (RBF) and the finite-difference (FD) method. The association of the two techniques leads to a meshless algorithm that does not requires the linearization of the nonlinear terms. First, the partial differential equation is transformed into a system of ordinary differential equations (ODEs) using radial kernels. Then, the ODE system is solved by means of an ODE solver of higher-order. It is shown that the proposed method is stable. In order to illustrate the validity and the efficiency of the technique, five problems are tested and the results compared with those provided by other schemes.

Keywords: nonlinear wave phenomen; RBF; local RBF-FD; stability

1. Introduction

Nonlinear waves are important phenomena in scientific research. Due to that reason, a number of models have been proposed to describe their behavior. Indeed, we find a variety of mathematical descriptions of wave dynamics, such as the Rosenau, regularized-long wave (RLW), and Korteweg-de Vries (KdV) equations [1–8]. The KdV equation has been applied in the description of dynamical effects such as longitudinal astigmatic, ion sound, and magnetic fluid waves [4–9]. The convergence properties, existence and the regularity of solutions of KdV-type equation have been discussed in [10–12].

Kaya and Aassila calculated the explicit solutions of the KdV equation with an initial condition by using the Adomian decomposition method [13]. Özer and Kutluay applied an analytical–numerical method to the KdV equation [14]. The RLW equation was developed by Peregrine, as an alternative to the classical KdV formulation in order to investigate the behavior of the solution [15,16]. Benjamin et al. proved the existence and uniqueness of the solution of the RLW model and determined its exact expression subject to restrictions in the initial and boundary conditions [2]. The RLW is also adopted in the modeling of long waves with small amplitudes on the water surface [17]. A noteworthy feature of the RLW problem is that the collision between two solitary waves results either in sinusoidal solutions, or in secondary solitary waves [18]. Since the KdV cannot describe wave–all and wave–wave interactions, another model, known as the Rosenau equation, was proposed by Rosenau to describe the dynamics behavior of dense discrete systems [7]. Zuo studied the solitary wave and periodic solutions for the Rosenau-KdV model [6]. Barreto et al. discussed the existence of solutions of the

Rosenau formulation with the plus sign in the advection-like term in moving domains by means of the Galerkin, multiplier, and energy estimate techniques [3].

Hereafter, we propose a numerical method for the initial value problem of the general Rosenau-KdV-RLW equation [19–24],

$$u_t + \alpha u_x + \beta (u^p)_x + \gamma u_{xxx} - \mu u_{xxt} + \delta u_{xxxxt} = 0, \tag{1}$$

with the initial condition

$$u(x,0) = f(x) \tag{2}$$

and boundary conditions

$$u(a,t) = u(b,t) = 0, \; u_x(a,t) = u_x(b,t) = 0, u(a,t) = u(b,t) = 0, \; u_{xx}(a,t) = u_{xx}(b,t) = 0, \tag{3}$$

where $u = u(x,t)$, is a real-valued function, the real constants α, β, γ and μ are non-negative, $p \geq 2$ is a positive integer, and $f(x)$ is a given smooth function.

Lemma 1. *(See [25].) The following conservative properties for the initial value problem* (1) *hold*

$$Q(t) = \int_a^b u(x,t) dx = \int_a^b u(x,0) dx = Q(0) \tag{4}$$

and

$$E(t) = \int_a^b (u^2 + c u_x^2 + u_{xx}^2) dx = ||u||_{L^2}^2 + c||u_x||_{L^2}^2 + c||u_{xx}||_{L^2}^2 = E(0), \tag{5}$$

where $Q(0)$ and $E(0)$ are constants depending on the initial conditions.

When $-a \gg 0$ and $b \gg 0$, the initial boundary value problem (1)–(3) is consistent and, the boundary condition (3) is reasonable [26]. Some particular cases of Equation (1) occur:

- if $\alpha = 0, \beta = 0.5, \gamma = 1, \mu = 0, \delta = 0$ and $p = 2$, then expression (1) is the KdV equation [14,27–29],

$$u_t + u u_x + u_{xxx} = 0;$$

- if $\alpha = 1, \beta = 0.5, \gamma = 0, \mu = 0, \delta = 1$ and $p = 2$, then expression (1) is the Rosenau equation [30–32],

$$u_t + \alpha u_x + u u_x + u_{xxxt} = 0;$$

- if $\alpha = 1, \beta = 0.5, \mu = 1, \gamma = 0, \delta = 0$ and $p = 2$, then expression (1) becomes the RLW equation [33]

$$u_t + u_x + u u_x - u_{xxt} = 0;$$

- if $\alpha = 1, \beta = 0.5, \gamma = 1, \mu = 0, \delta = 1$ and $p = 2$, then expression (1) is the Rosenau-KdV equation [6,34,35]

$$u_t + u_x + u u_x + u_{xxx} + u_{xxxxt} = 0;$$

- if $\alpha = 1, \beta = 1, \gamma = 0, \mu = 1$, and $\delta = 1$, then expression (1) is the generalized Rosenau-RLW model [26]

$$u_t + u_x + (u^p)_x - u_{xxt} + u_{xxxxt} = 0;$$

- if $p = 2, p = 3$ or $p \geq 4$, then expression (1) represents the classical, modified, and general Rosenau-RLW equations, respectively.

In recent years, various analytical and numerical methods have been used to approximate the solution of the initial boundary value problem (1)–(3). Razborova et al. presented a theoretical approach based on the Ansatz method for the Rosenau-KdV-RLW equation [9]. Later, Razborova et al.

used a semi-inverse Variational Principle to retrieve a single solitary wave solution [22]. Additionally, Razborova et al. and Sanchez et al. discussed the solutions of the perturbed Rosenau-KdV-RLW equation [23,24]. Wongsaijai et al. constructed a three-level weighted average implicit finite difference (FD) technique [19]. Pan et al. presented a C-N pseudo-compact conservative numerical scheme based on the FD technique [20]. Fernández and Ramos investigated a three-point compact method with fourth-second accuracy [21]. Wang et al. and Hu and Wang formulated FD schemes with linear three-level [31] and high-accuracy conservative [33] characteristics, respectively. Wongsaijai et al. proposed a compact FD technique [26] and Pan et al. developed a linear-implicit FD for the usual Rosenau-RLW equation [25,36]. Zheng et al. presented an average linear FD technique [34]. Mittal et al. implemented a numerical method based on the collocation of quintic B-splines over finite elements [37]. Hu et al. considered a second-order conservative FD scheme [38]. Ari et al. adopted a meshless kernel-based approach of lines [39]. Foroutan et al. developed a modified Chebyshev rational approximation [40]. Wang et al. advanced a three-level linear conservative FD [41], while Wongsaijai et al. came with a mass-preserving scheme, namely, a nonlinear algorithm based on a modification of the FD [42].

In this paper, we use the local meshless radial basis function (RBF) for solving the general Rosenau-KdV and the Rosenau-RLW models. Section 2 formulates and discusses the local meshless RBF based on the finite difference (RBF-FD) technique for discretizing Equation (1). Section 3 provides five numerical examples and compares the results with those of other schemes proposed in the literature. Finally, Section 4 presents the concluding remarks.

2. The RBF-FD Collocation Method

A mesh-free (or meshless) method adopts an algebraic system of equations for the complete domain without requiring a pre-defined mesh discretization of the domain and its boundary [43,44]. Mesh-free techniques are used to approximate scattered data, since generating meshes is one of the most laborious tasks of mesh-based numerical processes. Indeed, a mesh-free technique provides a low-cost alternative to schemes involving finite volume, finite difference, finite element, multivariate splines, and wavelets, all requiring node connectivity. Meshless techniques eliminate the mesh generation step and a collection of scattered data can be used. The RBF is one of the most widely used meshless techniques and reveals good performance in case of multidimensional scattered data interpolation [43,44].

Given a set of scattered node data $X_C = \{x_1, \ldots, x_N\} \subseteq \mathbb{R}^n$ and the corresponding function values $u_i = u(x_i)$, $i = 1, 2, \ldots, N$, the RBF interpolant is represented in the form

$$u(x) \simeq S(x) = \sum_{j=1}^{N} \alpha_j \phi_j(x, c), \tag{6}$$

where $\{\alpha_j\}_{j=1}^{N}$ are unknown coefficients, $\phi_j(x, c) = \phi(\|x - x_j\|_2, c)$, $j = 1, \ldots, N$, are RBF with shape parameter c, and the operation $\|\cdot\|_2$ represents the Euclidean norm [44,45]. Some popular choices of RBF include the linear, Cubic, Multiquadric (MQ), Gaussian (GA), and thin-plate spline (TPS) versions with dependence r, r^3, $\sqrt{c^2 + r^2}$, $\exp(-cr^2)$, and $r^4 \ln(r)$, respectively, where $r = \|x - x_j\|_2$. The coefficients $\{\alpha_j\}_{j=1}^{N}$ of Equation (6) are computed by imposing interpolation conditions $S(x_i) = u_i$, $i = 1, \ldots, N$. The relation (6) can be written in the following matrix form

$$A_\phi \alpha = f, \tag{7}$$

where

$$\alpha = \begin{bmatrix} \alpha_1 \\ \alpha_2 \\ \vdots \\ \alpha_N \end{bmatrix}, \quad f = \begin{bmatrix} f_1 \\ f_2 \\ \vdots \\ f_N \end{bmatrix}, \quad A_{\phi,ij} = \phi_j(x_i), \quad i,j = 1,\ldots,N.$$

The non-singularity of the associated linear system was proven in [46]. The main pros of the RBF collocation method when solving PDEs are its simplicity, easy application to different PDEs, and efficiency for solving problems involving complex domains. On the other hand, the major con of this method is related to the problem of full matrices. These matrices are strongly sensitive to the shape parameter c selected in the RBF and, therefore, they become difficult to solve in problems where we have too many unknowns. This problem arises from the fact that using the RBF interpolation increases the condition numbers of the related matrices for a large number of nodes. This occurs particularly when one selects inadequate data centers and uses basic functions that are infinitely smooth, such as the MQ, with extreme values of the shape parameter c [45].

The notation of local differentiation is popular in the RBF literature, particularly for time-dependent PDEs. The local radial basis function (RBF) generated by finite differences (RBF-FD), raised considerable interest owing to the structure of their differentiation and interpolation matrices [47,48]. It is possible to control the degree of sparsity of the differentiation and interpolation matrices produced by the local RBF. This sparsity can take advantage of parallelism and solve large problems [49,50]. The local RBFs have also been employed to reduce the model order. In some situations, researchers have found that the local RBF technique can produce the same degree of accuracy as the global RBF technique with a smaller mesh size [49–53]. Although small mesh sizes result in smaller ODE systems, the overall accuracy is maintained. Interested readers can find examples of the application of local RBFs to problems in the geosciences in [54,55]. Garshasbi et al. used the RBF collocation method for approximating the shallow water model named the Camassa–Holm equation [56]. Uddin connected the RBF to the pseudo-spectral method, known as RBF-PS method to approximate the equal width equation [57]. Nikan et al. solved numerically the nonlinear KdV-Benjamin-Bona-Mahony-Burgers (KdV-BBM-B) with the help of the RBF-PS [58]. Dehghan and Shafieeabyaneh addressed the RLW and extended Fisher-Kolmogorov (EFK) equations using local meshless RBF-FD [59]. Ebrahimijahan and Dehghan proposed a numerical technique for solving the nonlinear generalized BBBM-B and RLW equations based on the integrated RBF [60]. Rashidinia et al. implemented the local RBF-FD meshless method for generalized Korteweg-de Vries-Burgers [61] and Kawahara [62] equations.

Let us consider that $I_i = \{x_{i_1}, x_{i_2}, \ldots, x_{i_{n_i}}\}$ is a stencil of x_i. In the local RBF-FD collocation method, the linear differential operator $\mathcal{L}$ at every point can be approximated only the stencil instead of applying the complete number of point, i.e.,

$$\mathcal{L}u(x_i) = \sum_{j=1}^{n_i} w_j u(x_{i_j}), \tag{8}$$

where $x_i = x_{i_1}$ is the center point of stencil I_i. Figure 1 gives an example of a domain with 9 grid points and a stencil size of $n_i = 4$. At the point x_3, the $n_i - 1 = 3$ nearest neighbors are used in the computation. Figure 2 shows the sparsity patterns for $N = 50$ for two stencil sizes $n_i = 11$ and $n_i = 15$.

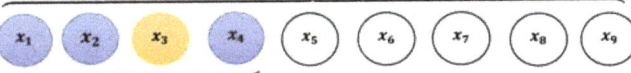

Figure 1. Illustration of one-dimensional case of stencil.

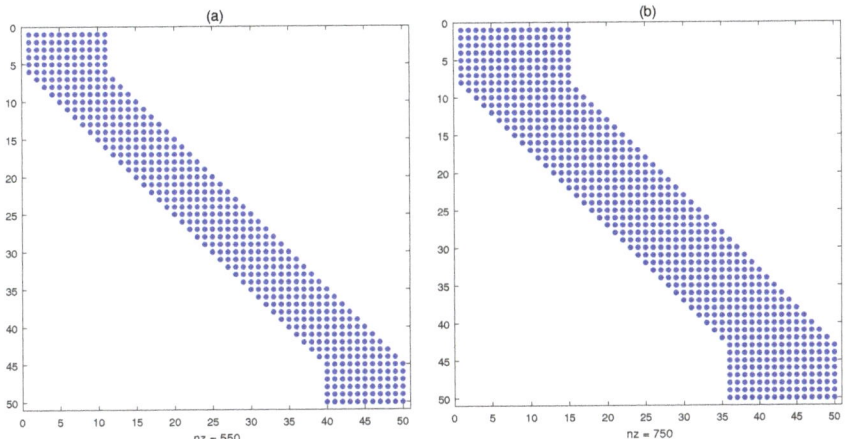

Figure 2. Sparsity patterns for $N = 50$ and two stencil sizes $n_i = 11$ (a) and $n_i = 15$ (b).

By deriving the RBF expansion of $u(x)$ in Equation (8), the weighted differences of stencil node can be obtained from the system as:
$$A_\phi w = l, \tag{9}$$

where

$$w = \begin{bmatrix} w_1 \\ w_2 \\ \vdots \\ w_{n_i} \end{bmatrix}, \quad l = \begin{bmatrix} \mathcal{L}\phi_{i_1}(x)|_{x=x_i} \\ \mathcal{L}\phi_{i_2}(x)|_{x=x_i} \\ \vdots \\ \mathcal{L}\phi_{i_{n_i}}(x)|_{x=x_i} \end{bmatrix}, \quad A_{\phi,ij} = \phi_{i_i}(x_{i_j}), \quad i,j = 1,\ldots,n_i. \tag{10}$$

Indeed, it is necessary to solve a small-sized linear system with a conditionally positive definite coefficient matrix in each stencil. The weighted differences of the stencil nodes $w_1, w_2, \ldots, w_{n_i}$ can be determined from the above system.

The first, second and third order derivatives can be approximated with the help of the function values at a set of n_i nodes (including x_i) in the stencil of x_i. That is, we can write

$$\frac{\partial u^k(x)}{\partial x}\bigg|_{x=x_i} = \sum_{j=1}^{n_i} w_{i,j}^{x,1} u^k(x_j^i) = \mathbf{W}_x \mathbf{u}, \tag{11}$$

$$\frac{\partial^2 u^k(x)}{\partial x^2}\bigg|_{x=x_i} = \sum_{j=1}^{n_i} w_{i,j}^{x,2} u^k(x_j^i) = \mathbf{W}_{xx} \mathbf{u}, \tag{12}$$

$$\frac{\partial^3 u^k(x)}{\partial x^3}\bigg|_{x=x_i} = \sum_{j=1}^{n_i} w_{i,j}^{x,3} u^k(x_j^i) = \mathbf{W}_{xxx} \mathbf{u}, \tag{13}$$

where $w_{i,j}^{x,l}$ represents the weighted differences of stencil node for the order derivatives $l = \{1, 2, 3\}$.
We can obtain the following semi-discrete system by considering the notations above as

$$\mathbf{u}' + \alpha \mathbf{W}_x \mathbf{u} + \beta \mathbf{W}_x (\mathbf{u}^p) - \mu \mathbf{W}_{xx} \mathbf{u}' + \gamma \mathbf{W}_{xxx} + \delta \mathbf{W}_{xxxx} \mathbf{u}' = 0. \tag{14}$$

The above equation can be represented as

$$(\mathbf{I} - \mu \mathbf{W}_{xx} + \delta \mathbf{W}_{xxxx}) \mathbf{u}' = -\alpha \mathbf{W}_x \mathbf{u} - \beta \mathbf{W}_x (\mathbf{u}^p) - \gamma \mathbf{W}_{xxx} \mathbf{u}. \tag{15}$$

We must note that the matrices $\mathbf{A} = -\alpha \mathbf{W}_x - \beta \mathbf{W}_x - \gamma \mathbf{W}_{xxx}$ and $\mathbf{B} = \mathbf{I} - \mu \mathbf{W}_{xx} + \delta \mathbf{W}_{xxxx}$ are time-independent. We conclude that

$$\mathbf{u}' = \mathbf{W}\mathbf{u}, \tag{16}$$

where $\mathbf{W} = \mathbf{B}^{-1}(-\alpha \mathbf{W}_x - \beta \mathbf{W}_x(\mathbf{u}^{p-1}) - \gamma \mathbf{W}_{xxx})$. Equation (16) is of the form

$$\mathbf{u}' = F(\mathbf{u}). \tag{17}$$

Equation (17) is an ODE with respect to $\mathbf{u}$ and it can be solved by means of an ODE solver in MATLAB such as ode113 or ode45. Let $\tau = T/M$ and $t_n = n\tau$, for $n = 0, 1, \ldots, M$, so that the mesh $\{t_n : n = 0, 1, \ldots, M\}$ is uniform. The initial solution u_0 is the starting vector. The package ode45 is an explicit Runge-Kutta of order 4(5) formula of the Dormand–Prince pairs [63]. The ode45 is a one-step solver that computes u_{t_n} given the solution at the preceding time point $u_{t_{n-1}}$. On the other hand, the ode113 is a variable-order Predict-Evaluate-Correct-Evaluate solver of the Adams–Bashforth–Moulton type [64]. This solver might be more efficient than the ode45 for close tolerances and, in particular, when the ODE file function is particularly expensive to evaluate. A multi-step solver, such as the ode113, needs the solutions corresponding to more than one preceding time point for calculating the current solution. Hereafter, we calculate the differentiation matrices, expressed by $\mathbf{W}_x$, $\mathbf{W}_{xx}$ and $\mathbf{W}_{xx}$, only one time outside the time-stepping operation. Additionally, merely matrix-vector multiplications are required within the time-stepping operation.

2.1. Stability Analysis

The method of lines represents the idea of using the FD method in the time direction t to solve a coupled system of ODEs. The numerical stability of the method of lines is investigated by a rule of tumb. The method of lines is stable if the eigenvalues of the (linearized) spatial discretization operator, scaled by τ, lie in the stability region of the time-discretization operator [57,65]. One defines the stability region as the portion of a multifaceted plane consisting of eigenvalues which result in the generation of bounded solutions. The coefficient matrix eigenvalues determine the stability of Equation (16) [66]. Hence, we need only to demonstrate that every eigenvalue $Re(\lambda_i)$ belonging to the coefficient matrix has a non-positive real term $Re(\lambda_i)$, where λ_i, $i = 1, 2, \ldots, N$, represents of the matrix eigenvalues. In other words, for all $i = 1, 2, \ldots, N$, we must have $Re(\lambda_i) \leq 0$ for obtaining stable solutions. The reader is referred to [66] for further details. In order to investigate the stable and unstable eigenvalue ranges of the Rosenau-KdV-RLW model, one must compute the eigenvalues belonging to the matrix $\mathbf{W}$, which are scaled by τ.

3. Computational Results and Comparisons

This section considers five test problems assessing the effectiveness and accuracy of the proposed method for various values of h, τ and c. To measure the accuracy of method in comparison with the exact solution, we compute the following error norms:

$$L_\infty = \max_{1\leq j\leq M-1} |u^{exact}(x_j, T) - u(x_j, T)|,$$

$$L_2 = \left(h \sum_{j=1}^{N} (u^{exact}(x_j, T) - u(x_j, T))^2 \right)^{\frac{1}{2}},$$

$$RMS = \left(\frac{1}{N} \sum_{j=1}^{N} (u^{exact}(x_j, T) - u(x_j, T))^2 \right)^{\frac{1}{2}},$$

where u and u^{exact} denote the numerical solution and exact solution, respectively. In addition, the invariants of motion are evaluated by

$$Q = \int_a^b u(x,t) dx = \sum_{i=1}^{N} u_i,$$

$$E = \int_a^b (u^2 + cu_x^2 + u_{xx}^2) dx = \sum_{i=1}^{N} \left(u_i^2 + c(u_x)_i^2 + (u_{xx})_i^2 \right).$$

It should be noted that the Gaussian function is used as a basis and the computations were performed in MATLAB R2016a with a computer system having a configuration including Intel(R) Core(TM) i5-2330 CPU 3.60 GHz and 8.00G RAM.

Example 1. *Let us consider the general Rosenau-KdV-RLW model (1) in the case of $\alpha = \mu = \gamma = 1$, $\beta = 0.5$, $p = 2$ and $\delta = 0$ in the spatial interval $x \in [-70, 100]$. The exact solution is $u(x,t) = k_{11} \operatorname{sech}^4[k_{12}(x - k_{13}t)]$ [38,41], where*

$$k_{11} = -\frac{35}{24} + \frac{35}{12}\sqrt{313}, \quad k_{12} = \frac{1}{24}\sqrt{-26 + 2\sqrt{313}}, \quad k_{13} = \frac{1}{2} + \frac{\sqrt{313}}{26}.$$

Table 1 lists the approximation errors in terms of L_∞, L_2 and RMS with $\tau = 0.01$ and $n_i = 489$. Table 2 compares the obtained results with those provided by the techniques described in [38,41]. It is seen that the errors obtained by the proposed technique are inferior to the others. Figure 3 depicts the motion of the single solitary wave with $h = \tau = 0.125$ over the spatial intervals $x \in [-40, 60]$ (left) and $x \in [-70, 100]$ (right) at final times $T \in \{0, 30, 40\}$. We verify that the single solitons move to the right at a constant speed and preserve their amplitude and shape with increasing time as anticipated. Figure 4 represents the absolute errors L_∞ at final times $T \in \{0, 30, 40\}$. Figure 5 portraits the eigenvalues of the linearized differentiation operator **A** and **B** (left and right panels, respectively) with $N = 100$. We observe that the eigenvalues calculated for **A** and **B** are zero or have negative values. The eigenvalues belonging to the linearized differentiation operators are real and negative or are complex with a negative real term. Hence, the stability of the proposed system for this case is proven.

Example 2. *Let us consider the general Rosenau-KdV-RLW model (1) in the case of $\alpha = \beta = \mu = \gamma = 1$, $p = 5$ and $\delta = 0$ over the spatial interval $x \in [-60, 90]$. The exact solitary wave solution is $u(x,t) = k_{21} \operatorname{sech}[k_{22}(x - k_{23}t)]$, where [34,41]*

$$k_{21} = \sqrt[4]{\frac{4}{15}(-5 + \sqrt{34})}, \quad k_{22} = \frac{-5 + \sqrt{34}}{3}, \quad k_{23} = \frac{5 + \sqrt{34}}{10}.$$

Table 3 reports the L_∞, L_2 and RMS errors with $\tau = 0.01$ and $n_i = 489$. Table 4 compares the results with those obtained by the techniques described in [34,41]. It is clear that the results of the new method are considerably more accurate. Table 5 illustrates the conservative law of the discrete energy E. Figure 6 depicts the motion of single solitary wave with $h = \tau = 0.125$ (left) and $h = \tau = 0.0625$

(right) over the spatial interval $x \in [-60, 90]$ at final times $T \in \{0, 10, 40\}$. The single solitons move to the right at a constant speed preserving their amplitude and shape. Figure 7 represents the absolute error L_∞ at final times $T = \{30, 40\}$. Figure 8 plots the eigenvalues of the matrices **A** and **B** (left and right panels, respectively) with $N = 100$. The eigenvalues calculated for **A** are negative values. For what concerns **B**, they are zero or have negative values. Therefore, the stability of the proposed system is confirmed.

Table 1. The L_∞, L_2 and RMS errors with $\tau = 0.01$ $N = 100$, $n_i = 589$ and $c = 1.8$ for Example 1.

Method	T	L_∞	L_2	RMS
RBF-FD	5	1.7556×10^{-11}	4.2037×10^{-11}	3.2241×10^{-12}
RBF-FD	10	3.4832×10^{-11}	8.3499×10^{-11}	6.4057×10^{-12}
RBF-FD	15	5.1114×10^{-11}	1.2388×10^{-10}	9.5038×10^{-12}
RBF-FD	20	6.6317×10^{-11}	1.6066×10^{-10}	1.2320×10^{-11}
RBF-FD	25	7.9818×10^{-11}	4.7209×10^{-10}	1.4929×10^{-11}
RBF-FD	30	9.2357×10^{-11}	2.2541×10^{-10}	1.7273×10^{-11}
RBF-FD	35	1.0459×10^{-10}	2.5348×10^{-10}	1.9441×10^{-11}
RBF-FD	40	1.1660×10^{-10}	2.7268×10^{-10}	2.0910×10^{-11}

Table 2. The L_∞ and L_2 errors under different mesh steps $h = \tau$ at $T = 40$ for Example 1.

	Method	c	N	n_i	L_∞	L_2
$h = \tau = 0.2$						
	RBF-FD	1.65	850	801	3.8494×10^{-12}	9.7017×10^{-12}
	[41]	—	850	—	7.8920×10^{-4}	—
$h = \tau = 0.1$						
	RBF-FD	3.65	1700	1689	3.2235×10^{-12}	2.8677×10^{-11}
	[41]	—	1700	—	1.8771×10^{-4}	—
	[38]	—	1700	—	1.1314×10^{-3}	—
$h = \tau = 0.05$						
	RBF-FD	5.60	3400	2971	2.3648×10^{-10}	2.9115×10^{-9}
	[41]	—	3400	—	2.8359×10^{-4}	—
	[38]	—	3400	—	4.6987×10^{-5}	—

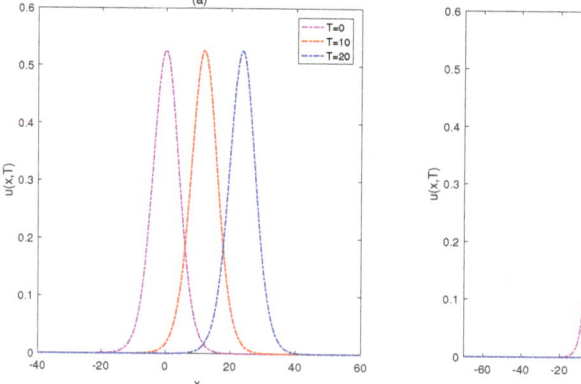

Figure 3. Motion of the single solitary wave with $\tau = h = 0.05$, at various times over the intervals: $x \in [-40, 60]$ (**a**) and $x \in [-70, 100]$ (**b**) for Example 1.

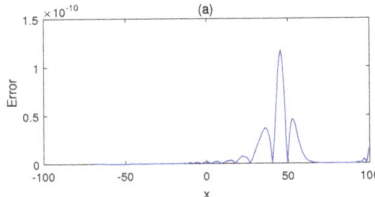
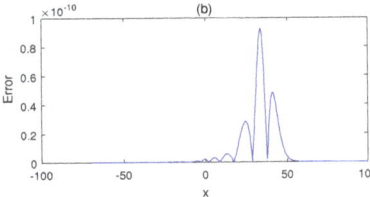

Figure 4. The absolute error L_∞ with $\tau = h = 0.05$, at final times $T = 30$ (**a**) and $T = 40$ (**b**) over the interval $x \in [-70, 100]$ for Example 1.

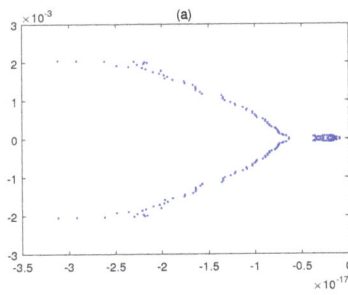
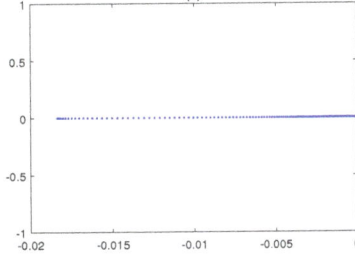

Figure 5. The eigenvalues of **A** (**a**) and **B** (**b**) for $N = 1000$, $n_i = 589$ and $c = 1.08$ in Example 1.

Table 3. The L_∞, L_2 and RMS errors with $\tau = 0.01$, $N = 900$ and $n_i = 489$ for Example 2.

Method	T	c	L_∞	L_2	RMS
RBF-FD	5	1.55	1.3384×10^{-8}	8.7070×10^{-8}	1.5499×10^{-9}
RBF-FD	10	1.55	1.5966×10^{-8}	3.6942×10^{-8}	3.0163×10^{-9}
RBF-FD	15	3.10	1.7030×10^{-8}	5.1031×10^{-8}	4.1667×10^{-9}
RBF-FD	20	2.90	1.7257×10^{-8}	6.3340×10^{-8}	5.1717×10^{-9}
RBF-FD	25	1.80	1.7608×10^{-8}	7.3863×10^{-8}	6.0309×10^{-9}
RBF-FD	30	3.10	3.5768×10^{-8}	9.2721×10^{-8}	7.5706×10^{-9}
RBF-FD	35	3.10	1.8542×10^{-7}	2.3769×10^{-7}	1.9407×10^{-8}
RBF-FD	40	1.55	9.5039×10^{-7}	1.0022×10^{-6}	8.1827×10^{-8}

Table 4. The L_∞ and L_2 errors under different mesh steps $h = \tau$ at $T = 40$ for Example 2.

	Method	c	N	n_i	L_∞	L_2
$h = \tau = 1/4$						
	RBF-FD	2.10	600	569	1.7483×10^{-8}	1.2820×10^{-7}
	[41]	–	600	–	1.7999×10^{-2}	–
	[34]	–	600	–	9.2311×10^{-3}	–
$h = \tau = 1/8$						
	RBF-FD	3.10	1600	869	1.7234×10^{-8}	1.7947×10^{-7}
	[41]	–	1600	–	4.5680×10^{-3}	–
	[34]	–	1600	–	2.3321×10^{-3}	–
$h = \tau = 1/16$						
	RBF-FD	6.80	2400	1541	1.6718×10^{-8}	2.4817×10^{-7}
	[41]	–	2400	–	1.1469×10^{-3}	–
	[34]	–	2400	–	5.8475×10^{-4}	–

Table 5. The energy E under different mesh steps $\tau = h$ for Example 2.

	Method	T	c	N	n_i	E
$h = \tau = 1/4$						
	RBF-FD	10	2.6	600	235	6.211055573870
	[41]	10	-	600	-	6.221349804819
	RBF-FD	20	2.6	600	235	6.211055573872
	[41]	20	-	600	-	6.221349804820
	RBF-FD	30	2.6	600	235	6.211055573871
	[41]	30	-	600	-	6.221349804820
	RBF-FD	40	2.6	600	235	6.211055573869
	[41]	40	-	600	-	6.221349804820
$h = \tau = 1/8$						
	RBF-FD	10	3.1	1200	869	6.216240094383
	[41]	10	-	1200	-	6.221405877565
	RBF-FD	20	3.1	1200	869	6.216240094388
	[41]	20	-	1200	-	6.221405877551
	RBF-FD	30	3.1	1200	869	6.216240094391
	[41]	30	-	1200	-	6.221405877549
	RBF-FD	40	3.1	1200	869	6.216240094396
	[41]	40	-	1200	-	6.216240094397
$h = \tau = 0.0625$						
	RBF-FD	10	6.8	2400	1541	6.218832354561
	[41]	10	-	2400	-	6.221419928242
	RBF-FD	20	6.8	2400	1541	6.218832354855
	[41]	20	-	2400	-	6.221419928522
	RBF-FD	30	6.8	2400	1541	6.218832354657
	[41]	30	-	2400	-	6.221419928339
	RBF-FD	40	6.8	2400	1541	6.218832354213
	[41]	40	-	2400	-	6.221419928294

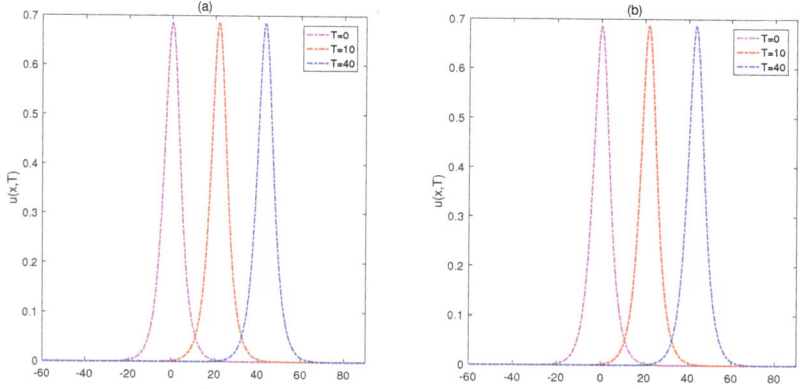

Figure 6. Motion of the single solitary wave with $h = \tau = 0.125$ (**a**) and $h = \tau = 0.0625$ (**b**) over the interval $x \in [-60, 90]$ at final times $T \in \{0, 10, 40\}$ for Example 2.

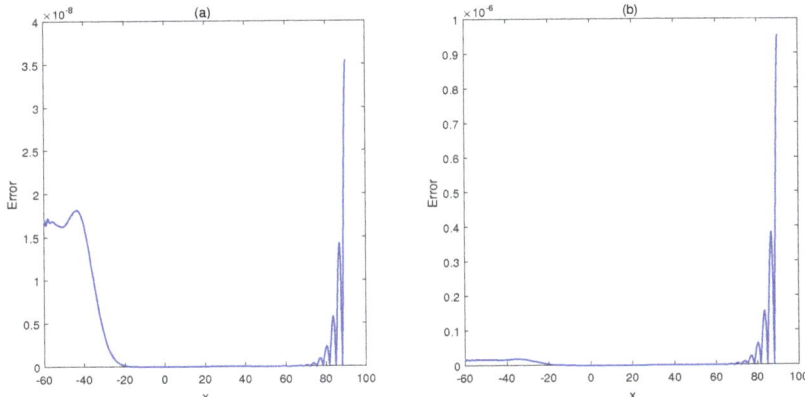

Figure 7. The absolute error L_∞ with $\tau = h = 0.05$, at final times $T = 30$ (**a**) and $T = 40$ (**b**) over the interval $x \in [-60, 90]$ for Example 2.

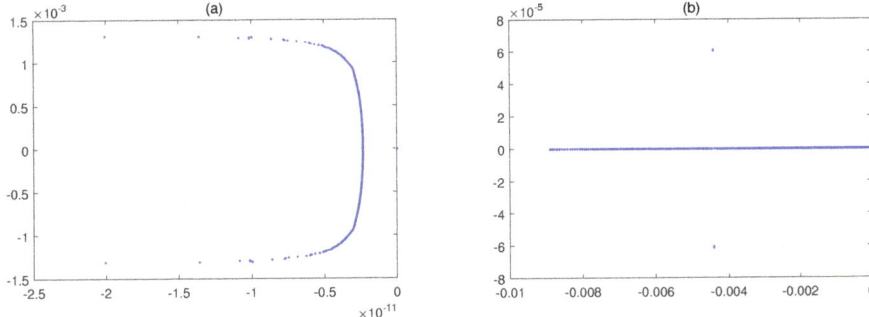

Figure 8. The eigenvalues of **A** (**a**) and **B** (**b**) for $N = 600$, $n_i = 235$ and $c = 1.12$ in Example 2.

Example 3. *We consider the general Rosenau-KdV-RLW model (1) corresponding to the case* $\alpha = \mu = 1$, $\beta = 0.5$, $p = 2$, $\mu = 0.1$ *and* $\delta = 0$ *over the spatial interval* $x \in [-40, 100]$. *The exact solution is* $u(x,t) = k_{31} \operatorname{sech}^4[k_{32}(x - k_{33}t)]$, *where [19]*

$$k_{31} = -\frac{5}{456}(25 - 13\sqrt{457}), \; k_{32} = \left(\frac{-13 + \sqrt{457}}{288}\right)^{1/2}, \; k_{33} = \frac{241 + 13\sqrt{457}}{266}.$$

Table 6 compares the results of the proposed method with those resulting from the schemes in [19,41]. The computational efficiency is clearly superior to the performance exhibited by the other schemes. Figure 9 plots the motion of single solitary wave with $h = \tau = 0.5$, (left) $h = \tau = 0.25$ (right) over the spatial interval $x \in [-40, 100]$ at final times $T \in \{0, 30, 40\}$. The peak of the solitary waves remains the same during the simulation. Figure 10 shows the eigenvalues of the matrices **A** and **B** (left and right panel, respectively) with $N = 100$. The eigenvalues calculated for **A** and **B** have zero or negative values. Hence, the stability of the proposed system for this case is confirmed.

Table 6. The L_∞ and L_2 errors under different mesh steps $h = \tau$ for Example 3.

	Method	T	c	N	n_i	L_∞	L_2
$h = \tau = 1/2$							
	RBF-FD	30	1.71	280	241	5.3379×10^{-1}	2.1555×10^0
	[19]	30	-	280	-	9.8675×10^{-1}	2.5784×10^0
$h = \tau = 1/4$							
	RBF-FD	30	2.90	560	431	6.5718×10^{-2}	3.5432×10^{-1}
	[41]	30	-	560	-	6.9960×10^{-1}	1.8662^0
	[19]	30	-	560	-	9.8675×10^{-1}	2.9434×10^0
$h = \tau = 1/8$							
	RBF-FD	30	5.40	1120	881	4.2035×10^{-2}	1.8697×10^{-2}
	[41]	30	-	1120	-	1.9713×10^{-1}	5.1866×10^{-1}
	[19]	30	-	1120	-	5.1920×10^{-2}	8.0563×10^{-1}

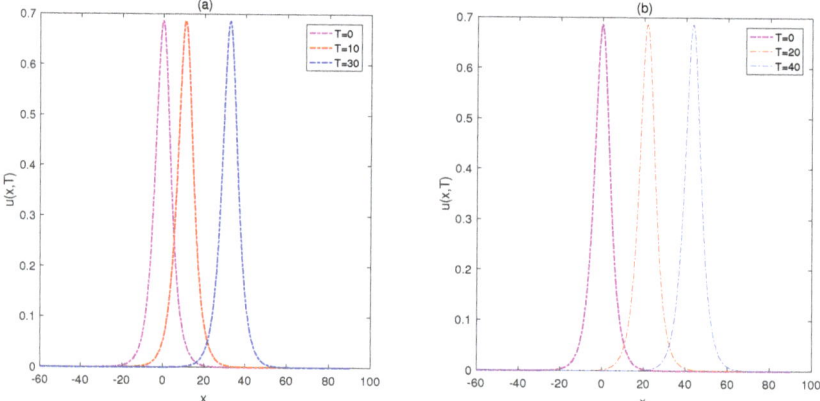

Figure 9. Motion of the single solitary wave with $h = \tau = 0.5$ (**a**) and $h = \tau = 0.25$ (**b**) over the interval $x \in [-40, 100]$ at final times $T \in \{0, 10, 30\}$ (**a**) and $T \in \{0, 20, 40\}$ (**b**) for Example 3.

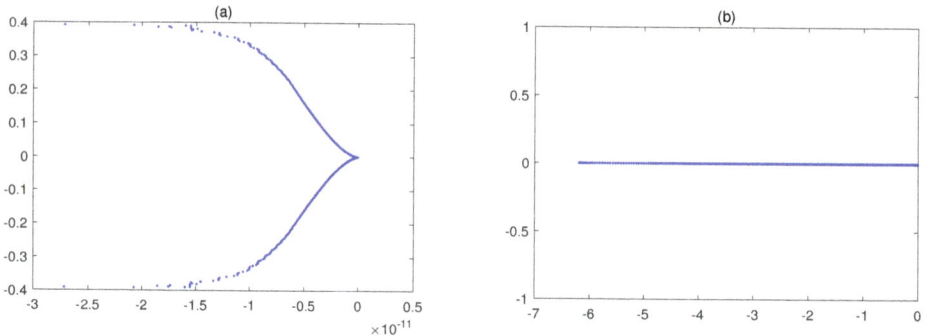

Figure 10. The eigenvalues of **A** (**a**) and **B** (**b**) for $N = 5600$, $n_i = 431$ and $c = 1.14$ for Example 3.

Example 4. Let us consider the general Rosenau-KdV-RLW model (1) in the case of $\alpha = \mu = 1$, $\beta = 1$, $p = 2$, $\gamma = 0$ and $\delta = 1$ over the spatial interval $x \in [-50, 150]$ [19,25,33,42]. The exact solution is $u(x, t) = k_{41} \operatorname{sech}^4[k_{42}(x - k_{43}t)]$, where

$$k_{41} = \frac{15}{19}, \quad k_{42} = \frac{\sqrt{13}}{26}, \quad k_{43} = \frac{169}{133}.$$

Table 7 compares the results of proposed method with those obtained with other schemes [19,25,33,42]. In this case, the accuracy of the method is slightly better than those achieved with the rest. Figure 11 depicts the motion of the single solitary wave with $h = \tau = 0.4$ (left) and $h = \tau = 0.2$ (right) over the spatial interval $x \in [-50, 150]$ at final times $T \in \{8, 16, 24, 32\}$. The crest of the soliton clearly remains the same during the simulation. Figure 12 plots the eigenvalues of the matrices **A** and **B** (left and right panels, respectively) with $N = 100$. The eigenvalues calculated for **A** are negative values, while for **B** they have zero or negative values. Hence, the stability of the proposed system for this case is verified.

Table 7. The L_∞ and L_2 errors under different mesh steps $h = \tau$ with $N = 250$ and $n_i = 181$ at $T = 24$ for Example 4.

	Method	T	c	N	n_i	L_∞	L_2
$h = \tau = 0.8$							
	RBF-FD	24	0.35	250	181	1.2281×10^{-11}	4.7975×10^{-11}
	[42]	24	-	250	-	3.09410×10^{-4}	7.78402×10^{-4}
	[25]	24	-	250	-	9.06883×10^{-4}	2.42851×10^{-1}
	[33]	24	-	250	-	1.16717×10^{-1}	3.11658×10^{-1}
	[19]	24	-	250	-	7.56362×10^{-3}	2.03287×10^{-2}
$h = \tau = 0.4$							
	RBF-FD	24	0.65	500	381	1.3151×10^{-11}	5.2620×10^{-11}
	[42]	24	-	500	-	1.87205×10^{-5}	4.73034×10^{-5}
	[25]	24	-	500	-	2.48437×10^{-4}	6.58790×10^{-2}
	[33]	24	-	500	-	3.27045×10^{-2}	8.62872×10^{-2}
	[19]	24	-	500	-	1.82402×10^{-3}	4.88759×10^{-3}
$h = \tau = 0.2$							
	RBF-FD	24	1.40	1000	631	1.3472×10^{-11}	5.4481×10^{-11}
	[42]	24	-	100	-	1.16521×10^{-6}	2.94078×10^{-6}
	[25]	24	-	1000	-	6.36404×10^{-3}	1.68468×10^{-2}
	[33]	24	-	1000	-	8.43616×10^{-3}	2.21942×10^{-2}
	[19]	24	-	1000	-	4.52324×10^{-4}	1.21311×10^{-3}
$h = \tau = 0.1$							
	RBF-FD	24	2.60	2000	1831	1.4086×10^{-11}	5.9844×10^{-11}
	[42]	24	-	2000	-	7.27778×10^{-8}	1.83776×10^{-7}
	[25]	24	-	2000	-	1.12985×10^{-4}	4.23946×10^{-3}
	[33]	24	-	2000	-	3.02978×10^{-4}	5.59422×10^{-3}
	[19]	24	-	2000	-	2.21651×10^{-3}	3.02978×10^{-3}

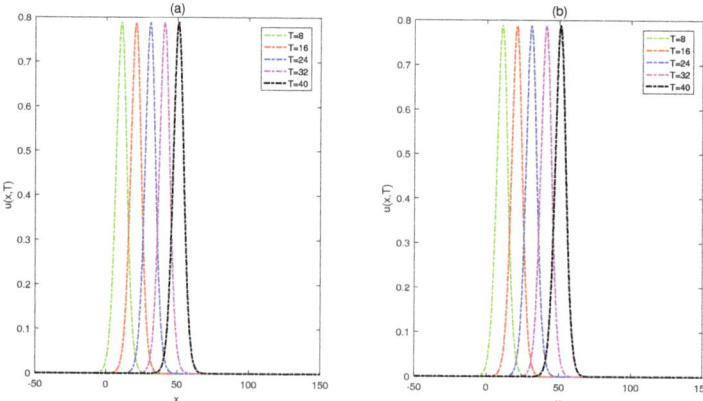

Figure 11. Motion of the single solitary wave with $h = \tau = 0.4$ (**a**) and $h = \tau = 0.2$ (**b**) over the interval $x \in [-50, 150]$ at final times $T \in \{8, 16, 24, 32\}$ for Example 4.

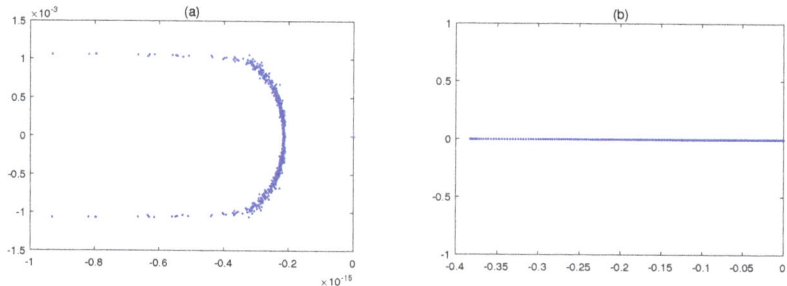

Figure 12. The eigenvalues of **A** (**a**) and **B** (**b**) for $N = 1000$, $n_i = 631$, $p = 8$ and $c = 0.95$ in Example 4.

Example 5. *Consider the general Rosenau-KdV-RLW model (1) with parameters as $\alpha = \beta = \mu = \delta = 1$ and $\gamma = 0$, in two spatial intervals, namely $x \in [-60, 120]$ and $x \in [-30, 120]$. The exact solution is given by*

$$u(x,t) = \exp(k_{51}) \operatorname{sech}^{\frac{4}{p-1}}[k_{52}(x - k_{53}t)],$$

where

$$k_{51} = (\ln[(p+3)(3p+1)(p+1)]/[2(p^2+3)(p^2+4p+7)])/(p-1),$$
$$k_{52} = \frac{p-1}{\sqrt{4p^2+8p+20}},$$
$$k_{53} = (p^4 + 4p^3 + 14p^2 + 20p + 25)/(p^4 + 4p^3 + 10p^2 + 12p + 21).$$

The initial boundary value problem (1)–(3) includes the following conservative quantities:

$$I_1 = \frac{1}{2}\int_a^b u\, dx = \frac{h}{2}\sum_{i=1}^{N} u_i,$$

$$I_2 = \frac{1}{2}\int_a^b (u^2 + u_x^2 + u_{xx}^2)\, dx = \frac{h}{2}\sum_{i=1}^{N}\left(u_i + (u_x)_i^2 + (u_{xx})_i^2\right),$$

related to mass and energy. The quantities I_1 and I_2 are applied to measure the conservation properties of the present method, calculated by means of the trapezoidal rule for the Rosenau-RLW equation.

Tables 8 and 9 compare the results of the proposed method with those obtained from the schemes presented in [26,36,37,39]. It can be observed that the computational results are clearly better than the others and that the invariants I_1 and I_2 remain constant during the simulation. Figure 13 plots the motion of the single solitary wave for various p at $T = \{0, 30, 60\}$ in the spatial interval $x \in [-60, 120]$. The single solitons move to the right at a constant speed and conserve their amplitudes and shapes. Figure 14 shows the eigenvalues of the linearized differentiation operator **A** and **B** (left and right panels, respectively) with $N = 100$. The eigenvalues calculated for **A** and **B** are zero, or have negative values. Therefore, the stability of the proposed system for this case is confirmed.

Table 8. The L_∞, L_2 and RMS errors and the invariants I_1 and I_2 with $N = 1500, n_i = 1089, c = 2.6$ and $\tau = 0.01$ in the spatial interval $x \in [-30, 120]$ for Example 5.

Method	T	L_∞	L_2	RMS	I_1	I_2
$p = 2$						
RBF-FD	10	4.2666×10^{-7}	1.1117×10^{-6}	9.0769×10^{-8}	1.89238729	0.53169648
[37]	10	7.6292×10^{-6}	1.8132×10^{-5}	—	1.89765990	0.53317753
RBF-FD	20	4.5738×10^{-7}	5.3007×10^{-6}	1.3686×10^{-7}	1.89238729	0.53169648
[37]	20	9.0949×10^{-6}	2.2513×10^{-5}	—	1.89766149	0.53317753
RBF-FD	30	4.6844×10^{-7}	6.5742×10^{-6}	1.6975×10^{-7}	1.89238729	0.53169648
[37]	30	1.0274×10^{-5}	2.5463×10^{-5}	—	1.89766306	0.53317753
RBF-FD	40	4.7437×10^{-7}	7.6096×10^{-6}	1.9648×10^{-7}	1.89238729	0.53169648
[37]	40	1.1378×10^{-5}	2.8139×10^{-5}	—	1.89766459	0.53317753
RBF-FD	50	4.7692×10^{-7}	8.4995×10^{-6}	2.1946×10^{-7}	1.89238729	0.53169648
[37]	50	1.2447×10^{-5}	3.0753×10^{-5}	—	1.89766608	0.53317753
$p = 3$						
RBF-FD	10	3.9146×10^{-6}	3.1606×10^{-5}	8.1606×10^{-7}	2.66518850	1.11037761
[37]	10	2.1569×10^{-5}	4.9409×10^{-5}	—	2.67262472	1.11347058
RBF-FD	20	4.2260×10^{-6}	4.9004×10^{-5}	1.2653×10^{-6}	2.66518850	1.11037761
[37]	20	2.7517×10^{-5}	6.5313×10^{-5}	—	2.67264006	1.11347058
RBF-FD	30	4.3421×10^{-6}	6.1274×10^{-5}	1.5821×10^{-6}	2.66518850	1.11037761
[37]	30	3.3326×10^{-5}	7.9999×10^{-5}	—	2.67265504	1.11347058
RBF-FD	40	4.4063×10^{-6}	7.1244×10^{-5}	1.8395×10^{-6}	2.66518850	1.11037761
[37]	40	3.9091×10^{-5}	9.4787×10^{-5}	—	2.67266966	1.11347058
RBF-FD	50	4.4481×10^{-6}	7.6096×10^{-6}	2.0648×10^{-6}	2.66518850	1.11037761
[37]	50	4.4846×10^{-5}	1.0984×10^{-4}	—	2.67268415	1.11347058
$p = 6$						
RBF-FD	10	3.3603×10^{-4}	8.0524×10^{-4}	6.5626×10^{-5}	3.97819339	1.91229616
[37]	10	3.1032×10^{-4}	6.5998×10^{-4}	—	3.99024365	1.91764461
RBF-FD	20	3.6994×10^{-4}	1.3622×10^{-3}	1.1107×10^{-4}	3.97819339	1.91229616
[37]	20	3.1897×10^{-4}	1.1382×10^{-3}	—	3.99024365	1.91764461
RBF-FD	30	3.8386×10^{-4}	1.7513×10^{-3}	1.4281×10^{-4}	3.97819339	1.91229616
[37]	30	3.2836×10^{-4}	1.4631×10^{-3}	—	3.99172706	1.91764489
RBF-FD	40	3.9219×10^{-4}	2.0639×10^{-3}	1.6841×10^{-4}	3.97819339	1.91229616
[37]	40	3.4181×10^{-4}	1.7187×10^{-3}	—	3.99458409	1.91764541
RBF-FD	50	3.9792×10^{-4}	2.3398×10^{-3}	1.9060×10^{-4}	3.99484237	1.91634750
[37]	50	3.4127×10^{-4}	1.9368×10^{-3}	—	3.99597486	1.91764566

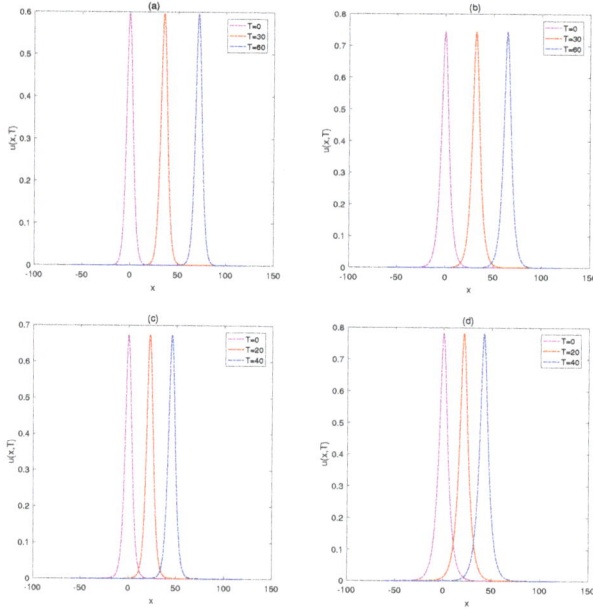

Figure 13. Motion of the single solitary wave for $p = 3$ (**a**), $p = 6$ (**b**), $p = 4$ (**c**), and $p = 8$ (**d**) at final times $T \in \{0, 30, 60\}$ (**a**,**b**) and $T \in \{0, 20, 40\}$ (**c**,**d**) in the spatial interval $x \in [-60, 120]$ for Example 5.

Table 9. The L_∞ and L_2 errors and the quantities Q and E with $N = 360, n_i = 295, c = 0.55$ and $\tau = 0.1$ in the spatial interval $x \in [-60, 120]$ for Example 5.

	Method	L_∞	L_2	Q	E
$p = 4$					
	RBF-FD	4.1402×10^{-10}	2.1363×10^{-9}	6.248401	2.859729
	[39]	1.3784×10^{-4}	9.3510×10^{-4}	6.266377	2.868226
	[39]	1.0310×10^{-5}	2.3550×10^{-5}	6.265844	2.867735
	[39]	2.9706×10^{-4}	6.6954×10^{-4}	6.265806	2.867684
	[39]	4.2250×10^{-4}	1.1045×10^{-3}	6.265992	2.867617
	[26]	1.7112×10^{-3}	4.4788×10^{-3}		
	[36]	2.7871×10^{-2}	7.4517×10^{-2}		
$p = 8$					
	RBF-FD	2.7865×10^{-6}	1.4924×10^{-5}	9.745127	4.722011
	[39]	1.3784×10^{-4}	3.8078×10^{-4}	9.742126	4.735346
	[39]	2.9490×10^{-5}	7.5220×10^{-5}	9.742181	4.735225
	[39]	6.2856×10^{-4}	1.7039×10^{-3}	9.742146	4.735302
	[39]	4.7892×10^{-4}	1.2762×10^{-3}	9.742227	4.735082
	[26]	1.6189×10^{-3}	4.3184×10^{-3}		
	[36]	2.9534×10^{-2}	8.0373×10^{-2}		
$p = 16$					
	RBF-FD	9.1964×10^{-4}	4.8646×10^{-3}	17.167390	8.372094
	[39]	4.4109×10^{-4}	2.3334×10^{-3}	17.168699	8.375376
	[39]	4.4493×10^{-4}	2.3199×10^{-3}	17.169258	8.375400
	[39]	5.3860×10^{-4}	3.0231×10^{-3}	17.172776	8.375393
	[39]	2.2709×10^{-3}	7.6218×10^{-3}	17.116828	8.375272
	[26]	1.1875×10^{-3}	3.5725×10^{-3}		
	[36]	2.2547×10^{-2}	6.1304×10^{-2}		

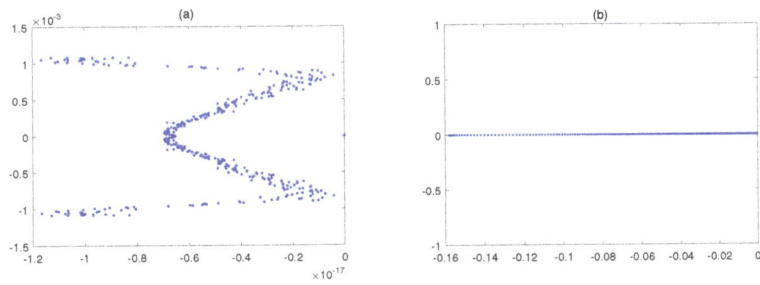

Figure 14. The eigenvalues of **A** (a) and **B** (b) for $N = 500$, $n_i = 111$ and $c = 1.26$ in Example 5.

4. Conclusions

We adopted the local meshless RBF-FD to calculate the approximate numerical solutions of the general nonlinear Rosenau-RLW equation without performing any linearization or transformation of the equation. In order to demonstrate the accuracy of the proposed numerical technique, the error invariants and error norms were computed, and the results were compared with others available in the literature. The local RBF-FD technique was verified to be remarkably accurate. In conclusion, the method is sufficiently accurate and fast due to its limited computational load.

Author Contributions: All authors contributed equally to this paper. All authors have read and agreed to the published version of the manuscript.

Funding: This research received no external funding.

Acknowledgments: The authors are thankful to the respected reviewers for their valuable comments and constructive suggestions towards the improvement of the original paper.

Conflicts of Interest: The authors declare that there is no conflict of interest regarding the publication of this article.

References

1. Korteweg, D.J.; De Vries, G. On the change of form of long waves advancing in a rectangular canal, and on a new type of long stationary waves. *Lond. Edinb. Dublin Philos. Mag. J. Sci.* **1895**, *39*, 422–443.
2. Benjamin, T.B.; Bona, J.L.; Mahony, J.J. Model equations for long waves in nonlinear dispersive systems. *Philos. Trans. R. Soc. Lond. Ser.* **1972**, *272*, 47–78.
3. Barreto, R.K.; De Caldas, C.S.; Gamboa, P.; Limaco, J. Existence of solutions to the Rosenau and Benjamin-Bona-Mahony equation in domains with moving boundary. *Electron. J. Differ. Equ.* **2004**, *2004*, 1–12.
4. Ramos, J.I. Explicit finite difference methods for the EW and RLW equations. *Appl. Math. Comput.* **2006**, *179*, 622–638.
5. Zhang, L. A finite difference scheme for generalized regularized long-wave equation. *Appl. Math. Comput.* **2005**, *168*, 962–972.
6. Zuo, J.M. Solitons and periodic solutions for the Rosenau–KdV and Rosenau–Kawahara equations. *Appl. Math. Comput.* **2009**, *215*, 835–840.
7. Rosenau, P. A quasi-continuous description of a nonlinear transmission line. *Phys. Scr.* **1986**, *34*, 827.
8. Cui, Y.; Mao, D.k. Numerical method satisfying the first two conservation laws for the Korteweg–de Vries equation. *J. Comput. Phys.* **2007**, *227*, 376–399.
9. Razborova, P.; Moraru, L.; Biswas, A. Perturbation of dispersive shallow water waves with Rosenau-KdV-RLW equation and power law nonlinearity. *Rom. J. Phys.* **2014**, *59*, 658–676.
10. Coclite, G.M.; di Ruvo, L. A singular limit problem for conservation laws related to the Rosenau–Korteweg-de Vries equation. *J. Math. Pures Appl.* **2017**, *107*, 315–335.
11. Mendez, A.J. On the propagation of regularity for solutions of the fractional Korteweg-de Vries equation. *J. Differ. Equ.* **2020**, *269*, 9051–9089.

12. Benia, Y.; Scapellato, A. Existence of solution to Korteweg–de Vries equation in a non-parabolic domain. *Nonlinear Anal.* **2020**, *195*, 111758.
13. Kaya, D.; Aassila, M. An application for a generalized KdV equation by the decomposition method. *Phys. Lett. A* **2002**, *299*, 201–206.
14. Özer, S.; Kutluay, S. An analytical–numerical method for solving the Korteweg–de Vries equation. *Appl. Math. Comput.* **2005**, *164*, 789–797.
15. Peregrine, D.H. Long waves on a beach. *J. Fluid Mech.* **1967**, *27*, 815–827.
16. Peregrine, D. Calculations of the development of an undular bore. *J. Fluid Mech.* **1966**, *25*, 321–330.
17. Bona, J.; Bryant, P.J. A mathematical model for long waves generated by wavemakers in non-linear dispersive systems. In *Mathematical Proceedings of the Cambridge Philosophical Society*; Cambridge University Press: Cambridge, UK, 1973; Volume 73, pp. 391–405.
18. Abdulloev, K.O.; Bogolubsky, I.; Makhankov, V.G. One more example of inelastic soliton interaction. *Phys. Lett. A* **1976**, *56*, 427–428.
19. Wongsaijai, B.; Poochinapan, K. A three-level average implicit finite difference scheme to solve equation obtained by coupling the Rosenau–KdV equation and the Rosenau–RLW equation. *Appl. Math. Comput.* **2014**, *245*, 289–304.
20. Pan, X.; Wang, Y.; Zhang, L. Numerical analysis of a pseudo-compact CN conservative scheme for the Rosenau-KdV equation coupling with the Rosenau–RLW equation. *Bound. Value Probl.* **2015**, *2015*, 65.
21. Apolinar-Fernández, A.; Ramos, J.I. Numerical solution of the generalized, dissipative KdV–RLW–Rosenau equation with a compact method. *Commun. Nonlinear Sci. Numer. Simul.* **2018**, *60*, 165–183.
22. Razborova, P.; Ahmed, B.; Biswas, A. Solitons, shock waves and conservation laws of Rosenau-KdV-RLW equation with power law nonlinearity. *Appl. Math. Inf. Sci.* **2014**, *8*, 485.
23. Razborova, P.; Kara, A.H.; Biswas, A. Additional conservation laws for Rosenau–KdV–RLW equation with power law nonlinearity by Lie symmetry. *Nonlinear Dyn.* **2015**, *79*, 743–748.
24. Sanchez, P.; Ebadi, G.; Mojaver, A.; Mirzazadeh, M.; Eslami, M.; Biswas, A. Solitons and other solutions to perturbed Rosenau-KdV-RLW equation with power law nonlinearity. *Acta Phys. Pol. A* **2015**, *127*, 1577–1586.
25. Pan, X.; Zhang, L. On the convergence of a conservative numerical scheme for the usual Rosenau–RLW equation. *Appl. Math. Model.* **2012**, *36*, 3371–3378.
26. Wongsaijai, B.; Poochinapan, K.; Disyadej, T. A Compact Finite Difference Method for Solving the General Rosenau–RLW Equation. *Int. J. Appl. Math.* **2014**, *44*, 192–199.
27. Dutykh, D.; Chhay, M.; Fedele, F. Geometric numerical schemes for the KdV equation. *Comput. Math. Math. Phys.* **2013**, *53*, 221–236.
28. Noon, N.J. Fully discrete formulation of Galerkin-Partial artificial diffusion finite element method for coupled Burgers' problem. *Int. J. Adv. Appl. Math. Mech.* **2014**, *1*, 56–75.
29. El-Sayed, M.; Moatimid, G.; Moussa, M.; El-Shiekh, R.; Al-Khawlani, M. New exact solutions for coupled equal width wave equation and (2+1)-dimensional Nizhnik-Novikov-Veselov system using modified Kudryashov method. *Int. J. Adv. Appl. Math. Mech.* **2014**, *2*, 19–25.
30. Park, M.A. Pointwise decay estimates of solutions of the generalized Rosenau equation. *J. Korean Math. Soc.* **1992**, *29*, 261–280.
31. Wang, M.; Li, D.; Cui, P. A conservative finite difference scheme for the generalized Rosenau equation. *Int. J. Pure Appl. Math.* **2011**, *71*, 539–549.
32. Karakoc, S.B.G.; Ak, T. Numerical simulation of dispersive shallow water waves with Rosenau-KdV equation. *Int. J. Adv. Appl. Math. Mech.* **2016**, *3*, 32–40.
33. Hu, J.; Wang, Y. A high-accuracy linear conservative difference scheme for Rosenau–RLW equation. *Math. Probl. Eng.* **2013**, *2013*, 423718, doi:10.1155/2013/870291.
34. Zheng, M.; Zhou, J. An average linear difference scheme for the generalized Rosenau-KdV equation. *J. Appl. Math.* **2014**, 202793, doi:10.1155/2014/202793.
35. Esfahani, A. Solitary wave solutions for generalized Rosenau-KdV equation. *Commun. Theor. Phys.* **2011**, *55*, 396–398.
36. Pan, X.; Zhang, L. Numerical simulation for general Rosenau–RLW equation: An average linearized conservative scheme. *Math. Probl. Eng.* **2012**, 517818, doi:10.1155/2012/517818.
37. Mittal, R.C.; Jain, R.K. Numerical solution of General Rosenau–RLW Equation using Quintic B-splines Collocation Method. *Commun. Numer. Anal.* **2012**, cna-00129. doi: 10.5899/2012/cna-00129

38. Hu, J.; Xu, Y.; Hu, B. Conservative linear difference scheme for Rosenau-KdV equation. *Adv. Math. Phys.* **2013**, 423718, doi:10.1155/2013/423718.
39. Arı, M.; Dereli, Y. Numerical solutions of the general Rosenau-RLW equation using meshless kernel based method of lines. *J. Phys. Conf. Ser.* **2016**, *766*, 012030, doi:10.1088/1742-6596/766/1/012030.
40. Foroutan, M.; Ebadian, A. Chebyshev rational approximations for the Rosenau-KdV-RLW equation on the whole line. *Int. J. Anal. Appl.* **2018**, *16*, 1–15.
41. Wang, X.; Dai, W. A three-level linear implicit conservative scheme for the Rosenau–KdV–RLW equation. *J. Comput. Appl. Math.* **2018**, *330*, 295–306.
42. Wongsaijai, B.; Mouktonglang, T.; Sukantamala, N.; Poochinapan, K. Compact structure-preserving approach to solitary wave in shallow water modeled by the Rosenau–RLW equation. *Appl. Math. Comput.* **2019**, *340*, 84–100.
43. Fasshauer, G.E. *Meshfree Approximation Methods with Matlab*; World Scientific Publishing Company: Singapore, 2007; Volume 6.
44. Wendland, H. *Scattered Data Approximation*; Cambridge University Press: Cambridge, UK, 2005.
45. Buhmann, M.D. *Radial Basis Functions: Theory and Implementations*; Cambridge University Press: Cambridge, UK, 2003; Volume 12.
46. Micchelli, C.A. Interpolation of scattered data: Distance matrices and conditionally positive definite functions. In *Approximation Theory and Spline Functions*; Springer: Berlin/Heidelberg, Germany, 1984; pp. 143–145.
47. Shu, C.; Ding, H.; Yeo, K. Local radial basis function-based differential quadrature method and its application to solve two-dimensional incompressible Navier–Stokes equations. *Comput. Methods Appl. Mech. Eng.* **2003**, *192*, 941–954.
48. Tolstykh, A.; Shirobokov, D. On using radial basis functions in a "finite difference mode" with applications to elasticity problems. *Comput. Mech.* **2003**, *33*, 68–79.
49. Sarra, S.A. A local radial basis function method for advection–diffusion–reaction equations on complexly shaped domains. *Appl. Math. Comput.* **2012**, *218*, 9853–9865.
50. Su, L. A radial basis function (RBF)-finite difference (FD) method for the backward heat conduction problem. *Appl. Math. Comput.* **2019**, *354*, 232–247.
51. Nikan, O.; Machado, J.T.; Golbabai, A. Numerical solution of time-fractional fourth-order reaction-diffusion model arising in composite environments. *Appl. Math. Model.* **2020**, *81*, 819–836.
52. Nikan, O.; Jafari, H.; Golbabai, A. Numerical analysis of the fractional evolution model for heat flow in materials with memory. *Alex. Eng. J.* **2020**, *59*, 2627–2637.
53. Nikan, O.; Machado, J.T.; Avazzadeh, Z.; Jafari, H. Numerical evaluation of fractional Tricomi-type model arising from physical problems of gas dynamics. *J. Adv. Res.* **2020**, *25*, 205–216.
54. Bollig, E.F.; Flyer, N.; Erlebacher, G. Solution to PDEs using radial basis function finite-differences (RBF-FD) on multiple GPUs. *J. Comput. Phys.* **2012**, *231*, 7133–7151.
55. Flyer, N.; Lehto, E.; Blaise, S.; Wright, G.B.; St-Cyr, A. A guide to RBF-generated finite differences for nonlinear transport: Shallow water simulations on a sphere. *J. Comput. Phys.* **2012**, *231*, 4078–4095.
56. Garshasbi, M.; Khakzad, M. The RBF collocation method of lines for the numerical solution of the CH-γ equation. *J. Adv. Res. Dyn. Control Syst.* **2015**, *4*, 65–83.
57. Uddin, M. RBF-PS scheme for solving the equal width equation. *Appl. Math. Comput.* **2013**, *222*, 619–631.
58. Nikan, O.; Golbabai, A.; Nikazad, T. Solitary wave solution of the nonlinear KdV-Benjamin-Bona-Mahony-Burgers model via two meshless methods. *Eur. Phys. J. Plus* **2019**, *134*, 367.
59. Dehghan, M.; Shafieeabyaneh, N. Local radial basis function–finite-difference method to simulate some models in the nonlinear wave phenomena: Regularized long-wave and extended Fisher–Kolmogorov equations. *Eng. Comput.* **2019**, 1–21. doi:10.1007/s00366-019-00877-z.
60. Ebrahimijahan, A.; Dehghan, M. The numerical solution of nonlinear generalized Benjamin-Bona-Mahony-Burgers and regularized long-wave equations via the meshless method of integrated radial basis functions. *Eng. Comput.* **2019**, 1–30. doi:10.1007/s00366-019-00811-3.
61. Rashidinia, J.; Rasoulizadeh, M.N. Numerical methods based on radial basis function-generated finite difference (RBF-FD) for solution of GKdVB equation. *Wave Motion* **2019**, *90*, 152–167.
62. Rasoulizadeh, M.N.; Rashidinia, J. Numerical solution for the Kawahara equation using local RBF-FD meshless method. *J. King Saud Univ.-Sci.* **2020**, *32*, 2277–2283.

63. Dormand, J.R.; Prince, P.J. A family of embedded Runge-Kutta formulae. *J. Comput. Appl. Math.* **1980**, *6*, 19–26.
64. Shampine, L.; Gordon, M. *Computer Solution of Ordinary Differential Equations. The Initial Value Problems*; W. H. Freeman: New York, NY, USA, 1975. doi:10.1093/comjnl/19.2.155.
65. Trefethen, L.N. *Spectral Methods in MATLAB*; SIAM: Philadelphia, PA, USA, 2000; Volume 10.
66. Jain, M.K. *Numerical Solution of Differential Equations*; Wiley Eastern: New Delhi, India, 1979.

© 2020 by the authors. Licensee MDPI, Basel, Switzerland. This article is an open access article distributed under the terms and conditions of the Creative Commons Attribution (CC BY) license (http://creativecommons.org/licenses/by/4.0/).

Article

Event-Based Implementation of Fractional Order IMC Controllers for Simple FOPDT Processes

Cristina I. Muresan [1], Isabela R. Birs [1,2,*] and Eva H. Dulf [1,3,*]

1. Automation Department, Technical University of Cluj-Napoca, Memorandumului 28, 400114 Cluj-Napoca, Romania; Cristina.Muresan@aut.utcluj.ro
2. Research Group of Dynamical Systems and Control, Faculty of Engineering and Architecture, Ghent University, EEDT Decision & Control, Flanders Make Consortium, Tech Lane Scrience Park 125, B-9052 Ghent, Belgium
3. Physiological Controls Research Center, Óbuda University, H-1034 Budapest, Hungary
* Correspondence: Isabela.Birs@aut.utcluj.ro (I.R.B.); Eva.Dulf@aut.utcluj.ro (E.H.D.)

Received: 2 August 2020; Accepted: 14 August 2020; Published: 17 August 2020

Abstract: Fractional order calculus has been used to generalize various types of controllers, including internal model controllers (IMC). The focus of this manuscript is towards fractional order IMCs for first order plus dead-time (FOPDT) processes, including delay and lag dominant ones. The design is novel at it is based on a new approximation approach, the non-rational transfer function method. This allows for a more accurate approximation of the process dead-time and ensures an improved closed loop response. The main problem with fractional order controllers is concerned with their implementation as higher order transfer functions. In cases where central processing unit CPU, bandwidth allocation, and energy usage are limited, resources need to be efficiently managed. This can be achieved using an event-based implementation. The novelty of this paper resides in such an event-based algorithm for fractional order IMC (FO-IMC) controllers. Numerical results are provided for lag and delay dominant FOPDT processes. For comparison purposes, an integer order PI controller, tuned according to the same performance specifications as the FO-IMC, is also implemented as an event-based control strategy. The numerical results show that the proposed event-based implementation for the FO-IMC controller is suitable and provides for a smaller computational effort, thus being more suitable in various industrial applications.

Keywords: fractional order IMC; first order plus dead-time processes; event-based implementation; numerical simulations; comparative closed loop results

1. Introduction

Fractional calculus has been reaching a larger part of the research community due to the numerous advantages it has. The increasing interest is mainly due to the ability to capture essential dynamics in physical phenomena. This is seconded by the demonstrated ability of fractional order controllers to meet more design specifications and provide for overall increased robustness and performance. Several researchers have used fractional order tools to model more accurately viscoelastic phenomena [1], aerodynamics [2], structural engineering [3], non-Newtonian characteristics in blood [4,5], type 1 diabetes [6], diffusion phenomena in magnetic resonance imaging [7], post-exposure prophylaxis model in HIV [8], epidemic models for infectious diseases [9], biochemical phenomena [10], etc.

In terms of fractional order control, the starting point is the generalization to arbitrary orders of the proportional-integral-derivative (PID) controller, as proposed in [11]. Ever since then, a manifold of papers have been published, presenting various modifications of the original fractional order PID (FO-PID) controller, various tuning methods and improvements. The key idea is that the generalization of the PID to a fractional order provides more flexibility in improving the system control performance [12,13]. Several enhancements for FO-PID controllers were proposed. An optimal FO-PID controller was proposed and tuned based on particle swarm optimization [14]. Designs based on phase and gain margin specifications are quite abundant [12,15,16], and quite frequently the design is based on ensuring the iso-damping property [12,16–18]. Tuning is usually performed in the frequency domain, but time domain approaches were also considered [19,20]. Autotuning methods for fractional order PID controllers were also proposed [21–23]. Some rather recent review papers on fractional order controllers can be found in [24–26] and provide an insight into fractional order control of different types. An excellent review paper on fractional order controllers, including their most widely used continuous and discrete approximation methods, as well as their digital and analogue implementation methods. The paper also presents the Matlab toolboxes that facilitate the use of fractional order calculus in modeling and control. At the same time, it clearly pinpoints the advantages and disadvantages of using fractional order calculus in control engineering [27].

For time delay systems, including first order plus dead time (FOPDT) processes, several approaches have been introduced and developed over the years. A recent review paper regarding the approaches for these types of system is given in [28]. Alternative control strategies based on fractional order calculus for variable time delay systems are proposed in [29]. For significant delays, a Smith predictor (SP) scheme can be useful. The fractional order controller design in this SP control scheme is based on several approaches. One method proposes a modified SP structure, where the tuning procedure is based on Bode's ideal transfer function and the internal model control (IMC) principle. The resulting control system is robust to changes in the process parameters [30]. A similar design for a fractional order PI controller in a SP control structure, also based on Bode's ideal transfer function, is presented in [31]. The analytical tuning rules are derived in the frequency domain and applied to various types of processes. The advantages of the method rely on a simple design scheme and a straightforward method, which can be easily implemented in the process industry. The SP control structure is also used as a means for comparing various fractional order controllers for a heat diffusion system in [32]. The research offers valuable insight into the performance of the proposed fractional order control algorithms. In [33], a time domain approach is considered for the design of fractional order controllers in a SP structure. Only two parameters need to be tuned, which simplifies considerably the design procedure. The tuning rules are derived based on an ideal closed-loop transfer function, with performance specifications imposed as overshoot and settling time.

One of the simplest tuning rules for integer order PID controllers, as well as for FO-PID controllers, highly suitable for time delay systems, is the IMC methodology. This consists in the simple inversion of the invertible part of the process model and in the addition of a properly selected filter. This method has also been tackled by researchers. For the design of a fractional order IMC controller (FO-IMC), the most widely used approach is based on using a modified fractional order filter [34]. Some tuning methods are based on the Ziegler–Nichols approach [35], Taylor series [36], dominant pole placement method [37]. Other approaches are based on frequency domain specifications, such as phase and gain margins [38–42]. Such an approach is also preferred in this research.

The FO-IMC control strategy proposed in this paper is based on specifying a certain gain crossover frequency, to ensure a specific closed loop settling time, as well as phase margin criteria to ensure a certain closed loop overshoot. The tuning rules are exemplified for first order plus dead-time processes. To implement the resulting fractional order controllers, an efficient method is used, namely the non-rational transfer function (NRTF) approximation method [43]. This allows for a more accurate approximation of the process dead-time and ensures an improved closed loop response [38]. The design is suitable for all types of FOPDT processes, but can be easily extended to higher order processes or even fractional order transfer functions [44]. The research is focused on controlling processes where CPU, bandwidth allocation, and energy usage are limited [45]. In this context, the idea of event-based control is a natural solution for controller implementation [46,47]. Such an approach has been only recently introduced to fractional order PID controllers [46,47]. The novelty of this paper resides in introducing an event-based methodology for FO-IMC controllers. The method is entirely original compared to [46,47], where a standard fractional order PID type controller is tuned according to some frequency domain specifications. The IMC methodology presented in the current manuscript is not used in the actual tuning. Then, the event-based implementation of the fractional order PID controller in [46] is based on direct discretization methods that use direct fractional order mappers of the fractional integrator and differentiator of the fractional order PID controller. The event-based algorithm relies then on a generalization of the standard direct discretization methods for fractional order elements, where the sampling period is considered as a variable parameter (it depends on an event being triggered). The study implements the proposed strategy entirely in the control signal generator, using a single function to compute the fractional order control value. The novelty of the current manuscript, apart from a different tuning procedure and a different fractional order controller type, is based on a two-step implementation of the event-based fractional order controller. Firstly, the fractional order controller determined based on the FO-IMC methodology is decomposed into an integer order PI and a fractional order filter. Then, the NRTF approach is used to determine a standard discrete-time approximation for the fractional order filter. The remaining PI controller is implemented in an event-based approach. Numerical results are provided for lag and delay dominant FOPDT processes. For comparison purposes, an integer order PI controller, tuned according to the same performance specifications as the FO-IMC, is also implemented as an event-based control strategy. The numerical results show that the proposed event-based implementation for the FO-IMC controller is suitable and provides for a reduction in the resources used for computing the control signal.

The paper is structured as follows. In Section 2, the proposed tuning procedure for the FO-IMC controllers is detailed. This is a novel approach, being based on the NRTF approximation method. Then, the event-based algorithm for the FO-IMC controller is described and the NRTF approach is briefly presented. Section 3 presents the results obtained for lag and delay dominant FOPDT processes, in terms of reference tracking, disturbance rejection, and robustness. Comparative results are also given. Section 4 includes a brief discussion of the previously presented results, while the last section concludes the research.

2. Materials and Methods

2.1. Tuning the FO-IMC Controller

The following mathematical representation of the processes is considered:

$$H_p(s) = \frac{k}{Ts+1} e^{-\tau s}, \qquad (1)$$

where k is the process gain, T is the time constant, τ is the time delay, and s is the Laplace variable. The transfer function in (1) is generally used to model various types of processes, such as thermal, chemical, biomedicine systems. For these types of processes, a FO-IMC control strategy is proposed, as indicated in Figure 1, where $H_c(s)$ is the equivalent controller in a standard feedback structure, $H_m(s)$ is the process model, d—is the disturbance signal, y—the output signal, r—the reference signal.

The design of the FO-IMC controller is based on the inversion of the process model. For time delay processes, the time delay cannot be inverted and needs to be approximated, using either first Taylor series or Padé approximations [39]. Current research has shown that using these widely used approximation methods leads to poorer closed loop results [38], compared to a new approach based on the non-rational transfer function (NRTF) approximation method [43]. The new approach to tuning FO-IMC controllers for FOPDT processes in (1) is detailed next.

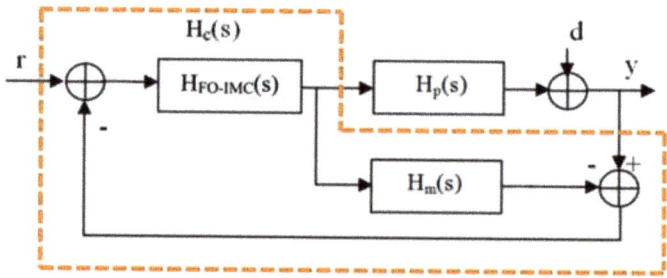

Figure 1. FO-IMC closed loop control scheme.

The proposed fractional order IMC (FO-IMC) controller is given by:

$$H_{FO-IMC}(s) = \frac{Ts+1}{k} \frac{1}{\lambda s^\alpha + 1}, \qquad (2)$$

where $\alpha \in (0, 2)$ is the fractional order and λ is the FO-IMC filter time constant. For $\alpha = 1$, the classical IMC controller is obtained. The limiting interval of the fractional order α is chosen such that the fractional order operation has a physical relevance from the control action point of view, as presented in [11].

Simple computations based on the diagram in Figure 1 lead to the following transfer function for the equivalent controller:

$$H_c(s) = \frac{Ts+1}{k(\lambda s^\alpha + 1 - e^{-\tau s})}. \qquad (3)$$

Notice the direct occurrence of the time delay in the denominator of (3). This can be further written as an integer order PI controller in series with a fractional order filter:

$$H_f(s) = \frac{s}{\lambda s^\alpha + 1 - e^{-\tau s}}. \qquad (4)$$

To tune the parameters of the FO-IMC controller, simple tuning methods can be used. In this particular approach, the phase margin and gain crossover frequency specifications are employed. These two performance specifications refer to the loop transfer function:

$$H_l(s) = H_p(s) \cdot H_c(s) = \frac{1}{\lambda s^\alpha + 1 - e^{-\tau s}} e^{-\tau s}. \qquad (5)$$

To meet the phase margin constraint, the phase equation is used:

$$\angle H_l(j\omega_c) = -\pi + PM, \qquad (6)$$

where ω_c is the desired gain crossover frequency and PM is the desired phase margin. The phase margin is a direct measure for the stability and robustness of a system. The larger the PM is, the more robust the overall closed loop system becomes. The selection of the gain crossover frequency is based on the maximization of the delay margin associated to the closed loop system, according to:

$$\omega_c = \frac{PM}{\tau_m - \tau}, \qquad (7)$$

where τ_m is the maximum time delay that would make the process in (1) unstable. To meet the gain crossover frequency constraint, the modulus equation is used:

$$|H_l(j\omega_c)| = 1. \qquad (8)$$

Then, replacing (1) and (3) in (6) and (8), leads to the following system of equations:

$$\begin{cases} \lambda \dfrac{\tan(\pi-PM-\tau\omega_c)-\tan(\pi-PM-\tau\omega_c)\cos(\tau\omega_c)-\sin(\tau\omega_c)}{\omega_c^a \sin\left(\frac{a\pi}{2}\right)-\omega_c^a \tan(\pi-PM-\tau\omega_c)\cos\left(\frac{a\pi}{2}\right)}, \\ \lambda^2 \omega_c^{2\alpha} + 2\lambda\omega_c^a\left[\cos\left(\frac{a\pi}{2}\right)-\cos\left(\frac{a\pi}{2}+\tau\omega_c\right)\right] - 2\cos(\tau\omega_c)+1 = 0. \end{cases} \quad (9)$$

The tuning of the FO-IMC controller is completed when the system of nonlinear equations in (9) is solved [12,17]. To implement the equivalent controller, an event-based algorithm is preferred. Such an approach leads to a smaller computational effort [46] and is more suitable in various industrial applications [46].

2.2. Event-Based Algorithm for FO-IMC Controllers

The equivalent controller for a standard feedback loop, as usually encountered in industrial applications, obtained based on the IMC methodology is given in (3). To implement this controller, an alternative form is preferred, as mentioned previously, with an integer order PI controller in series with the fractional order filter in (4). The new mathematical model for this fractional order equivalent controller is given as:

$$H_c(s) = C(s)H_f(s) = \frac{Ts+1}{ks}H_f(s) = \frac{T}{k}\left(1+\frac{1}{Ts}\right)H_f(s), \quad (10)$$

where $C(s)$ is the PI controller and $H_f(s)$ is the fractional order filter in (4).

Figure 2 presents the event-based paradigm, consisting of three components: process data measurement (data acquisition), event detector, and control input generator.

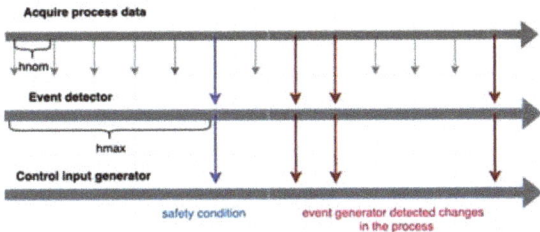

Figure 2. Basic paradigm of an event-based controller implementation [46].

The process output is measured at each sampling period h_{nom}, chosen according to standard discretization rules. The measured output data is transferred into the event detector. The main task of the detector is to decide whether an event has occurred, and in this case to trigger the control input generator. The event detector implements a function that optimizes the control process [47–49]. One of the most widely used event detection rules is based on computing the error signal and verifying whether it lies within a predefined range $[-\Delta_e, \Delta_e]$:

$$|e(t) - e(t - h_{act})| \geq \Delta_e, \quad (11)$$

where h_{act} denotes the elapsed time since the triggering of the previous event, $e(t)$ is the current error signal, and $e(t - h_{act})$ is the error at the previous event. Apart from the event triggering condition in (11), a safety condition is also used such that:

$$h_{act} \geq h_{max}, \quad (12)$$

where h_{max} is the maximum time between two consecutive events [50].

Once an event has been triggered, either when (11) or (12) occurs, the control input generator computes a new value for the control signal, according to a predefined algorithm. Since the computation of the control signal value occurs at variable sampling instants, the control law is represented by a discrete-time control algorithm where the sampling time parameter is considered as a variable parameter. The algorithm proposed in this paper is detailed below.

The overall Simulink implementation of such an event-based algorithm is given in Figure 3, where the blocks stand for: *nrtf_fo_filter* implements the NRTF approximation of the fractional order filter in (4), event-detector implements both the event detection part, as well as the control input generator for the integer order PI controller. The fractional order filter in (4) is used to filter the error signal. This occurs at every sampling period h_{nom}. The filtered error signal is then fed to the event-detector which implements the function that triggers of the control input generator. The latter is based on the standard PI controller, C(s) in (10). Thus, in the proposed approach, the PI controller is implemented in an event-based manner.

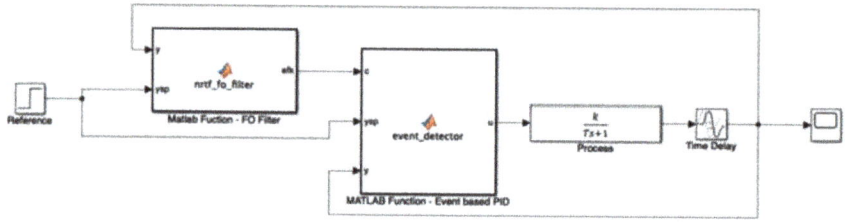

Figure 3. Simulink implementation of an event-based fractional order control algorithm.

Once the tuning of the FO-IMC has been performed and the two parameters, λ and α, determined, the fractional order filter in (4) is implemented in a discrete-time approximation based on the NRTF method, while the PI controller is implemented as an event-based algorithm. The control signal $U(s)$ of the PI controller $C(s)$ in (10) is computed based on:

$$U(s) = k_p\left(1 + \frac{1}{T_i s}\right) + E_f, \qquad (13)$$

where k_p and T_i are the proportional gain and integral time constant and $E_f(s)$ is the Laplace transform of the filtered error signal, at the output of the fractional order filter in (4):

$$H_f(s) = \frac{E_f(s)}{E(s)}, \qquad (14)$$

where $E(s)$ is the Laplace transform of the error signal $e(t)$ defined as $E(s) = Y_{sp}(s) - Y(s)$, $Y_{sp}(s)$ is the Laplace transform of the reference signal, $Y(s)$ is the Laplace transform of the measured output signal. For the particular case of FOPDT processes, the PI controller parameters are given as: $k_p = \frac{T}{k}$ and $T_i = T$, as resulting from (10) and (13). The event-based implementation of the PI controller can be achieved based on (13).

2.3. Comparisons with an Event-Based PI Controller

To compare the results, a classical integer order PI controller is designed for the same process in (1). The choice of the PI controller is based on the same number of parameters, as in the case of the FO-IMC controller, which allows for a similar tuning approach based on ensuring a certain gain crossover frequency and phase margin. The PI controller transfer function is given as:

$$C_{PI}(s) = k_p\left(1 + \frac{1}{T_i s}\right), \tag{15}$$

where k_p and k_i are the proportional and integral gains. The same performance specifications are used, as in the design of the FO-IMC controller. In this case, the loop transfer function is given by:

$$H_l(s) = H_p(s) \cdot C_{PI}(s) = k_p\left(1 + \frac{1}{T_i s}\right)\frac{k}{Ts + 1}e^{-\tau s}. \tag{16}$$

To meet the phase margin constraint, the phase Equation in (6) is used, leading to:

$$T_i = \frac{\tan\left(-\frac{\pi}{2} + PM - \angle H_p(j\omega_c)\right)}{\omega_c}, \tag{17}$$

where ω_c is the desired gain crossover frequency and PM is the desired phase margin. To meet the gain crossover frequency constraint, the modulus equation in (8) is used, leading to:

$$k_p = \frac{T_i \omega_c}{|H_p(j\omega_c)|\sqrt{T_i^2 \omega_c^2 + 1}}. \tag{18}$$

Thus, the tuning of the PI controller is complete, with the k_p and k_i parameters uniquely determined based on (17) and (18). To implement the PI controller, the event-based algorithm as proposed in [46] will be used. For a fain comparison, the parameters of the event-based algorithm will be similar to those used in the event-based implementation of the FO-IMC controller.

2.4. A Brief Overview of the NRTF Approximation Approach

Various discrete-time approximation methods for fractional order systems have been proposed over the years, including direct and indirect approaches [51]. One of the advantages of direct methods lies in the expedite approximation of fractional order systems as discrete-time higher order transfer functions. Most of the existing methods deal with the direct approximation of simple fractional order elements, such as the fractional order integrator or first order filter [52]. The NRTF method has been proposed as a means to offer a discrete-time approximation of low order for any type of non-rational transfer function, including complex fractional order elements and time delays [43]. The method consists of four steps, as detailed briefly below. A more detailed analysis and comparisons of the NRTF approach with other methods can be found in [43].

Step 1: The following generating function is used to replace the Laplace variable s in the fractional order system:

$$w(z^{-1}) = \frac{1 + \delta}{T_s} \frac{1 - z^{-1}}{1 + \delta z^{-1}}, \tag{19}$$

where $\delta \in [0, 1]$ is a shaping knob and T_s is the sampling period. To decrease the phase error between the approximation and the actual fractional order system, the parameter δ should be selected to be large, while a smaller value of δ decreases the magnitude error [43]. As fractional order systems have unlimited memory, the approximation is only possible within a certain limited frequency range. During this step, the maximum frequency boundary ω_h has to be specified, according to the Nyquist sampling theorem, with $\omega_h = \frac{\pi}{T_s}$. The approximation of the fractional order system will then be valid in an interval defined as $(0, \omega_h)$.

Step 2: The frequency response of the discrete-time fractional order system obtained in Step 1 is computed. To achieve this, the discrete-time operator z is replaced with $e^{j\omega T_s}$, where $\omega = \frac{2\pi}{T_s N_s}\begin{bmatrix} 0 & 1 & 2 & \ldots & \frac{N_s}{2} \end{bmatrix}$ is a vector of equally spaced frequencies and N_s is also a tuning knob. For a good approximation in the low frequency range, N_s should be large. The result of this second step consists in a vector of frequency response values of the fractional order discrete time transfer function.

Step 3: The inverse fast Fourier transform (FFT) algorithm is used to calculate the impulse response of the discrete-time fractional order system:

$$g[n] = \frac{1}{N_S}\sum_{k=0}^{N_S-1} G[k]e^{+j\frac{2\pi}{N_S}nk}, \; n = 0,1,2,\ldots,N_S-1, \qquad (20)$$

with $G[k]$ denoting the frequency response of the original fractional order system. The result of this step is a vector (20) containing N_s impulse response values.

Step 4: The Steiglitz–McBride approach [53] is used to determine a rational discrete-time transfer function with a similar impulse response as obtained from the inverse FFT in Step 3. The order N of the approximation has to be specified. The larger N is, the better the approximation. This also results in a higher order discrete-time transfer function. A compromise should be considered. The result of this step is the final discrete-time integer order transfer function of the form:

$$G(z^{-1}) = \frac{c_0 + c_1 z^{-1} + \ldots + c_N z^{-N}}{d_0 + d_1 z^{-1} + \ldots + d_N z^{-N}}, \qquad (21)$$

where $c_0, c_1 \ldots c_N$ and $d_0, d_1 \ldots d_N$ are coefficients computed according to the SteiglitzMcBride approach.

The step-by-step design procedure for a FO-IMC controller, as proposed in this paper, is detailed below, along with the event-based implementation.

Step 1: Select the desired PM and τ_m. Compute ω_c based on (7).

Step 2: Solve (9) to determine the FO-IMC controller parameters, λ and α.

Step 3: Compute the PI controller parameters according to $k_p = \frac{T}{k}$ and $T_i = T$ and the fractional order filter as indicated in (4), for the equivalent fractional order controller as described in (3).

Step 4: Select the parameters of the NRTF approximation method for the discrete-time approximation of the fractional order filter in (4): N, ω_h, and δ. The sampling period T_s is indirectly obtained as $\omega_h = \frac{\pi}{T_s}$.

Step 5: Select the parameters for the event-based implementation of the PI controller: h_nom = T_s, ∆_e, and h_max. The event-based algorithm is implemented based on two functions: an event detector (Figure 4) and a control signal generator (Figure 5), as follows.

```
function u = event_detector(ysp, y)

% define PI parameters Kp, Ti

global hact hnom Delta_e hmax es e u_last
e = ysp - y;
hact = hact + hnom;

    if (abs(e - es) > Delta_e || hact >= hmax)
        u = control_input_generator(ysp, y, Kp, Ti);
        u_last = u;
    else
        u = u_last;
    end

end
```

Figure 4. Event detector function.

```
function u = control_input_generator(ysp, y, Kp, Ti)

global hact y_old ui es e

up = Kp * (ysp - y);
u = up + ui;

ui = ui + Kp/Ti*hact*(ysp-y);

y_old = y;
hact = 0;
es = e;

end
```

Figure 5. Control input generator function.

3. Results

This section presents the main results obtained. The design is specific for FOPDT processes. Two different types will be discussed, the lag dominant and the delay dominant process. In both cases, reference tracking, disturbance rejection, and robustness to gain variations are considered as simulation tests. Reference tracking tests have been included in order to show the efficiency of the event-based controller in terms of setpoint trailing, while disturbance rejection results are considered in order to demonstrate the ability of the event-based controller to cope with external disturbances. Only step disturbance signals have been considered. As the results show, the event-based FO-IMC controller ensures better closed loop results compared to the event-based integer order PI controller, despite both controller being tuned and implemented in a similar fashion. Robustness tests are included to demonstrate that a fractional order IMC controller is intrinsically more robust to gain uncertainties, compared to a traditional integer order PI controller, even though robustness is not directly tackled in the design.

3.1. The Lag Dominant FOPDT Process

The following process model is considered for the first case study:

$$H_p(s) = \frac{1}{4s+1}e^{-s}. \tag{22}$$

To design the FO-IMC controller based on the tuning method described in Section 2.1, a phase margin PM = 85° and a gain crossover frequency ω_c = 0.3 rad/s are imposed. The solution of (9) yields λ = 1.95 and α = 0.85, with the FO-IMC controller given as:

$$H_{FO-IMC-NRTF}(s) = \frac{4s+1}{1} \frac{1}{1.95s^{0.85}+1}, \quad (23)$$

while the equivalent controller is computed as:

$$H_c(s) = \frac{4s+1}{1.95s^{0.85}+1-e^{-s}}. \quad (24)$$

In this case, the parameters of the PI controller are k_p = 4 and T_i = 4. The fractional order filter:

$$H_f(s) = \frac{s}{1.95s^{0.8475}+1-e^{-s}} = \frac{E_f(s)}{E(s)}, \quad (25)$$

is implemented as a discrete-time transfer function based on the NRTF approach with the order N = 7, δ = 1 and sampling period T_s = 0.1 s. For the event-based PI control algorithm, the following parameters are used: $h_{nom} = T_s$ = 0.1 s, h_{max} = 0.5, and Δ_e = 0.1.

For comparison purposes, a PI controller is also designed for the same performance specifications. The parameters are obtained according to (17) and (18), leading to:

$$C_{PI}(s) = 1.3841\left(1+\frac{1}{6.3715s}\right). \quad (26)$$

The controller in (26) is then implemented in an event-based algorithm with the same parameters as in the case of the FO-IMC controller.

The reference tracking, disturbance rejection and robustness results are given in Figures 6–9.

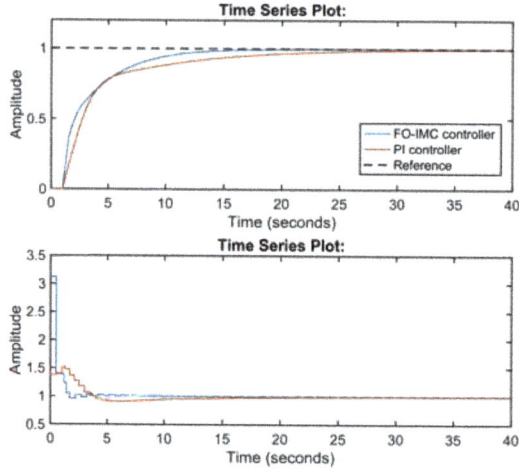

Figure 6. Comparison between the closed loop systems with the event-based FO-IMC and the classical proportional integral PI controllers for reference tracking (output signal in the upper plot, input signal in the lower plot).

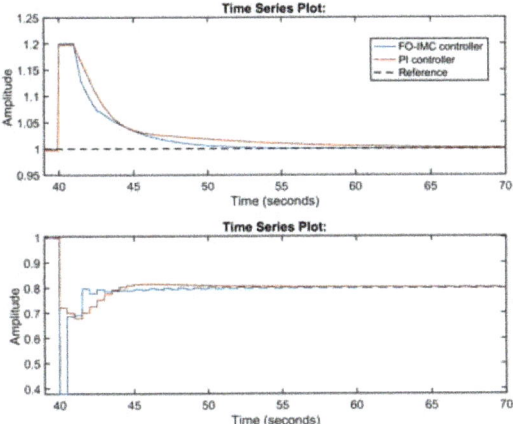

Figure 7. Comparison between the closed loop systems with the event-based FO-IMC and the PI controllers for a 0.2 disturbance rejection (output signal in the upper plot, input signal in the lower plot).

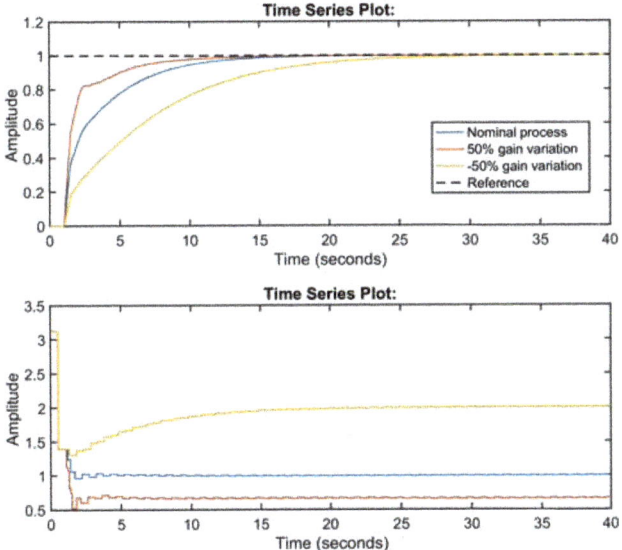

Figure 8. Robustness validation of the event-based FO-IMC controller (output signal in the upper plot, input signal in the lower plot).

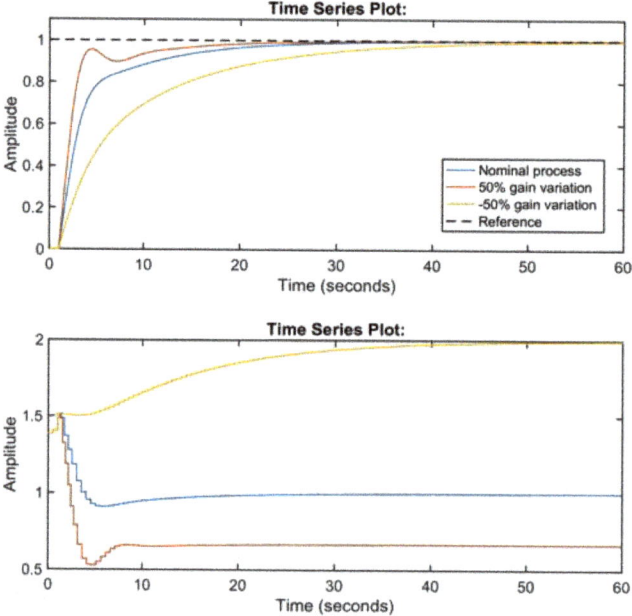

Figure 9. Robustness validation of the event-based PI controller (output signal in the upper plot, input signal in the lower plot).

3.2. The Delay Dominant FOPDT Process

The following process model is considered for the second case study:

$$H_p(s) = \frac{2}{s+1} e^{-2s}. \tag{27}$$

To design the FO-IMC controller based on the tuning method described in Section 2.1, a phase margin PM = 80° and a gain crossover frequency ω_c = 0.3 rad/s are imposed. The solution of (9) yields λ = 0.88 and α = 0.62, with the FO-IMC controller given as:

$$H_{FO-IMC-NRTF}(s) = \frac{s+1}{2} \frac{1}{0.88s^{0.62}+1}, \tag{28}$$

while the equivalent controller is computed as:

$$H_c(s) = \frac{s+1}{2(0.88s^{0.62}+1-e^{-2s})}. \tag{29}$$

In this case, the parameters of the PI controller are k_p = 0.5 and T_i = 1. The fractional order filter:

$$H_f(s) = \frac{s}{0.88s^{0.62}+1-e^{-2s}} = \frac{E_f(s)}{E(s)}, \tag{30}$$

is implemented as a discrete-time transfer function based on the NRTF approach with the order N = 5, δ = 0.5 and sampling period T_s = 0.1 s. For the event-based PI control algorithm, the following parameters are used: $h_{nom} = T_s$ = 0.1 s, h_{max} = 0.5 and Δ_e = 0.1.

For comparison purposes, a PI controller is also designed for the same performance specifications. The parameters are obtained according to (17) and (18), leading to:

$$C_{PI}(s) = 0.3430\left(1 + \frac{1}{2.9055s}\right). \quad (31)$$

The controller in (31) is then implemented in an event-based algorithm with the same parameters as in the case of the FO-IMC controller.

The reference tracking, disturbance rejection, and robustness results are given in Figures 10–13.

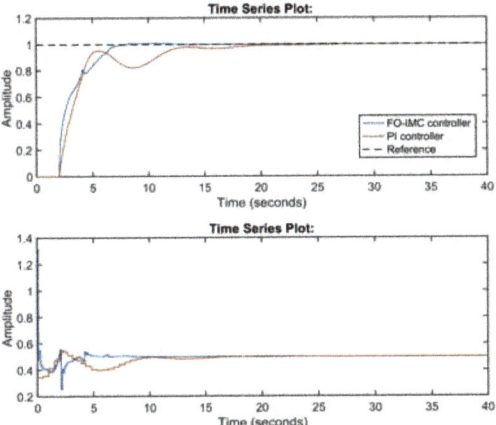

Figure 10. Comparison between the closed loop systems with the event-based FO-IMC and the PI controllers for reference tracking (output signal in the upper plot, input signal in the lower plot).

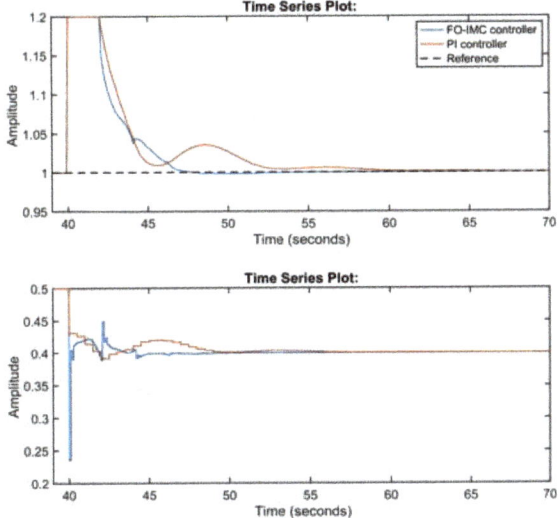

Figure 11. Comparison between the closed loop systems with the event-based FO-IMC and the PI controllers for a 0.2 disturbance rejection (output signal in the upper plot, input signal in the lower plot).

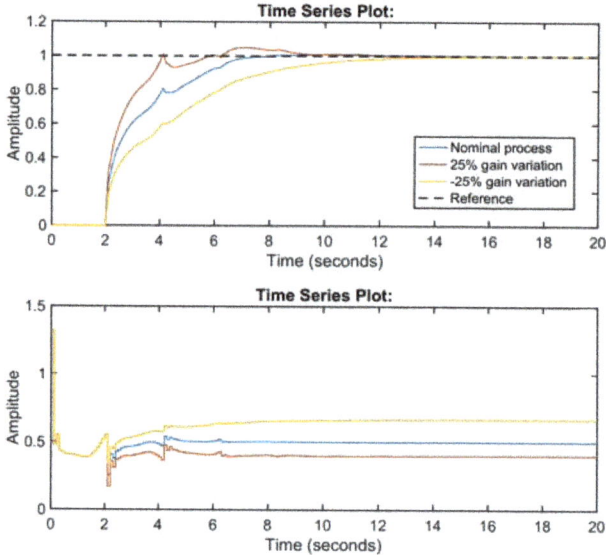

Figure 12. Robustness validation of the event-based FO-IMC controller (output signal in the upper plot, input signal in the lower plot).

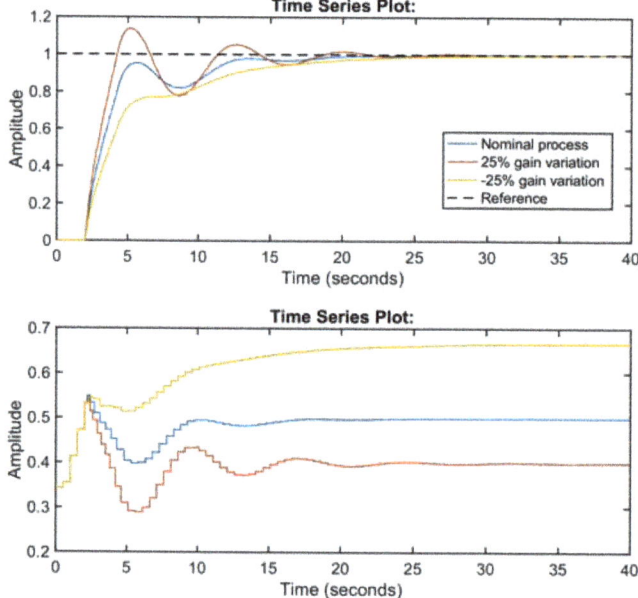

Figure 13. Robustness validation of the event-based PI controller (output signal in the upper plot, input signal in the lower plot).

4. Discussion

A performance comparison regarding the two different event-based controllers is presented in Table 1, for the lag dominant process. Figure 6 depicts a comparison regarding reference tracking with the event-based FO-IMC controller and the event-based PI controller. As indicated here, the FO-IMC achieves a faster settling time, compared to the PI: 13 s, compared to nearly 25 s. The drawback is that the FO-IMC control signal is twice as large, but solely during one event. Both controllers manage to achieve this without any overshooting. Over the 40 s seconds simulation time, the event-based FO-IMC controller requires 82 control signal computations; similarly for the event-based PI controller. In a classical discrete-time approximation, both type of controllers would have needed 400 computations. This leads to an overall reduction of the resources used of 79.5%. Improved disturbance rejection results are also visible in the case of the FO-IMC controller, compared to the PI controller, as indicated in Figure 7. The settling time for the FO-IMC is approximately 8 s, compared to 16 s for the PI. In terms of control signal computations, there are 61 for the event-based FO-IMC implementation, compared to 60 for the event-based PI implementation and 300 for the standard discrete-time implementation. This leads to 79–80% reduction in the resources used. The robustness validation of the event-based FO-IMC controller is given in Figure 8, whereas the robustness tests for the event-based PI controller are indicated in Figure 9. Although none of the control strategies were designed specifically to ensure the robustness to gain variations, the simulation results in Figures 8 and 9 show that for ±50% gain variations, the event-based controllers manage to maintain 0 overshoot. A comparison between the two shows that the event-based PI controller has better chances of turning the closed loop into an underdamped response for larger positive gain variations.

Table 1. Performance comparison between the event-based FO-IMC and event-based PI controller for the lag dominant FOPDT process.

Test Scenario	FO-IMC Control			PI Control		
	Overshoot (%)	Settling Time (s)	Control Computations	Overshoot (%)	Settling Time (s)	Control Computations
Reference tracking	0	13	82	0	25	82
Disturbance rejection		8	61		16	60
Robustness assessment						
50% gain variation	0	7		0	15	
−50% gain variation	0	23		0	40	

A performance comparison regarding the two different event-based controllers is presented in Table 2, for the delay dominant process. In this case, Figure 10 depicts a comparison regarding reference tracking with the event-based FO-IMC controller and the event-based PI controller. The event-based FO-IMC controller, proposed in this manuscript, achieves zero overshoot and a 7 s settling time. The event-based PI controller achieves no overshoot, but with an oscillatory response, as well as a larger settling time of 18 s. In terms of control effort, this is larger for the FO-IMC, compared to the PI. A comparison regarding resources used over 40 s simulation shows that the event-based FO-IMC requires 103 computations of the control signal, whereas the event-based PI controller needs solely 83. This accounts for 74% reduction in the resources used for the event-based FO-IMC and a slightly better 79% for the event-based PI controller, compared to the standard discrete-time implementation (with 400 computations over the 40 s simulation period). In terms of disturbance rejection, the comparative simulation results given in Figure 11 demonstrate the efficiency of the FO-IMC compared to the PI, with a settling time of 7 s, compared to 12 s. In this case also, the event-based FO-IMC requires 87 computations of the control signal, higher than the 61 for the event-based PI controller. The simulation period considered was 30 s, with a standard discrete-time implementation requiring 300 control signal computations overall. In this case, 71% improvement in the resources used is achieved with the event-based FO-IMC implementation and 80% with the event-based PI controller, compared to the standard discrete-time implementation. The robustness validation of the event-based FO-IMC controller is given in Figure 12, whereas the robustness tests for the event-based PI controller are indicated in Figure 13. Although none of the control strategies were designed specifically to ensure the robustness to gain variations, the simulation results in Figure 12 show that better robustness can be achieved by using the event-based FO-IMC controller, compared to the event-based PI controller. The latter has oscillatory response and higher overshoot and nearly twice settling time.

The results obtained for the event-based FO-IMC controller show that this can be considered as a viable option for controlling processes where resources, bandwidth allocation, energy usage are limited. Furthermore, as the simulation results show, the event-based FO-IMC control strategy offers better results in terms of reference tracking, disturbance rejection, and robustness for both lag and delay dominant systems. The robustness of the event-based FO-IMC controller is significantly improved compared to the event-based PI controller, for delay dominant systems. This is an aspect intrinsic to the IMC methodology. The sole disadvantage of the event-based FO-IMC controller is that there is an increase in the control effort and in the number of events that require the computation of the control signal, compared to the event-based PI controller.

Table 2. Performance comparison between the event-based FO-IMC and event-based PI controller for the delay dominant FOPDT process.

Test Scenario	FO-IMC Control			PI Control		
	Overshoot (%)	Settling Time (s)	Control Computations	Overshoot (%)	Settling Time (s)	Control Computations
Reference tracking	0	7	103	0	18	83
Disturbance rejection		7	87		12	61
Robustness assessment						
25% gain variation	10	9		15	20	
−25% gain variation	0	12		0	20	

Further research includes the modification of the FO-IMC tuning procedure to an optimization routine where the control effort is also directly tackled, as well as the robustness property. Additionally, an experimental validation is to be considered.

5. Conclusions

Fractional order calculus has been used to provide for a generalization of the IMC. Such an approach is usually considered for the control of dead time processes. In this paper, FOPDT processes are considered, including delay and lag dominant ones. The design is based on a new approximation approach, the NRTF method, for the equivalent controller in an IMC loop and on two performance specifications, the gain crossover frequency and the phase margin. As it has been previously demonstrated by the authors, the NRTF method allows for a more accurate approximation of the process dead-time and ensures an improved closed loop response.

The implementation of the final fractional order controller is usually a challenging task, since higher order transfer functions are used to approximate the dynamics of the original controller. In situations where CPU, bandwidth allocation, and energy usage are limited, resources need to be efficiently managed. In this paper, a solution for this is proposed, in terms of an event-based implementation of the FO-IMC controller. Such an approach, has only been recently proposed for fractional order PID-type controllers, but not for other types/structure of fractional order controllers. The originality of the approach consists in a two-step implementation. The equivalent fractional order controller, as obtained according to the proposed FO-IMC approach, is divided into an integer order PI controller and a fractional order filter. Then, the NRTF approach is used to determine a standard discrete-time approximation for the fractional order filter. The remaining PI controller is implemented in an event-based approach. Numerical results are provided for lag and delay dominant FOPDT processes. For comparison purposes, an integer order PI controller, tuned according to the same performance specifications as the FO-IMC, is also implemented as an event-based control strategy. The numerical results show that the proposed event-based implementation for FO-IMC controller is suitable and provides for better reference tracking, disturbance rejection, and robustness, compared to the integer order event-based PI controller, as well as a smaller computational effort compared to a standard discrete-time implementation, thus being more suitable in various industrial applications where resources need to be drastically limited.

Author Contributions: Conceptualization, C.I.M. and I.R.B.; methodology, C.I.M.; software, I.R.B.; validation, I.R.B.; formal analysis, E.H.D.; investigation, C.I.M.; resources, I.R.B.; data curation, I.R.B.; writing—original draft preparation, C.I.M.; writing—review and editing, C.I.M. and E.H.D.; visualization, I.R.B.; supervision, C.I.M.; project administration, C.I.M.; funding acquisition, E.H.D. All authors have read and agreed to the published version of the manuscript.

Funding: D.E.H. was funded by Hungarian Academy of Science, Janos Bolyai Grant (BO/00313/17) and the ÚNKP-19-4-OE-64 New National Excellence Program of the Ministry for Innovation and Technology. C.I.M. and I.B. have been supported by a grant of the Romanian National Authority for Scientific Research and Innovation, CNCS/CCCDI-UEFISCDI, project number PN-III-P1-1.1-TE-2016-1396, TE 65/2018.

Conflicts of Interest: The authors declare no conflict of interest. The funders had no role in the design of the study; in the collection, analyses, or interpretation of data; in the writing of the manuscript, or in the decision to publish the results.

References

1. Ionescu, C.M.; Machado, J.A.; De Keyser, R. Modeling of the lung impedance using a fractional-order ladder network with constant phase elements. *IEEE Trans. Biomed. Circuits Syst.* **2011**, *5*, 83–89. [CrossRef] [PubMed]
2. Muresan, C.I.; Folea, S.C.; Birs, I.R.; Ionescu, C.M. A Novel Fractional Order Model and Controller for Vibration Suppression in Flexible Smart Beam. *Nonlinear Dyn.* **2018**, *93*, 525–541. [CrossRef]
3. Muresan, C.I.; Dulf, E.H.; Prodan, O. A Fractional Order Controller for Seismic Mitigation of Structures Equipped with Viscoelastic Mass Dampers. *J. Vib. Control* **2016**, *22*, 1980–1992. [CrossRef]
4. Ionescu, C.M.; Birs, I.; Copot, D.; Muresan, C.I.; Caponetto, R. Mathematical modeling with experimental validation of viscoelastic properties in non-Newtonian fluids. *Philos. Trans. R. Soc. A* **2020**, *378*, 20190284. [CrossRef]
5. Birs, I.; Copot, D.; Muresan, C.I.; Nascu, I.; Ionescu, C. Identification for Control of Suspended Objects in Non-Newtonian Fluids. *Fract. Calc. Appl. Anal.* **2019**, *22*, 1378–1394. [CrossRef]
6. Carvalho, A.R.M.; Pinto, C.M.A.; de Carvalho, J.M. Fractional Model for Type 1 Diabetes. In *Mathematical Modelling and Optimization of Engineering Problems. Nonlinear Systems and Complexity*; Machado, J., Özdemir, N., Baleanu, D., Eds.; Springer: Cham, Switzerland, 2020; Volume 30.
7. Magin, R.L.; Karani, H.; Wang, S.; Liang, Y. Fractional Order Complexity Model of the Diffusion Signal Decay in MRI. *Mathematics* **2019**, *7*, 348. [CrossRef]
8. Pinto, C.M.A.; Carvalho, A.R.M.; Baleanu, D.; Srivastava, H.M. Efficacy of the Post-Exposure Prophylaxis and of the HIV Latent Reservoir in HIV Infection. *Mathematics* **2019**, *7*, 515. [CrossRef]
9. Kumar, S.; Ahmadian, A.; Kumar, R.; Kumar, D.; Singh, J.; Baleanu, D.; Salimi, M. An Efficient Numerical Method for Fractional SIR Epidemic Model of Infectious Disease by Using Bernstein Wavelets. *Mathematics* **2020**, *8*, 558. [CrossRef]
10. Dulf, E.-H.; Vodnar, D.C.; Danku, A.; Muresan, C.-I.; Crisan, O. Fractional-Order Models for Biochemical Processes. *Fractal Fract.* **2020**, *4*, 12. [CrossRef]
11. Podlubny, I. Fractional-order systems and $PI^\lambda D^\mu$-controllers. *IEEE Trans. Autom. Control* **1999**, *44*, 208–214. [CrossRef]
12. Monje, C.A.; Chen, Y.; Vinagre, B.M.; Xue, D.; Feliu-Batlle, V. *Fractional-Order Systems and Controls: Fundamentals and Applications*; Springer: Cham, Switzerland, 2010.
13. Folea, S.; Muresan, C.I.; De Keyser, R.; Ionescu, C. Theoretical Analysis and Experimental Validation of a Simplified Fractional Order Controller for a Magnetic Levitation System. *IEEE Trans. Control Syst. Technol.* **2016**, *24*, 756–763. [CrossRef]
14. Li, X.; Wang, Y.; Li, N.; Han, M.; Tang, Y.; Liu, F. Optimal fractional order PID controller design for automatic voltage regulator system based on reference model using particle swarm optimization. *Int. J. Mach. Learn. Cyber.* **2017**, *8*, 1595–1605. [CrossRef]
15. El-Khazali, R. Fractional-order $PI^\lambda D^\mu$ controller design. *Comput. Math. Appl.* **2013**, *66*, 639–646.

16. Monje, C.A.; Vinagre, B.M.; Feliu, V.; Chen, Y. Tuning and auto-tuning of fractional order controllers for industry applications. *Control Eng. Pract.* **2008**, *16*, 798–812. [CrossRef]
17. Muresan, C.I.; Folea, S.; Mois, G.; Dulf, E.H. Development and implementation of an FPGA based fractional order controller for a DC motor. *Mechatronics* **2013**, *23*, 798–804. [CrossRef]
18. Ionescu, C.M.; Dulf, E.H.; Ghita, M.; Muresan, C.I. Robust controller design: Recent emerging concepts for control of mechatronic systems. *J. Frankl. Inst.* **2020**, *357*, 7818–7844. [CrossRef]
19. Das, S.; Pan, I.; Das, S.; Gupta, A. A novel fractional order fuzzy PID controller and its optimal time domain tuning based on integral performance indices. *Eng. Appl. Artif. Intell.* **2012**, *25*, 430–442. [CrossRef]
20. Castillo-Garcia, F.J.; Feliu-Batlle, V.; Rivas-Perez, R.; Sanchez, L. Time Domain Tuning of a Fractional Order PIα Controller Combined with a Smith Predictor for Automation of Water Distribution in Irrigation Main Channel Pools. *IFAC Proc. Vol.* **2011**, *44*, 15049–15054. [CrossRef]
21. Maamir, F.; Guiatni, M.; El Hachemi, H.M.S.M.; Ali, D. Auto-tuning of fractional-order PI controller using particle swarm optimization for thermal device. In Proceedings of the 4th International Conference on Electrical Engineering, Boumerdes, Algeria, 13–15 December 2015; pp. 1–6. [CrossRef]
22. De Keyser, R.; Muresan, C.I.; Ionescu, C.M. A novel auto-tuning method for fractional order PI/PD controllers. *ISA Trans.* **2016**, *62*, 268–275. [CrossRef]
23. Juchem, J.; Muresan, C.I.; De Keyser, R.; Ionescu, C.M. Robust fractional-order auto-tuning for highly-coupled MIMO systems. *Heliyon* **2019**, *5*, e02154. [CrossRef]
24. Soukkou, A.; Belhour, M.C.; Leulmi, S. Review, Design, Optimization and Stability Analysis of Fractional-Order PID Controller. *Int. J. Intell. Syst. Appl.* **2016**, *8*, 73. [CrossRef]
25. Leng, B.Y.; Qi, Z.D.; Shan, L.; Bian, H.J. Review of Fractional Order Control. 2014. Available online: https://www.scientific.net/AMR.1049-1050.983 (accessed on 14 August 2020).
26. Shah, P.; Agashe, S. Review of fractional PID controller. *Mechatronics* **2016**, *38*, 29–41. [CrossRef]
27. Dastjerdi, A.A.; Vinagre, B.M.; Chen, Y.Q.; HosseinNia, S.H. Linear fractional order controllers; A survey in the frequency domain. *Annu. Rev. Control* **2019**, *47*, 51–70. [CrossRef]
28. Birs, I.; Muresan, C.I.; Nascu, I.; Ionescu, C. A Survey of Recent Advances in Fractional Order Control for Time Delay Systems. *IEEE Access* **2019**, *7*, 30951–30965. [CrossRef]
29. Copot, D.; Ghita, M.; Ionescu, C.M. Simple Alternatives to PID-Type Control for Processes with Variable Time-Delay. *Processes* **2019**, *7*, 146. [CrossRef]
30. Bettayeb, M.; Mansouri, R.; Al-Saggaf, U.; Mehedi, I.M. Smith Predictor Based Fractional-Order-Filter PID Controllers Design for Long Time Delay Systems. *Asian J. Control* **2016**, *19*, 587–598. [CrossRef]
31. Vu, T.N.L.; Lee, M. Smith predictor based fractional-order PI control for time-delay processes. *Korean J. Chem. Eng.* **2014**, *31*, 1321–1329. [CrossRef]
32. Jesus, I.S.; Machado, J.T. Fractional Control with a Smith Predictor. *J. Comput. Nonlinear Dyn.* **2011**, *6*, 31010–31014. [CrossRef]
33. Safaei, M.; Tavakoli, S. Smith predictor based fractional-order control design for time-delay integer-order systems. *Int. J. Dyn. Control* **2018**, *6*, 179–187. [CrossRef]
34. Maâmar, B.; Rachid, M. IMC-PID-fractional-order- filter controllers design for integer order systems. *ISA Trans.* **2014**, *53*, 1620–1628. [CrossRef]
35. Valerio, D.; Sa da Costa, J. Tuning of fractional PID controllers with Ziegler–Nichols-type rules. *Signal Process.* **2006**, *86*, 2771–2784. [CrossRef]
36. Abadi, M.R.R.M.; Jalali, A.A. Fractional order PID controller tuning based on IMC. *Int. J. Inf. Technol. Control Autom. IJITCA* **2012**, *2*, 21–35.
37. Lei, S.; Zhao, Z.; Zhang, J. Design of fractional order smith predictor controller for non-square system. In Proceedings of the 2016 12th World Congress on Intelligent Control and Automation (WCICA), Guilin, China, 12–15 June 2016. [CrossRef]
38. Muresan, C.I.; Birs, I.; De Keyser, R. An Alternative Design Approach for Fractional Order Internal Model Controllers for Time Delay Systems. *Int. J. Control* **2020**. under review.
39. Muresan, C.I.; Dutta, A.; Dulf, E.H.; Pinar, Z.; Maxim, A.; Ionescu, C.M. Tuning algorithms for fractional order internal model controllers for time delay processes. *Int. J. Control* **2016**, *89*, 579–593. [CrossRef]
40. Arya, P.P.; Chakrabarty, S. IMC based Fractional Order Controller Design for Specific Non-Minimum Phase Systems. *IFAC-PapersOnLine* **2018**, *51*, 847–852. [CrossRef]

41. Jain, S.; Hote, Y.V. Fractional order IMC controller via order reduction and CRONE principle for Load frequency control. In Proceedings of the 2018 8th International Conference on Power and Energy Systems (ICPES), Colombo, Sri Lanka, 21–22 December 2018. [CrossRef]
42. Dulf, E.H. Simplified Fractional Order Controller Design Algorithm. *Mathematics* **2019**, *7*, 1166. [CrossRef]
43. De Keyser, R.; Muresan, C.I.; Ionescu, C.M. An efficient algorithm for low-order discrete-time implementation of fractional order transfer functions. *ISA Trans.* **2018**, *74*, 229–238. [CrossRef]
44. Muresan, C.I.; Ionescu, C.M. Generalization of the FOPDT Model for Identification and Control Purposes. *Processes* **2020**, *8*, 682. [CrossRef]
45. Volanova, R.; Visioli, A. *PID Control in the Third Millennium*; Springer: Cham, Switzerland, 2012.
46. Birs, I.; Nascu, I.; Ionescu, C.M.; Muresan, C.I. Event-based fractional order control. *J. Adv. Res.* **2020**. [CrossRef]
47. Birs, I.; Muresan, C.I.; Ionescu, C.M. An event based implementation of a fractional order controller on a non-Newtonian transiting robot. In Proceedings of the European Control Conference (ECC), Saint Petersburg, Russia, 12–15 May 2020; pp. 1436–1441.
48. Heemels, W.P.; Johansson, K.H.; Tabuada, P. An introduction to event-triggered and self-triggered control. In Proceedings of the IEEE Conference on Decision and Control, Maui, HI, USA, 10–13 December 2012; pp. 3270–3285. [CrossRef]
49. Castilla, M.; Bordons, C.; Visioli, A. Event-based state-space model predictive control of a renewable hydrogen-based microgrid for office power demand profiles. *J. Power Sources* **2020**, *450*. [CrossRef]
50. Liu, Q.; Wang, Z.; He, X.; Zhou, D.H. A survey of event-based strategies on control and estimation. *Syst. Sci. Control Eng.* **2014**, *2*, 90–97. [CrossRef]
51. Petras, I. Tuning and implementation methods for fractional-order controllers. *Fract. Calc. Appl. Anal.* **2012**, *15*, 282–303. [CrossRef]
52. Li, Y.; Sheng, H.; Chen, Y.Q. Analytical impulse response of a fractional second order filter and its impulse response invariant discretization. *Signal Process* **2011**, *91*, 498–507. [CrossRef]
53. Steiglitz, K.; McBride, L.E. A Technique for the Identification of Linear Systems. *IEEE Trans. Autom. Control* **1965**, *10*, 461–464. [CrossRef]

© 2020 by the authors. Licensee MDPI, Basel, Switzerland. This article is an open access article distributed under the terms and conditions of the Creative Commons Attribution (CC BY) license (http://creativecommons.org/licenses/by/4.0/).

MDPI
St. Alban-Anlage 66
4052 Basel
Switzerland
Tel. +41 61 683 77 34
Fax +41 61 302 89 18
www.mdpi.com

Mathematics Editorial Office
E-mail: mathematics@mdpi.com
www.mdpi.com/journal/mathematics

www.ingramcontent.com/pod-product-compliance
Lightning Source LLC
LaVergne TN
LVHW070252100526
838202LV00015B/2211

9 783036 538471